WORLD SCIENTIFIC SERIES ON NONLINEAR SCIENCE

Discrete Systems with Memory

WORLD SCIENTIFIC SERIES ON NONLINEAR SCIENCE

Editor: Leon O. Chua
University of California, Berkeley

Series A.   **MONOGRAPHS AND TREATISES***

*To view the complete list of the published volumes in the series, please visit:
http://www.worldscibooks.com/series/wssnsa_series.shtml

WORLD SCIENTIFIC SERIES ON
NONLINEAR SCIENCE

Series A   Vol. 75

Series Editor: Leon O. Chua

Discrete Systems with Memory

Ramon Alonso-Sanz

Polytechnic University of Madrid

World Scientific

NEW JERSEY · LONDON · SINGAPORE · BEIJING · SHANGHAI · HONG KONG · TAIPEI · CHENNAI

Published by

World Scientific Publishing Co. Pte. Ltd.

5 Toh Tuck Link, Singapore 596224

USA office: 27 Warren Street, Suite 401-402, Hackensack, NJ 07601

UK office: 57 Shelton Street, Covent Garden, London WC2H 9HE

British Library Cataloguing-in-Publication Data

A catalogue record for this book is available from the British Library.

World Scientific Series on Nonlinear Science, Series A — Vol. 75
DISCRETE SYSTEMS WITH MEMORY

ISBN-13 978-981-4343-63-3
ISBN-10 981-4343-63-3

Printed by FuIsland Offset Printing (S) Pte Ltd. Singapore

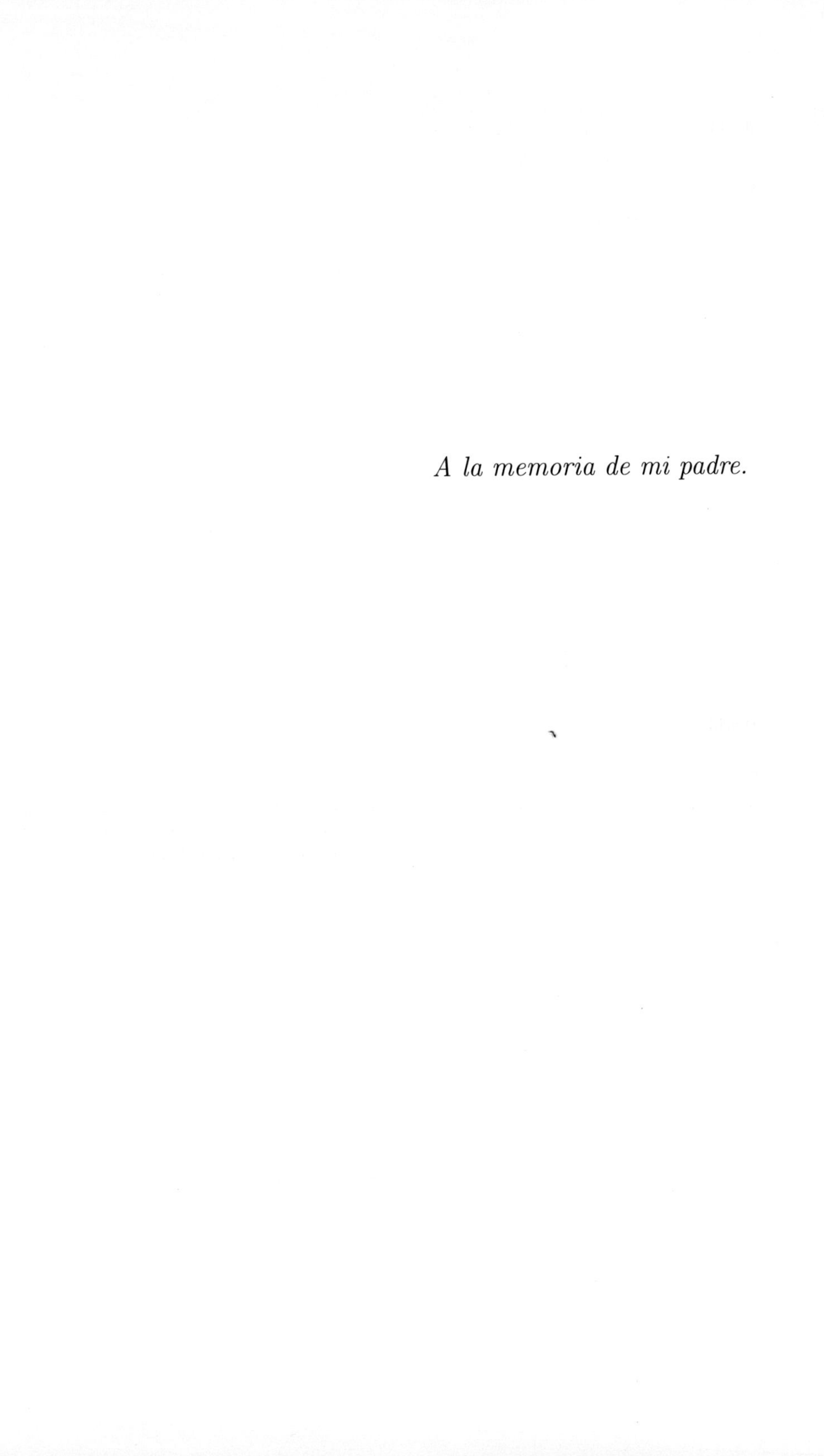

A la memoria de mi padre.

Preface

In conventional discrete systems, the new configuration depends solely on the configuration at the preceding time step. This book considers an extension to the standard framework of dynamical systems by taking into consideration past history in a simple way: the mapping defining the transition rule of the system remains unaltered, but it is applied to a certain summary of past states. The book focuses on the study of systems discrete -par excellence-, i.e., with every component, space, time and state variable, being discrete. These discrete universes are know as cellular automata in their more structured forms, and Boolean networks in a more general way. Thus, the mappings which define the rules of these dynamical systems are not formally altered in this book when implementing memory, but they are applied to cells (or nodes) that exhibit trait states computed as a function of their own previous states. So to say, cells -canalize- memory to the mapping.

After an introductory chapter, memories of average type are implemented in Chapter 2, whereas other types of memory are scrutinized in Chapter 3. A study is made of the effect of memory in systems with asynchronous updating and in one-dimensional automata with probabilistic rules in Chapter 4. The capacity of CA endowed with memory as random number generators is studied in Chapter 5. Although most of the automata studied here have two states, the case of three states is also present, in Chapter 6. Chapters 7 and 8 deal with reversibility scrutinized with respect to memory, the former with the so called Fredkin's rule, and the latter with block (or partitioned) automata.

In the generalization of the basic CA paradigm, known as structurally dynamic cellular automata, the connections between the cells are allowed to change according to rules similar in nature to the state transition rules associated with the conventional CA. These automata are taken into consideration in Chapter 9.

Automata on networks and proximity graphs, together with automata with different rules in their nodes and non-local connectivities, commonly referred to as Boolean networks, are considered in Chapter 10. The case of coupled layers is treated in Chapter 11.

Systems that remain discrete in space and time, but not in the state variable, are taken into account to some extent in Chapter 12. Chapter 13 is devoted to the study of spatial games, particular attention is paid in this chapter to the spatialized prisoner's dilemma with the players endowed with memory of both strategies and payoffs.

The fairly natural memory implementation mechanism adopted in the book, of straightforward computer codification, allows for an easy systematic study of the effect of memory in discrete dynamical systems. This may inspire some useful ideas in using discrete systems as a tool for modeling phenomena with memory. This task has been traditionally attacked by means of differential, or finite-difference, equations, with some (or all) continuous component. In contrast, full discrete models are ideally suited to digital computers. Thus, it seems plausible that further study on discrete systems with memory should prove profitable, and may be possible to paraphrase T.Toffoli [389] in presenting discrete systems with memory *as an alternative to (rather than an approximation of)* integro-differential equations in modeling phenomena with memory. Besides their potential applications, discrete systems with memory have an aesthetic and mathematical interest on their own, which the book illustrates with worked figures and detailed examples.

I am indebted to Dr. Andrew Adamatzky (UWE) who encouraged me to write this book. I wish to thank my friend Mike Talbot (ex. BIOSS) for the correction of an early draft of this text.

Ramón Alonso-Sanz

Universidad Politécnica de Madrid
ETSI Agrónomos (Estadística)
28040, Madrid, Spain
ramon.alonso@upm.es

Contents

Chapter 1

Cellular Automata and memory

1.1 Cellular Automata

Cellular Automata (CA) are discrete, spatially explicit extended dynamic systems. A CA system is composed of a grid of adjacent cells, arranged as a regular d-dimensional lattice (d is in most cases 1 or 2), which evolves in discrete time steps. Each cell is characterized by an internal state whose value belongs to a finite set of size k. The updating of these states is done simultaneously according to a common local transition rule involving only the neighbors of each cell. Thus, if $\sigma_i^{(T)}$ is taken to denote the state value of cell i at time-step T, the cell values evolve by iteration of the mapping:

$$\sigma_i^{(T+1)} = \phi\big(\{\sigma_{j \in \mathcal{N}_i}^{(T)}\}\big) \,, \quad \forall i,$$

where ϕ is an arbitrary function which specifies the cellular automaton *rule* operating on the cells in the neighborhood ($\mathcal{N}$) of the cell i. A formal definition of cellular automaton may be found elsewhere, for example, in [391] or [288]. It is to be stressed here that the *cells* in the contexts considered in this book are just the *bricks* of an oversimplified microworld which do not try to emulate real particles as in Molecular Dynamics [339].

Cellular automata are discrete *par excellence* as every component, space, time and state variable, is discrete. This perfectly fits the features of digital computers, enabling exact computation. The synchronicity of the updating mechanism, the regular topologies and the locality of interactions make the CA paradigm ideally suited for parallel computers.

CA are often described as a counterpart to partial differential equations [389], capable of describing continuous dynamical systems. Roughly speaking, there are two main levels of study of a natural system, corresponding to the scale of observation: the microscopic and the macroscopic. Often, the complexity of the macroscopic world appears to be disconnected from

1

that of the microscopic world, although the former is driven by the latter: the microscopic details are lost when the whole system is seen through a macroscopic filter. CA works at a microscopic level which drives the macroscopic behavior: the idea is not to try to describe a complex system from *above* - to describe it using difficult equations- but to simulate it by interaction of cells following easy rules, often formulated in a natural language (with minimal or *soft* mathematical demands) easily translated in a computer programming language, as postulated by the Artificial Intelligence community. In other words: not to describe a complex system with complex equations, but to let the complexity *emerge* by interaction of simple individuals following simple rules. Thus, cellular automata appear as an invaluable tool to study (idealized) complex systems [200, 291].

From the theoretical point of view, CA were introduced in the late 1940's by John von Neumann [308] and Stanislaw Ulam. One can say that the *cellular* part comes from Ulam, and the *automata* part from von Neumann. But CA were shunted into a sidings for a couple of decades [390], so did not reach the general public until 1970, when Martin Gardner published an account of John Conway's Game of "Life" in *Scientific American* [161, 333]. *Life* was destined to become the most famous CA and an inspiration to a generation of Artificial Life researchers.

A number of people at MIT began studying CA beyond *Life* during the 1970s. Probably the most influential figure there was Edward Fredkin, who around 1980 formed the *Information Mechanics Group at MIT* along with Tommaso Toffoli, Norman Margolus and Gérard Vichniac. By 1984, Toffoli and Margolus had nearly perfected the CAM-6 cellular automaton machine, a special computer designed for the lightning-quick execution of CA, and were generating some publicity. In addition, in the middle eighties (perhaps the *golden age* of CA), Stephen Wolfram was publishing numerous articles about CA (compiled in [425]) which "hooked" a number of researchers on modern CA (in particular after [420], a landmark review paper largely responsible for the resurgence of interest in CA in the eighties). In these articles, Wolfram suggested that many physical processes that seem random are, in fact, the deterministic outcome of computations that are simply so convoluted that they cannot be compressed into shorter form and predicted in advance (he spoke of these computations as *incompressible*).

During the following years, CA were developed and used in many different fields. The requirements for the application of the CA approach to real problems (connecting different levels of detail) enlarged the basic paradigm, leading to systems related to CA (mere *extensions* in some cases) such as inhomogeneous (Section 10.2), asynchronous (Chapter 4), continuous state (Chapter 12), Lattice-gas[1] or macroscopic[2] automata. Today, some authors use more comprehensive terms, such as *cellular networks* [341, 342] or *grid-based* models [419] in order to be freed of the restrictions that the CA paradigm imposes.

A vast body of literature has been produced on these topics[3], to a great extent devoted to many different unrelated areas of physics. Thus, the book by Chopard and Droz [110] might serve as a text-book for physics. The book by Ilachanski [208] is also intended mainly for a physicist audience, while [228] explores the use of CA in modeling chemical phenomena. Some modern books on statistical mechanics, e.g., [416], incorporate CA as a tool to study systems that are far from equilibrium. It has been argued that CA, intimately related to discrete statistical models, will play an important part elucidating basic ideas and general principles of statistical mechanics. Conversely, Rujan [347] also studies the usefulness of statistical physics methods to describe the properties of probabilistic CA. The book edited by Dieckmann *et al.* [124] frames CA in the context of (ecological) spatial analysis, while [120] focuses on biological issues. The book by L.J. Schiff [355] appears directed to a general audience, maybe as a text-book on CA, whereas the *Encyclopedia of Complexity* [288] stands up as an updated printed reference, compiling numerous entries dealing with particular aspects of CA. The book by Ilachinski [208] contains an extensive bibliography and provides a listing of CA resources on the WWW. Last but not least, the book by S.Wolfram [426] received the special attention of the CA community, with early reviews such as [101] and [165].

But the aim of this book is not to present a comprehensive review of the avatars of the CA approach but rather to enlarge the paradigm in a relatively unexplored direction: the consideration of past states (history) in the application of the CA rules (as explained in the following section). To achieve this, we start from a simple scenario: two-dimensional ($d = 2$)

---

[1]In which the update is split into two parts: collision and propagation, intended to guarantee propagation of quantities while keeping the proper updating rules (collision) simple [427].

[2]State variables refer to macroscopic quantities; cell dimension is *larger* [413].

[3]The biennial *Conference on CA for Research and Industry* (ACRI, from its original Italian acronym) aims to present an international forum for researchers who are active in the CA field, as well as for those interested in evaluating the possibility of applying them in their own fields. The *Journal of Cellular Automata* publishes since 2005 an special issue devoted to the annual AUTOMATA workshop, a reunion held since 1995. Outstandingly, the freeware DDLab [431] is a useful tool, in permanent upgrading, for studying finite discrete networks.

CA with two possible state values ($k = 2$) at each site: $\sigma \in \{0, 1\}$, with rules operating on nearest neighbors ($r = 1$). This is the case of Fig. 1.1, in which a cell becomes (or remains) alive if any cell in its neighborhood is alive, but becomes (or remains) dead otherwise. The perturbation in Fig. 1.1 spreads as fast as possible, i.e., at the *speed of light*.

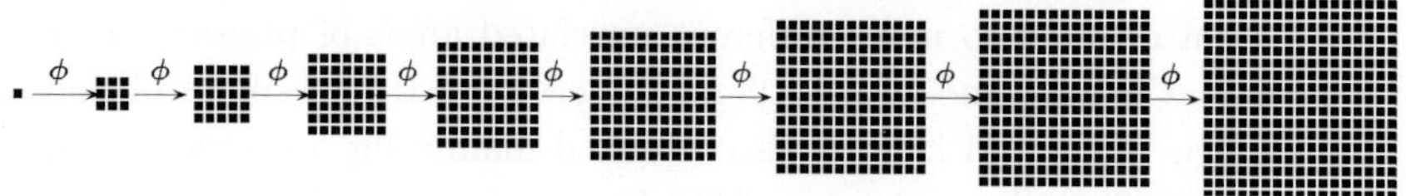

Fig. 1.1   The ahistoric 2D *speed of light* starting from a single active cell.

1.2   Memory

The standard framework of CA can be extended to the consideration of past states by implementing memory capabilities in cells, so that the general form of transition rule stated in the previous section becomes:

$$\sigma_i^{(T+1)} = \phi\big(\{s_{j \in \mathcal{N}_i}^{(T)}\}\big) , \quad \forall i,$$

with $s_j^{(T)}$ being a state function of the series of states of the cell j up to time-step T:

$$s_j^{(T)} = s\big(\sigma_j^{(1)}, \ldots, \sigma_j^{(T-1)}, \sigma_j^{(T)}\big)$$

Thus in CA with memory, while the mappings ϕ remain unaltered, historic memory of all past iterations is retained by featuring each cell by a summary of its past states. So to say, cells *canalize* memory to the map ϕ.

As an example: $s_i^{(T)} = mode(\sigma_i^{(1)}, \ldots, \sigma_i^{(T)})$, with $s_i^{(T)} = \sigma_i^{(T)}$ in case of a tie: $card\{1\} = card\{0\}$. Figure 1.2 shows the effect of *mode* memory on the speed of light starting as in Fig. 1.1. Memory exerts a characteristic inertial effect, which here might be described in terms of *punctuated equilibrium*: i.e., long periods of quiescence in the patterns of live cells, altered by changes that take place at well-defined steps (the punctuation marks), which in Fig. 1.2 are the powers of two time-steps when the automaton fires a new perimeter of live cells. The duration of the stable periods tends to increase with T . A large number of rules evolve essentially in this *punctuated equilibrium* manner in the *most frequent* memory scenario.

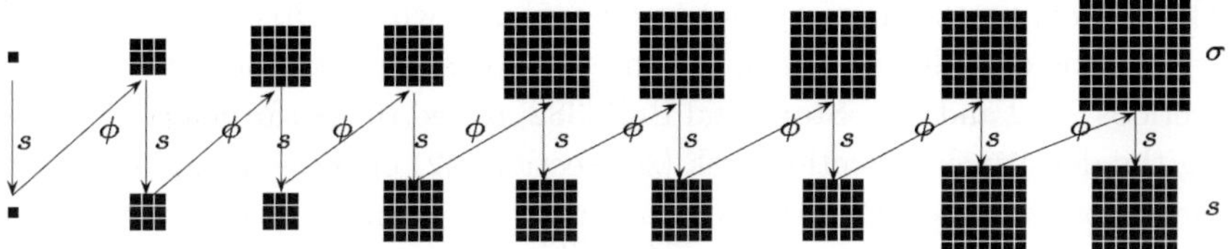

Fig. 1.2   The automaton of Fig. 1.1 with majority memory.

We will refer to conventional CA, in which the new state of a cell depends on the neighborhood configuration solely at the preceding time step, as ahistoric, albeit more precisely they are memory-one systems. Namely, the transition rule in previous section may be written as: $\sigma_i^{(T)} = \phi(\{\sigma_{j \in \mathcal{N}_i}^{(T-1)}\})$.

Disclaimer

The memory mechanism considered here differs from that of other CA with memory reported in the literature, often referred as *higher order* (in-time) CA^4. These, in most cases, explicitly alter the function ϕ and incorporate memory by directly determining the new configuration in terms of the configurations at previous time-steps. Thus, in second order in time (memory of capacity two) rules, the transition rule operates as: $\sigma_i^{(T+1)} = \Phi(\{\sigma_{j \in \mathcal{N}_i}^{(T)}\}, \{\sigma_{j \in \mathcal{N}_i}^{(T-1)}\})$. *Double* memory (in transition rule and in cells) can be implemented as: $\sigma_i^{(T+1)} = \Phi(\{s_{j \in \mathcal{N}_i}^{(T)}\}, \{s_{j \in \mathcal{N}_i}^{(T-1)}\})$. Particularly interesting is the reversible formulation: $\sigma_i^{(T+1)} = \phi(\{\sigma_{j \in \mathcal{N}_i}^{(T)}\}) \ominus \sigma_i^{(T-1)}$. Reversible CA with memory are studied in Chapter 7.

Some authors define rules with *memory* as those with dependence in ϕ on the state of the cell to be updated [427, 223]. So one-dimensional rules with no *memory* adopt the form: $\sigma_i^{(T+1)} = \phi(\sigma_{i-1}^{(T)}, \sigma_{i+1}^{(T)})$. Memory is not here identified with *delay*, i.e., referring cells exclusively to their state values a number of time-steps in the past [343]. So, for example, the cell to be updated may be referenced not at T but at $T - 1$: $\sigma_i^{(T+1)} = \phi(\sigma_{i-1}^{(T)}, \sigma_i^{(T-1)}, \sigma_{i+1}^{(T)})$ [250]. Again, the mapping function is not *extended*, for example, to consider the influence of cell i at time $T - 1$: $\sigma_i^{(T+1)} = \psi(\sigma_i^{(T-1)}, \sigma_{i-1}^{(T)}, \sigma_i^{(T)}, \sigma_{i+1}^{(T)})$, as done in [435].

Memory can be implemented in a *partial* way, thus memory is operative either only in the cell to be updated: $\sigma_i^{(T+1)} = \phi(\sigma_{i-1}^{(T)}, s_i^{(T)}, \sigma_{i+1}^{(T)})$ [344], or only in the cells of the neighbourhood: $\sigma_i^{(T+1)} = \phi(s_{i-1}^{(T)}, \sigma_i^{(T)}, s_{i+1}^{(T)})$.

---

^{4}See, for example, [423],p.118; [208],p.43; or class MEMO in [7],p.7 .

This kind of partial memory, not to be confused with limited trailing memory, will be considered only in passing when dealing with the LIFE rule in Sections 2.2.2 and 3.3 . Stone and Bull [382] tested the performance in solving the density classification task (see section 8.2) of these partial memory variations, together with that of implementing memory in just one of the neighbors, i.e., $\sigma_i^{(T+1)} = \phi\big(s_{i-1}^{(T)}, \sigma_i^{(T)}, \sigma_{i+1}^{(T)}\big)$.

The use of the locution *associative* memory usually refers, when used in the CA context, to the study of configuration attractors [160, 267], which are argued by A.Wuensche [430] to constitute the network's global states contents addressable memory in the sense of Hopfield [203].

To the best of our knowledge the study of the effect of memory on CA has been rather neglected . Thus, for example, Wuensche and Lesser ([429],p.15) just mention the possibility of *historical time reference*, which is excluded from their general study as "it would result in a qualitatively different behavior", whereas S.Wolfram [424] in the context of higher order linear CA, refers to "somewhat involved analysis not performed here".

Finally, it is not intended here to emulate *human* memory, i.e., the associative, pattern matching, highly parallel function of human memory. The aim is just to *store* the past, or just a part of it, to make it *work* in the dynamics. Thus, *working storage* might replace here the use of the term *memory*, avoiding the anthropomorphic, and rather unavoidable, connotations of the word memory.

Chapter 2

Average type memory

2.1 Average memory

Historic memory can be weighted by applying a geometric discounting process in which the state $\sigma_i^{(T-\tau)}$, obtained τ time steps before the last round, is actualized to $\alpha^\tau \sigma_i^{(T-\tau)}$, α being the *memory factor* lying in the $[0,1]$ interval. This well known mechanism fully takes into account the last round ($\alpha^0 = 1$), and tends to *forget* the older rounds.

Every cell will be featured by the rounded weighted mean of all its past states, so the memory mechanism is implemented in two steps at time-step T:

(i) The unrounded weighted mean (m) of the states of every cell is computed first:

$$m_i^{(T)}(\sigma_i^{(1)}, \ldots, \sigma_i^{(T)}) = \frac{\sigma_i^{(T)} + \sum_{t=1}^{T-1} \alpha^{T-t} \sigma_i^{(t)}}{1 + \sum_{t=1}^{T-1} \alpha^{T-t}} \equiv \frac{\omega_i^{(T)}}{\Omega(T)}$$

(ii) Then, the trait state s is obtained by rounding the obtained m by comparing it to the landmark 0.5 (if $\sigma \in \{0,1\}$), assigning the last state in case of an equality to this value, so that:

$$s_i^{(T)} = \mathcal{H}(m_i^{(T)}) = \begin{cases} 1 & if \quad m_i^{(T)} > 0.5 \\ \sigma_i^{(T)} & if \quad m_i^{(T)} = 0.5 \\ 0 & if \quad m_i^{(T)} < 0.5 \end{cases}.$$

The choice of the memory factor α simulates the long-term or remnant memory effect: the limit case $\alpha = 1$ corresponds to a memory with equally

7

weighted records (*full* memory, equivalent to the *mode* if $k = 2$), whereas $\alpha \ll 1$ intensifies the contribution of the most recent states and diminishes the contribution of the more remote states (short-term working memory). The choice $\alpha = 0$ leads to the ahistoric model. This memory implementation will be referred to as α-memory.

It is remarkable that this geometric memory mechanism is not *holistic* but *cumulative* in its demand for knowledge of past history: the whole $\{\sigma_i^{(t)}\}$ series needs not be known to calculate the term $\omega_i^{(T)}$ of the memory *charge* $m_i^{(T)}$, while to (sequentially) calculate $\omega_i^{(T)}$ one can resort to the already calculated $\omega_i^{(T-1)}$ and compute: $\omega_i^{(T)} = \alpha \omega_i^{(T-1)} + \sigma_i^{(T)}$. Consequently, only one number per cell needs to be stored. This positive property is accompanied by the drawback of any weighted average memory : it computes with real numbers, which is not in the realm of proper CA, that works only with integer arithmetics.

Computationally it is a saving if instead of calculating $m_i^{(T)}$ for every cell, we calculate the numerator $\omega_i^{(T)}$ all across the lattice and compare these figures to the factor $\dfrac{1}{2}\Omega(T) = \dfrac{1}{2}\sum_{t=1}^{T} \alpha^{T-t} = \dfrac{1}{2}\dfrac{\alpha^T - 1}{\alpha - 1}$.

Initially, $s_i^{(1)} = \sigma_i^{(1)}$, $s_i^{(2)} = \sigma_i^{(2)}$. After $T = 3$, history does not alter the series ($s_i^{(3)} = \sigma_i^{(3)}$) if $\sigma_i^{(1)} = \sigma_i^{(2)} = \sigma_i^{(3)}$, if $\sigma_i^{(2)} = \sigma_i^{(3)}$, or $\sigma_i^{(1)} = \sigma_i^{(3)}$ [51]. But the scenario may change if $\sigma_i^{(1)} = \sigma_i^{(2)} \neq \sigma_i^{(3)}$: $m(0,0,1) = \dfrac{1}{\alpha^2 + \alpha + 1} = 0.5 \equiv m(1,1,0) = \dfrac{\alpha^2 + \alpha}{\alpha^2 + \alpha + 1} = 0.5 \equiv \alpha^2 + \alpha - 1 = 0$ $\Leftrightarrow \alpha_3 = 0.61805$. Thus, provided that $\alpha > \alpha_3$, cells with state histories 001 and 110 will be assigned following $T = 3$ the trait states 0 and 1 respectively, instead of the last 1 and 0.

In the most unbalanced scenario, $\sigma_i^{(1)} = ... = \sigma_i^{(T-1)} \neq \sigma_i^{(T)}$, it holds that: $m = 0.5 \Rightarrow \alpha_T^T - 2\alpha_T + 1 = 0$ [2.1], where α_T holds for the critical value of α below which memory has no effect in simulations up to time-step T:

$$
\begin{cases}
m(0,0,\ldots,0,1) = \dfrac{1}{2} \equiv \qquad 1 \qquad = \dfrac{1}{2}\dfrac{\alpha^T - 1}{\alpha - 1} \\[4mm]
m(1,1,\ldots,1,0) = \dfrac{1}{2} \equiv \dfrac{\alpha^T - 1}{\alpha - 1} = \dfrac{1}{2}\dfrac{\alpha^T - 1}{\alpha - 1}
\end{cases}
\Rightarrow \alpha_T^T - 2\alpha_T + 1 = 0
$$

At $T = 4$, it is $\alpha_4^4 - 2\alpha_4 + 1 = 0 \Rightarrow \alpha_4 = 0.5437$. When $T \to \infty$, the equation [2.1] becomes: $-2\alpha_\infty + 1 = 0$, thus, in the $k = 2$ scenario, α-memory is not effective if $\alpha \leq 0.5$.

2.2 Two-dimensional lattices

2.2.1 *Totalistic rules*

In *totalistic* rules the value of a site depends only on the sum of the values of its neighbors and not on their individual values: $\sigma_i^{(T+1)} = \phi\left(\sum_{j \in \mathcal{N}_i} \sigma_j^{(T)} \right)$, that becomes with memory: $\sigma_i^{(T+1)} = \phi\left(\sum_{j \in \mathcal{N}_i} s_j^{(T)} \right)$. It has been argued that totalistic rules exhibit behavior characteristic of all CA.

A simple example is the *parity* rule: $\sigma_i^{(T+1)} = \sum_{j \in \mathcal{N}_i} s_j^{(T)} \mod 2$, in words: cell alive if the number of neighbors is odd, dead on the contrary case. Figure 2.1 shows the effect of memory on the parity rule starting from a single live cell in the Moore neighborhood, i.e., the eight nearest-neighbors in the Euclidean tessellation plus the cell itself. In Fig. 2.1, as stated in the previous section as a general fact,

- Memory has no effect up to $T = 3$
- The simulations corresponding to $\alpha = 0.6$ or below show the ahistoric pattern at $T = 4$, whereas memory leads to a pattern different to the ahistoric at $T = 4$[1].
- The pattern at $T = 5$ for $\alpha = 0.54$ and $\alpha = 0.55$ differ when $\alpha \geq 0.7$.

Non-low levels of memory tend to freeze the dynamics from the early time-steps, e.g. over 0.54 in Fig. 2.1. In the particular case of full memory small oscillators of short range in time are frequently generated, such as the period-two oscillator that appears as soon as $T = 2$ in Fig. 2.1. The group of evolution patterns shown in the [0.503,0.54] interval of α variation of Fig. 2.1, might not be expected to be generated by the parity rule, because they are too *sophisticated* for this simple rule. On the contrary, the evolution patterns with very small memory, $\alpha = 0.501$, resemble those of the ahistoric model in Fig. 2.1. But this similitude breaks later on, as Fig. 2.2 reveals: from $T = 19$, the parity rule with $\alpha = 0.501$ memory evolves to produce patterns notably different to the ahistoric ones. These patterns tend to be framed in squares of size not more than $T \times T$, whereas in the

---

[1] That of $T = 2$ in the particular case of Fig. 2.1: after $T = 3$, not only the new outer live cells at $T = 3$ are featured as dead, but also the outer live cells at $T = 2$ are featured as dead cells after $T = 3$, so the only cell featured as alive after $T = 3$ is the central cell, which leads at $T = 4$ to the same pattern as after the initial scenario.

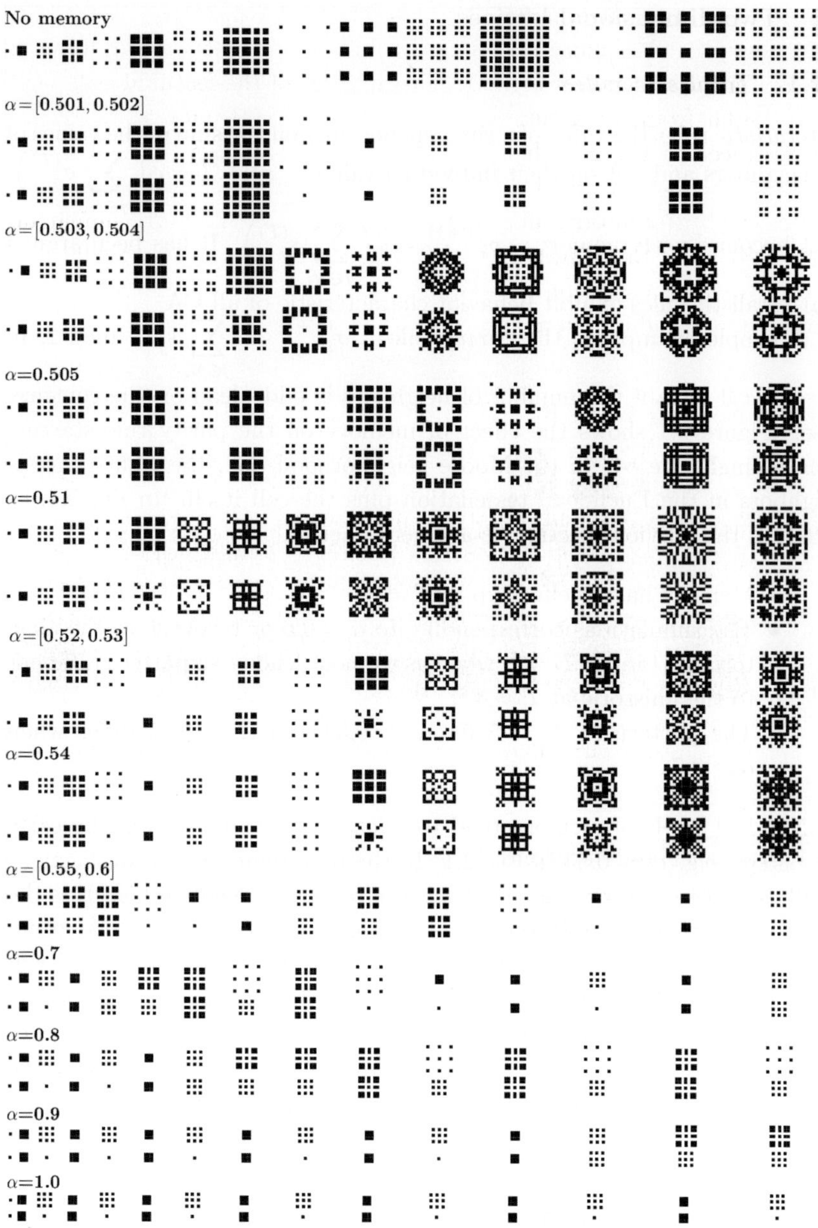

Fig. 2.1   The 2D parity rule with memory up to $T = 15$.

ahistoric case, the patterns tend to be framed in $2T \times 2T$ square regions, so even very small memory induces a very notable reduction in the affected cell area in the scenario of Fig. 2.1. The patterns of the featured cells tend not to be far from the actual ones, albeit examples of notable divergence can be traced in Fig. 2.1. In the very small memory scenario of Fig. 2.1, memory has no effect up to $T = 9$, when the pattern of featured live cells reduces to the initial one; afterwards both evolutions are fairly similar up to $T = 18$, but at this time step both kinds of patterns differ markedly, and from then on the evolution patterns in Fig. 2.2 notably diverge from those generated in the ahistoric model.

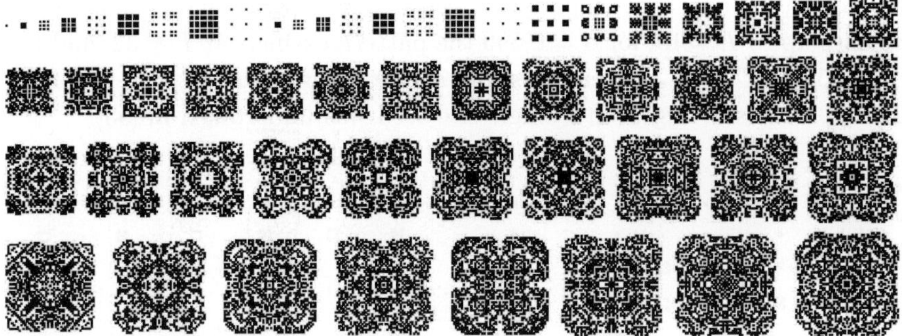

Fig. 2.2   The 2D parity rule with $\alpha = 0.501$ memory starting from a single site live cell up to $T = 55$.

Two-state, two-dimensional totalistic rules are characterized by a sequence of binary values $(\beta_\mathbf{s})$ associated with each of the possible values of the sum $(\mathbf{s})$ of the cell state values of the neighbors of a given cell. The rules are conveniently specified by a decimal integer, to be referred to as their rule number $\mathcal{R}$. It is $s \in \{0, 9\}$ when operating in the Moore neighborhood, in which case,

$$(\beta_9, \beta_8, \beta_7, \beta_6, \beta_5, \beta_4, \beta_3, \beta_2, \beta_1, \beta_0)_{binary} \equiv \sum_{\mathbf{s}=0}^{9} \beta_\mathbf{s} 2^\mathbf{s} \; decimal \; = \mathcal{R}$$

The rule number of this kind of totalistic rules ranges from 0 to $1023 = \mathrm{VR}_2^{10}\text{-}1$. *Quiescent* rules do not transform a dead cell with all neighbors dead into a live cell. The binary specification of a quiescent rule ends with a 0, its decimal rule number is even. This is the case of the parity rule 682 (1010101010) in Fig. 2.1.

To give consideration to historic memory in two-dimensional CA tends to confine the disruption generated by a single live cell. As a rule, full memory tends to generate oscillators, and less historic information retained, i.e. smaller α value, implies an approach to the ahistoric model in a rather smooth form. But the transition which decreases the memory factor from $\alpha = 1.0$ (full memory) to $\alpha = 0.5$ (ahistoric model), is not always regular, and some kind of *erratic* effect of memory can be traced. Thus, as an example, rule 514 (1000000010) in Fig. 2.3 provokes extinction when α in [0.7,0.8], but not in [0.9,1,0]; when in [0.52,0.54] but not if $\alpha = 0.60$, when in [0.501,0.502], but not in [0.503,0.504]. In the [0.501,0.502] interval memory has no effect up to $T = 9$, as in Fig. 2.2. But here the effect is so drastic that a notable pattern with 148 live cells in $T = 8$ is unexpectedly followed by extinction two time-steps later. Beyond the range of Fig. 2.3, which reaches up to $T = 19$, for $\alpha = 0.599$ the pattern extincts at $T = 32$, and for $\alpha = 0.51$ at $T = 34$. A general study of the effect of memory on totalistic 2D CA rules starting from a single site seed was undertaken in [53].

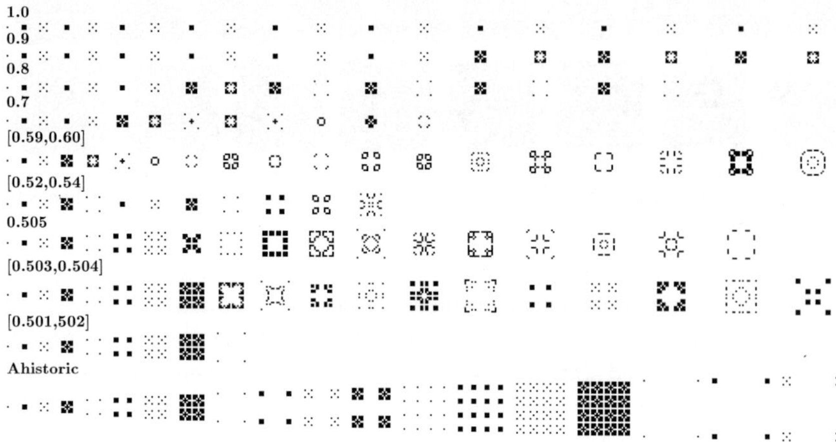

Fig. 2.3   The rule 514 (1000000010) with memory from a single site live cell.

In between the two types of effects of memory aforementioned, those gradual and erratic, for some rules like the two-dimensional parity rule 682 (1010101010), memory does not produce unexpected extinctions, but there are notable discontinuities in its effect, such as the the sharp restraining of the dynamics when the memory factor is over 0.54, compared to the rich evolution under this parameter value. Despite its formal simplicity, the parity rule exhibits complex behavior [219], which maintains the density and changing rate at high levels, close to 0.5, as can be seen in Fig. 2.8 regarding the density of the one-dimensional analog parity rule 150.

Figures 2.4 and 2.5 show the effect of memory on the parity rule in the hexagonal and triangular tessellations respectively, when starting from a small block of live cells. In both scenarios, full memory induces a period-two oscillator as early as at $T = 4$.

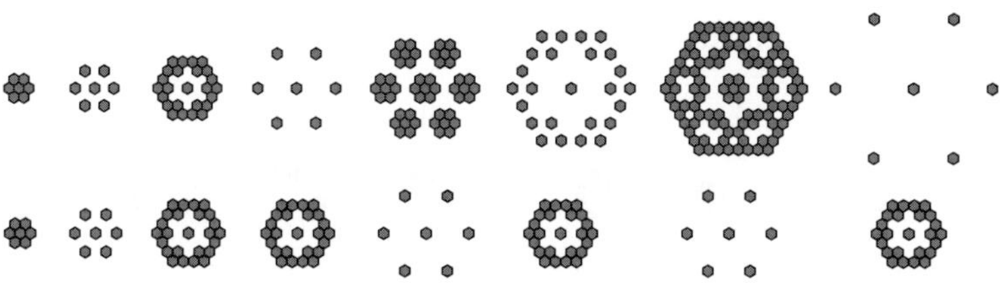

Fig. 2.4   The parity rule in the hexagonal tessellation. Ahistoric (up) and full memory models.

Fig. 2.5   The parity rule in the triangular tessellation.

2.2.2   *LIFE*

The well-know LIFE rule is described in words as: a cell that was dead will come alive if it is adjacent to exactly three live neighbors; a live cell will die unless it is adjacent to either two or three live neighbors [54]. Thus, LIFE is a semitotatistic rule of the form: $\sigma_i^{(T+1)} = \phi\Big(\sigma_i^{(T)}, \sum_{j \in \mathcal{N}_i} \sigma_j^{(T)}\Big),$

that becomes with memory: $\sigma_i^{(T+1)} = \phi\left(s_i^{(T)}, \sum_{j \in \mathcal{N}_i} s_j^{(T)}\right)$ [19].

Stable forms under LIFE remain, of course, stable with majority memory (see in [162] or in [426] p.964 the commonest stable forms). Also for those forms that reproduce in $T = 3$ the initial form will remain unaltered. So do the period-two (flip-flops) forms collected in Table 2.1. This is so because, starting from any of this kind of structures, the most frequent and actual states are coincident, i.e., $\sigma_i^{(T)} = s_i^{(T)}$, $\forall$ i,T.

Table 2.1   Period-two oscillators in LIFE.

But in general, the inertial (or conservative) effect of majority memory dramatically changes the dynamics of LIFE. Thus, the *vividness* that some small clusters exhibit in LIFE, has not been detected in LIFE with full memory. Thus, for example, the r-pentomino ▗▛ stabilizes at T=29 as ▗▙▖ , and the *glider* (a translating oscillator that moves across the lattice [75]) in LIFE does not *glide* at all in LIFE with memory as shown in Table 2.2 at its first time steps. The *glider* stabilizes from $T = 45$, very close to its initial position, as the *tub*: ✦ .

Table 2.2   The *glider* in ahistoric (upper) and full memory (lower) LIFE.

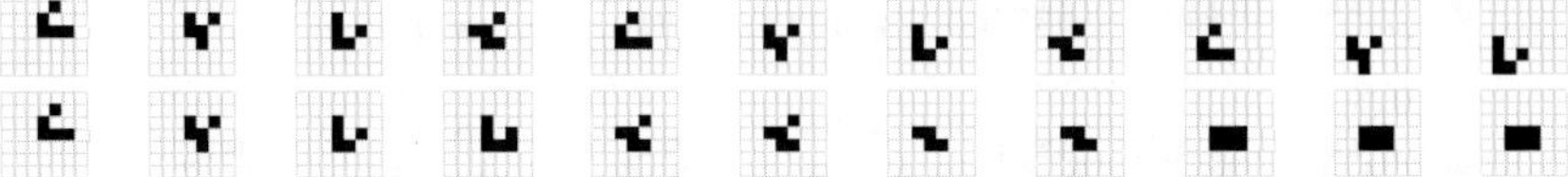

As the size of a configuration increases, often live clusters tend to persist with a higher number of live cells in LIFE with memory than in the ahistoric formulation. Thus, the 8×8 square in Table 2.3 does not extinguish but generates a period-two oscillator.

A single *mutant* appearing in a stable rich configuration of live cells (*agar*) can lead to its destruction in the ahistoric model, whereas its effect tends to be restricted to its proximity with memory. Thus, the stable agars in Table 2.4 suffer an inexorable destruction without memory due to a sole error in its central part. A period-two oscillator is generated with full memory (lower) in both examples, as soon as at T=2. The effect of

Table 2.3 The 8×8 square in ahistoric (up) and full memory (down) LIFE.

short-term memory on the agars of Table 2.4 is shown in Table 3.4.

Table 2.4 The effect of a virus in a tub (up) and a block (down) agar in LIFE. Ahistoric (upper) and full memory (lower) models.

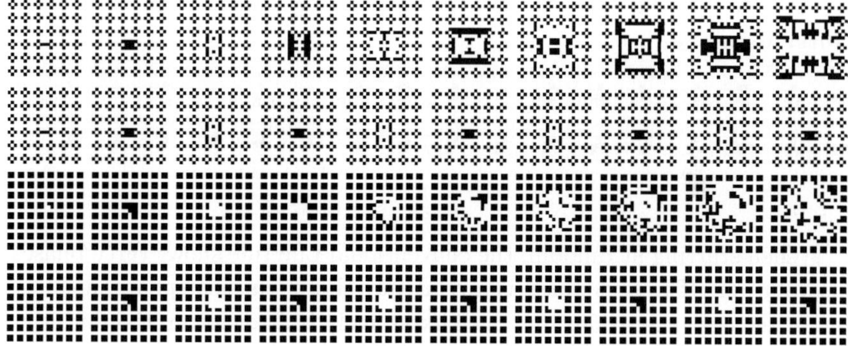

Due to the inertial effect associated with majority memory, finding configurations that can grow without limit when endowed with this kind of memory is a rather challenging goal that we have not achieved. J.H.Conway conjectured that no finite configuration can grow without limit regarding LIFE in 1970 and offered a \$ 50 prize for the first proof or disproof. We support the same enterprise with a prize of 50 euro in what respect LIFE with majority memory. Sorry, no updated prize amount. Restraining the range of memory to the last states may facilitate this task, but the solution found by W.Gosper, the glider-gun [114], is no longer valid, as shown in Section 3.3 with short-term majority memory.

Partial memory can be implemented in LIFE either as inner memory:

$$\sigma_i^{(T+1)} = \phi\left(s_i^{(T)}, \sum_{j \in \mathcal{N}_i} \sigma_j^{(T)}\right),$$

or as outer memory:

$$\sigma_i^{(T+1)} = \phi\left(\sigma_i^{(T)}, \sum_{j \in \mathcal{N}_i} s_j^{(T)}\right).$$

Table 2.5 shows the evolution of the glider in both scenarios, with cells endowed with full memory. In the former, the dynamic ceases at T = 78 producing the stable (*beehive*) configuration: , in the latter, Table 2.5 shows how a period-two oscillator is generated at T = 11. The evolution of the glider with partial τ=3 majority memory is given in Table 3.13.

Table 2.5   The effect of partial full memory in the LIFE glider.

Inner memory

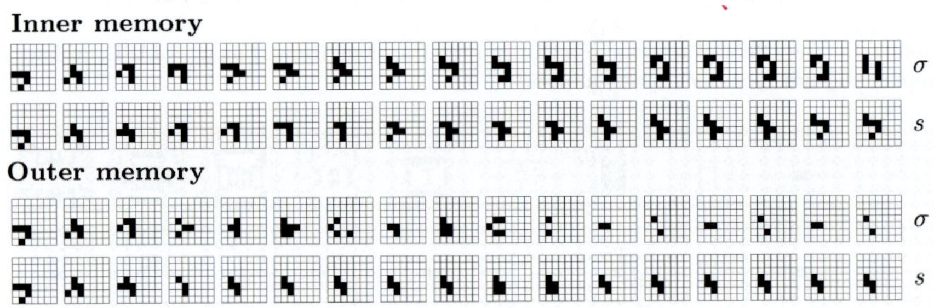

Outer memory

When starting at random, the conventional Life rule shows three phases of evolution: The first phase is a relatively short transient phase, at most ten or tens of generations, in which excessively high or low initial densities adjust themselves; the second phase may last for thousands of generations in which nothing seems to be definite; followed by the third and final phase in which isolated groups of cells go through predictable cycles of evolution. The left panel of Table 3.11 shows the adjustment in density from initial values ranging from 10 % to 80 % up to plateaus around a 10 % of live cells.

A simple measure of the degree of the activity of a given automaton is the measure of its *changing rate* :

$$\mathsf{c}(T) = \frac{1}{N{\times}N} \sum_{i=1}^{N}\sum_{j=1}^{N} \sigma_{i,j}^{(T-1)} \oplus \sigma_{i,j}^{(T)} \quad , \quad \mathsf{c}(1) = \frac{1}{N{\times}N} \sum_{i=1}^{N}\sum_{j=1}^{N} \sigma_{i,j}^{(1)} .$$

Table 2.6   The effect of memory on the density (red) and changing rate (blue) in the LIFE rule starting at random.

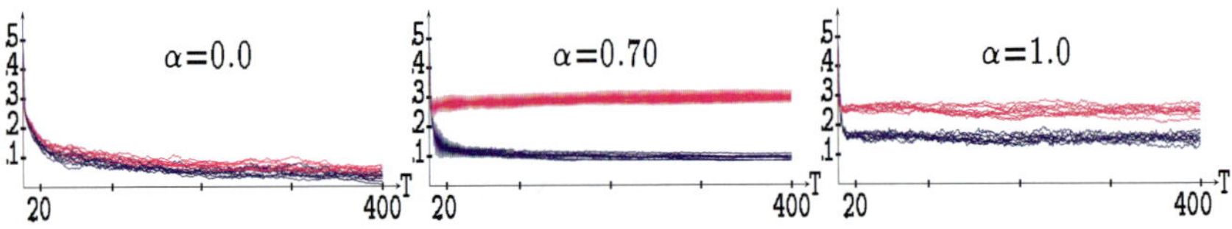

Table 2.6 shows how the inertial effect of memory restrains the decrease in density and changing rate characteristic in the ahistoric LIFE formulation. Nine different simulations starting with a 50 % of live cells randomly distributed across a 100×100 lattice are shown in Table 2.6. Table 2.7 shows the patterns at $T=400$ in one of such simulations.

Table 2.7   LIFE starting at random at $T=100$.

| Ahistoric | $\alpha=0.7$ | $\alpha=1.0$ |

2.3   One-dimensional layers

2.3.1   *Elementary rules*

Elementary rules are one-dimensional ($d = 1$) CA with two possible values ($k = 2$) at each site ($\sigma \in \{0,1\}$), with rules operating on nearest neighbors ($r = 1$). Following Wolfram's notation, these rules are characterized by a sequence of binary values (β) associated with each of the eight possible triplets $\left(\sigma_{i-1}^{(T)}, \sigma_i^{(T)}, \sigma_{i+1}^{(T)}\right)$:

| 111 | 110 | 101 | 100 | 011 | 010 | 001 | 000 |
| β_1 | β_2 | β_3 | β_4 | β_5 | β_6 | β_7 | β_8 |

The *rule number* of elementary CA, $\mathcal{R} = \sum_{i=1}^{8} \beta_i 2^{8-i}$, ranges from 0 to 255.

Legal rules are *reflection symmetric* (so that 100 and 001 as well as 110 and 011 yield identical values), and *quiescent* ($\beta_8 = 0$). These restrictions leave 32 possible *legal* rules of the form: $\beta_1\beta_2\beta_3\beta_4\beta_2\beta_6\beta_40$.

The computer code in Table 2.8 generates the ahistoric and α-memory patterns of the parity rule 150, with $\alpha=1.0$ in the instance shown.

Table 2.8   A $MATLAB^{\textregistered}$program for rule 150 with and without memory.

```
T=13; N=2*T+1; rule=150; alpha=1.0;
for memo=0:1
    [SIGMA,w,W]=init(N,T,alpha);
    for t=1:T;   S=SIGMA;ST(t,:)=SIGMA;
        if(memo==1&t>2)
            for i=1:N
                if(w(i)<W(t))S(i)=0;end
                if(w(i)>W(t))S(i)=1;end
            end
        end
        [SIGMA]=TRANSITION(S,N,rule);
        w=alpha*w+SIGMA;
    end;   subplot(1,2,memo+1);imagesc(ST);
end
function [SIGMA]=TRANSITION(S,N,rule)
    SIGMA(1)=RULES(S(N),S(1),S(2),rule);
    for i=2:N-1
        SIGMA(i)=RULES(S(i-1),S(i),S(i+1),rule);
    end
    SIGMA(N)=RULES(S(N-1),S(N),S(1),rule);
function new=RULES(l,c,r,rule)
    switch {rule}
    case {150};new=mod(l+c+r,2)
    end
function [SIGMA,w,W]=init(N,T,alpha)
    SIGMA(1:N)=0;c=(N+1)/2;SIGMA(c:c)=1;
    W(1:T)=1;for t=2:T; W(t)=alpha*W(t-1)+1;end;W=W/2;w=0;
```

Figure 2.6 shows the spatio-temporal patterns of elementary legal rules affected by memory when starting from a single live cell[2]. The evolution of the cellular automata at successive time steps is shown at successive horizontal lines. Patterns are shown up to $T = 63$, with the memory factor varying from 0.6 to 1.0 by 0.1 intervals, and adopting also values close to the limit of its effectiveness: 0.5.

As a rule, the transition from the $\alpha = 1.0$ (fully historic) to the ahistoric scenario is fairly gradual, so that the patterns become more expanded as less historic memory is retained (smaller α). Rules 50, 122, 178,250, 94, and 222,254 are paradigmatic of this smooth evolution. Rules 126 (studied in [212]) and 182 also present a gradual evolution, although their patterns with high levels of memory models hardly resemble the historic ones. But the non-smooth effect of memory is also present in Fig. 2.6: *i*) with rule 150 is sharply restrained at $\alpha = 0.6$, *ii*) the important rule 54 expires in [0.8,0.9], but not with full memory, *iii*) the rules in the group {18,90,146,218} become

---

[2]Evolving from a single active site, history does not affect the simple class-1 and 2 rules. History does affect the remaining 16 legal rules: *i*) the kind of legal class-3 simple rules which copies the initial 1 to generate a uniform structure which expands by one site in each direction on each time step (exemplified by 50, 122 and, of course, 254), and *ii*) the *complex* (non-simple) legal rules [52].

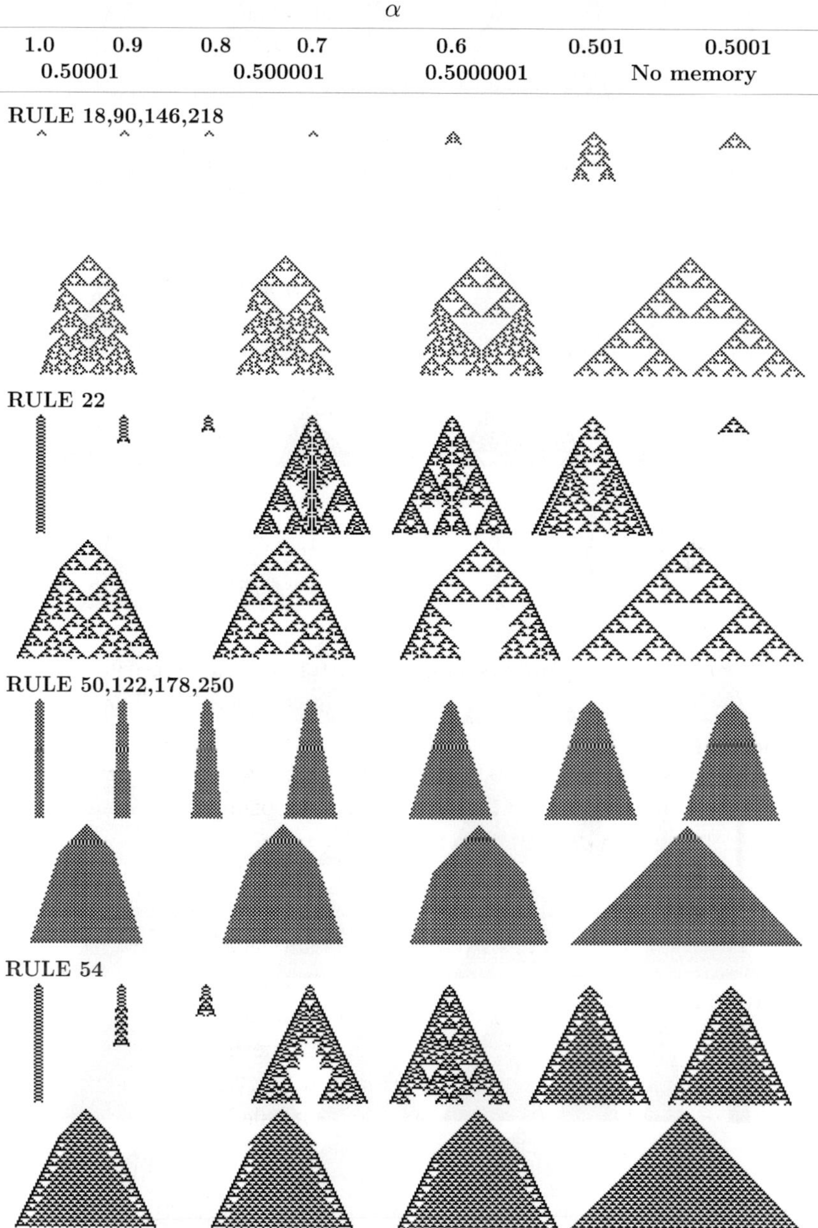

Fig. 2.6   Elementary, legal rules with memory from a single site live cell.

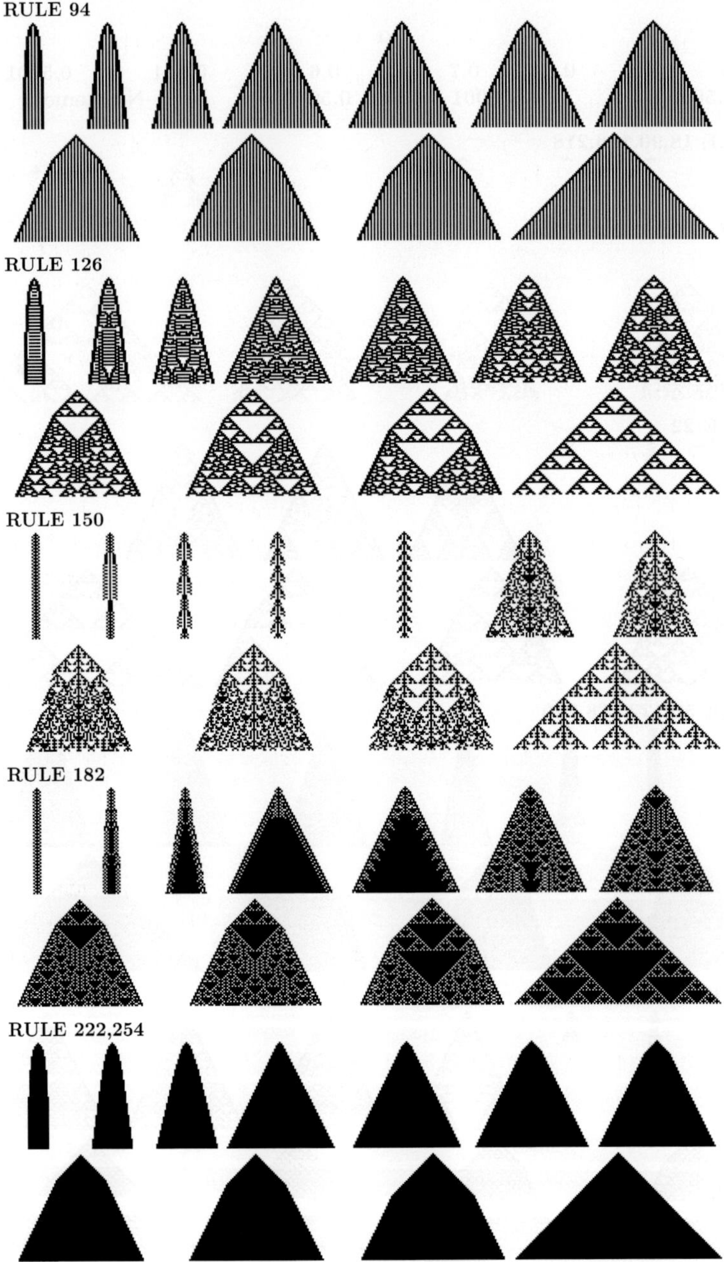

Fig. 2.6   (*continued*)

extinct from $\alpha = 0.5001$[3], and *iv*) rule 22 becomes extinct for $\alpha = 0.5001$, not in 0.501, 0.6, and 0.7, again extinguished at 0.8 and 0.9, and finally generate an oscillator with full memory. A rather erratic bhavior.

As noted elsewhere [168, 215], rules 18, 22, 122, 146 and 182 *simulate* rule 90 in that their behavior coincides when restricted to certain spatial subsequences. Starting with a single site live cell, the coincidence fully applies in the historic model for rules 90, 18 and 146. Rule 22 shares with these rules the extinction for high α values, with the notable exception of no extinction in the fully historic model. Rules 122 and 182 diverge in their behavior: there is a gradual decrease in the width of evolving patterns as α is higher, but they do not reach extinction.

Random starting

Figure 2.7 shows the effect of memory on elementary, legal rules when starting at random: the values of sites are initially uncorrelated and chosen at random to be 0 (*blank*) or 1 (*gray*) with probability 1/2. The pictures show also the differences in patterns resulting from reversing the center site value. The *damaged* region is enhanced with black pixels, corresponding to the site values that differed among the patterns generated with the two initial configurations. Patterns are shown up to $T = 60$, in a line of size 129. The memory factor varying from 0.6 to 1.0 by 0.1 intervals.

Periodic boundary conditions are imposed on the edges all across the simulations in this book. It has been argued that periodic boundary conditions tend to minimize the finite-size effects [89, 186].

Some legal rules are unaffected by memory when starting from a random initial configuration. That is the case of the *simple* [420] rules which evolve to the null state in the ahistoric model (e.g., rules 0, 32 or 72), the *identity* rule 204, or the *majority* rule 232. Other rules are minimally affected by memory. For example, rules 250 and 254 which soon blacken the space, or those which serve as *filters* (Wolfram's Class II) exemplified by rule 36.

Only the nine legal rules which generate non-periodic patterns in the ahistoric scenario are significantly affected by memory. Figure 2.7 shows the evolution patterns of these rules, studied by Grassberger [168] and Jen [215], among others, in the ahistoric scenario.

---

[3]Memory kills the evolution for these rules already at $T=4$ for α values over α_3. Thus over 0.6 in Fig. 2.6, the actual spatio-temporal pattern at $T=3$ is , consequently all the cells, even the two outer cells alive at $T=3$, are featured as dead, which means extinction at $T=4$.

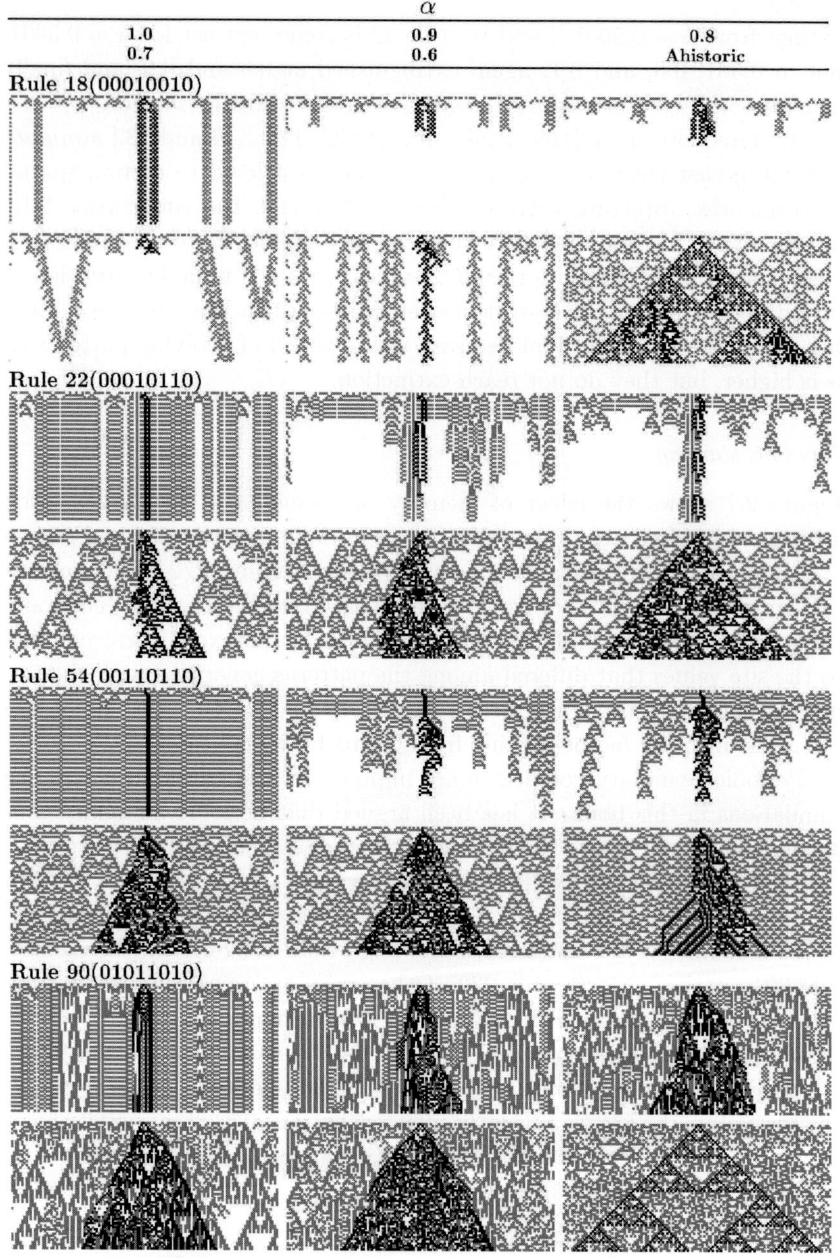

Fig. 2.7   Elementary, legal rules with memory starting at random.

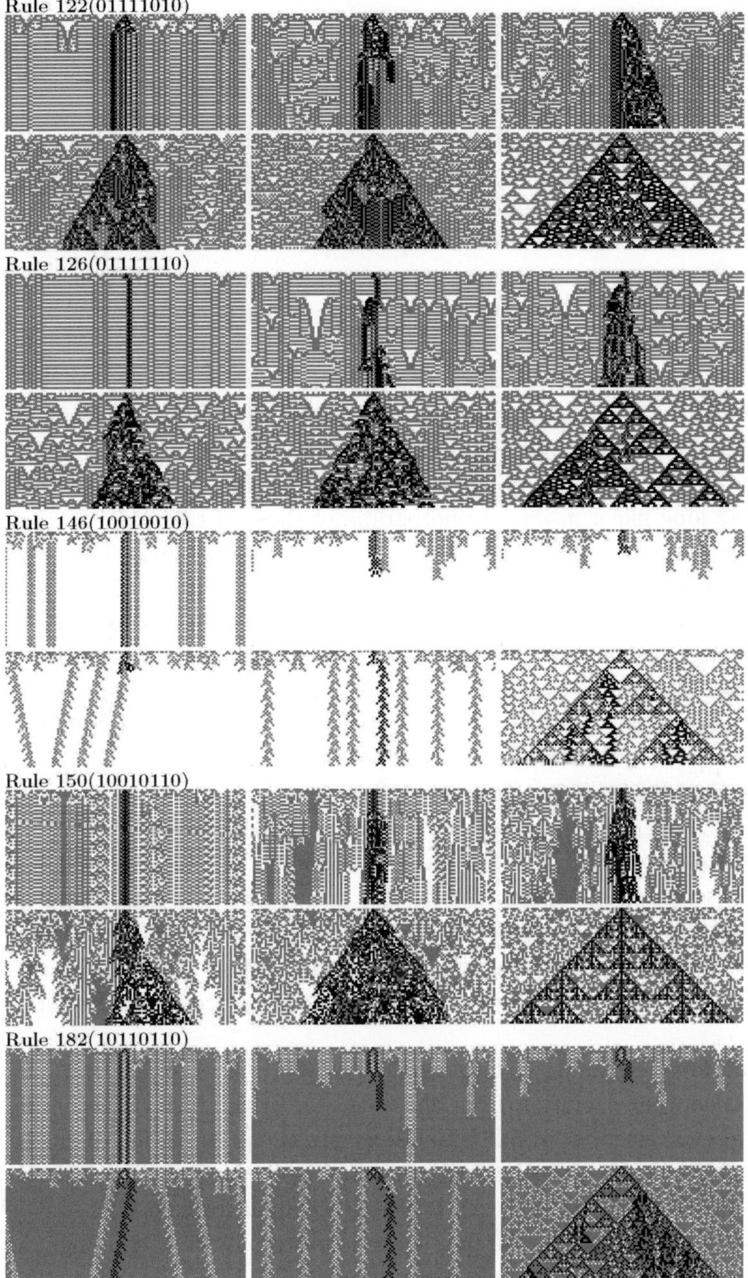

Fig. 2.7 (*continued*)

Rule 18 (00010010) allows only dead cells with exactly one living neighbor to become alive. All living cells die. The seemingly simple rule 18 was one of the first rules carefully analyzed [169], following its *intriguing* properties [212]. History has a dramatic effect on rule 18. Even at the low value of $\alpha = 0.6$, the appearance of the spatio-temporal pattern completely changes: a number of isolated periodic structures are generated, far from the distinctive inverted triangles world of the ahistoric pattern. For $\alpha = 0.7$, the live structures are fewer, advancing the extinction found in [0.8,0.9]. In the fully historic model, only simple periodic patterns of live cells survive.

Rule 146 is affected by memory in much the same way as rule 18. This is because, although their rule numbers are relatively distant, their binary configurations differ only in their β_1 value. The spatio-temporal patterns of rule 182 and its equivalent rule 146 are reminiscent, though those of rule 182 look like a *negatives* photograph of those of rule 146.

The effect of memory on rules 22 and 54 is similar. Their spatio-temporal patterns in $\alpha = 0.6$ and $\alpha = 0.7$ keep the essentials of the ahistoric, although the inverted triangles become enlarged and tend to be more sophisticated in their *basis*. A notable discontinuity is found for both rules as the value of the memory factor increases, so with $\alpha = 0.8$ and $\alpha = 0.9$ only a few simple structures survive for both rules. But unexpectedly, the patterns of the fully historic scenario differ markedly from the others, showing a high degree of synchronization. The behavior of rule 54 in the ahistoric model has been featured to some extent as transitional between very simple Wolfram's class I and II rules and chaotic Class III (see next section for a description of Wolfram's classes). Thus, rule 54 appears among the two one-dimensional rules (with Rule 110) that seem to belong to Wolfram's (*complex*) Class IV. Rule 54 (and 110) is one of the rules having a complexity index $\kappa = 2$, which is the *threshold of complexity* after Chua [112].

The four remaining chaotic legal rules (90, 122, 126 and 150) show a much smoother evolution from the ahistoric to the historic scenario: no pattern evolves either to full extinction or to the preservation of only a few isolated persistent propagating structures (solitons). Rules 122 and 126 (close in terms of rule number), evolve in a similar form (particularly when comparing the ahistoric and fully historic patterns), showing a high degree of synchronization in the fully historic model.

Some spatio-temporal patterns of rules in the ahistoric model are reminiscent of the patterns of pigmentation observed on the shells of certain molluscs [426, 237]. Wolfram [[425],p.413] illustrates this idea by showing a natural *cone shell* with a pigmentation intended to be reminiscent of the pattern generated by rule 90 in the ahistoric scenario. But the clearings in the shell would suggest rule 22 for some value of α in between 0.6 and 0.7: with $\alpha = 0.6$ the clearings seems to be scarce, with $\alpha = 0.7$ the clearings appear to be excessive.

Density

Figure 2.8 shows the evolution of the average fraction of sites with value 1 at time T, density ρ_T, up to 200 time steps, starting with an initial density $\rho_0 = 0.5$. In order to improve plot resolution, the y-axes scopes have been tailored in this figure. The simulation is implemented for the same rules as in Fig. 2.7 (except rule 182), but with notably wider lattice: $N = 500$.

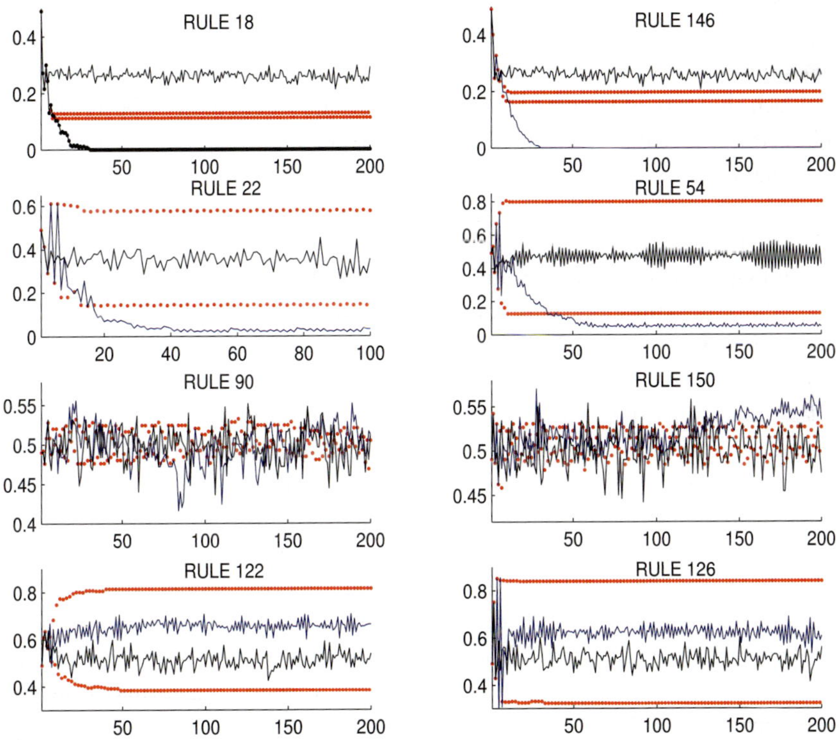

Fig. 2.8 Evolution of the density starting at random in elementary rules affected by memory. Plots code as: blue → full memory, black → $\alpha = 0.8$, undotted red dots → ahistoric model.

A visual inspection of the plots in Fig. 2.8, confirms the general features observed in the patterns in Fig. 2.7. Thus, starting with a disordered configuration of any non-zero density, the evolution of the density ρ_T according to rule 18 in the ahistoric model yields an asymptotic density $\rho_\infty = 1/4$. Figure 2.8 illustrates how this value is reached soon, and how history induces a depletion of the asymptotic density, null for $\alpha = 0.8$ and $\alpha = 0.9$. In the fully historic model, a smooth period-two density oscillator $\{0.105, 0.115\}$ is generated as early as at $T = 8$.

Rule 146 density plots resemble those of rule 18, but with a sharper period two oscillator in the fully historic model: $\{0.175, 0.204\}$. Rule 182 (not present in Fig. 2.8) yields $\rho_\infty = 3/4$ in the ahistoric model, the shape of its evolution density curves resembling the complement to 1 of its equivalent rule 146.

Quoting Grassberger [168], rule 22 is the only one of the elementary rules whose long-time behavior is not yet understood. This totalistic rule, referred as LIFE in one dimension in [303], has been used to illustrate the spontaneous generation of complex structure, in contrast to superficial evidence that would suggest that this simple rule would lead to fairly simple behavior [134, 168, 441]. Wolfram [420] resorts to simulations to report $\rho_\infty = 0.35 \pm 0.02$ for evolution with rule 22.

Rule 54, very notably absent in Wolfram's [420] considerations, again resembles rule 22 with regard to Fig. 2.8. For example, both rules present a very low density in the [0.8,0.9] interval. Wolfram [[425], Table 6] reports $\rho_\infty = 1/2$ for rules 122 and 126. In our simulation, ρ_T oscillates around values slightly over $1/2$ in the ahistoric model.

Wolfram [420] proved that for rule 90 it is $\rho_\infty = 1/2$, independent of the initial density ρ_0 (so long as $\rho_0 \neq 0$). With memory, ρ_T varies erratically around 0.5 in Fig. 2.8 in rules 90 and 150, without either periodic evolution or tendency to a fixed point, not even in the fully historic model. So the effect of memory on the linear rules 90 and 150 turns out to be atypical. Usually, memory depletes the density curves (sharply for rules 18 and 146, softly for rules 122 and 126), and in the fully historic model, synchronous behavior is frequently found (e.g., rules 122 and 126). This is not so with rules 90 and 150. The effect of memory on the multifractal properties of rule 90 has been studied in [350] by using discrete wavelet transforms, after the study made in [351] on these properties on memoryless linear elementary automata. Further analysis of multifractal properties of one-dimensional memoryless CA, with particular reference to rules 90, 105 and 150, has been reported in [302].

The high degree of synchronization visually appreciated for rules 22, 54, 122 and 126 in the fully historic model in Fig. 2.7, stands out also in Fig. 2.8 (un-joined dot plots). Synchronization in CA is not a trivial task since synchronous oscillation is a global property, whereas CA typically employs only local interactions; but the phenomenon of synchronous oscillations occurs in nature in fairly striking forms. The synchronization task, i.e., given any initial configuration, the CA must reach a final configuration, within a finite number of time steps, that oscillates between all $0s$ and all $1s$ at successive time steps, has been investigated by Das *et al.*[119] and Sipper [369], who conclude that rules 21 and 22 are the most effective synchronizer rules.

Difference patterns

Figure 2.7 and the figures in Appendix A show also the differences in patterns (DP) resulting from change in the value of its initial center site value (black pixels).

The perturbations in proper *chaotic* rules propagate to the right and left at a single (maximum) velocity at any time. This behavior illustrates the *butterfly effect* : a small perturbation grows, and finally rules the whole system. The velocity in the *damage spreading* is quantified by means of the left and right Lyapunov exponents (λ_L, λ_R) which measure the rate at which perturbations spread to the left and right, and are given by the slopes of the left and right boundary of the growth of the difference patterns. Thus, zero values indicate periodicity, whereas negative velocity indicates perturbation repair. The maximum λ attainable when $r = 1$ is $\lambda = 1$. The chaotic Class III rules in Fig. 2.7 reach this maximum value of λ. The Lyapunov exponents of ahistoric elementary CA are tabulated in [425].

As a rule, the effect of memory on the DP mimics that on the spatio-temporal patterns, so that the rule parallelisms found for the spatio-temporal patterns are again applicable to the DP. In the case of rule 18 for example, a periodic structure (with only four elements) appears in the fully historic scenario; the differences die out when $\alpha \in [0.7, 0.9]$, and the perturbation remains localized (again in the form of a periodic structure) when $\alpha = 0.6$. Finally, in the ahistoric model, the perturbation grows close to the *speed of light*: $\lambda_L = \lambda_R = 1$. Thus, an effect of memory on damage spreading comparable to that on the whole lattice and with that starting with a single site active cell.

The DP of rule 146 resembles that of rule 18, and evolves in a similar way. The DP of rule 182 resembles that of its equivalent rule 146. The 22 and 54 DP are again similar: the DP is constrained when history begins to actuate ([0.6,0.7]), and ceases, or is localized near the central site, at higher α values. This coincidence in the behavior agrees with their *complexity*: they are the only *legal* rules with no extreme and irrational Lyapunov exponents: $\lambda_L = \lambda_R \simeq 0.75$ for rule 22 and $\lambda_L = \lambda_R \simeq 0.55$ for rule 54 (see Table 6 in [425]). The group of rules 90, 122, 126 and 150 shows a fairly gradual evolution from the ahistoric to the historic scenario, so that the DP appear more constrained as more historic memory is retained, with no extinction for any α value.

The patterns with inverted triangles dominate the scene in the ahistoric difference patterns on Fig. 2.7 (the exception is the peculiar DP of rule 54), but history destroys this common appearance (and that of rule 54), even at the lowest value of α in the figure: $\alpha=0.6$. Thus, there is a sort of discontinuity implied in the consideration of historic memory (perhaps with the exception of rule 22) regarding the DP, which rule 90 might exemplify: memory, at the low rate $\alpha = 0.6$, destroys the structures characteristic of ahistoric DP. To avoid coined terms such as chaotic or random, the DP generated for $\alpha = 0.6$ could be described as *helter-skelter*. Regarding the central site specifically, for rules 22, 54, 90, 122, 126 and 150, the disruption induced by its initial reversion, is in some manner more unpredictable in the historic model with $\alpha = 0.6$ than in the ahistoric. Extreme examples are: rule 90 (after initial reversion, the original evolution is restored in the ahistoric model) and rule 150 (the initial reversion remains for ever).

The conclusions drawn from Fig. 2.7 are supported by the $N = 500$ simulations run for Fig. 2.8. Again, the central site has been reversed and two type of plots implemented to feature the DP: i) the Hamming distance (i.e., the number of non-zero site values of the difference patterns), and ii) the width of these patterns. From these plots (not included here to avoid graphical overloading) the overall conclusion drawn from Fig. 2.7 remains valid: memory implies a depletion in the damaged region (i) and in the speed of propagation of perturbation (ii).

Equivalence classes

In order to systematize the analysis of the DP, one can resort to the equivalence classes, formed under the negative, reflection and negative plus reflection transformations [429, 112]. Memory is expected to affect all the rules of an equivalence class in a similar way, as was already observed in legal class $\{146, 182\}$.

Figure A.1 shows the DP of the remaining rules equivalent to the legal ones, the symmetric but non-quiescent rules of the form: $(\beta_1\beta_2\beta_3\beta_4\beta_2\beta_6\beta_4 1)$. A fairly consistent correspondence in the apparency of the DP is observed between equivalent rules: $\{18, 183\}$, $\{22, 151\}$, $\{54, 147\}$, $\{90, 165\}$, $\{122, 161\}$ and $\{126, 129\}$. The DP of rule 150, which constitutes alone an equivalence class [332], resembles that of its complementary rule 105. Complementary rules assign complementary β values, i.e., $\overline{\beta_i} = 1 \ominus \beta_i$, so their number adds 255.

Figure 2.9 shows the DP of some asymmetric rules affected by memory in a lattice of size $N = 97$ up to $T = 90$. Damage reaches the borders in most of the ahistoric patterns of this figure, because the size $N = 97$ is not large enough for free expansion during $T = 90$ time steps (which would be feasible with $N = 181$). Recall that with $r = 1$, an initial sole mutation may affect the values of at most $2T$ sites after T time steps. The rules in Fig. 2.9 are grouped by equivalence classes and the left and right Lyapunov exponents of the lowest rule number of each class (minimal representative) in the ahistoric model are given after rule codes.

Rules in which almost all changes in initial configuration die out, and rules with $\lambda_{L,R} = 0$ are not greatly affected by memory (e.g., Rules 156 and 100). A considerable number of equivalence classes of asymmetric rules have $|\lambda_L| = |\lambda_R| = \lambda$ with $\lambda = \pm 1$ or $\lambda = -1/2$ as their minimal representative rule. These rules present a *diagonal* as DP in the ahistoric model, which is *rectified* (in the sense of having both Lyapunov exponents evolving to zero) and/or led to extinction by memory. But not always are either extinction or rectification of the trajectory of the perturbation achieved in a uniform way. Examples of unexpected evolution have been found in this context.

The important rule 110 [4], and the others of its equivalence class, may serve as a paradigmatic example of the expected effect of memory: the damage induced by the reversal of the initial central site value becomes more constrained as the memory factor increases, with no discontinuities in the preserving effect. The same applies to all the rules of the equivalence classes of the three that have one irrational Lyapunov exponent: rules 30, 45 and 106.

History however has an unexpected effect on most of the rules whose damage propagation direction alternates: $\lambda_L = (-1, 1), \lambda_R = (1, -1)$. Fig-

---

[4]This rule shows highly complex properties of information transmission, associated with particle-like structures [253]. Rule 110 is a universal Turing machine.

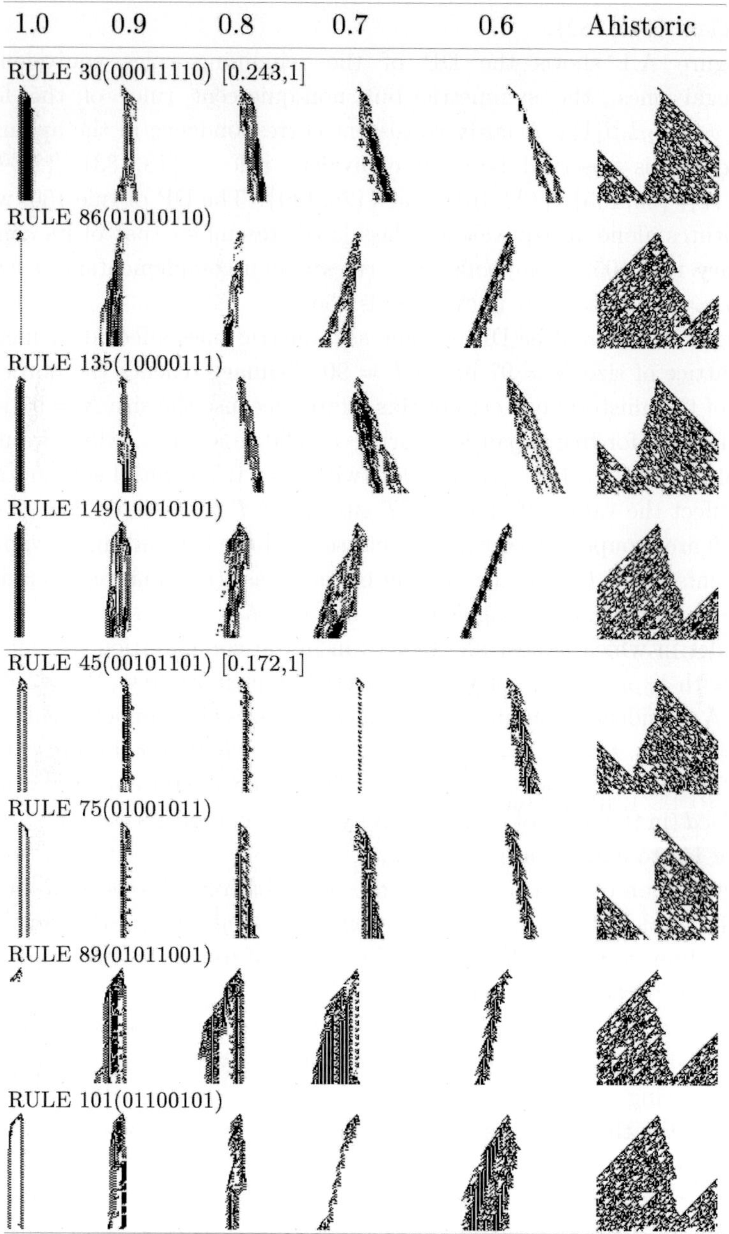

Fig. 2.9   Damage spreading of some asymmetric elementary rules.

RULE 106(01101010) [1,-0.134]

RULE 120(01111000)

RULE 169(10101001)

RULE 225(11100001)

RULE 110(01101110) [0.26-5,0.27-0.0]

RULE 124(01111100)

RULE 137(10001001)

RULE 193(11000001)

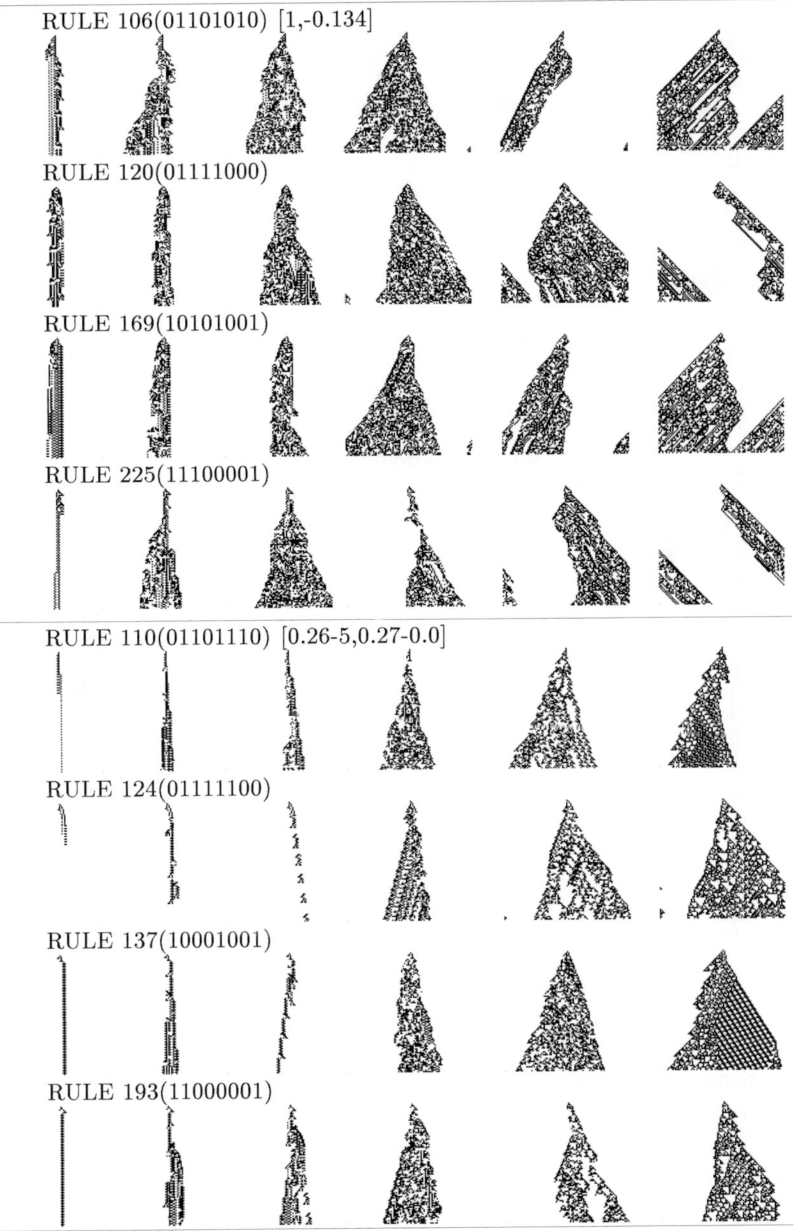

Fig. 2.9 (*continued*)

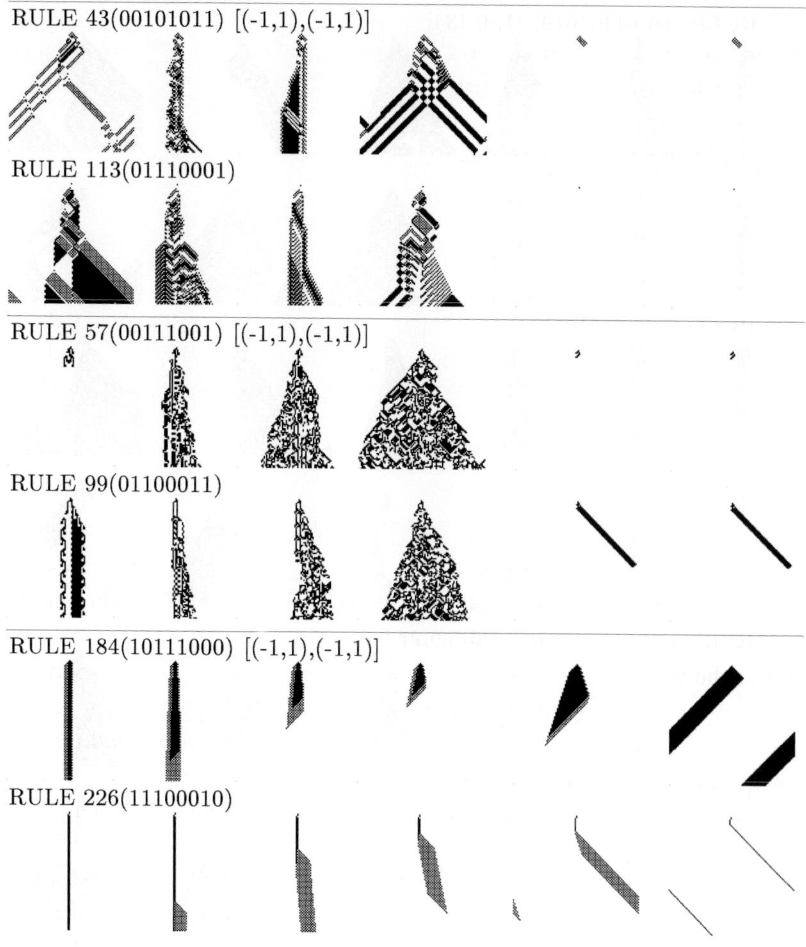

Fig. 2.9   (*continued*)

ure 2.9 shows the case of fully asymmetric rules 43, 57 and 184. Particularly curious are the DP generated for $\alpha > 0.6$ in Rules $\{43, 113\}$.

The DP of rules $\{184, 226\}$ are *rectified* when history is taken into account (that of rule 184 becomes extinct for α in [0.6,0.8]). The rules 184 and 226 have proved particularly effective in solving the *density* problem: to decide whether an arbitrary initial configuration contains a density of 1*s* above or below ρ_c, particularly $\rho_c = 0.5$. As in synchronization, the *density task* comprises a non-trivial computation for CA: again, density is a global property of a configuration, where small-radius CA rely solely on local interactions [5]. The effectivity of elementary rules with short-range memory in density discrimination is described in section 11.1, whereas that of block cellular automata is described in section 8.2.

2.3.2 *Nearest and next-nearest neighbors*

In the $r=2$ CA, the value of a given site depends on values of the nearest and next-nearest neighbors: $\sigma_i^{(T+1)} = \phi\big(\sigma_{i-2}^{(T)}, \sigma_{i-1}^{(T)}, \sigma_i^{(T)}, \sigma_{i+1}^{(T)}, \sigma_{i+2}^{(T)}\big)$. We will analyze here only $k=r=2$ *totalistic* rules: $\sigma_i^{(T+1)} = \phi\big(s_{i-2}^{(T)} + s_{i-1}^{(T)} + s_i^{(T)} + s_{i+1}^{(T)} + s_{i+2}^{(T)}\big)$, characterized by a sequence of binary values (β_s) associated with each of the six possible values of the sum (s) of the neighbors, conveniently specified by their *rule number* (R), ranging from 0 to 63:

$$\big(\beta_5, \beta_4, \beta_3, \beta_2, \beta_1, \beta_0\big)_{binary} \equiv \sum_{s=0}^{5} \beta_s 2^s \quad \underset{decimal}{} = R$$

Figure 2.10 shows the spatio-temporal patterns starting from a single live site, for all the quiescent rules sensitive to a sole live cell, i.e., the sixteen rules with $\beta_1 = 1$, thus with rule number from 2 to 62 by 4 intervals. The memory factor varies in these figures from 1.0 to 0.6 by 0.1 intervals in a first row of patterns, whereas in a second row, α has been stated to values close to the value that implies no memory effect, i.e., $\alpha = 0.5$. Evolution in Figure 2.10 is up to 36 time steps for high values of the memory factor and up to 217 time steps for low values of α. The patterns corresponding to the high α values, are zoomed compared to those corresponding low α values. The patterns shown are symmetric due to the exclusive consideration of totalistic rules.

Two main conclusions can be derived from Fig. 2.10: *i)* as an overall rule the patterns become more expanded as less historic memory is retained (smaller α), *ii)* the transition from the fully historic $(\alpha = 1.0)$ to the ahistoric scenario $(\alpha = 0.5)$ is gradual in most cases.

Rules 14, 22, 26, 30, 42, 46, 54, 58 and 62 are paradigmatic of smooth evolution from an expansive pattern in the ahistoric model to a narrow one in the fully historic, which does not resemble the ahistoric one.

Nevertheless, notable examples of discontinuity are found. Thus, rule 18 dies out sharply from $\alpha = 0.7$, in fact this is the only rule in which extinction is found in the fully historic model. Rule 2 dies out in [0.9,0.501], rule 6 in [0.9,0.6] and surprisingly in 0.501, rule 10 in 0.6, rule 34 in [0.8,0.501], rule 38, unexpectedly, in 0.5001, and rule 50 in 0.9 and in [0.7,0.6]. But none

---

[5]The density task has been related to the way in which stomatal apertures become synchronized into patches that exhibit richly complicated dynamics [329].

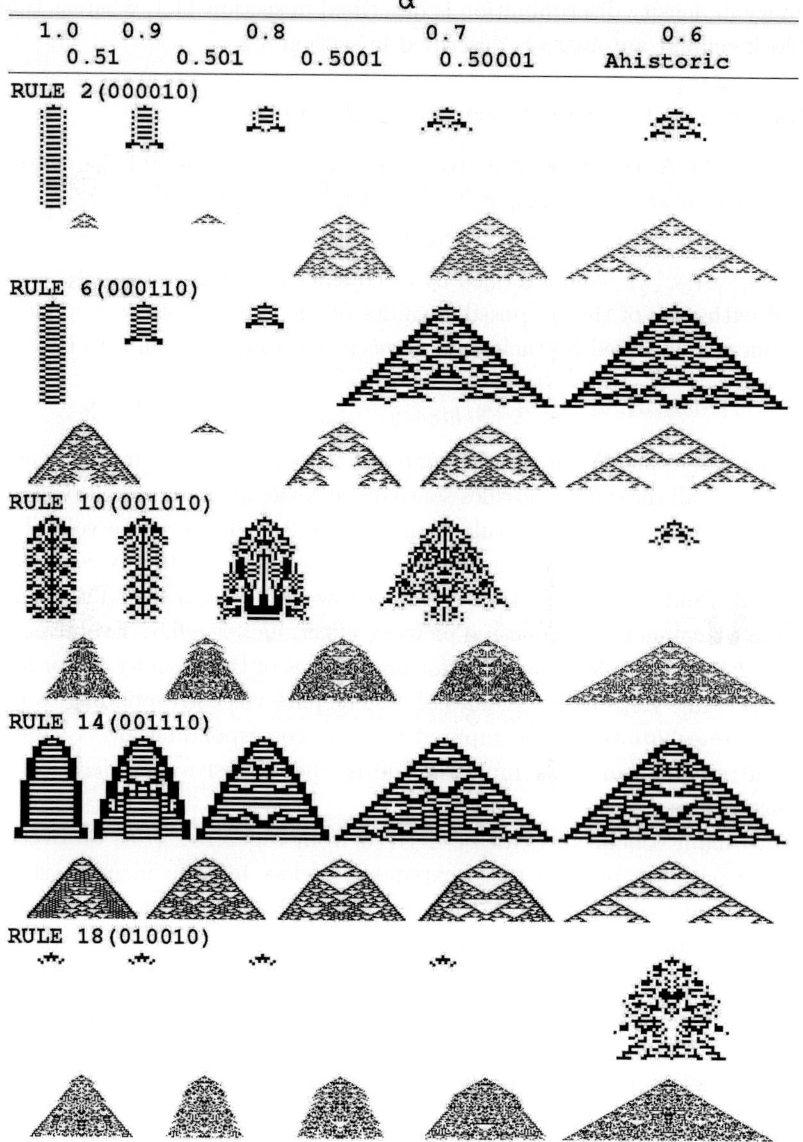

Fig. 2.10   Evolving patterns of totalistic, $k = r = 2$ quiescent rules starting from a single seed. The patterns corresponding to the high α values, presented in the first row of patterns, are zoomed.

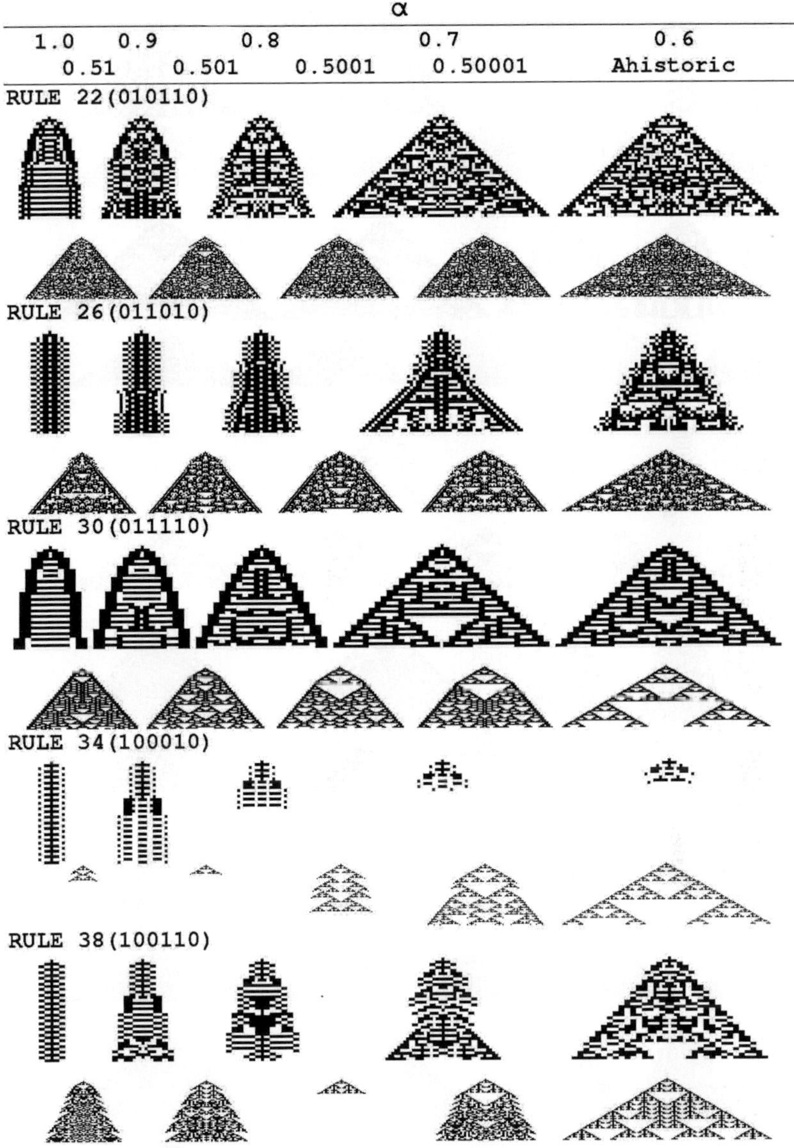

Fig. 2.10   (*continued*)

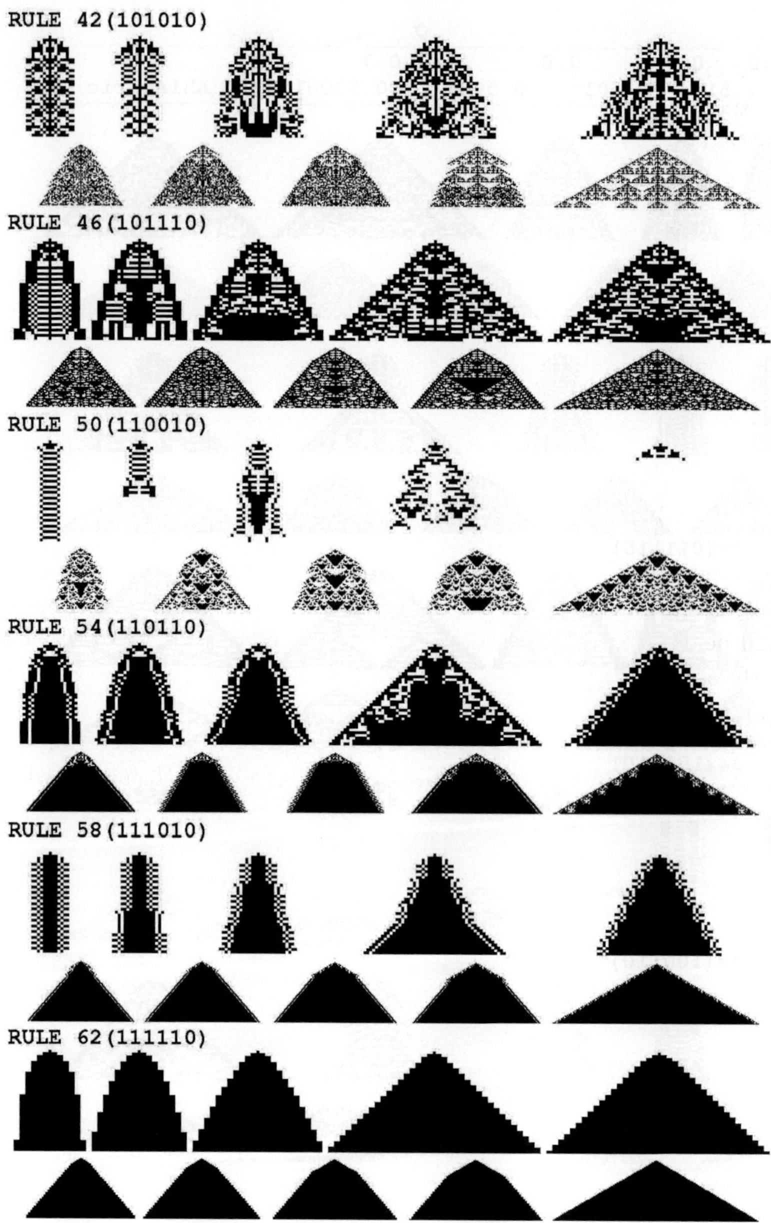

Fig. 2.10   (*continued*)

of these rules dies in the fully historic model. Most patterns in the fully historic model are oscillators; the others grow in the *punctuated equilibrium* way described in section 1.2.

S. Wolfram [423] classified the behavior of $r=k=2$ CA rules in four qualitative classes according to the fate of the evolving patterns:

Class I : evolution leads to a homogeneous state (realized for rules 0, 4, 16, 32, 36, 48, 54, 60 and 62).

Class II : evolution leads to a set of separate simple stable or periodic structures (rules 8, 24, 40, 56 and 58).

Class III : evolution leads to a chaotic pattern (rules 2, 6, 10, 12, 14, 18, 22, 26, 28, 30, 34, 38, 42, 44, 46 and 50).

Class IV : evolution leads to complex localized structures, sometimes long-lived (rules 20 and 52).

The Langton's $\lambda = \dfrac{\#\beta = 1}{\#\beta}$ parameter , has been proposed as a measure of the *entropy* inherent in a CA rule. Thus, rules with $\lambda \simeq 0.5$ tend to generate Class III behaviour, and rules with λ approaching 0.5 are candidates for exhibiting complex, i.e., class IV, behaviour.

Except for rules 54 and 62 (Class I) and rule 58 (Class II), the rules of Fig. 2.10 belong to Wolfram's Class III, that of *chaotic* rules, which consequently are activated by means of a single site cell. The two rules of Wolfram's Class IV are not active when starting from a single live cell, so they are not covered in Fig. 2.10. They are present in Fig. A.2 in Appendix A, elaborated to examine the behavior of Class III and IV when starting from a disordered configuration.

As a rule, the patterns in the fully historic model of Fig. A.2 might be classified as Class II: evolution leads to a set of separate simple stable or periodic structures. Half of the rules in Fig. A.2, rules 10, 12, 14, 22, 26, 28, 30, 38 and 42 , led from the chaotic spatio-temporal in the ahistoric model to the Class II one in the fully historic, in a fairly gradual way, though the graduality in the evolution of the patterns is rather problematic for some rules when comparing the fully historic pattern with the pattern of $\alpha = 0.9$. For example, rules 12, 14, 28 and, notably rule 30, present, at $\alpha = 0.9$, some triangular clearings that characterize the ahistoric patterns which are completely lost in the fully historic.

Albeit the norm is a gradual effect of memory in Fig. A.2, there are also present in it unexpected extinctions, such as in rule 2, which, in parallel with what happens starting with a single site seed, dies out soon in the $[0.6, 0.9]$ α interval, but not in the fully historic model. The schedule of extinction starting with a single site seed is reflected by its starting with a disordered configuration, but without an automatic translation. For example, rule 18, which dies out for any $\alpha \geq 0.7$ in Fig. 2.10, does not die out in the fully historic model nor for $\alpha = 0.9$ in Fig. A.2; Rule 6 does not go extinct in [0.8,0.9], neither does rule 10 in 0.6; rule 34 becomes extinct in [0.7,0.8], as in Fig. 2.10, but also for $\alpha = 0.9$, and not in 0.6. Rule 50 presents both in 0.8 and in 0.7 only a persistent structure. In most cases the aspect of the spatio-temporal patterns produced in these *discontinuities* is *complex*, showing in some cases persistent structures which closely resemble the archetypical evolving patterns generated by rule 20 in the ahistoric model

Figure A.2 shows also the differences in pattern (DP) produced by evolution from the disordered initial configuration resulting from change in the value of its initial center site. The *damaged* region is shown as black squares corresponding to the site values that differed among the patterns generated with the two initial configurations. In most cases, the perturbations in the ahistoric model propagate very rapidly to the right and left a great velocity at any time (*butterfly effect*). In the particular case of rule 30, the reversal of the central site value ($01001 \rightarrow 01101$) does not alter the initial output.

Thus the rules that are smoothly affected by the increase of the memory factor, tend to be affected in the same way in respect to the damage spreading, i.e., rules 10, 12, 22, 26, 28, 38 and 42. As an overall rule, historic memory produces a preserving effect regarding the damage induced by the reversal of a single site value. This result can be expected from Fig. 2.10, since this can be seen as a particular case of the initial alteration of a unique (central) site value, so there is notable parallelism in the qualitative evolution of patterns in Fig. 2.10 and in the damage spreading in Fig. A.2. For example, the disruption for rule 2 dies out in [0.9,0.6] but not in the fully historic model. But the parallelism seems to work better compared to what happens starting at random, particularly in respect to extinction. Thus, for rule 18 the disruption does not die out in the fully historic model, rule 6 does not die out in [0.8,0.9], nor rule 10 in 0.6, disruption on rule 34 dies out in [0.7,0.9] as starting at random (not for $\alpha = 0.9$ in Fig. 2.10). Historic memory has no significant effect on the complex rules 20 and 52 with the disruption confined already in the ahistoric model. Two rules serve as counter-examples in Fig. A.2: rule 14, in which a imperceptible disruption in the ahistoric model is to be compared with a notable damage in the other memory models (or at least a persistent one such as with full memory), and rule 20 whose gradation in the length of the damaged region is expected to be the opposite.

Chapter 3

Other memories

3.1 Average-like memory

A number of average-like memory mechanisms can readily be proposed by generalizing the expression of the memory charge of Section 2.1 as:

$$m_i^{(T)} = \frac{\displaystyle\sum_{t=1}^{T} \delta(t)\sigma_i^{(t)}}{\displaystyle\sum_{t=1}^{T} \delta(t)} \equiv \frac{\omega_i^{(T)}}{\Omega(T)}$$

Possible weight functions are, the exponential $\delta(t) = e^{-\beta(T-t)}$, $\beta \in \mathbb{R}^+$, it is $\alpha = e^{-\beta}$; and the inverse $\delta(t) = \alpha^{t-1}$, featured by $\delta(t) > \delta(t+1)$. The latter is adopted by Oprisan [317] in the form: $\delta(t) = \dfrac{1}{r^{t-1}}$ with the free parameter r being real and $r > 1$.

Among the possible choices of δ stand the weights $\delta(t) = t^c$ and $\delta(t) = c^t$, in which the larger the value of c, the more heavily is the recent past taken into account, and consequently the closer the scenario to the ahistoric one. Both weights allow for integer-based arithmetics ($\grave{a}$ la CA) comparing $2\omega^{(T)}$ to $\Omega(T)$ to get the trait states s (which is a clear computational advantage over the α-based model), and remain cumulative in respect to the charge: $\omega_i^{(T)} = \omega_i^{(T-1)} + \begin{Bmatrix} T^c \\ c^T \end{Bmatrix} \sigma_i^{(T)}$ [50]. Nevertheless, both weights share the same drawback: powers *explode* at high values of T, even for $c = 2$.

39

Integer-based t^c memory has no effect at $T=2$, nor has at $T=3$[1]. But memory may alter the dynamics at $T=4$. Thus, in the most unbalanced scenario:

$$\begin{cases} m(0,0,0,1) = \dfrac{4^c}{1+2^c+3^c+4^c} < \dfrac{1}{2} \\[2ex] m(1,1,1,0) = \dfrac{1+2^c+3^c}{1+2^c+3^c+4^c} > \dfrac{1}{2} \end{cases} \Rightarrow 4^c < 1+2^c+3^c$$

Thus, the trait state at $T=4$ is the reverse of the last one in the most unbalanced scenario if $c=1$, i.e., $\delta(t) = t$. Taking into account the formulas for the sum of powers of integers given in Table 3.1, in the most unbalanced scenario, the trait state reverses the last one, if $12T^2 < T(T+1)(2T+1)$ if $c=2$, what happens form $T=5$, and if $8T^3 < T^2(T+1)^2$ if $c=3$, what happens from $T=6$.

Table 3.1   Sum of the c-th powers of the first T integers

$c=1$	$c=2$	$c=3$	$c=4$
$\dfrac{T(T+1)}{2}$	$\dfrac{T(T+1)(2T+1)}{6}$	$\left(\dfrac{T(T+1)}{2}\right)^2$	$\dfrac{2T^6 + 6T^5 + 5T^4 - T^2}{12}$

For general c, it is [91, 264, 319, 328, 360, 446]:

$$\sum_{t=1}^{T} t^c = \frac{1}{c+1} \sum_{j=1}^{c+1} (-1)^{\delta_{jc}} \binom{c+1}{j} B_{c+1-j} T^j$$

where δ_{jc} is the Kronecker delta and B_i is the i-*th* Bernoulli number. The Bernouilli numbers are a sequence of rational number that verify: $\dfrac{x}{e^x - 1} = \sum_{j=0}^{\infty} \dfrac{B_j x^j}{n!}$, with $\dfrac{x}{abs(x)} < 2\pi$. Bernouilli numbers may be calculated by using the recursive formula: $\sum_{j=0}^{i} \binom{i+1}{j} = 0$. Thus, $m=1$: $\binom{2}{0} B_0 + \binom{2}{1} B_1 = 0 \rightarrow 1{\times}1 + 2B_1 = 0 \rightarrow B_1 = -1/2$. And, $m=2$: $\binom{3}{0} 1 + \binom{3}{1}(-1/2) + \binom{3}{2} B_2 = 0 \rightarrow 1 \times 1 - 3/2 + 3B_2 = 0 \rightarrow B_2 = 1/6$. Applying the general formula to the lowest values of c, thus $c=1$: $\sum_{t=1}^{T} t = \dfrac{1}{2}\left(-2B_1 T + B_0 T^2\right) =$

---

[1] This contrasts with what happens with α-memory, in which case memory switches the trait state of a cell at $T=3$ if $\sigma_i^{(1)} = \sigma_i^{(2)} \neq \sigma_i^{(3)}$ and $\alpha > 0.61805$.

$$\frac{1}{2}\big(-2(-1/2)T + 1 \times T^2 \big) . \ = \tfrac{1}{2}\big(T + T^2\big) , \text{ and } c{=}2: \sum_{t=1}^{T} t^2 = \frac{1}{3}\big(3B_2 T -$$

$$3B_1 T^2 + B_0 T^3\big) = \frac{1}{3}\big(3\tfrac{1}{6}T - 3(-1/2)T^2 + 1 \times T^3\big) = \frac{1}{6}\big(T + 3T^2 + 2T^3\big) .$$

Figure 3.1 shows the spatio-temporal patterns of the legal rules significantly affected by $\delta = t^c$ memory when starting from a single live site. The c factor varies in the figure from 0 (fully historic model) to 6 by 1 intervals, from 10 to 25 by 5 intervals. The ahistoric pattern is also shown in the figure. The main conclusions regarding α-memory derived fron Fig. 2.6 apply also to Figure 3.1: *i)* as an overall rule, the patterns become more expanded as less historic memory is retained (higher c), *ii)* the transition from the fully historic ($c = 0$) to the ahistoric scenario is gradual in most cases. Rules 50, 122, 178, 250, 94, and 222, 254 are paradigmatic of the smooth evolution. Rules 126 and 182 also present a gradual evolution, although the historic patterns do not resemble the ahistoric at all. In some cases the above transition presents a notable discontinuity. This applies for *i)* rules such as 22 and 54 which tend to die out in $1 < c < 10$ but generate oscillators in the fully historic model, and *ii)* the group of rules for 18, 90, 146, and 218 which exhibit extinction till $c = 25$ (with the exception of $c = 20$) and, abruptly, the typical ahistoric expansion. The equivalent rules 146 and 182 are not affected by memory in a similar way.

The weight c^t is not operative in the two-state scenario, but it becomes effective when $k > 2$ (see Chapter 6).

Another memory mechanism, similar to that used in the context of connection weights adjustments in neural networks, is one in which the distance between the state value and the actual one is adjusted with the so called *learning rate* β. Thus, in this memory implementation, referred to as Widrow-Hoff or β-memory, the trait state s is obtained by rounding the *expected* (unweighted) average value $m_i^{(T)} = \frac{1}{T}\sum_{t=1}^{T}\sigma_i^{(t)}$ incremented by the pondered discrepancy between the mean and the current state σ: $s_i^{(T)} = \mathcal{H}\big(m_i^{(T)} + \beta(\sigma_i^{(T)} - m_i^{(T)})\big)$. Thus, the β parameter controls the memory charge, with $\beta{=}1$ corresponding to the ahistoric model, and $\beta{=}0$ to full memory.

Tables 3.3 and 3.2 show the effect of decreasing values of β-memory (i.e., increasing memory charge) on the parity rule in the one-dimensional scenario. In both contexts, the elementary rule 150 and the $r{=}2$ parity rule, the inhibition of the spatio-temporal development due to the inertial

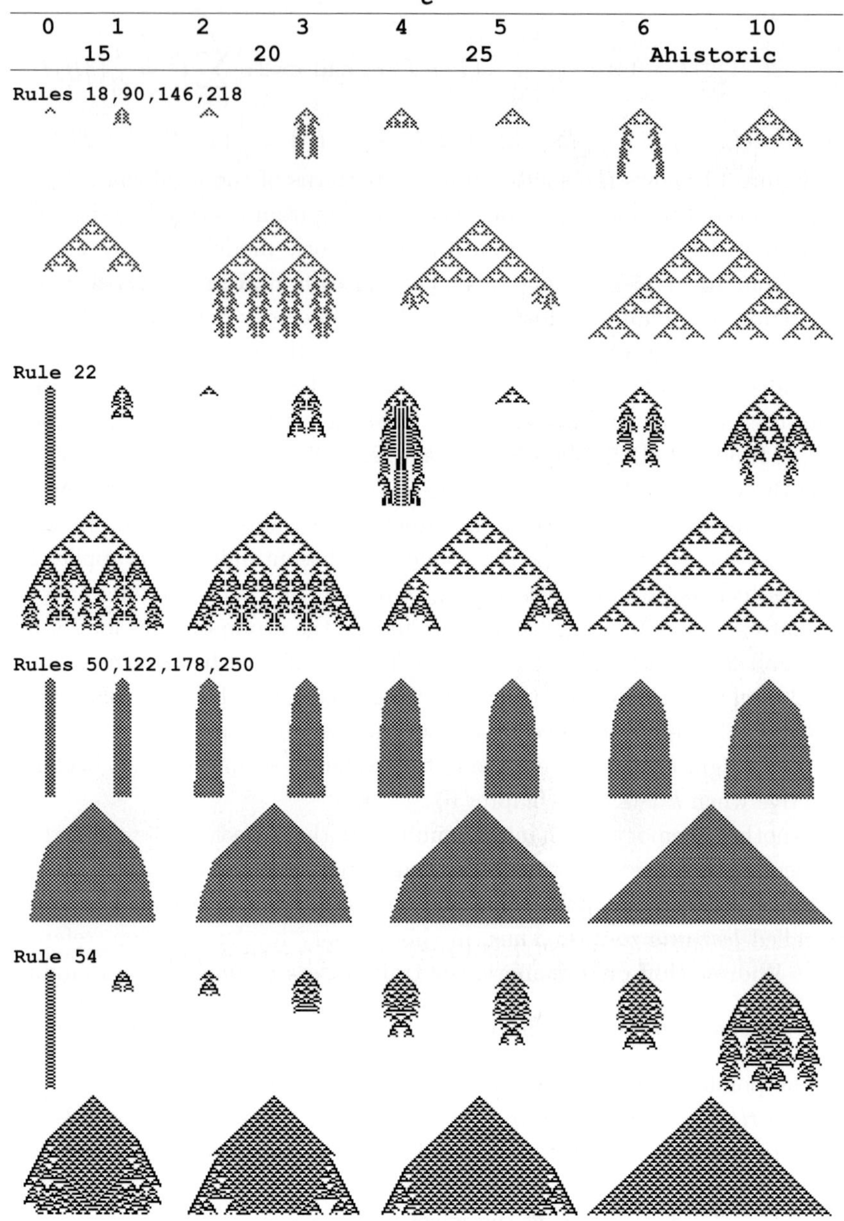

Fig. 3.1   Elementary rules with $\delta = t^c$ memory.

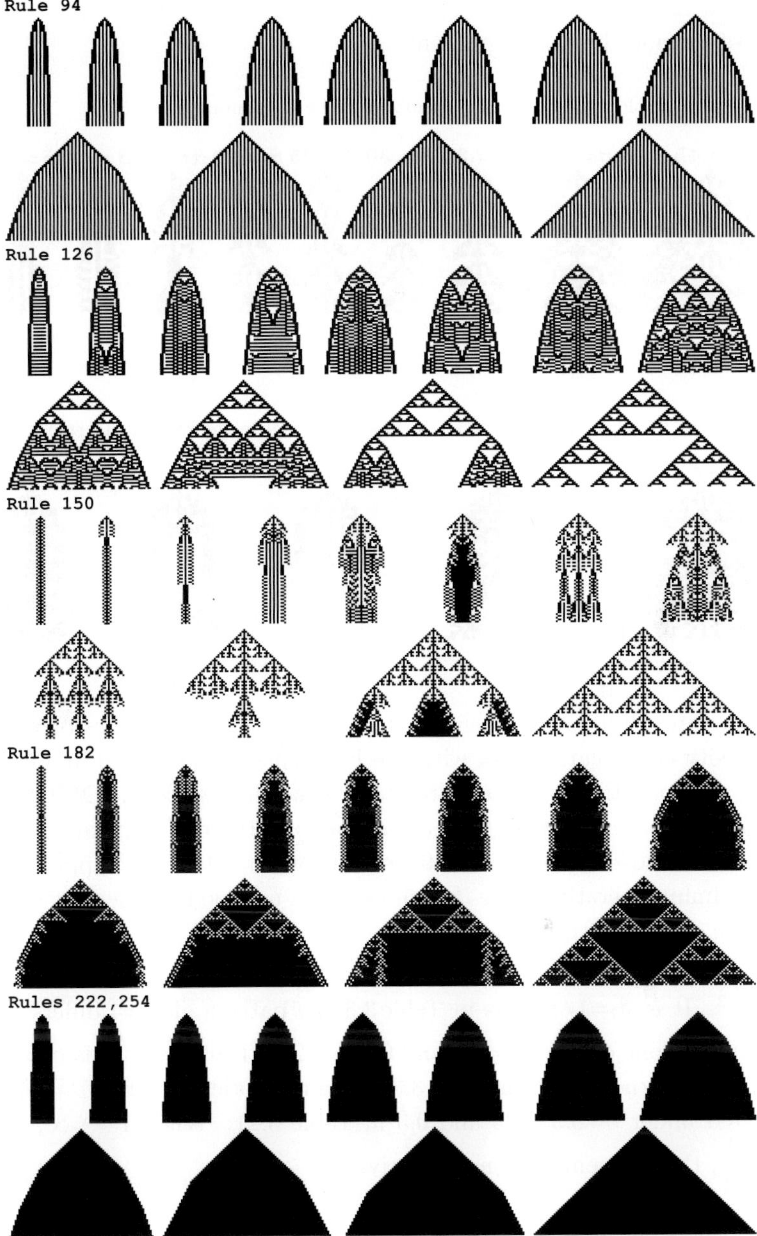

Fig. 3.1   (*continued*)

effect of memory becomes apparent.

Table 3.2   Effect of β-memory on the elementary rule 150.

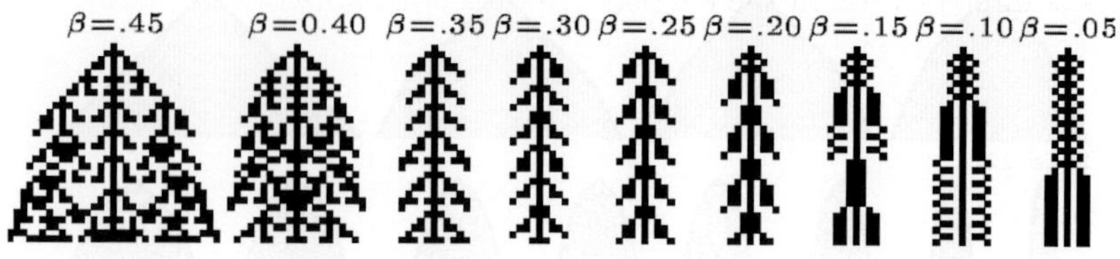

Table 3.3   Effect of β-memory from a single cell in the $r=2$ parity rule.

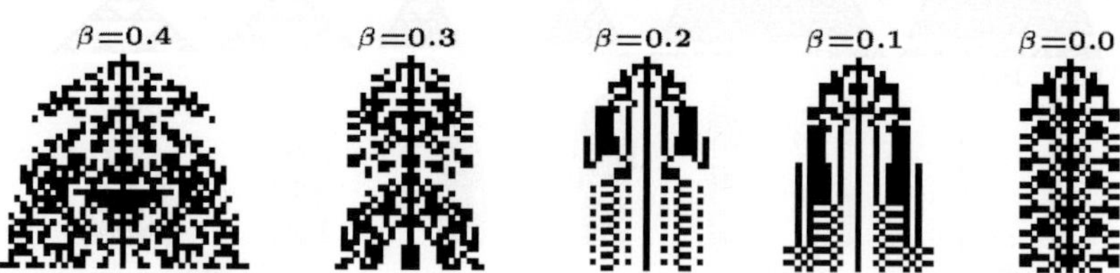

Stone and Bull have implemented β-memory in their studies regarding the density classification task with the ECA rule 184 [382] and in a evolutionary computation approach [383] to evolve solutions to the DCT in the one-dimensional $r=3$ context. In the latter case, the best performance (oscillating around $\epsilon=0.80$ for $n=149$) is found with values of β approaching the maximum operative value $\beta=0.5$, i.e., with not too high memory charge as in the α-memory implementation.

In the most unbalanced scenario, it holds that: $m = 0.5 \Rightarrow \beta_T = \frac{1}{2}\frac{T-2}{T-1}$ [2]. It is $\beta_3=1/4$, thus in Table 3.3 the pattern at $T=4$ differs from the ahistoric one only for $\beta=0.2$ and 0.1 (≤ 0.2 in Table 3.2). It is $\beta_4=1/3$, thus the pattern at $T=5$ in Table 3.3 is the ahistoric one when $\beta=0.4$ but is altered under $\beta=0.3$ (0.35 and 0.3 in Table 3.2). When $T \to \infty$, it is: $\beta_\infty = \frac{1}{2}$, thus β-memory is not effective if $\beta \geq 0.5$.

---

[2]
$$\left.\begin{cases} m(0,0,\ldots,0,1) = \dfrac{1}{T} + \beta\left(1-\dfrac{1}{T}\right) = \dfrac{1}{2} \\[2mm] m(1,1,\ldots,1,0) = \dfrac{T-1}{T}+\beta\left(0-\dfrac{T-1}{T}\right) = \dfrac{1}{2} \end{cases}\right\} \Rightarrow \beta_T = \frac{1}{2}\frac{T-2}{T-1}$$

3.2 Limited trailing memory

Limited trailing memory would keep memory of only the last τ time-steps. This is implemented in the context of average memory as:

$$m_i^{(T)}(\sigma_i^{(T-\tau+1)}, \ldots, \sigma_i^{(T)}) = \frac{\sum_{t=\mathsf{T}}^{T} \delta(t)\sigma_i^{(t)}}{\sum_{t=\mathsf{T}}^{T} \delta(t)}, \quad \text{with } \mathsf{T} = \max(1, T - \tau + 1).$$

Limiting the trailing memory would take the model closer to the ahistoric model ($\tau = 1$). In the geometrically discounted method, $\delta(t) = \alpha^{T-t}$, such an effect is more appreciable when the value of α is high, whereas at low α values (already close to the ahistoric model when memory is not limited) the effect of limiting the trailing memory is not so important. In the $k = 2$ context, τ must be at least three for memory to have effect. For $\tau = 3$, provided that $\alpha > \alpha_3 = 0.61805$, the memory mechanism turns out to be that of selecting the mode of the last three states, i.e., $s_i^{(T)} = mode(\sigma_i^{(T-2)}, \sigma_i^{(T)}, \sigma_i^{(T-1)})$, equivalent to the elementary *majority* rule 232 [46].

A seemingly natural choice of the depth of memory is that of the extent of the spatial rule. Thus, in the $r=2$ parity rule with memory in Fig. 3.2 the majority memory rule has length $\tau=5$.

Fig. 3.2 The ahistoric $r=2$ parity rule (left) and this rule memory of the majority of the last five states (center).

In the Boolean $k=2$ context, the majority function may be computed as: $mode(\sigma^{(1)}, \ldots, \sigma^{(T)}) = \chi\left(\frac{1}{2} + \frac{\sum_{t-1}^{T}\sigma_t}{T} - \frac{(-1)^{\sigma_T}}{3T}\right)$, where the indicator function χ is defined as: $\chi(x) = 1$ if > 1 and $\chi(x) = 0$ if < 1. In case of a tie, the term $-\frac{(-1)^{\sigma_T}}{3T}$ serves to break in favor of the last state value [233].

A study of the (neural) automaton $x_{t+1} = \mathcal{H}\left(\sum_{t=\mathsf{T}}^{T} \delta_t x_t - \theta\right)$ is made in [115]. In the paper by Layman [242], the weights (coupling factors) of the, also neural (but unlimited), type of memory are $\delta_t = c^{T-t}$, $0 < c < 1$.

No memory up to T=τ

A variant memory implementation of memory limited to the last τ time-steps is that of letting the automaton to evolve in the conventional form, i.e., with no memory, up to T=τ, and activate memory from this precise time-step. This variant is not adopted here, but in the example of Fig. 3.3.

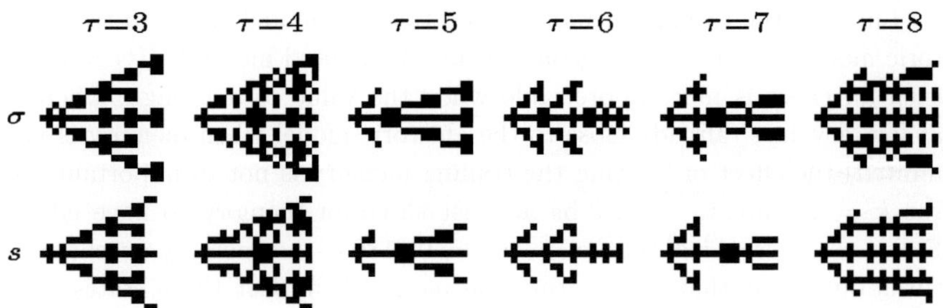

Fig. 3.3   The parity rule with τ-majority memory activated from T=τ.

Memory of depth τ activated at T=τ is implemented in the studies performed in [277] and [278]. These studies found interesting complex dynamics when implementing τ=8 and τ=4 majority memory on rules 30 and 126 respectively. The reference [276] deals with the chaotic rules 86 and 101. Particular attention is paid in these articles to the study of how gliders emerge and interact from random initial configurations.

A de Bruijn diagram is a graph-theoretic tool, originated in shift-register theory, where nodes are sequences of symbols from an alphabet (states of elements of a spatially extended system), and edges describe how the sequences may overlap. The edges of a de Bruijn diagram are naturally associated with the neighbourhoods of an automaton using the same symbols, which associates the links with a step of evolution in an equally natural fashion. de Bruijn diagrams can accurately predict complex propagating patterns, and intrinsic features of space-time dynamics of cellular automata [275, 440]. The articles [277] and [278] analyze the de Bruijn diagrams of rules 30 and 126. Rule 126 has been taken into account in the design of a cellular automaton using nonlinear memory resistors, or memristors [213], a circuit component which "remembers" changes in the current passing through it by changing its resistance.

3.3 Majority of the last three state memory

Memory of the most frequent of the last three states operates on the canonical example of Fig. 1.1 as shown in Fig. 3.4. This minor memory induces a softer inertial effect compared to that of unlimited trailing memory in Fig. 1.2 .

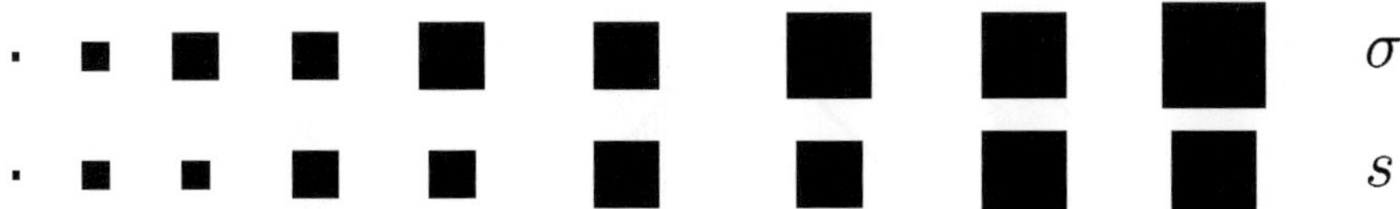

Fig. 3.4 The automaton of Fig. 1.1 with τ=3 majority memory.

Figure 3.5 shows the evolution of the legal rules affected by the mode of the last three state memory. As it is known, history has a dramatic effect on rules 18, 90, 146 and 218 as their pattern dies out as early as at $T = 4$. The case of rule 22 is particular: two *branches* are generated at $T = 17$ in the historic model; the patterns of the remaining rules in the historic model are much more reminiscent of the ahistoric ones, but, let us say, *compressed*. Majority of the last three state memory exerts in some rules in Fig. 3.5 an effect similar to that described in Fig. 3.4, i.e., the replication of every ahistoric pattern from T=3. That is the case of rules 94, 126, 150, and 222,225 .

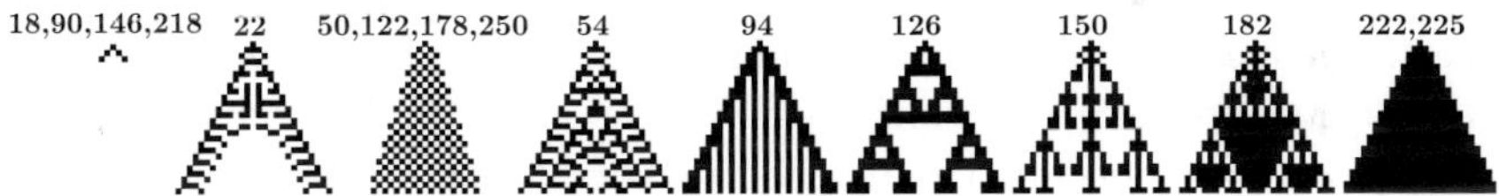

Fig. 3.5 Elementary legal rules significantly affected by the τ=3 majority memory. Evolution from a single site seed up to T=26 .

Figure 3.6 shows the effect of memory on some relevant [3] quiescent asymmetric rules. Evolution in both Figs. 3.5 and 3.6 is from a single site seed up to $T = 26$. Rule 2 *shifts* a single site live cell one space at every time-step in the ahistoric model; with memory the shift actuates just up to $T = 3$, and at $T = 4$ the pattern dies. This evolution is common to

---

[3]Either because the effect on them is representative of the effect on many others or because they have been particularly studied in the ahistoric scenario. For example, Rules 30 and 110.

all rules that just shift a single site cell without increasing the number of living cells at $T = 2$, this is the case of the *number conserving* rules 184 and 226. The patterns generated by rules 6 and 14 are *rectified* [4] by memory in such a way that the total number of live cells in the historic and ahistoric spatio-temporal patterns is the same [5]. Again, the historic patterns of the remaining rules in Fig. 3.6 seem, as a rule, like the ahistoric ones compressed.

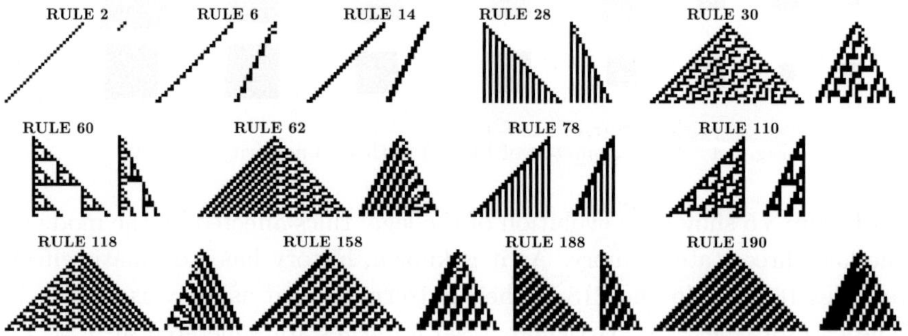

Fig. 3.6   Evolution of some quiescent asymmetric elementary rules affected by the $\tau{=}3$ majority memory.

Majority memory is much more in the realm of CA than the average memory because it does not use any number type other tha integer type. Thus majority memory allows for manipulation of symbols, avoiding any real computing/arithmetic. By the way, the mode memory mechanism has been implemented in models of opinion formation such as those proposed in [359] and in [348], in which agents resort to their individual history when *frustration* occurs, i.e., neighbors have different opinions and do not induce any updating.

Figures B.1, B.2, B.3 in Appendix B show the effect of the mode memory on elementary rules starting with the same random initial configuration with 371 sites up to $T = 150$.

Figure B.1 shows the evolution of the legal rules significantly affected by memory when starting from a random initial configuration: the nine legal rules which generate non-periodic patterns in the ahistoric scenario.

---

[4]In the sense of the lines in the spatio-temporal pattern having a slower slope.

[5]The other quiescent rules affected by memory in such a way are: 14, 20, 38, 52, 46, 84, 116, 134, 142, 148, 166, 174, 180, 212, 244, and 245.

The remaining legal rules are absent from Fig. B.1 because they are not significantly affected by memory, as explained in 2.3.1 . Figure B.1 shows also the differences produced in patterns (DP) when reversing the value of its initial center site, i.e., the *damaged* black pixels. Again, as in Fig 3.5, history has a dramatic effect on rule 18, completely changing the appearance of its spatio-temporal patterns: isolated periodic structures are generated, far from the distinctive inverted-triangle world of the ahistoric pattern. Reversing the central site cell generates a periodic localized perturbation in the historic model, whereas in the ahistoric model, the perturbation grows at the *speed of light*: $\lambda_L = \lambda_R = 1$. The effect of memory on rules 22, 90, 122, 126, and 150 is fairly similar: the spatio-temporal patterns preserve the inverted triangles characteristic of the ahistoric model and the DP are notably restrained, albeit rule 122 shows an exceptional extinction of the DP in the ahistoric model in the simulation of Fig. B.1 . The constraint of the DP due to memory is also found in rules 146 (in fact, extinction) and in 182, but their spatio-temporal patterns are much more altered. Also rule 54 shows an exceptional extinction of the DP in the ahistoric model in the simulation of Fig. B.1, whereas in the historic one the DP is a distinctive feature of this rule, notably altered by memory. The effect of mode memory on symmetric but not quiescent rules, that is not shown here, can, to a great extent, be described in terms of the effect on their equivalent legal rules. Thus, the pattern with memory of rule 183 resembles that of its equivalent (under the negative transformation) rule 18, 151 resembles that of 22, 147 that of 54, 165 that of 90, 161 that of 122, and 129 that of 126 (rule 150 is its own negative transformed).

Figures B.2 and B.3 show the spatio-temporal and DP of some of the asymmetric rules affected by memory in the initial scenario of Fig. B.1. Fig. B.2 deals with *semi-asymmetric* (either $\beta_2 \neq \beta_5$ or $\beta_4 \neq \beta_7$, but not both) rules; Fig. B.3 with *fully asymmetric* ($\beta_2 \neq \beta_5$ and $\beta_4 \neq \beta_7$) rules. Rules in these figures are grouped as belonging to the same equivalence class; often showing only the minimal representative rule of the class affected by memory. Rules not affected by memory tend to belong to the same equivalence class. Thus, for example, memory has no effect on any of the rules of the subclasses: $\{8, 64, 239, 253\}$, $\{12, 68, 207, 221\}$, $\{13, 69, 79, 93\}$, and $\{29, 71\}$.

The rule 110 and its three equivalent rules under the negative and reflection transformations (rules 124, 137 and 193), may serve as a paradigmatic example of the expected effect of memory: the preservation of the general aspect of the spatio-temporal pattern and the constraint in the damage

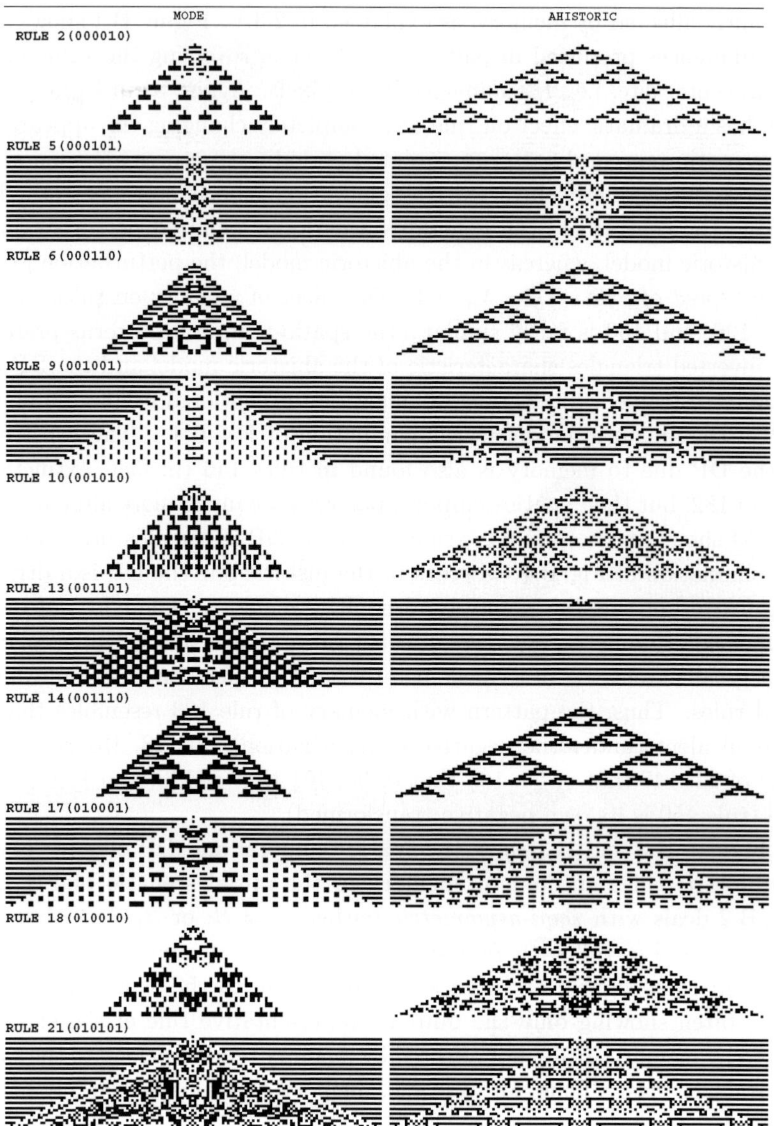

Fig. 3.7   Totalistic $k=r=2$ rules affected by $\tau=3$ majority memory. Evolution from a single live cell.

Fig. 3.7 (*continued*)

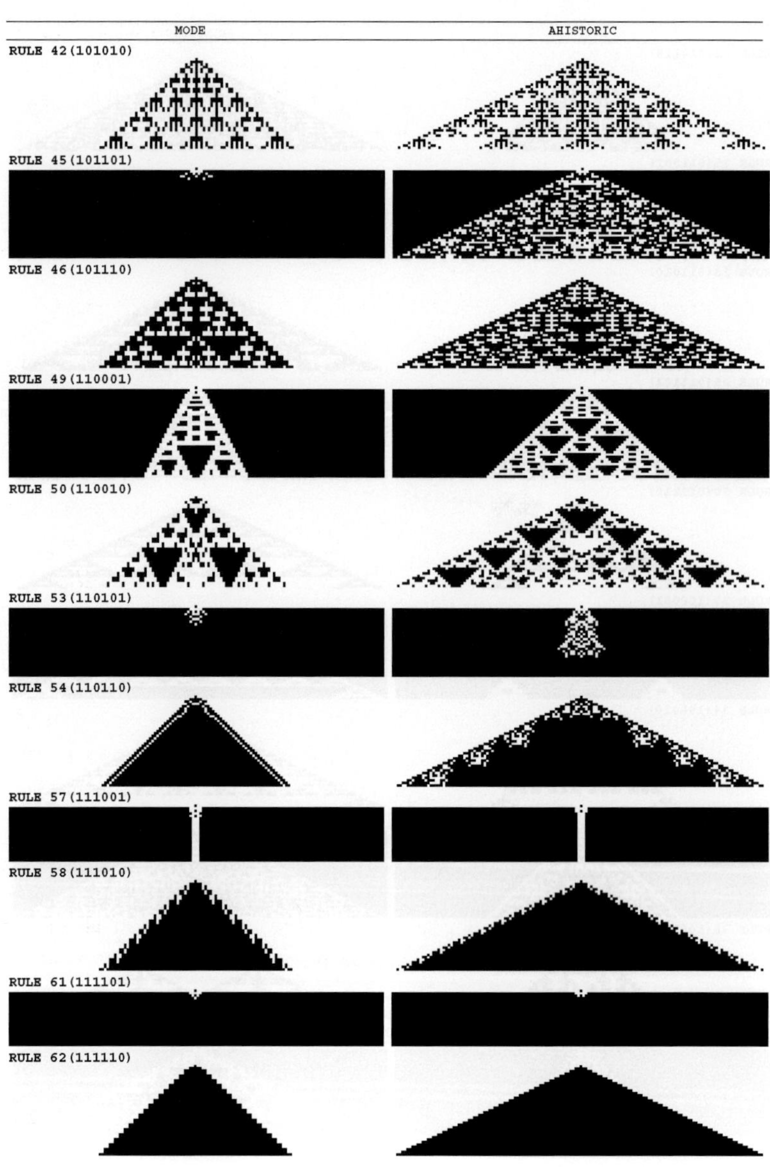

Fig. 3.7   (*continued*)

induced by the reversal of a single site. The same applies to the subclass $\{60, 102, 195, 153\}$ in Fig. B.3. History however has an unexpected effect on other rules. This is the case of the rule 184 and its equivalent rule 226 and of the rules 57 and 99; their similar aspect in their ahistoric spatio-temporal patterns are notably altered by memory. The extinction induced by memory on rule 2 was predictable from Fig. B.1. *Rectification* of lines (or more complex patterns) to a slower slope seems to be a common effect of memory in many rules, e.g., rules 26 and 10.

Figures 3.7 and B.4 in Appendix B deal with totalistic CA rules operating on the near and next-nearest neighbors ($r = 2$) scenario. Figure 3.7 shows the symmetric spatio-temporal patterns starting from a single live site up to $T = 37$, for all the $r = 2$ totalistic rules sensitive to a sole live cell, i.e., the sixteen rules with $\beta_1 = 1$, thus with rule number from 2 to 62 by 4 intervals. Two conclusions can be derived from Fig. 3.7: *i)* memory preserves the general aspect of the ahistoric patterns from a single live cell, *ii)* memory constrains the evolving patterns. Figure B.4 shows the evolving patterns of the quiescent rules of Fig. 3.7 (and of rules 20 and 52 that are not active when starting from a single live cell) starting at random, with the differences in pattern (DP) produced from a change in the value of its initial center site superimposed. The effects starting from a single site seed on the rules of the classes III and IV in Fig. 3.7 can be extended when starting at random (Fig. B.4): *i)* memory preserves the general aspect of the ahistoric patterns, and *ii)* the damage-spreading becomes constrained when memory enters. The latter conclusion is fully applicable to the chaotic rules, in which the perturbations in the ahistoric model propagate very rapidly to the right and left at great velocity at any time (*butterfly effect*) in the ahistoric model; in the particular case of the *complex* rules 20 and 52, their singular damage-spreading in the ahistoric model is not actually constrained in the model with memory.

Often, the conclusions reached starting from a sole active cell (as in Fig. 3.7) may be (qualitatively) extended to damage spreading from a single damaged cell when starting at random (as in B.4), as the former may be seen as a (very) particular form of damage spreading generated from the initial alteration of a unique site value.

Mean field

Cellular automaton evolution generates correlations between state values at different sites. Nevertheless, as a simple approximation, one may ignore these correlations, assuming each site independently to have value 1 with probability equal to the density $p = \rho$, and to have value 0 with probability equal to $q = 1 - p = 1 - \rho$.

Thus, in the so called mean-field theory, departing in one-dimensional layers from:

$$P^{(T+1)} = \sum_{j=1}^{8} \beta_j P\big((\sigma_l^{(T)}, \sigma_c^{(T)}, \sigma_r^{(T)})/j\big),$$

and assuming independence it is:

$$P^{(T+1)} = \sum_{j=1}^{8} \beta_j \big(P(\sigma_l^{(T)})P(\sigma_c^{(T)})P(\sigma_r^{(T)})/j\big)$$

With memory implemented in the dynamics:

$$P^{(T+1)} = \sum_{j=1}^{8} \beta_j \big(P(s_l^{(T)})P(s_c^{(T)})P(s_r^{(T)})/j\big)$$

Endowing cells with $\tau=3$ majority memory, and again assuming independence between the successive values in every cell, it is: $P(s^{(T)} = 1) = P(\sigma^{(T)} = 1)P(\sigma^{(T-1)} = 1)P(\sigma^{(T-2)} = 1) + P(\sigma^{(T)} = 1)P(\sigma^{(T-1)} = 1)P(\sigma^{(T-2)} = 0) + P(\sigma^{(T)} = 1)P(\sigma^{(T-1)} = 0)P(\sigma^{(T-2)} = 1) + P(\sigma^{(T)} = 0)P(\sigma^{(T-1)} = 1)P(\sigma^{(T-2)} = 1) = P^{(T)}P^{(T-1)}P^{(T-2)} + P^{(T)}P^{(T-1)}(1 - P^{(T-2)}) + P^{(T)}(1 - P^{(T-1)})P^{(T-2)} + (1 - P^{(T)})P^{(T-1)}P^{(T-2)}$.

Thus, for the ahistoric formulation of rule 90 it is: $P^{(T+1)} = P(\sigma_l^{(T)} = 1)P(\sigma_c^{(T)} = 1)P(\sigma_r^{(T)} = 0) + P(\sigma_l^{(T)} = 1)P(\sigma_c^{(T)} = 0)P(\sigma_r^{(T)} = 0) + P(\sigma_l^{(T)} = 0)P(\sigma_c^{(T)} = 1)P(\sigma_r^{(T)} = 1) + P(\sigma_l^{(T)} = 0)P(\sigma_c^{(T)} = 0)P(\sigma_r^{(T)} = 1) = p^2 q + pq^2 + qp^2 + q^2 p = 2p^2 q + 2pq^2 = 2pq(p + q) = 2pq$, with $p = P(\sigma^{(T)} = 1)$, $q = 1 - p$. The equation: $P^{(T+1)} = 2p(1 - p)$ has a fixed point at $p = 1/2 = \#_1(\text{R90})/8$.

Endowing $\tau=3$ majority memory in cells, and, in a further step of simplification, assuming that $P^{(T)} = P^{(T-1)} = P^{(T-2)} = p$, it is $P(s^{(T)}=1) = p^3 + 3p^2 q = p^2(1 + 2q)$, and $P(s^{(T)}=0) = q^3 + 3q^2 p = q^2(1 + 2p)$. The mean-field equation for rule 90 in this scenario: $P^{(T+1)} = 2p^2 q^2 (1 + 2q)(1 + 2p)$, preserves the fixed point $p = 1/2$. In general, the rules with fixed point $p = 1/2$ in their ahistoric formulation will preserve this fixed point when substituting p by $p^2(1 + 2q)$ in its mean-field equation, because $(1/2)^2(1 + 2(1/2)) = 1/2$. This is also the case of rule 150: $P^{(T-1)} = p^3 + 3p^2 q$, and of rule 30: $P^{(T-1)} = p^2 q + 3pq^2$. But rules with fixed point other than $p = 1/2$, do not preserve their fixed point in the foregoing scenario. Thus, rules 22 and 126, with $P^{(T-1)} = 3pq^2$ and $P^{(T-1)} = 3pq$, do not preserve their fixed points $p = 1 - 1/\sqrt{(3)}$, and $p = 2/3$. The same

results hold when endowing $\tau=3$ parity memory in cells, in which case:
$P(s^{(T)}=1) = p^3 + 3pq^2$, and, again, $(1/2)^3 + 3(1/2)(1-1/2)^2 = 1/2$.

Probabilistic approximations like the mean-field theory can also be used for other quantities, such as difference patterns. In general, such approximations tend to work better for systems in larger number of dimensions, where correlations tend to be less important.

Number conserving rules

The only non-trivial elementary rules conserving the number of active sites are the rule 184 (10111000) and its reflected rule 226 (11001000) [87, 154]. Extended examples of both rules are given in Fig. 11.12 and Fig. C.4. Rule 184 has been used as a simple model for traffic flow in a single line highway, and form the basis for many CA models of traffic flow with greater sophistication. Because of that, rule 184 is sometimes called the *traffic* rule.

When implementing majority memory, rules 184 and 226 lose the number conserving property, so that the pattern drifts to a fixed point of all 1s or all 0s depending upon whether within the configuration the number of 1s was superior to that of 0s or the contrary. The variation in the number of live cells in Fig. 11.13 and Fig. C.5 is low because the initial density is fairly close to the watershed 0.5, so that only by $T=1000$, the pattern completelly blackens. The cases of low and high initial densities in Figs. 11.14 show how the rule 184 with $\tau=3$ majority memory readily relaxes in both cases (as in the low density example of rule 226 in Fig. C.6) to a fixed point of all 0s or all 1s, correctly classifying the initial configuration.

The *majority* rule 232 together with the equivalent rules 184 and 226 have proved to be particularly effective among elementary rules in dealing with the density classification task. So far, it is to expecte that rules endowed with memory in cells in the form of the mode of the three last states acting in the temporal domain will produce good results in classifying density, as suggested here with respect to the elementary rules 184 and 226. Thus, it becomes apparent from Fig. 3.8 that rule 184 with $\tau=3$ very soon relaxes to the correct fixed point if the initial density is either $\rho\geq0.6$ or $\rho\leq0.4$. In the $[0.4, 0.6]$ interval the drift of density is much slower, but tending to the steady state that marks *correct* classification. This is so even for the particular cases that appear stabilized up to $T=300$, as shown below in Fig. 3.8. Only one of the plots that finally evolve upwards in the lower graph of Fig. 3.8, that red-marked, corresponds to an initial density under 0.5, albeit very close to this limit value.

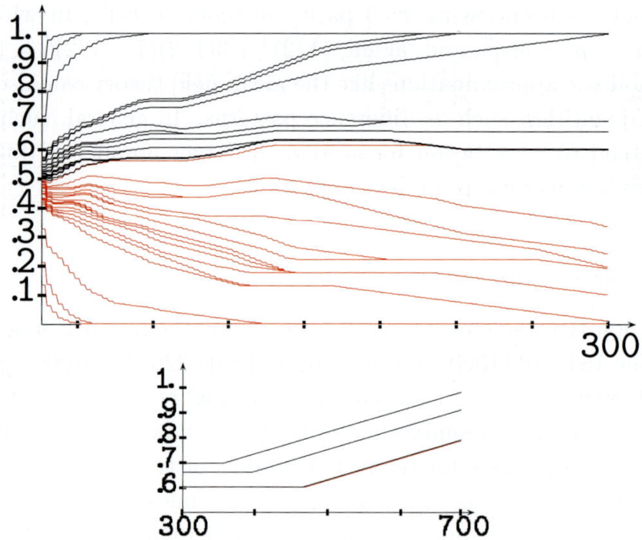

Fig. 3.8   Evolution of density under rule 184 with $\tau{=}3$ majority memory.   Further evolution of the instances stable up to $T{=}300$ is shown below.

In a simulation of a series of 10000 different initial densities uniformly covering the whole $[0.0, 1.0]$ interval, with every simulation run up to $T{=}1500$ in a $N{=}400$ layer, 114 densities were incorrectly classified[6]. Increasing the length of the trailing memory up to $T{=}5$ does not imply increasing the performance in the solution of the classification task. Beyond this trailing length, inertial effects oppose the drift to all 1s or all 0s intended in the density classification. It seems that $\tau{=}3$ as length of the majority memory is a good (and simple) choice regarding the density classification task.

The evolution of the rules 184 and 262 with memory (in Fig. 11.13 and Fig. C.5) is reminiscent to that of the **r=3** Gacs-Kurdyumov-Levin rule (GKL, [157, 167])[7], one of best density classifier rules.   The GKL rule

---

[6]Most of them, up to 71, were densities over 0.5 ($0.502{\times}11$, $0.505{\times}6$, $0.507{\times}6$, $0.510{\times}9$, $0.512{\times}10$, $0.515{\times}7$, 0.517, $0.520{\times}17$, $0.522{\times}7$, $0.525{\times}2$), thus 43 under this limit value ($0.463{\times}5$, 0.468, 0.470, 0.472, 0.475, $0.477{\times}2$, $0.480{\times}5$, 0.482, $0.485{\times}3$, $0.487{\times}2$, $0.490{\times}5$, $0.492{\times}4$, $0.495{\times}4$, $0.497{\times}9$).

[7]
$$\sigma_i^{(T+1)} = \begin{cases} mode\!\left(\sigma_i^{(T)}, \sigma_{i-1}^{(T)}, \sigma_{i-3}^{(T)}\right) \; if \; \sigma_i^{(T)} = 0 \\[2mm] mode\!\left(\sigma_i^{(T)}, \sigma_{i+1}^{(T)}, \sigma_{i+3}^{(T)}\right) \; if \; \sigma_i^{(T)} = 1 \end{cases}$$

Table 3.4   The effect of $\tau=3$ majority memory in the tub (up) and block (down) LIFE agars.

when applied to the same above reported series of initial densities. misclassified 160 densities. Both GKL and rules with $\tau=3$ mode memory have in common the presence of the mode operation, that seems to be in the origin of their excellent results in the density task.

Non-uniform or *hybrid* CA have been shown to be a good tool to solve the density task [369]. The CA with memory here may be also considered hybrid (or composite), but in space and time, implementing in the time context the most successful density task solver rule: the majority rule. It is expected that a synergic effect will emerge, so that rules in the standard context that poorly deal with the density task problem, when combined with the majority memory become good density solvers.

LIFE with $\tau=3$ majority memory

Short-term $\tau=3$ majority memory turns out to be much less effective in the control of the advance of the destruction of a stable rich configuration (*agar*) under LIFE [3]. This is exemplified in the comparison of Table 3.4 with Table 2.4. The minimal damage with full memory achieved in Table 2.4 is bigger with only $\tau=3$ memory in Table 3.4. But in any case, memory notably restrains the advance of the agar structure destruction without memory, as depicted in Table 2.4.

Table 3.5 shows how the glider evolves in LIFE with $\tau=3$ majority memory, at half the velocity as in the conventional LIFE rule from T=5, repeating twice every live structure. The r-pentomino generated as most-frequent state configuration following T=3 in Table 3.5 is, as in conventional LIFE, a *methulselah* when seeded as initial configuration, generating five $\tau=3$ memory gliders. The actual configuration at T=4 in Table 3.5 has also a notable activity from T=1, but does not produce gliders.

In most cases, the gliders interact in LIFE with $\tau=3$ majority memory as in conventional LIFE, though at half velocity, repeating every pattern. Tables 3.6 and 3.7 show examples of two gliders traveling in the same di-

Table 3.5   The glider in LIFE with τ=3 majority memory up to T=10.

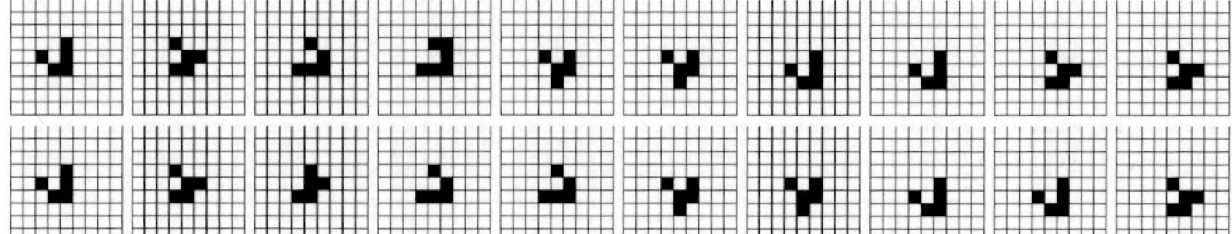

rection and in orthogonal directions respectively. In Table 3.8, four gliders collide. Conventional LIFE reaches extinction, a blinker (second scenario of Table 3.6), and a *loaf* (lower scenario of Table 3.6) as in Tables 3.6-3.8, but without the repetition of patterns from T=5.

Table 3.6   Collisions of gliders traveling in the same direction in τ=3 LIFE.

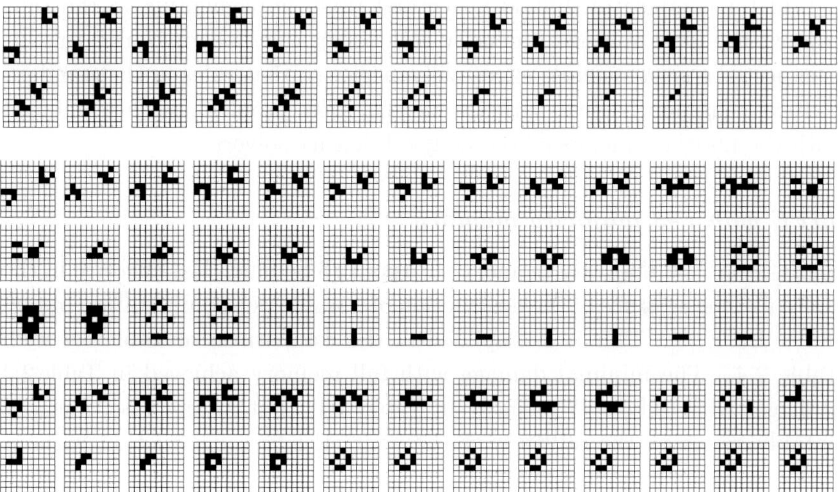

Table 3.9 shows a well-known glider-gun in conventional LIFE at T=70, and the corresponding configuration in LIFE with τ=3 majority memory. In the latter case, only one glider is generated up to T=70. Later on, a second glider gliding in parallel with that generated by T=70 emerges, but no more gliders are generated. Thus, the glider-gun with τ=3 majority memory is not stable and only shoots two *bullets*.

The important spaceships in conventional LIFE collected in Table 3.10, do not travel but either extinguish at T=22 in the *lightweight* upper case

Table 3.7   Glider collisions traveling in orthogonal directions in $\tau=3$ LIFE.

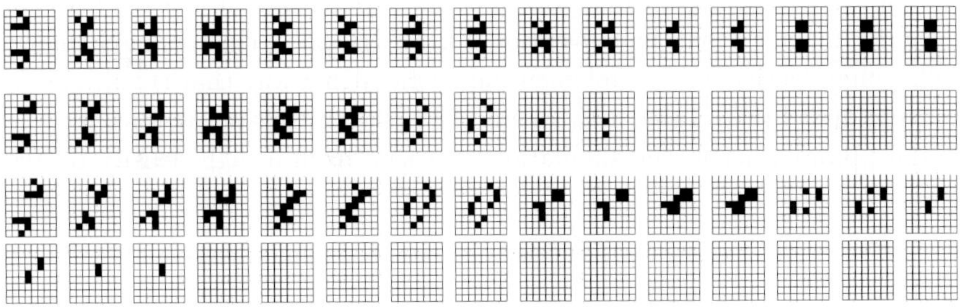

Table 3.8   A collision of four gliders in $\tau=3$ LIFE.

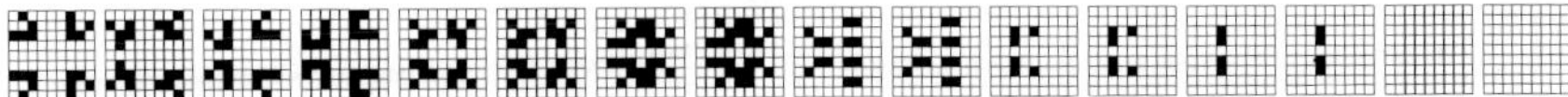

Table 3.9   A glider-gun at $T=70$ in conventional LIFE and with $\tau=3$ majority memory.

NO MEMORY   **TR-232**

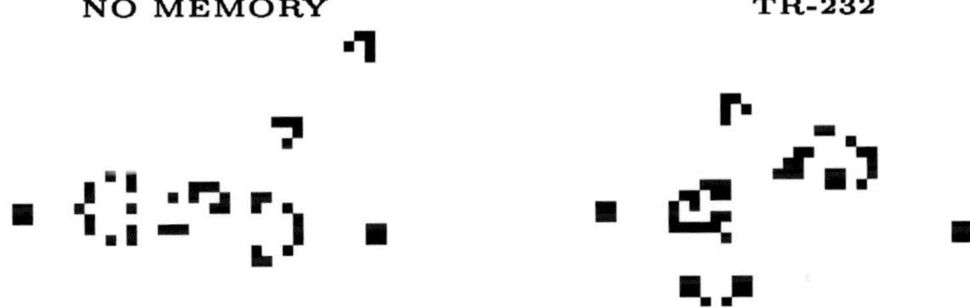

in the figure, or soon stabilize as in the *middle* and *heavy* configurations presented below. No mobile configurations, other that the glider, have been found with $\tau=3$ majority memory.

Table 3.10   Spaceships in LIFE with $\tau=3$ majority memory up to T=21.

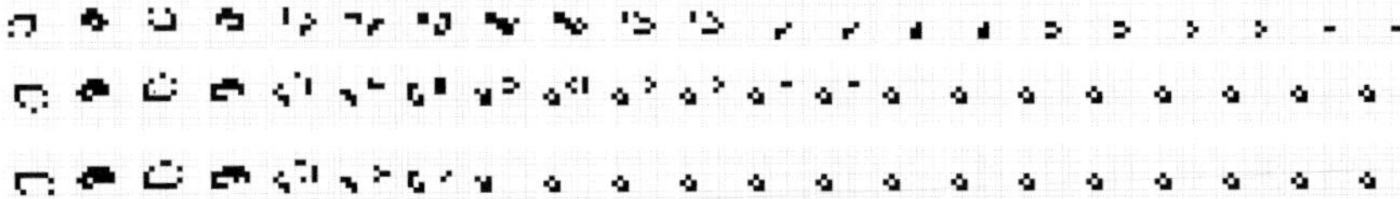

Table 3.11 shows the evolution of the density (red) and changing rate (blue) in conventional and $\tau=3$ majority memory LIFE, when starting from random configurations. In both cases, both parameters tend to decline, but

in the scenario with memory to a lower extent. Thus, the patterns starting from a random configuration tend to be more populated in LIFE with $\tau=3$ majority memory than in conventional LIFE. Besides, clusters of changing cells remain active when implementing memory in a way that is not seen in conventional LIFE. Table 3.12 zooms the first ten time-steps, showing the initial plummeting in the simulations starting from a high density of live cells.

Table 3.11   Density (red) and changing rate (blue) up to T=400 in LIFE.

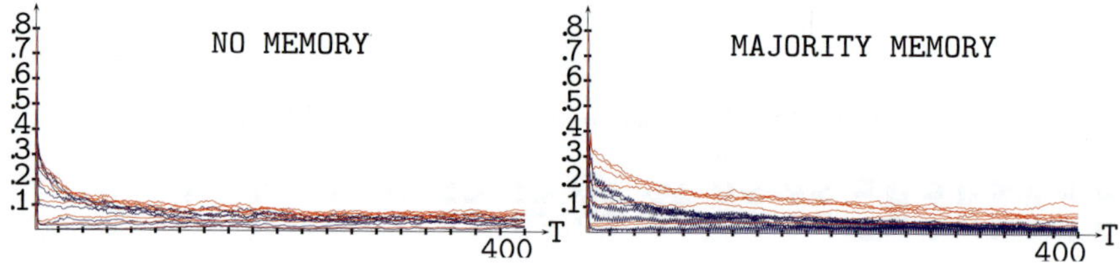

Table 3.12   Density and changing rate up to T=10 in LIFE.

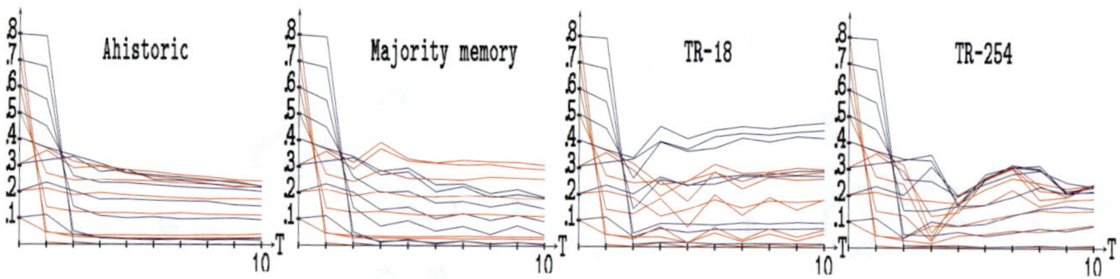

Table 3.13 shows the dynamics of the glider with partial $\tau=3$ majority memory, so the short-term variant of full memory dynamics in Table 2.5. Unlike with full memory, Table 3.13 shows that the glider soon dies with partial short-term majority memory.

Table 3.13   The glider in LIFE with partial $\tau=3$ majority memory.

3.4   Elementary rules as memory

Elementary CA rules (f) can in turn act as memory rules:
$$s_i^{(T)} = f\big(\sigma_i^{(T-2)}, \sigma_i^{(T)}, \sigma_i^{(T-1)}\big)$$
In this context, the spatial rules (ϕ) will be referred as S-rules, and the memory rules (f), actuating on time, as T-rules. In such a way that SXTY will refer to the spatial rule X actuating on cells featured by the memory rule Y. In figures, $+$Y refers to the temporal rule.

Figures 3.9 and 3.10 show the effect of memory on rules 90 and 150 starting from a single site live cell up to $T = 13$. The computer code in Table 3.14 produces the patterns for rules 150 and S150T150.

Table 3.14   A MATLAB$^{\circledR}$ program for rules 150 and S150T150 in Fig. 3.10.

```
      T=13; N=2*T+1; ruleS=150;ruleT=150;
      for memo=0:1
          [SIGMA]=init(N,T);
          for t=1:T;  S=SIGMA; ST(t,:)=SIGMA;
          if(memo==1&t>2)
              for i=1:N
        S(i)=RULES(ST(t-2,i),ST(t,i),ST(t-1,i),ruleT);
              end
          end
          [SIGMA]=TRANSITION(S,N,ruleS);
          end;  subplot(1,2,memo+1);imagesc(ST);
      end
      function [SIGMA]=TRANSITION(S,N,rule)
          SIGMA(1)=RULES(S(N),S(1),S(2),rule);
          for i=2:N-1
          SIGMA(i)=RULES(S(i-1),S(i),S(i+1),rule);
          end
          SIGMA(N)=RULES(S(N-1),S(N),S(1),rule);
      function new=RULES(l,c,r,rule)
          switch {rule}
          case {150}; new=mod(l+c+r,2)
          ...
          end
      function [SIGMA]=init(N,T)
          SIGMA(1:N)=0;c=(N+1)/2;SIGMA(c:c)=1;
```

Complementary memory rules have the same effect on rule 90 (regardless of the role played by the three last states in f) starting from a single seed. This is why Fig. 3.9 shows only the effect of memory rules up to $f=127$. A subset of rules does not affect rule 90; grouped as pair of complementary rules they are: (19,236), (28,228), (51,204), (59,196), (59,196), (68,187), (68,187), (76,179), (100,155) and (108,147). Rule 150 is affected by every memory rule except, of course, by the *identity* rule 204(11001100), which assigns $s_i = \sigma_i$ and therefore does not affect any rule. Most of the rules that extinguish starting from a single site seed, lead to extinction (e.g.,

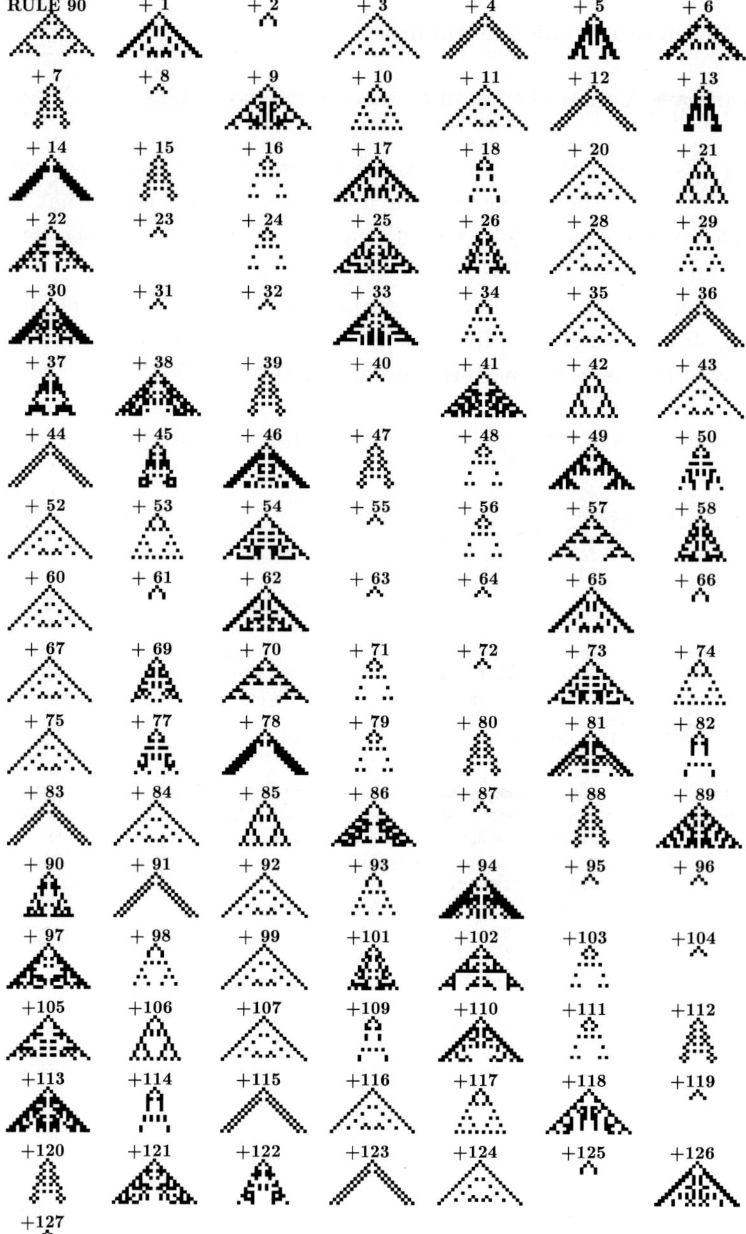

Fig. 3.9   The rule 90 with elementary rules as memory from a single seed.

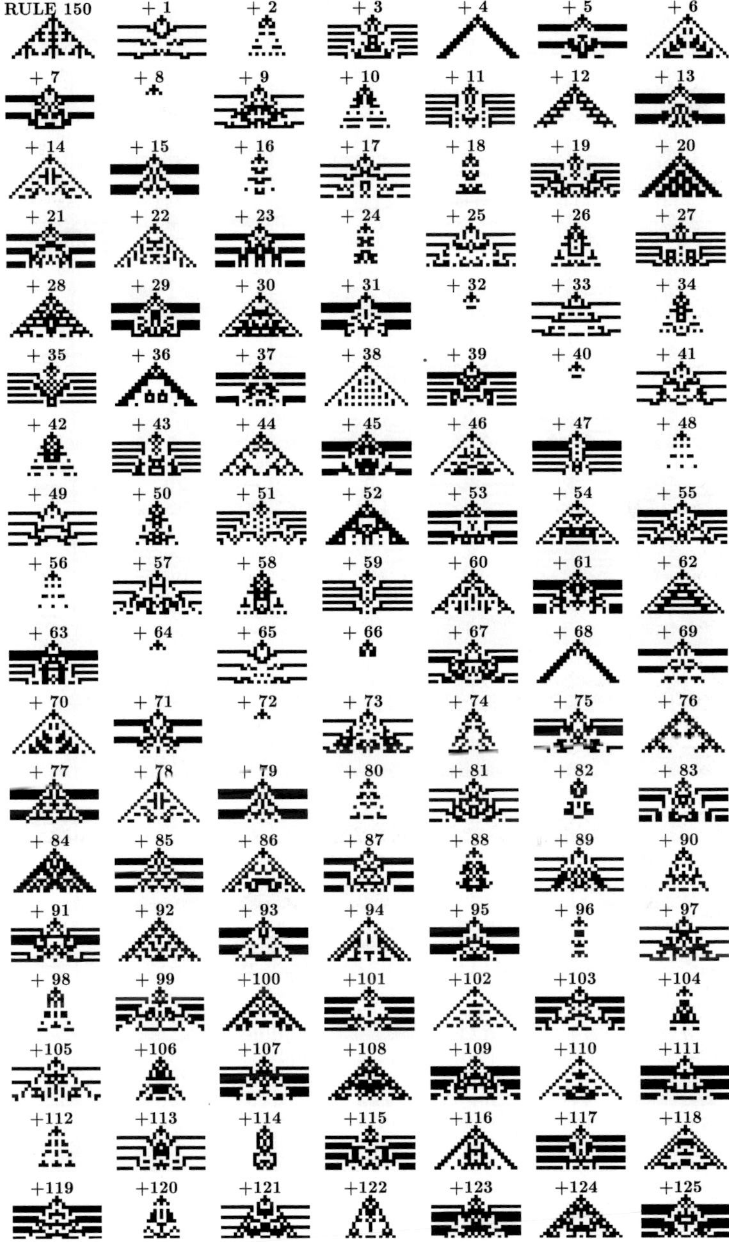

Fig. 3.10 The rule 150 with elementary memory rules from a single seed.

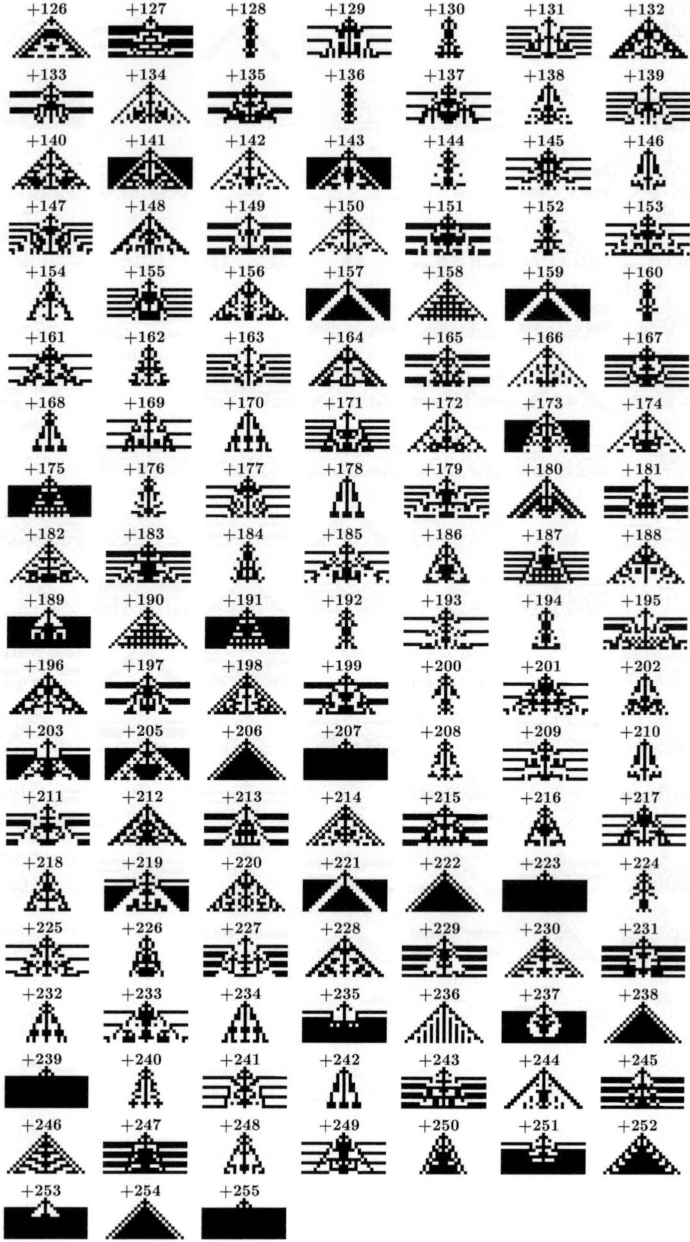

Fig. 3.10   (continued)

Table 3.15 Rules 90 and 150 starting with two adjacent live cells. Ahistoric patterns and patterns with rules 90 and 150 acting as memory.

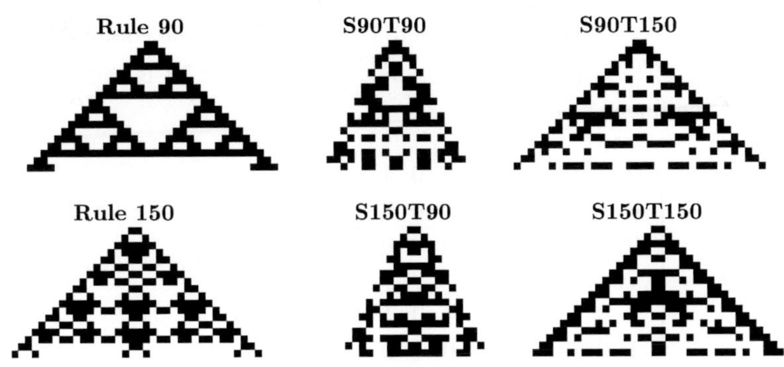

the legal rules 32, 72, 104, 128, 160, 200 and 232 as early as at $T = 4$), or produce patterns consisting of only two *branches* (e.g., S90T4) when acting as memory rules on rule 90. Rule 150 is much more resilient to extinction (only five rules lead to its extinction), or to the formation of two branches (S150T4,S150T68). Some patterns with memory are reminiscent of the ahistoric ones, e.g., S90T20 or S150T100, but as a normal effect, memory notably alters the spatio-temporal patterns. As a rule, no general relevant concordance can be traced between the effect of *all* rules in the same equivalence class. Anyway, *i*) reflected rules tend to produce similar memory effect, e.g., {8,64}, {13,69}, {29,71}, {30,86}, or {184,226}, *ii*) rules in some equivalence classes produce a similar effect when acting as memory rules, e.g., the important rule 110 and its three equivalent rules 124,137 and 193, or those rules in the class {60, 102, 153, 195}. The counterexample is seen in rules {19,55} regarding their effect on rule 90: rule 19 does not alter rule 90, rule 55 leads to its extinction at $T=4$.

Rules 90 and 150 are *linear* transition rules which employ only XOR logic, i.e., arithmetics performed modulo 2 in the two-states scenario. Thus, R150: $\sigma_i^{(T+1)} = \sigma_{i-1}^{(T)} \oplus \sigma_i^{(T)} \oplus \sigma_{i+1}^{(T)}$,and R90: $\sigma_i^{(T+1)} = \sigma_{i-1}^{(T)} \oplus \sigma_{i+1}^{(T)}$, with $\oplus$ denoting sum in $GF(2)$, i.e., XOR operation equivalent to modulo 2, when $k = 2$. Rules 90 and 150 (together with the the trivial 0 and 204) are the only linear legal rules [8]. Linear rules remain linear when cells are endowed with linear memory rules. Thus for example, S150T150 turns out:

---

[8]The eight elementary linear rules are: R0, R60: $\sigma_i^{(T+1)} = \sigma_{i-1}^{(T)} \oplus \sigma_i^{(T)}$, R90, R102 (mirror image of R60): $\sigma_i^{(T+1)} = \sigma_i^{(T)} \oplus \sigma_{i+1}^{(T)}$, R150, R170: $\sigma_i^{(T+1)} = \sigma_{i+1}^{(T)}$, R204: $\sigma_i^{(T+1)} = \sigma_{i-1}^{(T)}$, R204 (identity): $\sigma_i^{(T+1)} = \sigma_i^{(T)}$.

$$\sigma_i^{(T+1)} = (\sigma_{i-1}^{(T)} \oplus \sigma_{i-1}^{(T-1)} \oplus \sigma_{i-1}^{(T-2)}) \oplus (\sigma_i^{(T)} \oplus \sigma_i^{(T-1)} \oplus \sigma_i^{(T-2)}) \oplus (\sigma_{i+1}^{(T)} \oplus \sigma_{i+1}^{(T-1)} \oplus \sigma_{i+1}^{(T-2)})$$

, or in matrix terms: $\mathbf{C}^{(T+1)} = \mathbf{M}\big(\mathbf{C}^{(T)} \oplus \mathbf{C}^{(T-1)} \oplus \mathbf{C}^{(T-2)}\big) = \mathbf{MC}^{(T)} \oplus \mathbf{MC}^{(T-1)} \oplus \mathbf{MC}^{(T-2)}$, where $\mathbf{C}^{(T)}$ stands for the configuration at time-step T, $\mathbf{C}^{(T)} = \big(\sigma_1^{(T)}, \sigma_2^{(T)}, \ldots, \sigma_{N-1}^{(T)}, \sigma_N^{(T)}\big)'$, and $\mathbf{M}$ is the transition matrix of the linear rule. Thus,

$$\mathbf{M_{90}} = \begin{pmatrix} 0 & 1 & 0 & 0 & 0 & \ldots & 0 & 0 & 1 \\ 0 & 1 & 0 & 1 & 0 & \ldots & 0 & 0 & 0 \\ 0 & 0 & 1 & 0 & 1 & \ldots & 0 & 0 & 0 \\ \multicolumn{9}{c}{\dotfill} \\ 0 & 0 & 0 & 0 & 0 & \ldots & 1 & 0 & 1 \\ 1 & 0 & 0 & 0 & 0 & \ldots & 0 & 1 & 0 \end{pmatrix} \qquad \mathbf{M_{150}} = \mathbf{M_{90}} \oplus \mathbf{I}$$

Linear rules are *additive*: i.e., any initial pattern can be decomposed into the superposition of patterns from a single site seed. Each of these configurations can be evolved independently and the results superposed (module two) to obtain the final complete pattern. Mathematically, the distributive property holds for every pair of initial configurations $\mathbf{u}$ and $\mathbf{v}$: $\phi(\mathbf{u} \oplus \mathbf{v}) = \phi(\mathbf{u}) \oplus \phi(\mathbf{v})$.

The additivity of rules 90 and 150 is illustrated in Table 3.15 up to $T = 18$. The evolution patterns starting with two adjacent live cells as shown in this table coincide with the XOR superposed configuration of those evolved independently starting with a single seed, shown in Figs. 3.9 and 3.10.

Figures B.5 and B.6 show the effect of legal memory rules on rules 90 and 150, starting from the same random initial configuration with 211 sites up to $T = 150$. These figures also show superimposed the differences produced in patterns (DP) when reversing the value of its initial center site (darker pixels). Some features of the patterns in Figs. B.5 and B.6 could be predicted from Figs. 3.9 and 3.10. For example, rules 4, 32, 36, 72, 128, 132, 160, 164, and 200, when acting as memory rules on rule 90, lead either to extinction, to the formation of periodic patterns, or they produce soliton-like structures. Nevertheless, the prediction fails for rules such as 104 and 232, or 76, 108 and 236 which do not alter rule 90 starting from a single site seed but notably alter its pattern when starting at random. The effect of rules 4, 32, and 72 on rule 150 in Fig. B.6 agrees qualitatively with that on Fig. 3.10 . The constrained patterns of rules 128, 160, and 200 seem to advance their particular patterns in Fig. B.6 . Rule 36, which exemplifies the type of simple rules which serve as *filters* (Wolfram's Class II), turns out to be sophisticated when acting on rule 150. More predictable is the strong effect induced by the *complex* rule 54 on both rules.

Figure 3.11 shows the effect of some elementary rules acting as memory on the 2D parity rule implemented with the von Neumann neighborhood [23]. Figure 3.11 commences from a single site live cell, so that the (not shown) three first patterns are: ■ ✚ ■ ∴ ■ . In the case of rule 18 acting as memory rule, the actual pattern has not any cell alive at two time-steps in Fig. 3.11, which does not imply extinction after any of them. These kind of cataleptic episodes are unfeasible in the ahistoric context, but they are not rare when endowing cells with elementary rules as memory.

Fig. 3.11   The parity rule with elementary rules as memory. T=4-15.

LIFE with CA rules as memory

The computer code in Table 3.16 generates the spatio-temporal patterns of
LIFE with legal elementary CA rules as memory shown in Table 3.17.

Table 3.16   A MATLAB® program for LIFE with legal elementary rules as memory.

```
maxt=20;m=19;nh=(m+1)/2;
n=[m 1:m-1];e=[2:m 1];s=[2:m 1];w=[m 1:m-1]; % Indices for periodic boundary
for TR=1:254
    X(1:m,1:m)=0;X(nh-1:nh+1,nh-1:nh+1)=1; BN(1:8)=0;
    for ix=1:8
        rest=mod(TR,2);quot=(TR-rest)/2;BN(8-ix+1)=rest;TR=quot;
    end
    if(BN(8)==0&(BN(2)==BN(5))&(BN(4)==BN(7))
        for t=1:maxt
          if(t>3);X1=X2;X2=X3;X3=X;
          else; switch(t);case(1);X1=X;case(2);X2=X;case(3);X3=X;end
          end
          if(t>2)
            for i=1:m;for j=1:m
               S(i,j)=BN(8-(X(i,j)*2+X2(i,j)+X1(i,j)*4));
            end;end
            else;S=X;
            end
          figure(1);subplot(2,maxt,t);imagesc(1-X);colormap(gray)axis('off');axis image;
          figure(1);subplot(2,maxt,maxt+t);imagesc(1-S);axis('off');axis image
        N=S(n,:)+S(s,:)+S(:,e)+S(:,w)+S(n,e)+S(n,w)+S(s,e)+S(s,w);
        X=S;
        for j=1:m;for i=1:m
            if(S(i,j)==0&                N(i,j)==3  )X(i,j)=1;end
            if(S(i,j)==1&(~(N(i,j)==2|N(i,j)==3)))X(i,j)=0;end
        end;end
        end
    end
end;
```

Table 3.17 shows the evolving patterns from a 3×3 square, under conventional LIFE and in LIFE with elementary legal rules as memory rules.
In ahistoric LIFE, the 3×3 square becomes at $T=6$ a set of four blinkers that
generates a simple period-two oscillator, named *traffic light*. Examples of
nil patterns in the actual and trait series may also be traced in Table 3.17,
which does not imply extinction after any of them. These cataleptic patterns are more probable in early time-steps, though they also occur, to a
lesser extent, later in time.

Table 3.18 shows the evolution of the density of live cells (red) and of
the changing rate in LIFE with cells endowed with legal elementary rules
in a 100×100 lattice. Rule 18 seems to induce a fairly unexpected changing
rate when acting as memory rule. The seemingly simple rule 18 (00010010)
supports some kind of *intriguing* dynamics when acting as memory rule on

Table 3.17   The 3×3 square with elementary legal memory rules in LIFE.

Ahistoric

TR-4

TR-18

TR-22

TR-32,36

TR-50

TR-54

TR-72,76

TR-90

TR-94

TR-104

TR-108

TR-122

TR-126

TR-128,132

TR-146

Table 3.17   (*continued*)

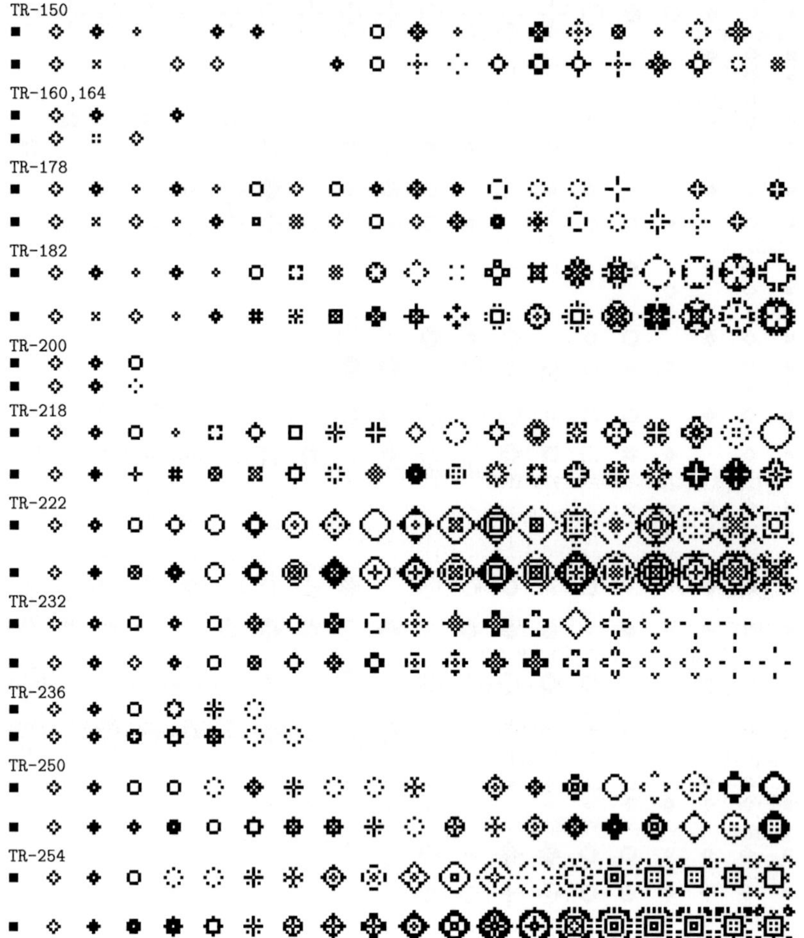

LIFE, in agreement with its own *intriguing* nature, as qualified in [212].

The main features of the dynamics in Table 3.17 are indicative of the dynamics starting at random as in Table 3.18. Thus, for example, rules 4, 32, 36, 72, 76, 128, 132, 160, 164 soon extinguish starting at random, as do in Table 3.17. Rule 104 acting as memory induces a strong decay in the activity of LIFE, that by T=20 is reduced to small active clusters in some simulations. The not shown rule 200, behaves much as rule 104 when acting as memory rule. Rule 108 is an exception as it does not induces extinction, as may be expected from Table 3.17. On the contrary, rule 108 activates LIFE in most simulations, with a case of very slow, though monotone, increase in both density and change rates, that by T=400 reaches only 10 %. As a norm, the rules that induce activity in Table 3.17 show also fairly high density and change rates in Table 3.18. This is so, for example, with the linear rules 90 and 150 and with rule 22 (00010110), considered as the one-dimensional LIFE analog, which induces a notable activity in LIFE. Rule 22 seems to corroborate this supporting, as rule 18 does, a high activity in LIFE when acting as memory rule.

In most cases in Table 3.18, the changing rate curve is above that of the density once they have reached their almost-steady levels. Rule 222 stabilizes both parameters into a fairly coincident rate close to 0.3. Oddly, rule 236 and, notably, rule 254 show the density curve above the changing rate.

The majority of the asymmetric quiescent rules that support activity when acting as memory rules in LIFE evolve to reach plateaus for both changing rate density, with the former above density, as in the most frequent evolution in Table 3.18. Just to cite an example, the density and changing rate of the important rule 110 evolves much like, say, rule 54 in Table 3.18. The *traffic* rule 184 does not support a permanent activity, and both parameters decay much as with majority memory in Table 3.11, albeit with changing rate above density. Non-quiescent memory rules tend to annihilate the LIFE dynamics: after the initial decay most trait states are the live one at $T=3$, which means almost extinction by overcrowding at $T=1$.

The main characteristics of the effect of elementary rules acting as memory, i.e., their *signature*, remain in lattices of smaller size than in that of 100×100 size chosen here. The main effect of decreasing the lattice size is the increase in the oscillation of both density and changing rates, which is fairly small in Table 3.18 once reached the (almost) steady levels. Table 3.19, shows how this applies for rule 254 acting as memory in 50×50 and 25×25 lattices: the most important characteristic of rule 254 acting as memory in LIFE is found in the simulations in the bigger lattice, i.e., changing rate above density, remains in the smaller ones.

Table 3.18   Density and changing rate in LIFE with elementary legal memory rules in a 100×100 lattice.

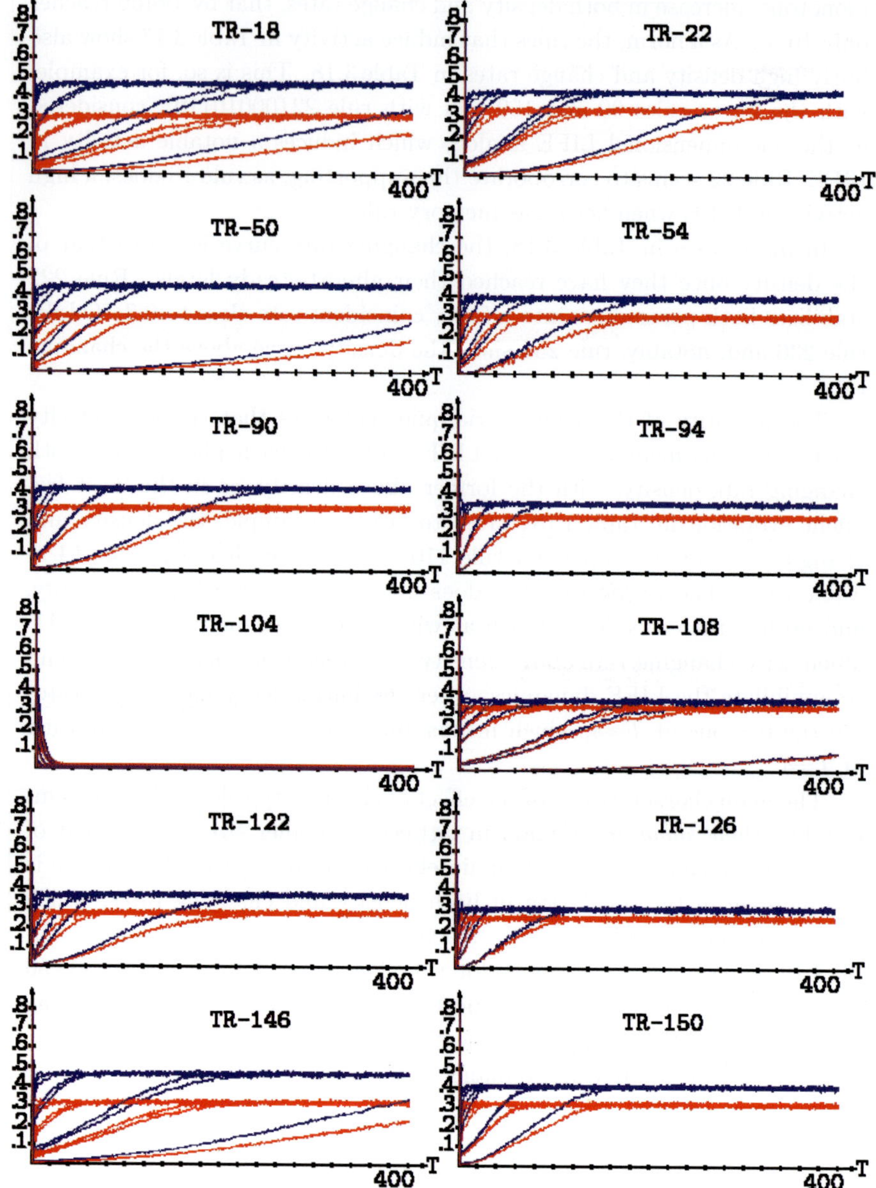

Table 3.18   (*continued*)

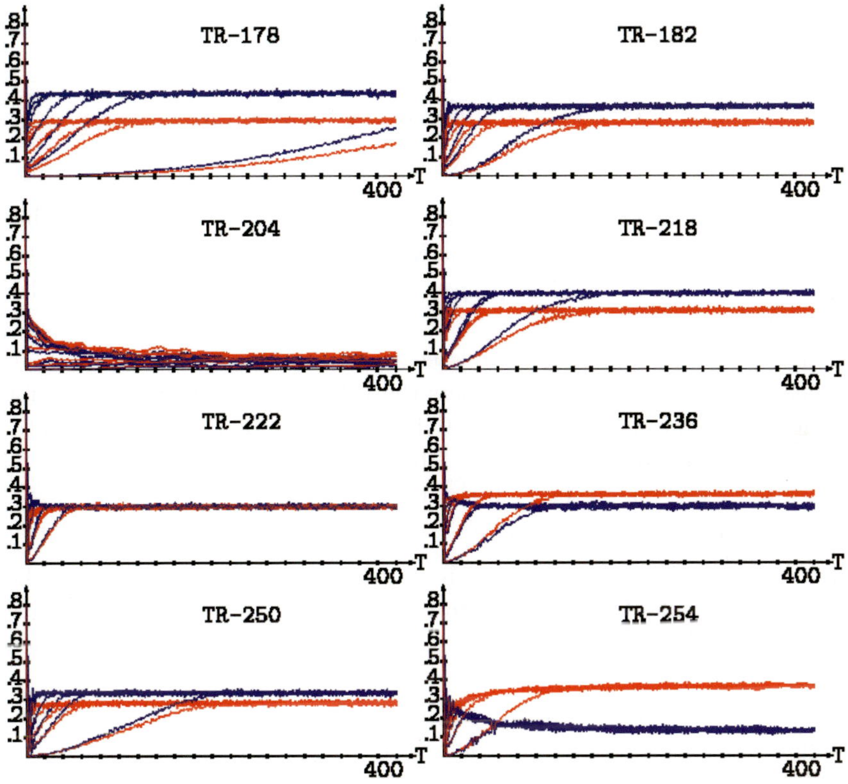

Table 3.19   Density (red) and changing rate (blue) in LIFE with rule 254 as memory. The lattice sizes are 50×50 in left panel and 25×25 in the right one.

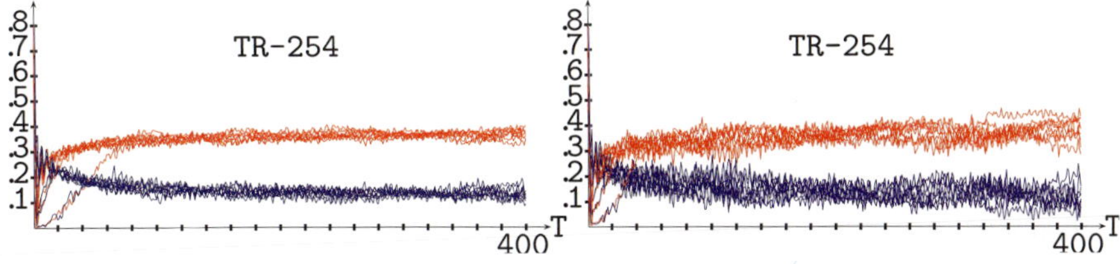

3.5   Minimal memory

The lowest degree of memory conceivable is that of featuring cells by Boolean functions of their last two states: $s_i^{(T)} = f\big(\sigma_i^{(T)}, \sigma_i^{(T-1)}\big)$ [37].

These mappings are characterized by a sequence of four binary values (β) associated with each of the four possible pairs $\left(\sigma_i^{(T)},\sigma_i^{(T-1)}\right)$ (a two-bit analogue of the codification in Section 2.3.1):

$$
\begin{array}{cccc}
11 & 10 & 01 & 00 \\
\beta_1 & \beta_2 & \beta_3 & \beta_4
\end{array}
\qquad \equiv \sum_{i=1}^{4} \beta_{\mathrm{s}} 2^{4-i} = \mathcal{R}
$$

The rule number varies in the $[0,15]$ interval, with $\mathcal{R} = 12\,(1100)$ being the identity rule $s_i^{(T)} = \sigma_i^{(T)}$, and $\mathcal{R} = 6\,(0110)$ being the parity rule: $s_i^{(T)} = \sigma_i^{(T-1)} \oplus \sigma_i^{(T)}$. The *proper* majority rule is $\mathcal{R}=8\,(1000)$ in this context.

Every two-bit rule may be related to an elementary three-bit one, in which the former operates regardless the value of the third bit. Thus, for example, the two-bit parity rule $6\,(0110)$ relates to the three bit rule 102:

$$
\begin{array}{cccccccc}
111 & 110 & 101 & 100 & 011 & 010 & 001 & 000 \\
0 & 1 & 1 & 0 & 0 & 1 & 1 & 0
\end{array},
$$

or to rule $60\,(00111100)$ with $\sigma_i^{(T-1)}$ playing the role of $\sigma_{i-1}^{(T)}$ instead of that of $\sigma_{i+1}^{(T)}$. In the same vein, the two-bit rule $14\,(1110)$ relates to the three-bit rule $238\,(11101110)$, or to rule $252\,(11111100)$ in the alternative case. These four three-bit rules, i.e., 60, 102, 238 and 252, are non-legal rules.

But the patterns generated by related rules in the two and three time-step memory scenarios are not coincident, because the dynamic differs initially at T=3. In the three-state memory scenario, memory becomes activated for the first time following T=3, thus the pattern at this time-step is the same as in the ahistoric model, whereas in the two-state memory scenario, memory is activated already following T=2, thus the pattern at T=3 is (likely) not the same as in the ahistoric model.

Rule $10\,(1010)$ acting as memory replicates twice every pattern, because: $s_i^{(T)} = (\sigma_i^{(T)} \cap \sigma_i^{(T-1)}) \cup (\overline{\sigma}_i^{(T)} \cap \sigma_i^{(T-1)}) = \sigma_i^{(T-1)} \cap (\overline{\sigma}_i^{(T)} \cup \sigma_i^{(T)}) = \sigma_i^{(T-1)}$. The two-bit rule 10 relates to the three-bit rules $170\,(10101010)$ and $240\,(11110000)$, i.e., the shift rules: $\sigma_i^{(T+1)} = \sigma_{i+1}^{(T)}$, $\sigma_i^{(T+1)} = \sigma_{i-1}^{(T)}$, when acting as spatial rules.

Figure 3.12 shows the effect of the parity of the last two state memory on rules 30, 90, and 150. Thus, the linear rules 90 and 150 become with $\tau=2$ parity memory (i.e., T6): $\sigma_i^{(T+1)} = (\sigma_{i-1}^{(T)} \oplus \sigma_{i-1}^{(T-1)}) \oplus (\sigma_{i+1}^{(T)} \oplus \sigma_{i+1}^{(T-1)})$, $\sigma_i^{(T+1)} = (\sigma_{i-1}^{(T)} \oplus \sigma_{i-1}^{(T-1)}) \oplus (\sigma_i^{(T)} \oplus \sigma_i^{(T-1)}) \oplus (\sigma_{i+1}^{(T)} \oplus \sigma_{i+1}^{(T-1)})$, or in matrix terms: $\mathbf{C}^{(T+1)} = \mathbf{M}\big(\mathbf{C}^{(T)} \oplus \mathbf{C}^{(T-1)}\big)$.

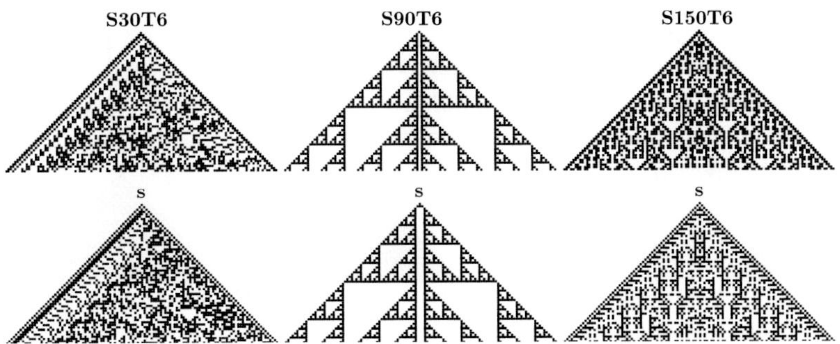

Fig. 3.12 The rules 30, 90 and 150 with parity of the last two state (rule 6) memory.
The evolving patterns of the featured (s) cells are shown below.

Figures 3.13 and 3.14 show the evolving patterns of some elementary
rules with memory of the last two states. Again, as in the $\tau = 3$ memory
context, complementary memory rules (now adding 15) have the same effect
on rule 90 and on the elementary complementary to rule 90, i.e., rule 165.
This is reflected in Fig. 3.13 and in Fig. B.7 in Appendix B, which shows
the effect of $\tau = 2$ memory on some elementary rules starting at random
up to $T = 100$.

An alternative mechanism that only demands an additional bit of memory per cell is that of keeping unlimited track of the sum of previous state
values, $s_i^{(T)} = \sigma_i^{(1)} \oplus \ldots \oplus \sigma_i^{(T-1)} \oplus \sigma_i^{(T)}$, as $s_i^{(T)} = s_i^{(T-1)} \oplus \sigma_i^{(T)}$. Lines 3
to 8 of the MATLAB code in Table 3.14 should be rewritten to implement
this idea as follows:

```
3              [SIGMA]=init(N,T);  S=SIGMA;
4              for t=1:T;ST(t,:)=SIGMA;
5-8               if(memo==1&t>1); S=mod(S+SIGMA,2); <- trait states
```

Figure B.8 in Appendix B shows the effect of such a kind of minimal
memory mechanism on some elementary rules, in a register of size 150
and up to $T = 60$. In the case of linear rules it holds that, $\mathbf{C}^{(T+1)} =
(\mathbf{M} + \mathbf{I})\mathbf{C}^{(T)}$ [9], consequently rule S150TUP evolves after $T = 2$ as rule 90
and S90TUP evolves as rule 150. This can be checked in Table 3.20, in
which the evolution from $T = 3$ is that of rule 90.

---

[9] $\mathbf{C}^{(2)} = \mathbf{M}\mathbf{C}^{(1)}$

$\mathbf{C}^{(3)} = \mathbf{M}(\mathbf{C}^{(2)} + \mathbf{C}^{(1)}) = \mathbf{M}\mathbf{C}^{(2)} + \mathbf{C}^{(2)}$

$\mathbf{C}^{(4)} = \mathbf{M}(\mathbf{C}^{(3)} + \mathbf{C}^{(2)} + \mathbf{C}^{(1)}) = \mathbf{M}\mathbf{C}^{(3)} + \mathbf{C}^{(3)}$

. . . .

$\mathbf{C}^{(T+1)} = \mathbf{M}\mathbf{C}^{(T)} + \mathbf{C}^{(T)} = (\mathbf{M} + \mathbf{I})\mathbf{C}^{(T)}$

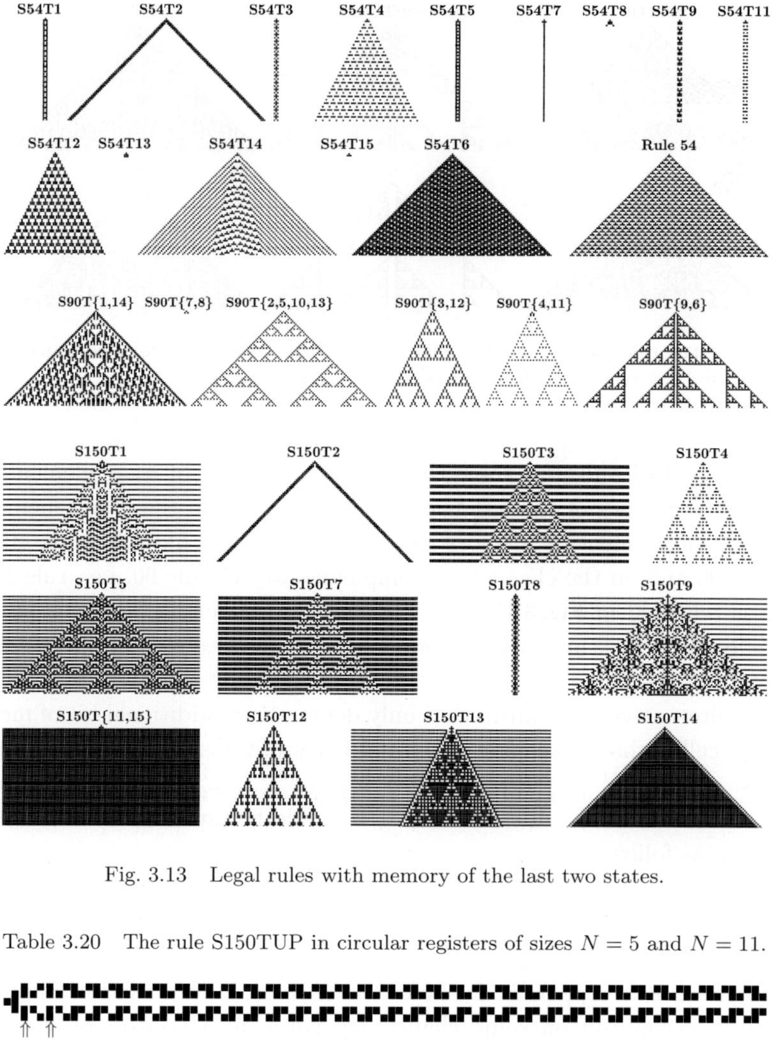

Fig. 3.13   Legal rules with memory of the last two states.

Table 3.20   The rule S150TUP in circular registers of sizes $N = 5$ and $N = 11$.

Endowing cells with $\tau=2$ majority memory in the mean-field approach it is:

$$P\big(s^{(T)} = 1\big) = P(\sigma^{(T)}=1)P(\sigma^{(T-1)}=0) + P(\sigma^{(T)}=0)P(\sigma^{(T-1)}=1) =$$
$$P^{(T)}(1 - P^{(T-1)}) + (1 - P^{(T)})P^{(T-1)}.$$

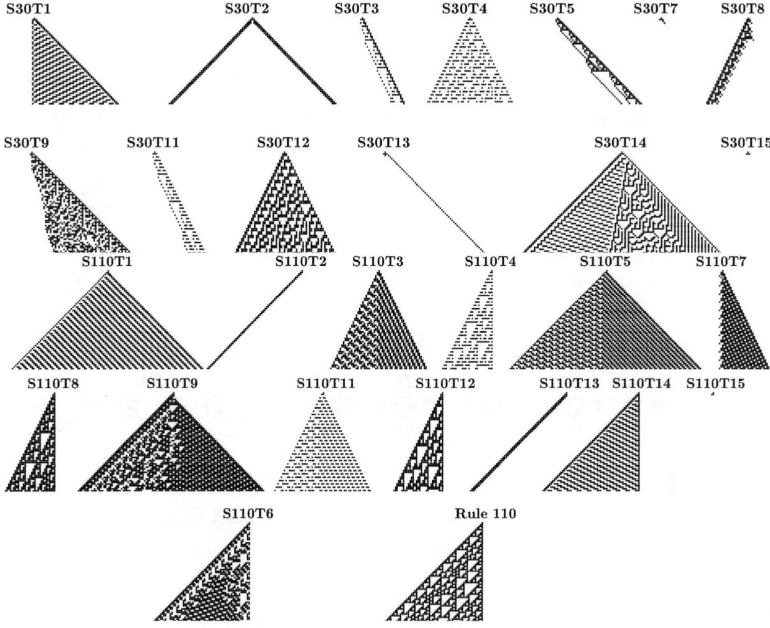

Fig. 3.14   No-legal rules with memory of the last two states.

In the $P^{(T)} = P^{(T-1)}$ scenario, the rules with fixed point $p = 1/2$ in their ahistoric formulation will preserve this fixed point when endowing cells with parity of the last two state memory, because when substituting p by $2p(1-p)$ it remains $2(1/2)(1-1/2) = 1/2$.

LIFE with minimal memory

Table 3.21 figure shows the collision of four gliders starting as in Table 3.8 in LIFE with the two-bit rule 10 (1010) as memory. As expected, rule 10 replicates every configuration. In the same vein, the glider-gun in Table 3.9 remains stable with rule 10 acting as memory, generating gliders as in the conventional ahistoric LIFE formulation, but at half the velocity.

Table 3.21   A collision of four gliders in $\tau{=}2$ rule 10 memory LIFE.

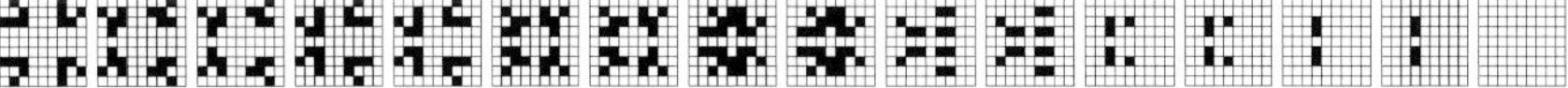

Table 3.22   The LIFE rule with minimal memory rules.

Table 3.22 shows the evolution of the initial 3×3 square in LIFE with two different minimal memory rules. The parity rule (T6) soon generates a period-three oscillator one of whose components is the nil-pattern, whereas T14, i.e., cell considered alive if was not dead in any of the two last time-steps, soon generates a sophisticated dynamic.

Table 3.23   Evolution of the 3×3 square in LIFE with the $\tau=2$ rule 14 memory. Density and changing rates up to T=400 (left), and pattern at T=400 in a 100×100 lattice.

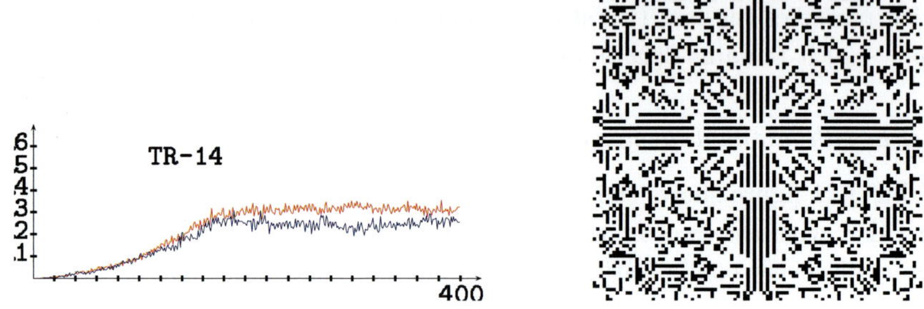

Accordingly to the expectation from Table 3.22, rule 14 leads to a fairly high density and change rates, close to 0.3, when starting from a 3×3 square in a 100×100 lattice, as shown in Table 3.23. This plateau is seemingly coincident with that reached starting at random, as shown in the TR-14 panel of Table 3.24. In fact both parameters slightly oscillate close to 0.3, with the density a bit greater than the changing rate. Density above changing rate relates the effect of the two-bit rule 14 to that of its qualitatively equivalent three-bit rule 254 .

On the contrary, the effect of rule 6 on LIFE when starting at random is not envisaged in Table 3.22. Thus, Table 3.24 shows a plateau for the density again circa 0.3, and a higher level of the changing rate, set at 0.4 .

Table 3.24   Density and changing rate up to T=200 in LIFE with two τ=2 memory rules in a 100×100 lattice.

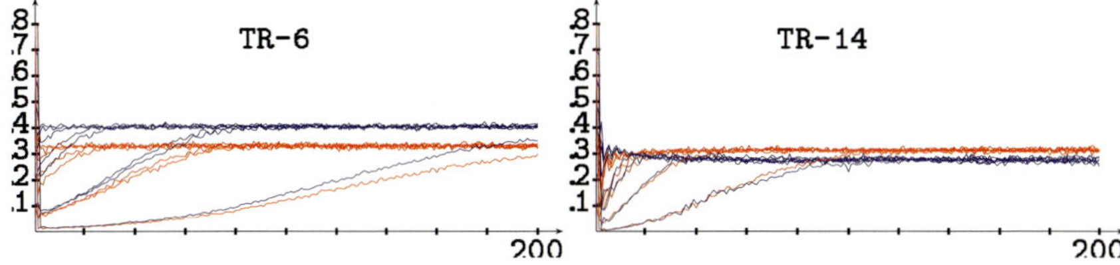

As already pointed in the three-time step memory scenario, decreasing the lattice size increases the oscillation of both density and changing rates, but the signature of the effect of the rule memory remains. Table 3.25 gives an example with the rules in Table 3.24. Incidentally, a simulation collapses in the TR-6 panel of Table 3.25

Table 3.25   Density and changing rate in LIFE with the two τ=2 memory rules of Table 3.24 in a 25×25 lattice.

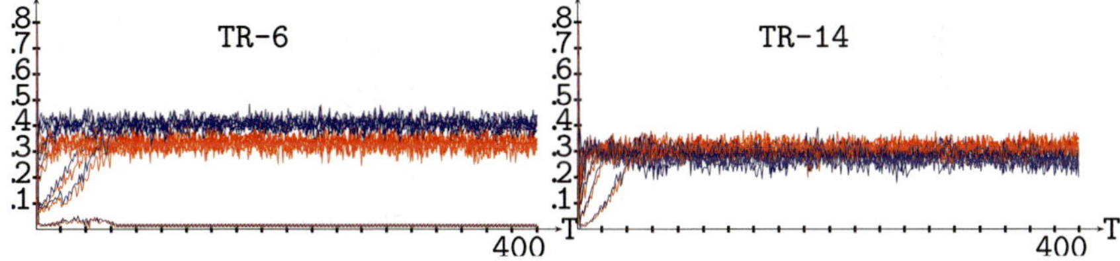

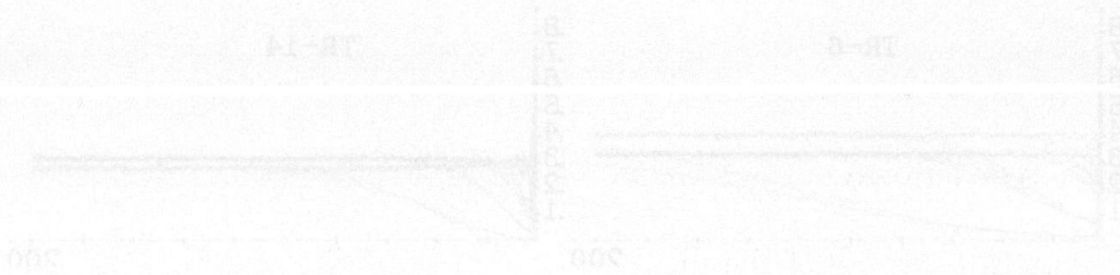

Chapter 4

Asynchrony and probabilistic rules

4.1 Asynchrony

Departing from one of the prominent features of cellular automata, that of the simultaneous updating of all cell states, asynchronous updating is allowed in one of the generalizations, or extensions, of the CA paradigm [263, 363]. In some applications involving rules with several components, some of then are applied in parallel, whereas others are applied asynchronously.

It has been advocated that asynchrony is more natural since there is no universal clock in Nature, but the controversy on synchrony *vs* asynchrony is an open question in spatially distributed systems in general, becoming a major question to be addressed regarding the CA methodology in the ecological contexts. Hogeweg [202] and others, have claimed that the simultaneous updating of all cells is at odds with the localness of interaction that is one of the strengths of CA. Frequently, the interesting structure seen in the evolution of a CA is, in fact, an artifact of the synchronous updating. What needs to be addressed is whether or not there are ecological systems for which universal updating is not an unwarranted assumption. Quoting Hogeweg [202]: "Because synchrony is obviously not a property of natural systems, it should be dropped from the formalism or treated with caution.". The study of the effect of memory in ecological systems undertaken in [187], also considers asynchrony in its stochastic transition rule. The issue is again commented in the context of the spatialized prisoner's dilemma in Chapter 13.

Asynchronism can be implemented in various forms, classified in [363] as: *i) time-driven* methods, where the updating of each single cell is governed by an exponentially distributed waiting time. This approach relates to that known as *Kinetic Monte Carlo* (KMC) simulations on a lattice, or

81

atomistic KMC, studied with memory in [281]. *ii) step-driven* methods, characterized by the absence of an explicit time variable, and in which an algorithm determines the order of evaluation of the cells. The update of a single cell is called a single *step*.

The step-driven method that consists in choosing the cell to be evaluated randomly with uniform distribution on the grid, i.e., a random choice with replacement, imposes the need for each cell having a counter of the number of the *successive* times which it has been updated. Thus, following the t_i-*th* update of the generic cell i, the cell will be endowed with weighted average memory $(s_i^{(t_i)})$ of its previous states by the application of the geometric mechanism of section 2.1:

$$m_i^{(t_i)} = \frac{\sigma_i^{(t_i)} + \alpha\omega_i^{(t_i-1)}}{\Omega(t_i)} \quad \Rightarrow \quad s_i^{(t_i)} = \begin{cases} 1 & if \quad m_i^{(t_i)} > 0.5 \\ \sigma_i^{(t_i)} & if \quad m_i^{(t_i)} = 0.5 \\ 0 & if \quad m_i^{(t_i)} < 0.5 \end{cases}.$$

The properties of elementary cellular automata with two types of asynchronous updating mechanisms have been studied by Ingerson and Buvel [209], which pay particular attention to Rule 22.

In the references by Fates and Morvan [138], and in [90, 246], at every time-step, each cell has a certain probability, the *synchrony rate p*, to be updated.

We will analyze in this section the effect that both memory with factor α and updating at synchrony rate p has on the LIFE rule. LIFE has proved to be very sensitive to the alterations in the updating scheme, so that asynchrony tends to *freeze* its evolution, eliminating its complex dynamics in the asynchronous model [83].

Figure 4.1 shows the patterns at $T{=}400$ after LIFE is started at random, in the ahistoric and three α-memory with four levels of synchrony rate scenarios. The patterns show the central 61×61 part of a $100{\times}100$ lattice (i.e., the same initial scenario as Table 2.7), marking the changing cells at $T{=}100$ as blue and green cells. Low levels of synchrony soon freezes the dynamics. This is so regardless of the presence or absence of memory, which is reflected in the qualitative resemblance of the four $p{=}0.1$ patterns shown in Fig. 4.1, characterized by domains of alternative dead and live horizontal and vertical stripes, with most of the scarce activity occurring at the domain boundaries.

Figure 4.2 shows the dynamics of the density and changing rate in the $p{=}0.5$ and $p{=}1.0$ scenarios of Fig. 4.1. It is shown in this figure how the main variation in these parameters is achieved early, let us say by $T{=}20$,

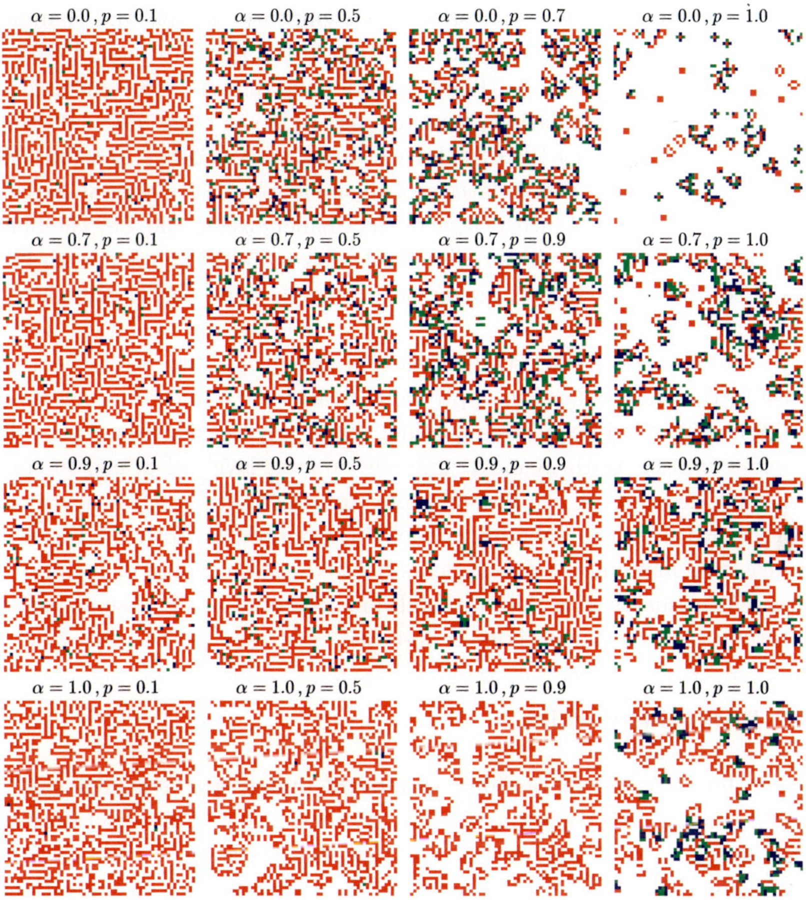

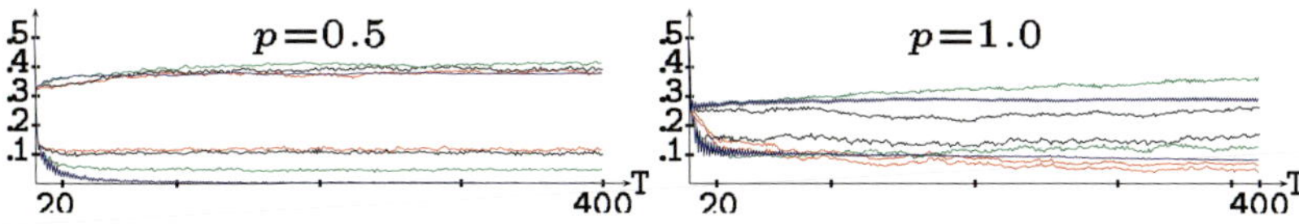

$\alpha = 0.0, p = 0.1$	$\alpha = 0.0, p = 0.5$	$\alpha = 0.0, p = 0.7$	$\alpha = 0.0, p = 1.0$
$\alpha = 0.7, p = 0.1$	$\alpha = 0.7, p = 0.5$	$\alpha = 0.7, p = 0.9$	$\alpha = 0.7, p = 1.0$
$\alpha = 0.9, p = 0.1$	$\alpha = 0.9, p = 0.5$	$\alpha = 0.9, p = 0.9$	$\alpha = 0.9, p = 1.0$
$\alpha = 1.0, p = 0.1$	$\alpha = 1.0, p = 0.5$	$\alpha = 1.0, p = 0.9$	$\alpha = 1.0, p = 1.0$

Fig. 4.1 Patterns in LIFE starting at random with memory factor α and asynchrony rate p. Code color : ■ : $1\to1$, ■ : $0\to1$, ■ : $1\to0$.

and afterwards only small corrections are foreseeable.

$p=0.5$ **$p=1.0$**

Fig. 4.2 Evolution of density and changing rate in LIFE starting at random with memory factors $\alpha=0.0$ (red), $\alpha=0.7$ (black), $\alpha=0.9$ (green), $\alpha=1.0$ (blue), and the synchrony rates indicated in each panel.

Figure 4.3 shows the variation in the asymptotic density and changing rate in LIFE with varying synchrony rate p in the ahistoric model, and in three models with memory. The two order parameters are computed at T=400 across nine simulations starting with different random assignments of a 50 % of live cells in a 100×100 lattice (black marked dots), in a 150×150 lattice (red), and in a 50×50 lattice. The simulation in the biggest lattice, the red dots, seems to *fit* with those of the smaller size 50×50, i.e. the black dots. The simulation in the smallest lattice size, 50×50, marked with blue dots, also follows fairly well bigger ones, though some outlier simulations appear at high p values.

In the ahistoric model, the increase in synchrony rate produces a decrease in the steady density, appreciable from $p \simeq 0.5$ and observed up to $p \simeq 0.9$ in Fig. 4.3. Beyond this p-value, the steady density seems to stabilize close to 0.1 . With memory, only the highest levels of p lead to some significant decrease of ρ . The changing rate seems to reach a maximum value at p=0.7 in the ahistoric model (an example is shown in Fig. 4.1), whereas this maximum value appears delayed as the charge of memory is increased, in such a way that with full memory only in the case of perfect synchrony (p=1), some activity is observed.

4.2 Probabilistic rules

Rather than introducing randomness into the updating scheme, one can instead include it directly in the cellular automaton rule. This generalization of the deterministic framework to the probabilistic scenario will enable us to study perturbations to deterministic automata, as well as transitional changes from one deterministic automata to another.

Introducing probabilities in the transition rules approaches the context to that of the Markov random fields, the basic tool in the study of *interacting particle systems* [254] . But in contrast with what happens mainly in these statistical mechanics scenarios, and in *cellular neural networks* [113, 126], time remains discrete in what follows.

Thus in the context of elementary rules, the binary β_i assignments may be replaced by probabilities p_i :

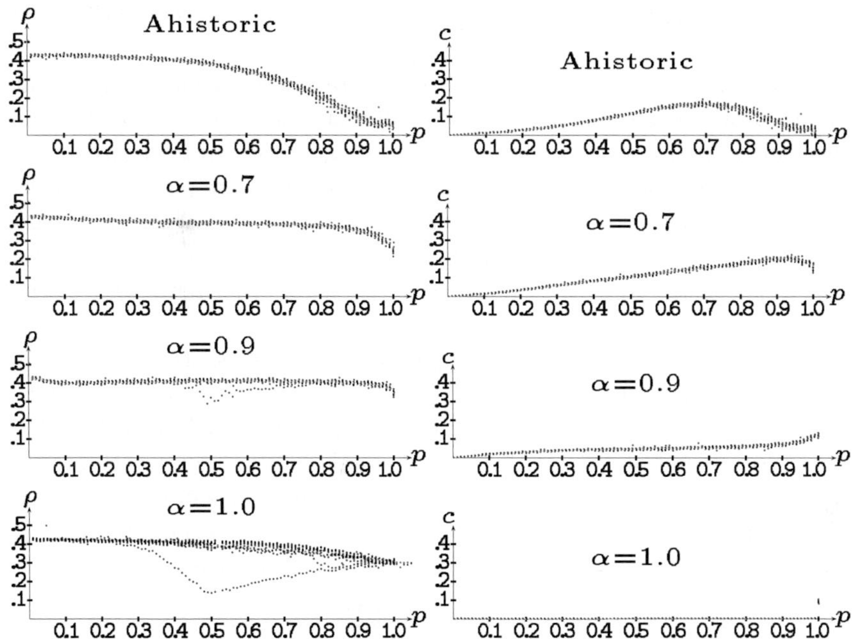

Fig. 4.3   Density (left) and changing rate (right) in LIFE with varying synchrony rate p in four α-memory scenarios. The results from nine simulations in a 100×100 lattice are marked black, that from a simulation in a 150×150 lattice are marked red, and that from a simulation in a 50×50 lattice are marked blue.

111	110	101	100	011	010	001	000	
β_1	β_2	β_3	β_4	β_5	β_6	β_7	β_8	$\beta \in \{0,1\}$
p_1	p_2	p_3	p_4	p_5	p_6	p_7	p_8	$p \in [0,1]$

Domany and Kinzel [127, 230] pioneered in the study of these probabilistic CA (PCA for short [1]). The statistical mechanics of PCA was studied by Lebowitz *et al.* [245] and Grinstein *et al.* [176]. Probabilistic CA rules are found in the simulation of hysteresis phenomena such as those in shape memory alloys, materials capable of large recoverable inelastic strain [399].

As in the deterministic scenario, memory can be embedded in PCA by featuring cells with a summary of past states:

$$\text{Conventional PCA:} \quad p = P\big(\sigma_i^{(T+1)} = 1/\sigma_{i-1}^{(T)}, \sigma_i^{(T)}, \sigma_{i+1}^{(T)}\big)$$

[1]The probabilistic reinforcement mechanism in [403–406] increases the probability of previously chosen deterministic rules. No probabilistic rules are taken into account in this evolutionary context.

$$\text{PCA with memory:} \quad p = P\big(\sigma_i^{(T+1)} = 1\big/ s_{i-1}^{(T)}, s_i^{(T)}, s_{i+1}^{(T)}\big)$$

Again, in the probabilistic scenario, cells are endowed with memory but the stochastic transition rules remain unaltered. Legal rules are now of the form $(p_1, p_2, p_3, p_4, p_2, p_6, p_4, 0)$. The five probabilities parameterize a five-dimensional hypercube, where the deterministic rules are placed in the 32 corners. Elementary CA quickly evolve into configurations with a density of non-null states fluctuating about a given mean (ρ) so the asymptotic density is considered in the literature as a natural order parameter in the phase space .

The simulations in this chapter were run for 1000 time steps on a lattice consisting of 500 sites and periodic boundary conditions. All the simulations start from the same random initial configuration, in which the value of each site is initially uncorrelated, and is taken to be 0 or 1 with probability 0.5 . All the PCA evolve with identical realizations of the stochastic noise (with equal sequence of random numbers). The asymptotic density was taken as the mean density over the last 100 time steps.

Table 4.1   Asymptotic densities of the deterministic rules 18, 72, and 90.

	α					
	1.0	0.9	0.8	0.7	0.6	0.5
$\rho_0 = 0.5$						
Rule 18 (00010010)	0.120	0.000	0.000	0.012	0.108	0.250
Rule 72 (01001000)	0.096	0.096	0.096	0.096	0.096	0.096
Rule 90 (01011010)	0.503	0.521	0.497	0.495	0.507	0.500
$\rho_0 = 0.9$						
Rule 18 (00010010)	0.004	0.000	0.000	0.000	0 004	0.131
Rule 72 (01001000)	0.016	0.016	0.016	0.016	0 016	0.016
Rule 90 (01011010)	0.384	0.522	0.502	0.500	0 500	0.496
$\rho_0 = 0.1$						
Rule 18 (00010010)	0.050	0.000	0.000	0.016	0 046	0.253
Rule 72 (01001000)	0.032	0.032	0.032	0.032	0 032	0.032
Rule 90 (01011010)	0.162	0.466	0.504	0.502	0 493	0.490

We explore the effect of memory in three different subsets. We study first the subset $(0, p_2, 0, p_4, p_2, 0, p_4, 0)$, delimited with corners being the deterministic rules 0, 18, 72 and 90. Petersen and Alstrom [331] study the transitions across the three regions apparent in the -no memory- graph of Fig.4.4: that with $\rho = 0^2$, that in which ρ has a cylindrical shape,

---

^{2}Excluding the singular rule 72. The introduction of even the slightest probabilities

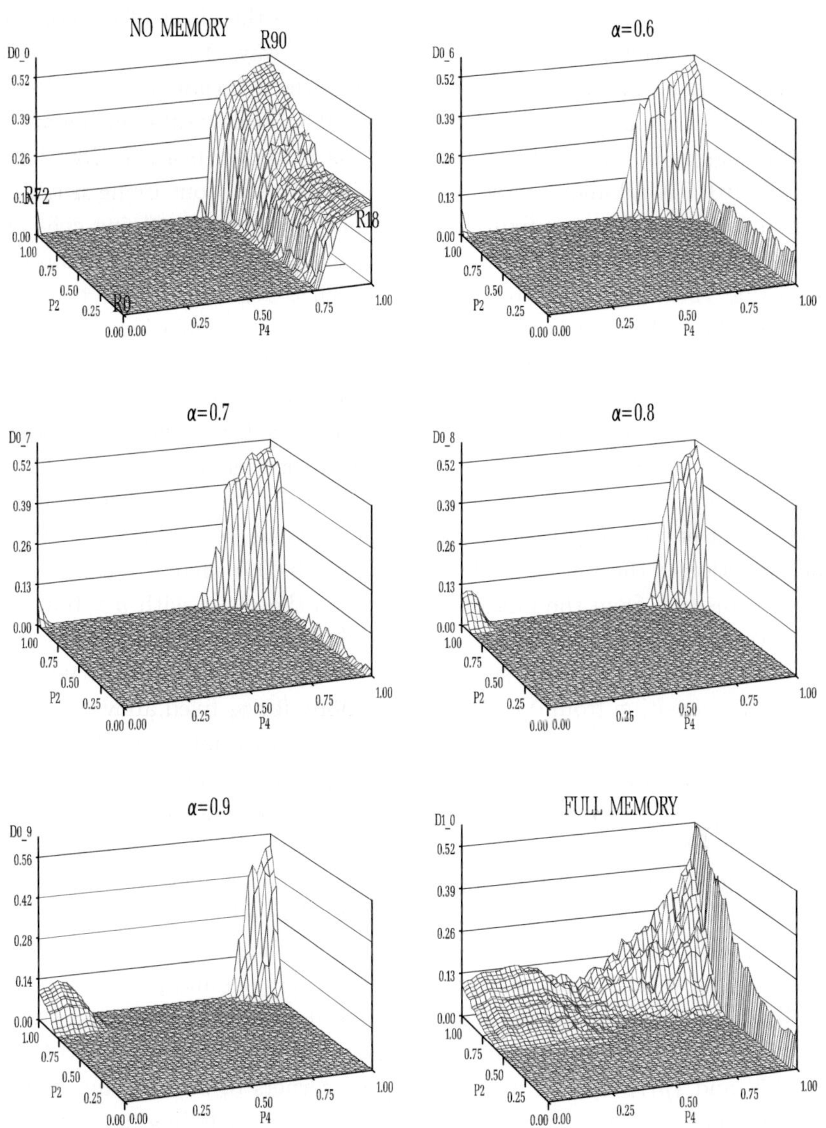

Fig. 4.4 The asymptotic density of the legal probabilistic CA rules of the form $(0, p_2, 0, p_4, p_2, 0, p_4, 0)$. The probabilities are varied from 0 to 1 by 0.02 intervals. The memory factor α is varied from 0.6 to 1.0 by 0.1 intervals.

lead to extinction.

and the region near rule 90. Let us first examine the effect of memory on these corners, which evolve in a rather different way. Table 4.1 shows the numerical values of the asymptotic densities of the deterministic rules 18, 72 and 90 under the label (initial density) $\rho_0 = 0.5$. Rule 0 dies out instantly. History has a dramatic effect on rule 18, as is known from section 2.3.1. Rule 72 (01001000) allows only living cells with exactly one living neighbor to remain alive . No new living cells are born, and the surviving cells are unaltered after the first iteration (see an example in [331]). Because of that, rule 72 is unaffected by memory, which is reflected in the constant value 0.096 in Table 4.1 (this value is 0.125 in the simulations in [331]). The totalistic rule 90 shows a much smoother evolution from the ahistoric to the historic scenario (Fig.2.7 in Appendix A) : no pattern evolves to full extinction nor to the preservation of only a few isolated persistent propagating structures (solitons). In the ahistoric scenario it is $\rho \simeq 1/2$ for rule 90, independently of the initial density $\rho_0 \neq 0$. When historic memory is taken into account, ρ varies little from this value (see Table 4.1). With the exception of the singular rule 72 (with $\rho = 0.096$) there is a continuous phase transition from the phase with $\rho = 0$ to the phase with $\rho > 0$ in the ahistoric model. This transition is observed when p_4 is approximately 0.8, the transition value p_4^c being smaller along the transition from 72 to 90 than along the transition from 0 to 18. Furthermore, for p_4 fixed above p_4^c there is a continuous phase transition from a phase with ρ fairly constant close to 0.26 to a phase with $\rho > 0.26$. For example, along the transition from R18 to R90 it is $\rho \simeq 0.26$ until $p_2 \simeq 0.49$. Above this value, ρ increases monotonically toward the value $\rho = 0.5$ at the corner R90. Memory tends to deplete the cylindrical region apparent near R18 in the ahistoric scenario, and to enhance that with $\rho > 0$ near R72. The effect of memory when starting with high and low initial densities is remarkably coincident with that starting with $\rho = 0.5$ as shown in [27] for $\rho = 0.9$ and $\rho = 0.1$.

The two following figures of this section deal with the subsets : $\big(0, 0, p_3, 1, 0, p_6, 1,0 \big)$, with deterministic corners the rules 18, 22, 50 and 54 in Fig.4.5, and $\big(p_1, p_2, p_1, p_2, p_2, 0, p_2, 0\big)$ with corners 0, 90, 160 and 250 in Fig.4.6 . The former is a very interesting subset, as including not only rule 18, but also the *complicated* (to avoid the coined term *complex*) rules 22 and 54. The rules of the latter subset are *peripheral*, i.e., with no dependence in ϕ of the state of the cell to be updated, $\sigma_i^{(T+1)} = \phi\big(\sigma_{i-1}^{(T)}, \sigma_{i+1}^{(T)}\big)$. and totalistic $\sigma_i^{(T+1)} = \phi\big(\sigma_{i-1}^{(T)} + \sigma_{i+1}^{(T)}\big)$. The equivalence between *peripheral* PCA and the problem of directed percolation is studied in [127]. Inciden-

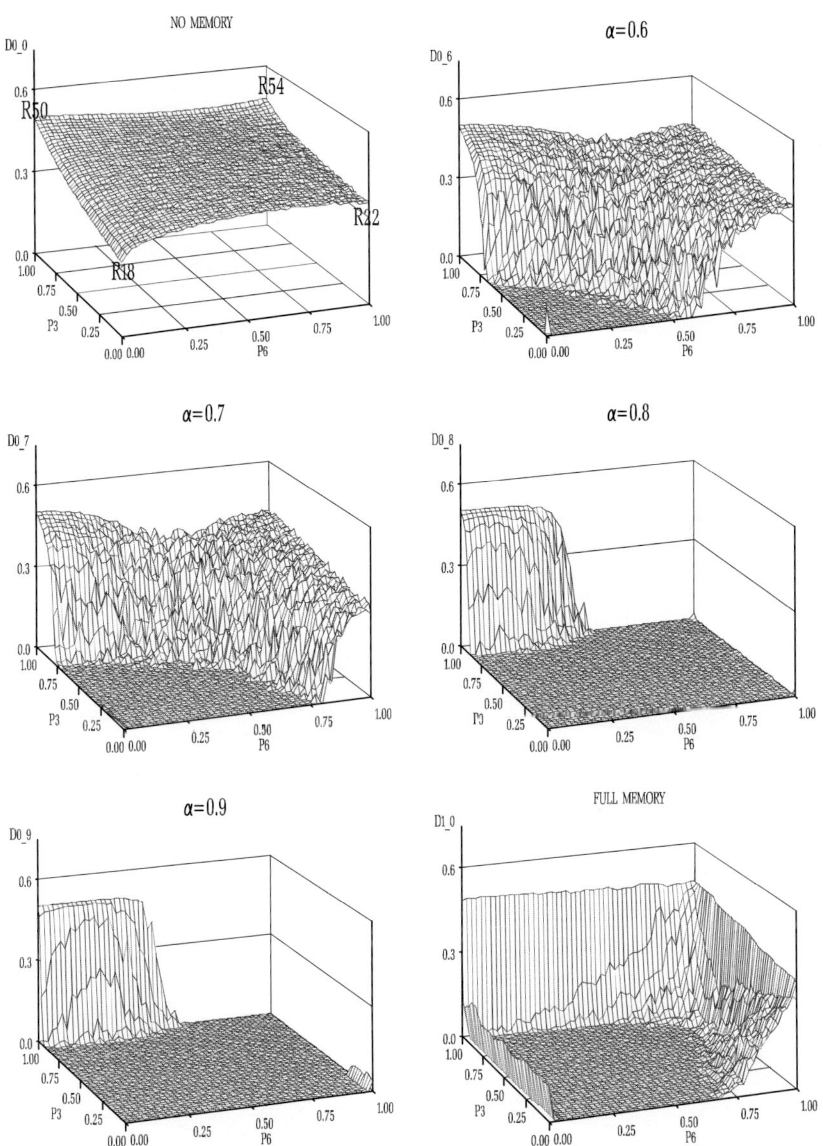

Fig. 4.5 The asymptotic density of the legal probabilistic CA rules of the form $(0, 0, p_3, 1, 0, p_6, 1, 0)$ in the scenario of Fig. 4.4.

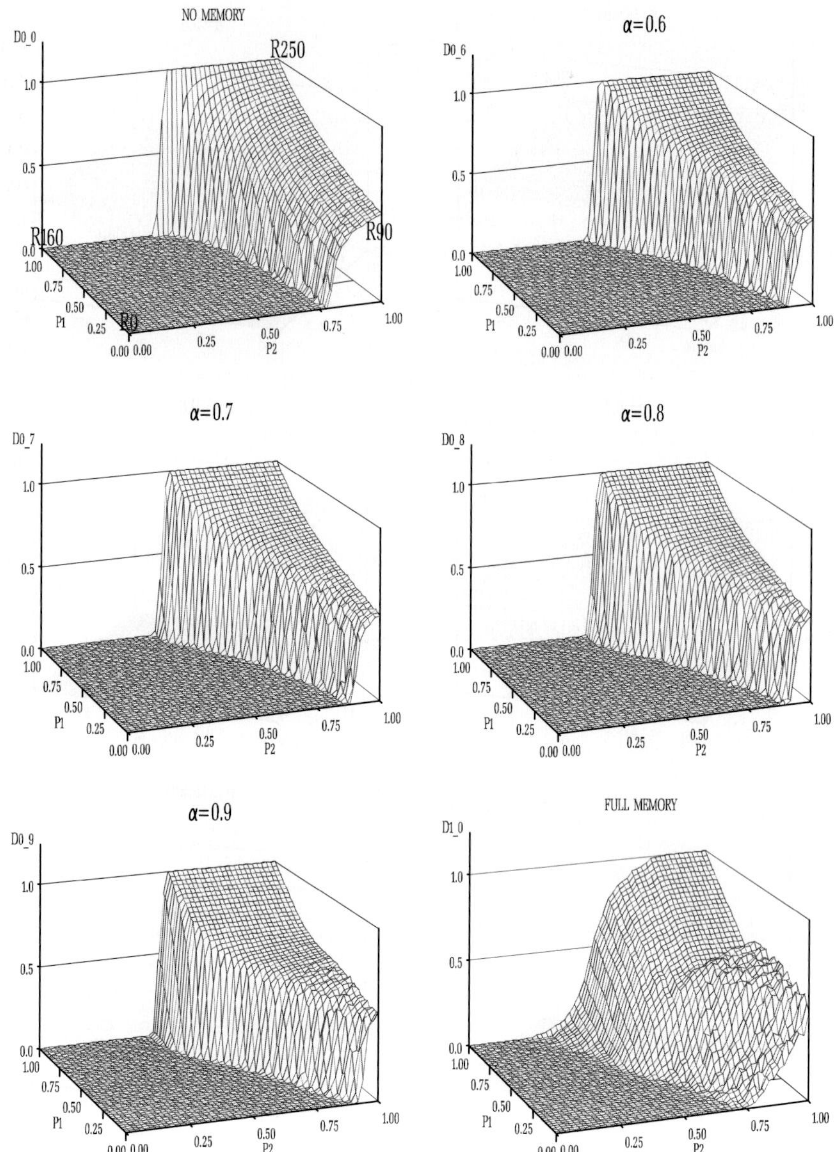

Fig. 4.6 The asymptotic density of the legal probabilistic CA rules of the form $(p_1, p_2, p_1, p_2, p_2, 0, p_2, 0)$ in the scenario of Fig. 4.4.

tally, rule 22 in the former subset is totalistic but not peripheral.

The effect of memory on the deterministic corners of the subset $(0,0,p_3,1,0,\ p_6,1,0)$ (see Appendix A), is also reflected in the asymptotic density graphs of Fig.4.5. Thus, the inhibition induced by memory on rules 18, 22 and 54 when α is in the $[0.8,0.9]$ interval is reflected in a clear depletion of the corresponding graphs in Fig.4.5, in which only the proximities of the corner R50 are not depleted. When α varies in the $[0.6,0.7]$ interval, the ρ value of rule 54 is not notably altered, thus depletion in the 0.6 and 0.7 graphs is restricted only to the R18 corner. The aspect of the full memory graph is fairly particular, with a dominant tendency to the null value in proper probabilistic rules, but not in the fairly deterministic ones (the sides).

The gradual effect of memory on rule 90 is reflected in a smooth effect of memory on the graphs of Fig.4.6. The smooth effect of memory on this peripheral and totalistic subset has been found also for other totalistic but non-peripheral subsets, in particular on the important PCA subsets studied by Kinzel [230]: $(p_2,p_2,p_2,p_4,p_2,p_4,p_2,0) \rightarrow 0,22,232,254$, $(p_1,p_2,p_2,p_2,p_2,p_2,p_2,0) \rightarrow 0,126,128,254$, and $(p_4,p_2,p_2,p_4,p_2,p_4,p_2,0) \rightarrow 0,104,150,254$.

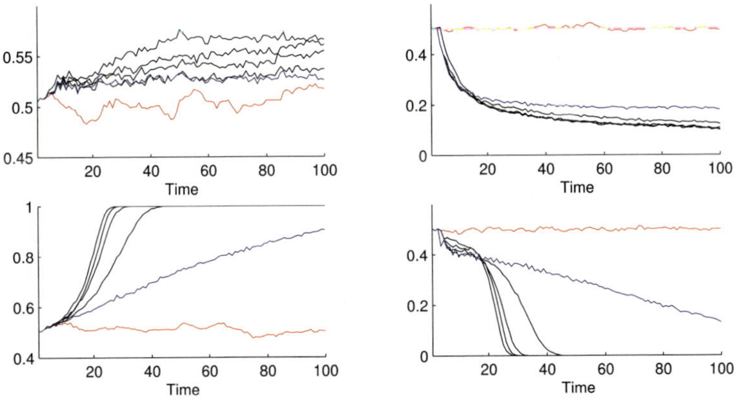

Fig. 4.7 Density (*left*) and changing rate (*right*) in the probabilistic 2D model described in text. Regular (upper) and random (lower) $K{=}4$ lattices and $\rho_o = 0.505$.

Intrinsic probability

When considering probabilistic totalistic automata, the transition probability depends only on the sum of states of the neighbors. A seemingly natural way of designing these probabilities is to make them equal to the ratio of this sum to the total number of cells connected. Thus, as an example, Fig.4.7 shows the effect of memory with this probabilistic rule in a 100×100 lattice with cells interacting with their four nearest-neighbours. No self-interaction is considered in the simulation of Fig.4.7, so the allowed probabilities are $\{0,1/4,2/4,3/4,1\}$. The two upper snapshots concern the canonical regular von Neumann neighborhood , i.e., the N-S-E-W nearest-neighbors in the Euclidean lattice. In this context, memory does not have a marked effect on the evolution of density, but does have on the rate of changing cells, around 0.5 in the ahistoric model, and notably depleted with memory. The two lower snapshots of Fig.4.7 refer to a lattice with cells connected at random (via rewiring, see section 10.1). In this scenario, the ahistoric model shows again fairly consistent dynamics of the density and changing rates around the 0.5 . But when memory is taken into account the evolution is dramatically altered. Thus, density tends to one in a slower way in the case of full memory. The effect of memory appears inverse to its intensity : higher α values seem to delay the tendency to dynamic collapse, and in the extreme case of full memory extinction is not reached in the time range shown in Fig.4.7. So, contrary to expectations, full memory dynamics are closer to any other to the ahistoric case in this figure.

The initial density in Fig.4.7 is $\rho_0 = 0.505$, a little bit over the landmark 0.500, in which case the density of the evolving patterns with memory keeps over 0.5 in the regular lattice scenario, and tends to 1.0 in the fully rewired context. In Fig.4.8 the initial density is $\rho_0 = 0.499$, just below 0.500, in which case the density of the dynamics with memory is under 0.5 in the regular lattice and extinction is the fate of cases the evolving patterns in the random lattice context. Memory seems to detect initial density around the critical 0.5 value, i.e. acts as a *density task* solver. This effect of memory has been checked in lattices with homogeneous degree of connectivity up to nine, and is also present in lattices with heterogeneous connectivity as shown in Fig.10.7.

Figure 4.9 shows the reported effect of memory in the probabilistic rule on a one-dimensional lattice with initial density below 0.50 . Thus, in the rewired scenarios extinction is the fate of with memory except with full memory, in which case the density is just depleted. In the regular lattice

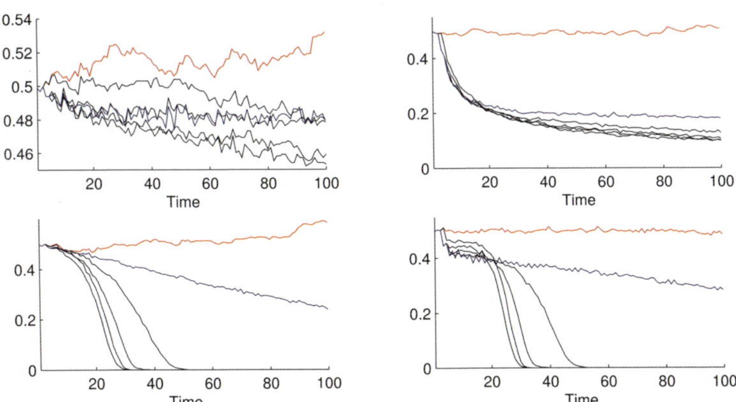

Fig. 4.8   Density (*left*) and changing rate (*right*) in scenario of Fig .4.7 but with initial density $\rho_o = 0.499$.

context, memory has not a notable effect on density, although the aspect of the spatio-temporal patterns are remarkably altered.

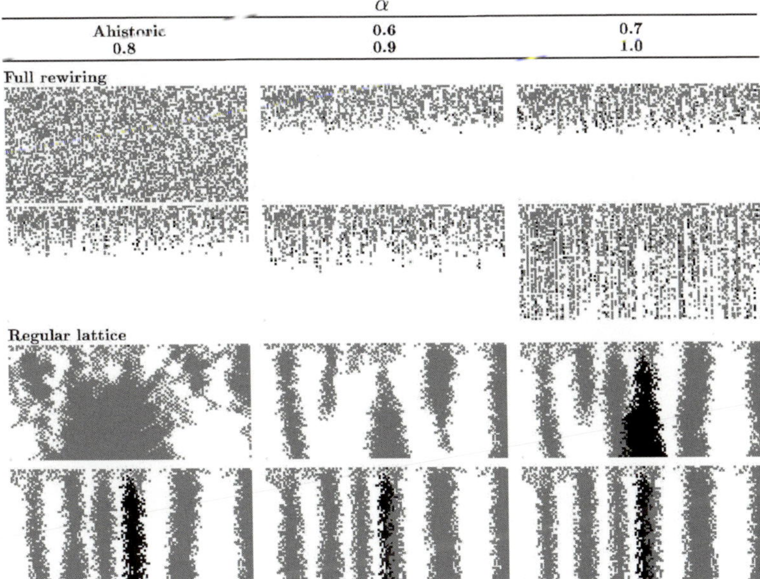

Fig. 4.9   Spatio-temporal patterns in the probabilistic rule on an one-dimensional lattice with four inputs. Initial density $\rho_o = 0.495$. Central $n = 300$ cells of a $N = 1000$ register.

Chapter 5

Cycles and random sequences

5.1 Cycles

An important feature in the characterization of CA is that of the length of the sequence of patterns produced, i.e., after how many time-steps does the sequence repeat.

Cycles in CA with memory are harder to find compared to the conventional ahistoric scenario, in which the mere repetition of a sole pattern marks the beginning of a cycle. This is not so in CA with memory. We will consider the case of two and three time-step memory, in which cases, two and three consecutive patterns have to be repeated to start a cycle.

As a simple example, Table 5.1 shows the dynamics up to $T = 100$ of the ahistoric rule 150 and that of S150T150 in small lattices of sizes $N = 5$ and $N = 11$ (with periodic boundary conditions), starting from a single live cell in its central site. The ahistoric evolution generates a period-three oscillator as soon as at $T = 4$ when $N = 5$; in the historic scenario, the first repetition of three consecutive patterns (in fact the three first ones) is achieved at $T = 66$, a figure larger than the total number of possible configurations: $2^5 = 32$. When $N = 11$, the oscillator is of period 33 in the ahistoric formulation, whereas no cycle is found up to $T = 100$ in the historic context. So, the cycles appear notably later, thus they are notably longer.

In Table 5.2 the first repetition of two consecutive patterns (again the two first ones) in the rule S150T6 occurs at $T = 16$ in the register of size $N = 5$, whereas with $N = 11$ this rule has period length 93, notably lower than the total number of different configurations, i.e., $2^{11} = 2048$, but notably longer than the period in the ahistoric context.

In general, in a CA constructed so that the new configuration depends on the preceding τ ones of the sequence, the maximum period conceivably

Table 5.1   The ahistoric rule 150 (upper), and rule S150T150 (lower) in circular registers of sizes $N = 5$ and $N = 11$.

Table 5.2   The rule 150 with parity of the last two state memory (S150T6) in circular registers of sizes $N = 5$ (upper) and $N = 11$ (lower).

attainable is $\left(2^{N}\right)^{\tau}$.

In order to circumvent the difficult analytical study of the cycles in the CA dynamics, the so called return map helps to visually detect its mere existence by plotting the points representing pairs of consecutive configurations. Thus, (x_T, x_{T+1}), where x_T is a real number representing the configuration at time-step T. Usually: $x_T = \sigma_1^{(T)} + \sum_{i=2}^{N} \sigma_i^{(T)} (0.5)^{i-1}$. Figure 5.1 shows the return maps of rules on 30, 90, 150 and S150T150 for a simulation with a common initial random configuration on a $N = 150$ lattice. The well known characteristic patterns of the ahistoric rules 30, 90 and 150 are completely changed compared to other ones of *random* aspect exemplified by that of S150T150 in Fig. 5.1 .

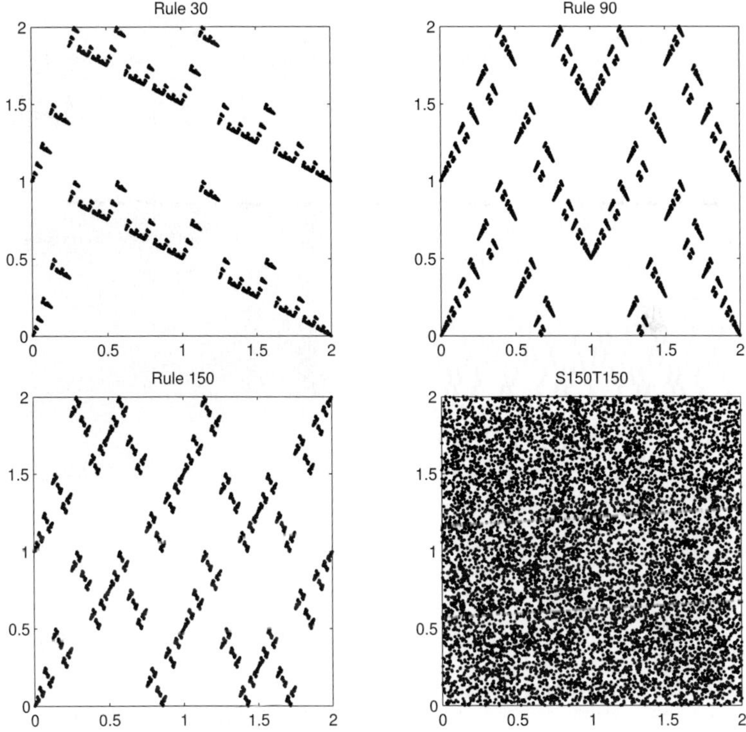

Fig. 5.1   Return maps for rules 30, 90, 150 and S150T150.

The correlation in state values induced by the local transition rule in conventional CA turns out to be weakened by the action of the temporal rule, so that there is a sort of random re-starting at every time-step. The two important characteristics of CA with memory reported here (longer cycles, correlation avoided) justify the potential value of CA with memory as good RNG, and accomplish, at least qualitatively, the commitment stated by Kunth [232],p.5, -random numbers should not be generated with a method chosen at random. Some theory should be used -.

Return maps may unveil partially hidden order in the CA dynamics as well as *random* features in the pattern sequence generation. Both aspects are illustrated in Fig. 5.2. The ahistoric and historic spatio-temporal patterns of these two $r = 2$ rules do not seem to qualitatively differ. But the ahistoric one encloses an internal order that appears not be present at all in the historic evolution.

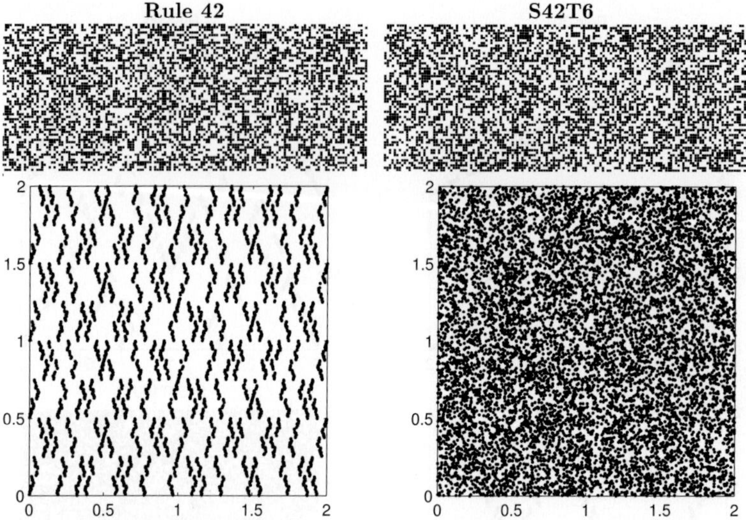

Fig. 5.2 Rules 42(101010) and S42T6. Spatio-temporal patterns (upper) and return maps (lower).

5.2 Random number generation by CA

Random number sequences can be found in a large number of applications ranging from technological (cryptography, communications, computer-based gaming, VLSI, hardware self-test (BIST)) to scientific large-scale statistical physics simulations or stochastic natural processes modeling [241].

Yet finding good random number generators (RNG) is a non-trivial task. Fairly recent studies have shown that CA is a promising technique for generating *pseudo*-random[1] numbers. This kind of generator has the advantage of being highly parallel [396] and thus easily scalable with a relatively tiny hardware cost. Moreover, due to its Boolean nature, CA are free of numerical errors derived from the finite precision of floating point representation of real numbers in computers. This facilitates the *portability*, so when a random number sequence has been generated on some particular machine it is easy to generate the same sequence on other

---

[1]Random sequences generated on a digital computer are usually referred as *pseudo*-random, as distinguished from true random numbers, resulting from some natural physical process.

Table 5.3   The first ten real numbers in the $[0, 1]$ interval generated by the ahistoric rule 150 and S150T150 in the N=5 register of Table 5.1 .

S150			S150T150		
00100	4	0.129	00100	4	0.129
01110	14	0.452	01110	14	0.452
10101	21	0.677	10101	21	0.677
00100	4	0.129	11111	31	1.000
01110	14	0.452	01110	14	0.452
10101	21	0.677	01110	14	0.452
00100	4	0.129	11111	31	1.000
01110	14	0.452	11111	31	1.000
10101	21	0.677	10101	21	0.677
00100	4	0.129	00100	4	0.129

machines also. This overcomes the main problem of the linear congruential technique, $x_{n+1} = (ax_n + c) \bmod m$, that involves a multiplication and a division (to obtain the remainder) making it both slow and expensive in the digital computer area.

Rules 30,90 and 150 have been considered for random number generation. Rules 90 and 150 operate *á la* congruential form, so it is to be expected that they perform good results in random number generation terms. The intriguing properties of rule 30 have been largely studied, among others, by S.Wolfram [426, 422]. These rules powered with any of them acting as memory produce better randomized series.

Hybrid CA using both rules 90 and 150 have been also implemented regarding RNG [204, 235, 387]. The CA with memory here may be termed also *hybrid*, but in space and time. In both scenarios, a synergic effect emerges, so that rules that separately can not be used as randomizers, when combined have very good statistical properties.

There are several methods of extracting random numbers from a cellular automaton, with our own choice based on simplicity : at each time-step a real number in the $[0, 1]$ interval is obtained by rating a sequence of adjacent digits, converted to an integer, to the maximum attainable. This mechanism of generation of real numbers is illustrated in Table 5.3, in the simple scenario of the register of N=5 cells shown in Table 5.1, with maximum integer attainable being $11111 \equiv 31$.

Testing randomness

In order to demonstrate the efficacy of a proposed RNG, it is usually subject to a battery of empirical and theoretical tests, among which the most well known are those described by D.Knuth [232].

Although there are compiled batteries of tests to deal with the issue of qualifying randomness[2], we opted for the robust set of four tests implemented by the Numerical Algorithms Group (NAG®, www.nag.co.uk). The four tests for randomness implemented by NAG (runs, pairs, triplets, and gaps), belong to the class of *Chi-square* tests, in which the final operative parameter to decide on rejection of the null hypothesis of randomness is that of the tail *probability* associated with the chi-square statistic (with the corresponding degree of freedom), i.e., the significance level. After Knuth, *good* results are in the (0.1,0.9) interval, ideally close to 0.5, with extremities in both sides representing non-satisfactory random sequences.

To assess the potential value of CA with memory as random number generator, one hundred simulations, i.e., starting from different initial random configurations, of one-dimensional registers with 1000 cells were run for 10000 time-steps. Every central sequence of 101 digits was transformed to a real number in the $[0, 1]$ as indicated above.

In the use of any RNG it is important of the length of the sequence produced (i.e., after how many numbers does the sequence repeat). Sampling a fairly high number of bits, 101 in the subsequent simulations, allows for cycles of long length. This is a desirable feature of a RNG, although the quality of the randomness of a sequence of numbers is not defined by it. Thus for example, the conventional (non-CA) linear congruential method produces sequences of at most length m. Generalizing the method so that x_{n+1} depends on x_{n-1} as well as on x_n, the maximum period is m^2. Thus the simple Fibonacci sequence $x_{n+1} = (x_n + x_{n-1}) \bmod m$ (1)   usually gives a period length greater that m. Since no multiplications are involved, this implementation has the advantage of being fast, but tests have shown that the numbers produced by the Fibonacci recurrence are definitely not satisfactorily random, and, quoting D.Knuth [232],p.26 : - at the present time the main interest in (1) as a source of random numbers is that it makes a nice "bad example" -.

Mean and standard deviations (sd) of the Chi-square $(\overline{\chi^2})$ and probability $(\overline{P})$ parameters are shown in Tables 5.4 and 5.5. These tables report results concerning rules 30,90, and 150 as well as those obtained by the application of the tests to 100 sequences of 10000 random numbers generated by the program provided by NAG (G05KAF) to obtain pseudo-random numbers from a uniform distribution in the $[0, 1]$ interval. These parameters, under the column headed NAG, act as reference parameters, as it is expected that the results obtained with good randomizers should be not

---

[2]Such as the suites : ENT (*www.fourmilab.ch/random/*), NIST (cscr/nist/gov), or maybe the most currently applied, DIEHART (*www.stat.fsu.edu/pub/diehard*).

far from them.

Table 5.4   Run test: Mean and standard deviation (*sd*) of Chi-square($\overline{\chi^2}$) and probability ($\overline{P}$) across 100 simulations of 10000 observations each.

	NAG	S30	+T30	+T90	+T150
χ^2	6.452	1732.083	19.917	9.543	6.437
sd	4.130	36.975	7.551	4.384	3.268
$\overline{P}$	0.481	0.000	0.039	0.256	0.452
sd	0.308	0.000	0.101	0.235	0.294

	S90	+T30	+T90	+T150	S150	+T30	+T90	+T150
χ^2	26.220	7.601	6.115	7.012	34.500	6.389	5.770	6.460
sd	17.558	4.836	3.562	3.880	19.925	3.939	2.905	4.019
$\overline{P}$	0.053	0.403	0.495	0.425	0.008	0.481	0.507	0.479
sd	0.143	0.295	0.289	0.294	0.023	0.303	0.264	0.280

Table 5.5   Gap test in the scenario of Table 5.4.

	NAG	S30	+T30	+T90	+T150
χ^2	8.653	111.824	10.269	15.062	10.731
sd	3.663	18.080	4.850	6.932	5.520
$\overline{P}$	0.508	0.000	0.422	0.224	0.410
sd	0.274	0.000	0.289	0.223	0.304

	S90	+T30	+T90	+T150	S150	+T30	+T90	+T150
χ^2	97.780	9.998	9.784	8.988	31.214	10.259	9.011	8.727
sd	20.618	4.615	4.582	3.877	10.982	5.531	4.447	3.732
$\overline{P}$	0.000	0.440	0.457	0.491	0.011	0.439	0.502	0.505
sd	0.000	0.285	0.304	0.287	0.031	0.301	0.292	0.275

It becomes apparent from Tables 5.4 and 5.5 that the rules without memory do not pass the tests: their probability parameter is extremely low. Rule 30 dramatically fails, rule 90 seems to have some efficiency only under the run test, and rule 150 merely produces probabilities slightly non-null in the run and gap tests. It seems that the *regularities*, the triangular features appreciated in their ahistoric patterns, are translated into some kind of tendencies that the tests for randomness detect. These poor results obtained by the conventional ahistoric rules are due to the correlation induced by extracting the values of a wide sequence of contiguous sites. To try to remove correlation, it is customary to sample only a rather limited number of sites (either adjacent or spaced), or only one as considered by

S.Wolfram regarding rule 30 [422].

On the contrary, rules 30, 90 and 150 with memory (with no obvious patterns in their space-time diagrams in Figs. B.5 and B.6), dramatically increase their performance, as their probability parameter under the four tests considered here turns out to be fairly close to 0.5, that of a genuine random sequence. The exception is S30T30 and S30T90.

Figure 5.3 shows the three dimensional grids of triplets of successive numbers for rules 30, 90, 150 and S150T150. Conrary to what happens without memory, the aspect of the distribution of points in the S150T150 (and in the not shown S30T150 and S30T150) plot is fairly *random*. This is not so when considering two-state memory, as shown in Fig. 5.4.

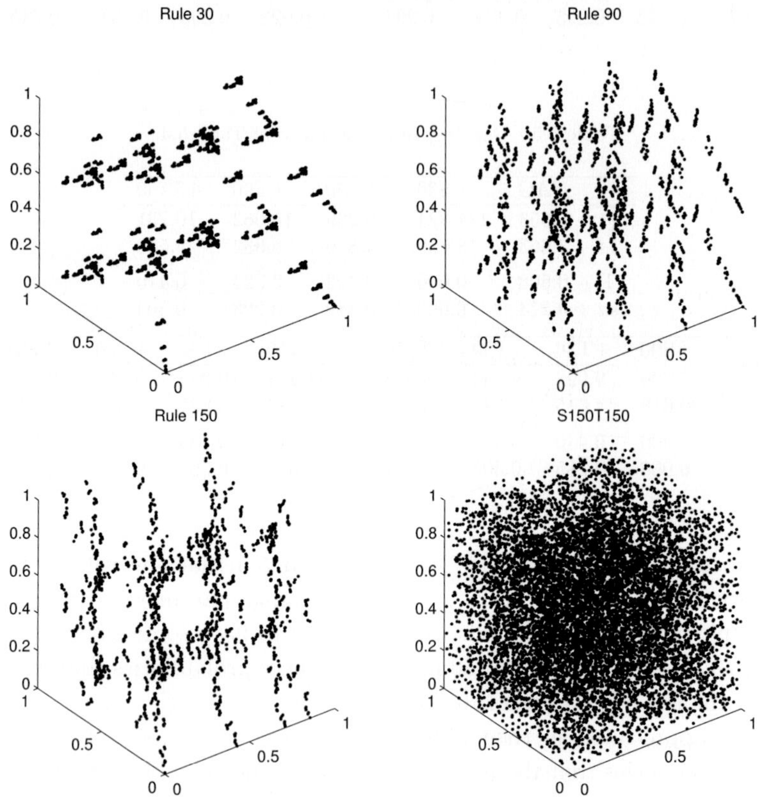

Fig. 5.3   Grids of triplets of successive numbers for rules 30, 90, 150 and S150T150 with $\tau = 3$.

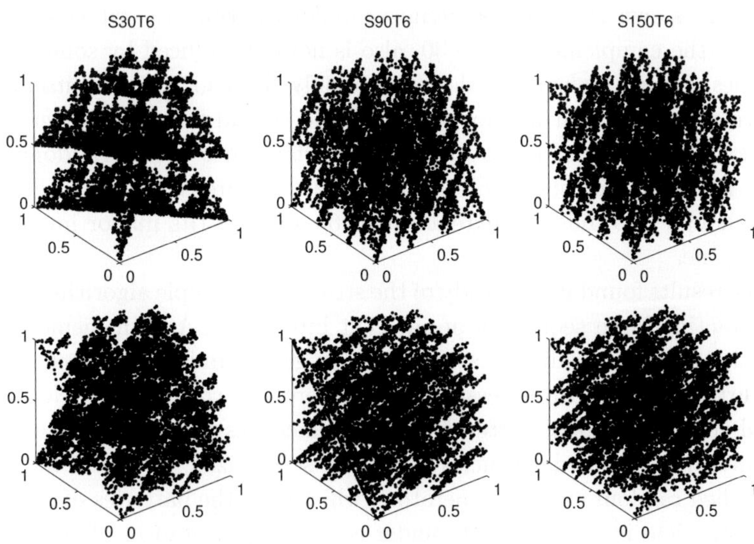

Fig. 5.4   Grids of triplets of successive numbers for rules S30T6, S90T6 S150T6 with $\tau = 2$. Two perspectives of the same data set are shown.

Rules 30, 90, and 150 are not uniquely propelled by themselves for random number generation purposed. Rules 60, 75, 105, 120, 165, 195, 210, and 225 seem also to induce randomness when acting as memory rules on them [40]. Reference to some of these rules may be traced in the specialized literature regarding RNG by CA. Thus, rule 45 is present in one of the hybrid CA studied in [204], and its complementary rule, i.e., rule 210, is also cited. Rule 60 (a totalistic linear rule much of the type of rule 90) is also reported in [422]. Rule 195 is the complementary rule of 60. Rules 120 and 225 belong to the same equivalence class. Rules 105, 165 and 225 are present in the coevolved CA combinations designed in [371], and rules 105 and 165 compose, together with rules 90 and 150, a self-programmable CA in [182]. So rules 225, 165 and 105 are complementary to 30, 90 and 150 respectively. It can be noted here that no attempt is made in this book to implement memory in *programmable* CA, first introduced in [306], which allow for dynamically alterable rules in cells. A random bit generator by means of Rule 90 employing an algorithm related to the automaton backward evolution has been proposed in [301].

Most of the above cited rules, in turn, when endowed with memory may perform good random sequence numbers [40]. The efficiency of the rules complementary to 150 and 90 is remarkable, i.e., that of rules 105, and 165 which on its own is a good randomizer under the run test criterion. Rule 225, i.e., the complementary to 30, also is notably induced by some memory rules. Linear transition rules employ only XOR logic, i.e., arithmetics performed modulo 2 in the two-state scenario, a kind of *congruential* computation that is expected to have good performance in the randomizing task. All the proper linear rules (60,90, 102, 150) are relevant as randomizers, though to a different extent: rules 60 and 102 (its mirror image) to a lesser one.

The results found in the study of the structurally simple algorithm based on extracting large sequences of adjacent bits from CA with memory capabilities in cells, qualify it as a plausibly good randomizer mechanism. But further studies are necessary to draw broader conclusions. Not only should the conventional batteries of randomness tests be applied, but also the potential *hidden* deficiencies[3] must be scrutinized, because even well established random number generators, which pass the batteries of random tests, can yield incorrect results under certain circumstances. It has been claimed that the origin of the failures can be traced to the special role of the zero in the algebra of finite fields [74]. In $\mathbb{Z}_2$ the zero element is essential, but already in $\mathbb{Z}_3$ it can be circumvented to generate perfectly balanced runs. This respect may be of interest when considering of three-state CA with memory [58].

The space-search suffers a true *combinatorial explosion* when composing spatial and temporal rules. Evolutionary techniques [180, 181, 337, 394] have been previously used to search for traditional rules performing qualified random sequences. We are currently exploring this approach for $r=2$, reversible, two-dimensional [395], and disordered CA [356] endowed with memory.

---

[3]Such as those found in [141] in cluster MC simulations of the 2D-Ising model with the Wolff algorithm. Basically, the bias consists in a tendendy to cluster zeros together.

$$\text{Chapter 6}$$

Three state automata

This chapter deals with one-dimensional CA of range $r = 1$ (the value of a given cell depends on the values of its nearest neighbors), with three possible values at each site ($k = 3$), which are coded as $\{0, 1, 2\}$, and noted as blank, black and gray in the figures.

6.1 Totalistic rules

This section considers *totalistic* rules: $\sigma_i^{(T+1)} = \phi\big(\sigma_{i-1}^{(T)} + \sigma_i^{(T)} + \sigma_{i+1}^{(T)}\big)$, characterized by a sequence of ternary values (β_s) associated with each of the seven possible values of the sum (s) of the neighbors:

$$\big(\beta_6, \beta_5, \beta_4, \beta_3, \beta_2, \beta_1, \beta_0\big)_{ternary} \equiv \sum_{s=0}^{6} \beta_s 3^s \underset{decimal}{=} \mathcal{R}$$

The *rule number* ($\mathcal{R}$) for $k = 3$ and $r = 1$ will range from 0 to 2186. Attention in this chapter is on *quiescent* rules ($\beta_0 = 0$) sensitive to a sole non-dead cell (σ_c), so rules with $\beta_1 > 0$ if $\sigma_c^{(1)} = 1$, or $\beta_2 > 0$ if $\sigma_c^{(1)} = 2$.

With three states, the rounding mechanism becomes:

$$s_i^{(T)} = \begin{cases} 0 & if\ m_i^{(T)} < 0.5 \\ \sigma_i^{(T)} & if\ m_i^{(T)} = 0.5 \\ 1 & if\ 0.5 < m_i^{(T)} < 1.5 \qquad [6.1] \\ \sigma_i^{(T)} & if\ m_i^{(T)} = 1.5 \\ 2 & if\ m_i^{(T)} > 1.5 \end{cases}$$

In order to study the effect of discounting memory, we consider the most unbalanced cell dynamics, either $\sigma_i^{(1)} = \sigma_i^{(2)} = \ldots = \sigma_i^{(T-1)} = 0$ and $\sigma_i^{(T)} = 2$, or $\sigma_i^{(1)} = \sigma_i^{(2)} = \ldots = \sigma_i^{(T-1)} = 2$ and $\sigma_i^{(T)} = 0$. In either case, memory will take effect when a cell with such dynamics is to be featured with a state value different to the last one. In the former scenario the m charge is to be compared to $3/2$, in the latter to $1/2$:

$$\begin{cases} m(0,0,\ldots,0,2) = \dfrac{3}{2} \equiv \dfrac{2}{\dfrac{\alpha^T-1}{\alpha-1}} = \dfrac{3}{2} \\[4em] m(2,2,\ldots,2,0) = \dfrac{1}{2} \equiv \dfrac{2\dfrac{\alpha^T-\alpha}{\alpha-1}}{\dfrac{\alpha^T-1}{\alpha-1}} = \dfrac{1}{2} \end{cases} \Rightarrow 3\alpha_T^T - 4\alpha_T + 1 = 0$$

In both cases, historic memory takes effect after time step T only if $\alpha > \alpha_T$, with $3\alpha_T^T - 4\alpha_T + 1 = 0$. For example, after $T = 2$, it is, $3\alpha_2^2 - 4\alpha_2 + 1 = 0 \Leftrightarrow \alpha_2 = 1/3$.

In the limit, it is: $\lim_{T\to\infty} 3\alpha_*^T - 4\alpha_* + 1 = 0 \equiv -4\alpha_* + 1 = 0 \Leftrightarrow \alpha_* = 0.25$. It is then concluded that memory does not affect the scenario if $\alpha \leq 0.25$. Thus, the value 0.25 of the memory factor becomes a bifurcation point that marks the transition to the ahistoric scenario.

In general, in CA with k states (termed from 0 to $k-1$),

$$s_i^{(T)} = \begin{cases} 0 & \text{if } m_i^{(T)} < 0.5 \\ \ldots & \ldots\ldots \\ \sigma_i^{(T)} & \text{if } m_i^{(T)} = \kappa - 0.5 \\ \kappa & \text{if } \kappa - 0.5 < m_i^{(T)} < s + 0.5 \; \kappa = 1, k-2 \\ \sigma_i^{(T)} & \text{if } m_i^{(T)} = \kappa + 0.5 \\ \ldots & \ldots\ldots \\ k-1 & \text{if } m_i^{(T)} > (k-2) + 0.5 \end{cases}$$

It is:

$$\begin{cases} m(0,0,\ldots,0,k-1) = \dfrac{3}{2k-3} \equiv \dfrac{k-1}{\dfrac{\alpha^T-1}{\alpha-1}} = \dfrac{2k-3}{2} \\[4em] m(k-1,\ldots,k-1,0) = \dfrac{1}{2} \equiv \dfrac{(k-1)\dfrac{\alpha^T-\alpha}{\alpha-1}}{\dfrac{\alpha^T-1}{\alpha-1}} = \dfrac{1}{2} \end{cases}$$

This leads to a general form of the characteristic equation: $(2k-3)\alpha_T^T - (2k-1)\alpha_T + 1 = 0$, which becomes $-2(k-1)\alpha_* + 1 = 0$ in the temporal limit. It is then concluded that memory does not affect the scenario if $\alpha \leq \alpha_*(k) =$

$\dfrac{1}{2(k-1)}$, the simplest particular cases being: $\alpha_*(2) = \dfrac{1}{2}$, $\alpha_*(3) = \dfrac{1}{4}$.

Computationally it is a saving if instead of calculating $m_i^{(T)}$ for every cell, we calculate $\omega_i^{(T)} = \sum_{t=1}^{T} \alpha^{T-t} \sigma_i^{(t)}$ all across the lattice and compare the ω figures to the factors $\dfrac{1}{2}\Delta(T)$ and $\dfrac{3}{2}\Delta(T)$. With $\alpha = 1$, the fully historic model is recovered: $m_i^{(T)} = \dfrac{1}{T} \sum_{t=1}^{T} \sigma_i^{(t)}$. In this scenario, instead of comparing $\omega_i^{(T)} = \sum_{t=1}^{T} \sigma_i^{(t)}$ to $\dfrac{T}{2}$ and $3\dfrac{T}{2}$ it is better to compare $2\omega_i^{(T)}$ to T and $3T$: this enables us to work only with integers.

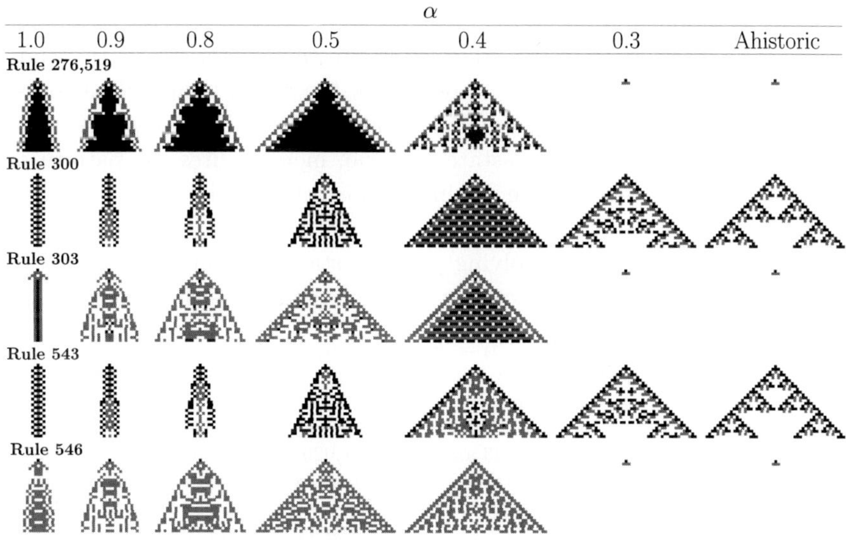

Fig. 6.1 Parity $k = 3$ rules starting from a single site seed with $\sigma = 1$. The black cells are at state 1, the gray ones at state 2.

Figure 6.1 shows the spatio-temporal patterns starting from a single live site seed with $\sigma = 1$, in quiescent ($\beta_0 = 0$) *parity* rules, i.e., rules with β_1, β_3 and β_5 non-null, and $\beta_2 = \beta_4 = \beta_6 = 0$. Patterns are shown up to $T = 26$, and the memory factor varies from 1.0 (fully historic) to 0.0 (ahistoric model); the pattern for $\alpha = 0.3$ is shown to test its proximity to the ahistoric one (recall that if $\alpha \leq 0.25$ memory takes no effect). Rule 273

(0101010) is not a proper $k = 3$ but a $k = 2$ rule, because no β is equal to 2 in its ternary description, it is equivalent to the *parity* rule in the $k = 2$ scenario, thus rule 150. Rule 516 (0201010) evolves as rule 273 starting with a single $\sigma = 1$ seed because in this scenario the sum of neighbors is never five and only $\beta_5 = 2$. For this reason, the spatio-temporal patterns of rules 273 and 516 are not in Fig. 6.1.

Historic memory acts on rules 300 and 543 in the characteristic *inhibition of growth* manner of rules with memory (growth phenomena in CA are studied in [171]). But the effect of memory on rules 276, 519, 303 and 546 is somewhat unexpected: they die out at $\alpha \leq 0.3$ but at $\alpha = 0.4$ the pattern expands [1], the expansion being inhibited (in Fig. 6.1) only at $\alpha \geq 0.8$.

Starting with a single site seed it can be concluded, regarding proper three-state rules, that, :

i) as an overall rule the patterns become more expanded as less historic memory is retained (smaller α),

ii) the transition from the fully historic to the ahistoric scenario tends to be gradual in regard to the amplitude of the spatio-temporal patterns, although their composition can differ notably, even at close α values.

iii) In contrast to the two-state scenario, memory fires the pattern of some three-state rules that die out in the ahistoric model, and no rule with memory dies out.

The similarities in the evolving patterns starting from a single seed in Fig. 6.1 are qualitatively reflected starting at random as shown with rules

---

[1] These quiescent rules have $\beta_2 = \beta_4 = \beta_6 = 0$ and $\beta_2 = 2$, thus their evolution in the ahistoric model is truncated at $T = 3$ in the ahistoric model after: ▪ → ▬ . Memory has effect on the two outer live cells (evolving $0 \rightarrow 2$) if $\alpha > 1/3$. This activating effect of memory is found in most of the *parity like* rules (rules with non null β value either β_1, β_3 or β_5) such as Rules 6, 33 and 60 ($000\beta_3020$), 249, 492 and 492 ($0\beta_500020$). Other rules activated by memory starting with a single $\sigma = 1$ seed, either from extinction or from short range perturbation in the ahistoric model, are the rules 87, 114 and 141 ($001\beta_3020$); 168, 195 and 222 ($002\beta_3020$); [thus the whole set of rules ($00\beta_4\beta_3020$) is activated by memory];

330, 357 and 384($011\beta_3020$); 411, 438 and 465($012\beta_3020$); 573, 600 and 627($021\beta_3020$);

654, 681 and 708($022\beta_3020$); 816, 843 and 870($101\beta_3020$); 897, 924 and 951($102\beta_3020$);

978, 1005 and 1032; 1059, 1086 and 1113; 1140, 1167 and 1194($11\beta_4\beta_3020$)

1221, 1248 and 1275; 1302, 1329 and 1356; 1383, 1410 and 1437($12\beta_4\beta_3020$)

1464, 1491 and 1518; 1545, 1572 and 1599; 1626, 1653 and 1680($20\beta_4\beta_3020$)

1707, 1734 and 1761; 1788, 1815 and 1842; 1869, 1896 and 1923($21\beta_4\beta_3020$)

1950, 1977 and 2204; 2031, 2058 and 2085; 2112, 2139 and 2166($22\beta_4\beta_3020$).

276, 300 and 519 in Fig. B.9 in the Appendix. The spatio-temporal patterns of rules 543 (very much resembling that of rule 300), 303 and 516 can be seen in [48].

The *parity* Rules 276, and 546 are clearly of Class I in the ahistoric model; rules 303 and 519 are of Class II (on the borderline with Class I: some rules, do not fit squarely into any of the four basic classes), whereas Rules 300, 516 and 543 belong to Class III, that of *chaotic* rules, which consequently are activated by means of a single site cell. As a rule, the patterns in the fully historic model might be classified as Class II: evolution leads to a set of separate simple stable or periodic structures. The ahistoric chaotic rules (300, 516 and 543) seems to progress from the ahistoric to the full memory patterns in a fairly gradual way, whereas the dissimilarities in patterns, appreciated already when starting with a single $\sigma = 1$ seed, in the rules 276, 303 and 519 in Fig. 6.1 are also present starting at random (see [48]), in which no rule with memory is led to extinction. What happens starting from a single seed turns out to be very informative about what happens starting at random. So, virtually all the rules *activated* by memory when starting from a single $\sigma = 1$ site (cited in footnote 1) are also *activated* (even at $\alpha = 0.3$) when starting at random. It is remarkable that all these rules (and many of the Class IV rules [2]) have $\beta_1 = 2$ and $\beta_2 = 0$. When $\beta_2 = 0$ but $\beta_1 = 1$ the effect of memory also tends to be remarkable even for low values of α, whereas if $\beta_2 > 0$ (e.g., rules 300 and 543 starting at random in [49]) the effect of memory tends to be less abrupt. These tendencies have of course exceptions, so that, for example, rules with $\beta_2 > 0$ can be greatly affected by memory, as for example with the simple Rules 18 (0000200) and 180 (0020200): their spatio-temporal pattern are fully altered at $\alpha = 0.3$, and become complex at $\alpha = 0.4$ and $\alpha = 0.5$.

Figure 6.2 shows the effect of $\delta = t^c$ memory on the *parity* rules when starting from a single live ($\sigma = 1$) site up the $T = 30$. The forgetting factor varies in Fig. 6.2 from 0 (fully historic model) up to 5. In Fig.5 of [48] time reaches $T = 62$ and c applies also to 6, 10, 15, 20, 25, closer to the ahistoric scenario. The conclusions drawn after Fig. 6.1 apply also in the $\delta = t^c$ memory scenario:

i) as an overall rule, the patterns become more expanded as less historic memory is retained (higher c),

---

[2]Wolfram [[426],p.948] includes the following rules in Class IV: 357, 438, 600, **792**, 924, **1038**, 1041, 1086, 1329, 1572, 1599, **1635**, **1659**, **1662**, 1815, **2007**, **2043**, **2049**. The rules marked in bold type in this list are not cited in the preceding footnote.

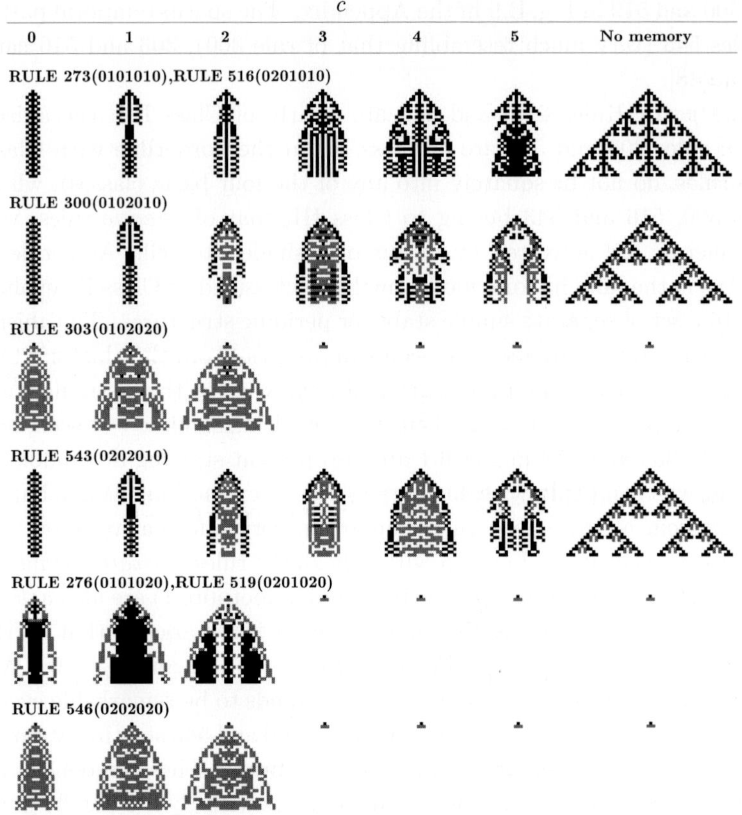

Fig. 6.2   Parity $k = 3$ rules with $\delta = t^c$ memory, from a single $\sigma = 1$ site for different values of the forgetting factor c.

ii) the transition from the fully historic ($c = 0$) to the ahistoric scenario is gradual in most cases. Rules 273, 516, 300, and 543 in Fig. 6.2 are paradigmatic of this smooth evolution.

iii) memory causes the generation of patterns for rules which die out in the ahistoric scenario. In Fig. 6.2, the patterns for Rules 303, 276, 519 and 546 do not die when $c \geq 2$.

Another integer-based weight memory mechanism, operative when $k = 3$, is $\delta = c^t$. For $c = 1$ the fully historic model is recovered, and again, the larger the value of c, the closer the scenario comes to the ahistoric one. It is $\Omega(T) = (c^{T+1} - c)/(c - 1)$. In the most unbalanced scenario, in a cell with state dynamics $\sigma^{(t)} = 0, t < T, \sigma^{(T)} = 2$, it is $2\omega = 4c^T$. Thus the last state (2) will feature the cell up to the T value in which $4c^T$ is below 3Ω.

For $c = 2$, with $\Omega(T) = 2^{T+1} - 2$, memory takes effect already at $T = 2$ (as for $c = 1$) : $4c^2 = 16 < 18 = 3\Omega(T)$, and being $16 > 6 = \Omega(T)$ it is $s^{(2)}(0,2) = 1$.

In Figures 6.3 and B.10 in Appendix B, two memory models are considered: the mode of the three last state memory (headed Mode) and a minimal memory of capacity two (headed $\tau = 2$) implemented as:

$$m_i^{(T)} = \frac{1}{2}\left(\sigma_i^{(T-1)} + \sigma_i^{(T)}\right)$$

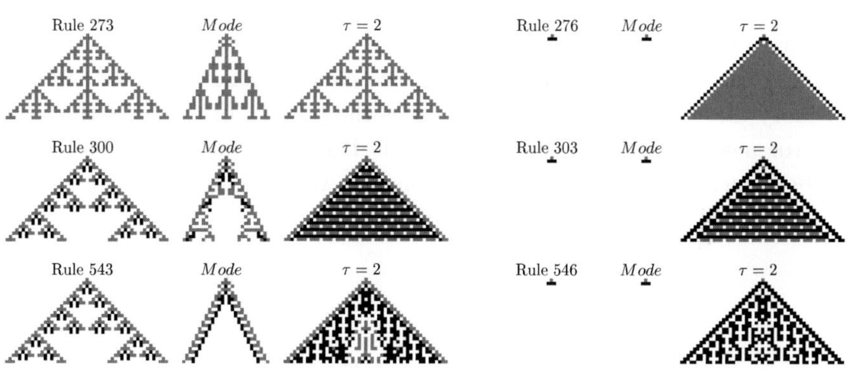

Fig 6.3   Parity $k = 3$ rules with mode of the last three states and average of the last two state memories.

This kind of mean memory mechanism actually only changes the trait cell after the sequences of states 02 and 20, featured as 1. Figure 6.3 shows the effect of memory on the spatio-temporal patterns starting from a single live site seed with $\sigma = 1$, in all the totalistic quiescent ($\beta_0 = 0$) *parity* rules, i.e., rules with β_1, β_3 and β_5 non null, and $\beta_2 = \beta_4 = \beta_6 = 0$. Again, the patterns in Fig. 6.3 are symmetric due to the consideration of totalistic rules. Rule 273 (0101010) is not a proper $k = 3$ but $ak = 2$ rule, because no β is equal to 2 in its ternary description. For this reason, its spatio-temporal pattern in the mean memory model coincides with the ahistoric one. In the mode memory model, Rule 273 shows in Fig. 6.3 the characteristic *inhibition of growth* already shown for its equivalent rule in the $k = 2$ scenario: the *parity* Rule 150. Rule 516 (0201010) evolves like Rule 273 starting with a single $\sigma = 1$ seed because in this scenario the sum of neighbors is never five and only $\beta_5 = 2$. Mode memory acts on rules 300 and 543 also inhibiting growth toward the formation of two *branches*. The remaining rules in Fig. 6.3 (276, 519, 303 and 546) are unaffected by the mode memory: they extinguish at $T = 3$, as in the ahistoric model.

But these rules show a rich dynamic in the mean memory model[3]. This activation under (mean) memory of rules that die at $T = 3$ in the ahistoric model is infeasible in the $k = 2$ scenario. Figure B.10 shows the evolving patterns of the $k = 3$ *parity* rules starting with the same initial configuration with values chosen at random as 0, 1 or 2 with probability 1/3. Again mode and mean memory models are considered in Fig. B.10. The similarities in the evolving patterns starting from a single seed in Fig. 6.3 are qualitatively reflected in Fig. B.10, so in the mode memory rules 276 and 519 are similar, the patterns of rules 300 and 543 are distinctive and rules 276, 303, 519 and 546 are unaffected or minimally affected by mode memory. All the rules in Fig. B.10 show a rich dynamic in the mean memory model, so no rule with memory dies out and mean memory fires the pattern of the rules that die out (or nearly so) in the ahistoric model. Rules 273 and 516 are affected by both memory models in a similar smooth way, not shown in Fig. B.10.

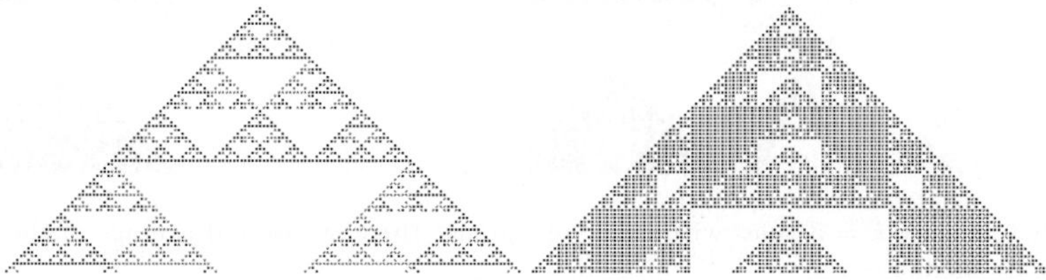

Fig. 6.4   Evolving patterns in a $k=3$ automaton with transition rule being that of the sum of nearest neighbors (left) and with this rule applied to cells endowed with memory of the sum of the last two states (right).

Figure 6.4 deals with an example of $k=3$ cells endowed with minimal memory. It shows the effect of featuring cells by the sum of their last two states on a cellular automaton with transition rule being the sum of the states of their nearest neighbors and the cell itself (sums performed modulo 3). It serves as final example of the notable alteration induced by memory, even of the shortest-term, in this context.

---

[3]These quiescent parity ($\beta_0 = \beta_2 = \beta_4 = \beta_6 = 0$) rules have $\beta_1 = 2$, thus their evolution in the ahistoric model is truncated at $T = 3$ in the ahistoric and mode memory model

$$\text{after:} \quad \begin{matrix} 1 & & \to & 222 \\ \downarrow & \uparrow & & \downarrow \\ 1 & \to & & 222 \end{matrix} \quad .$$ In the mean memory model extinction is avoided as the featuring

$$\text{after } T = 2 \text{ is} \quad \begin{matrix} 1 & & \to & 222 \\ \downarrow & \uparrow & & \downarrow \\ 1 & \to & & 121 \end{matrix} \quad .$$

6.2   Excitable systems

The effect of memory in excitable media has been recently studied in [145], [298] and [108], and in reaction-diffusion processes in particular in [361], [289], and [436]. Modeling in these references is made by means of differential equations supplemented with memory terms. In general, these terms are the time convolution of a linear operator applied to the unknown function with a suitable memory kernel [107].

In discrete *excitable* systems, the three states are featured: resting 0, excited 1 and refractory 2. State transitions from excited to refractory and from refractory to resting are unconditional, they take place independently of the cell's neighborhood state: $\sigma_i^{(T)} = 1 \rightarrow \sigma_i^{(T+1)} = 2$, $\sigma_i^{(T)} = 2 \rightarrow \sigma_i^{(T+1)} = 0$. Frequently, this kind of CA are termed the Greenberg-Hastings (GH) model, after the name of the authors that pioneered its study [172].

The GH approach has been also applied to the study of *slime mold* [148], a broad term describing fungi-like organisms that use spores to reproduce. They were formerly classified as fungi, but are no longer considered part of this group [92]. Their common name refers to part of some of these organism's life cycles where they can appear gelatinous (hence the name slime). Memory in this context has been addressed in [386, 349].

The excitation rule may adopt a kind of Pavlovian phenomenon of defensive inhibition [33]: when strength of stimulus applied to some parts of nervous system exceeds certain limit the system 'shuts down', this can be naively interpreted as an inbuilt protection of energy loss and exhaustion. To simulate the phenomenon of defensive inhibition we adopt interval excitation rules, developed in [1], and put as a resting cell becomes excited only if one or two of its neighbors are excited. If more than two neighbors are excited the defensive inhibition comes into action and prevents the cell from excitation: $\sigma_i^{(T)} = 0 \rightarrow \sigma_i^{(T+1)} = 1$ if $\sum_{j \in \mathcal{N}_i} (\sigma_j^{(T)} = 1) \in \{1, 2\}$.

An example of simple development is shown in Fig. 6.5 where configurations of defensive-inhibition CA starting from an excited singleton, and evolving up to $T = 15$; there the Moore neighborhood is adopted.

The defensive inhibition rule was studied in [5] amongst other interval excitation rules in a context of morphological and dynamical complexities. We demonstrate there that the rule exhibits quite a low morphological complexity amongst rules with lowest number of different cluster (i.e., clusters of excited and refractory states) sizes. However CA governed by the rule exhibits the longest transient periods. Calculating Langton's λ parameter

Fig. 6.5   The defensive inhibition CA rule from an excited singleton. Black cells are excited (state 1), gray cells are refractory (state 2).

shows that the rule stays toward the middle of the rule phases space, and occupies the general position of rules with complex behavior. A contradiction between morphological, i.e., *a posteriori*, and function-based λ, i.e., *a priori*, measurements of complexity sounds alarming. Moreover, the closest to the defensive inhibition rule – in the interval excitation universe – is the interval $[2, 2]$ rule is a 'kingdom of complexity and universality' [1]. Mobile localizations, in other terminologies called wave-fragments and gliders, are essential attributes of complexity. As you can see in Fig. 6.5, the rule is a distant analog of replicator rules.

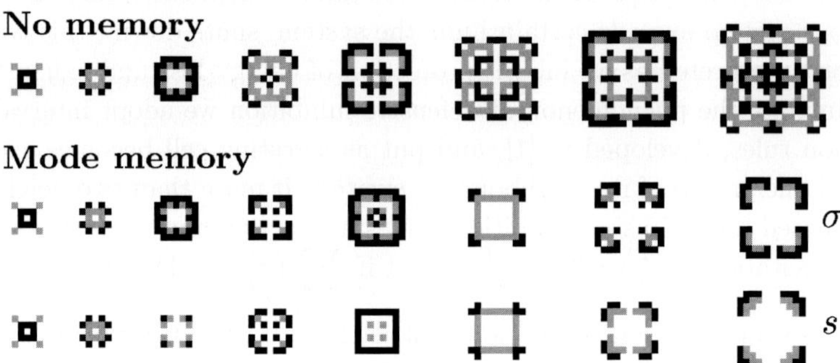

Fig. 6.6   Effect of mode memory on the defensive inhibition CA rule starting from a simple configuration.

The restraining effect of memory may be fatal starting with simple configurations such as the singleton of Fig. 6.5, which extinguishes at $T = 4$ with mode memory: The only cells not featured as resting up to $T = 3$ are those refractory at $T = 3$: , which implies immediate extinction.

This leads us to change the simple initial configuration for demonstration purposes to that of Fig. 6.6. The last series of evolving patterns of Fig. 6.6 shows the underlying patterns with cells featured by their most frequent state (s) along the last three time steps; these patterns generate the actual patterns (σ) with mode memory to their right above. As a general rule, memory tends to restrain the evolution as shown in the case of mode memory in Fig. 6.6. It is generally so from the beginning of the effective memory action, so at $T = 3$ the outer excited cells in the actual pattern evolution are not featured as excited but as resting cells, as this is the their most frequent state up to this time step (twice resting versus one excited). Typically, the series of evolving patterns with memory diverges from the ahistoric evolution already at $T = 4$. From this early time-step, the patterns with memory turn out to be less expanded, as shown in Fig. 6.6.

An exhaustive study of spatio-temporal excitation dynamics in two-dimensional automata with eight-cell neighborhood in made in [2]. In a general form, a resting cell becomes excited if the number of excited neighbors $(\#(1))$ belongs to the interval $[\theta_1, \theta_2]$, and that of refractory neighbors $(\#(2))$ belongs to the interval $[\delta_1, \delta_2]$. Thus, under the generic rule $R(\theta_1, \theta_2, \delta_1, \delta_2)$: $\sigma_i^{(T)} = 0 \rightarrow \sigma_i^{(T+1)} = 1$ if $\#(1)_i^{(T)} \in [\theta_1, \theta_2]$ and $\#(2)_i^{(T)} \in [\delta_1, \delta_2]$. Such a excitation mode is named *mutualistic* in [2] because for a cell to be excited not only excited neighbors but also neighbors in refractory states are required. These automaton rules may be regarded with *memory* as: $\sigma_i^{(T)} = 0 \rightarrow \sigma_i^{(T+1)} = 1$ if $\#(1)_i^{(T-1)} \in [\theta_1, \theta_2]$ and $\#(1)_i^{(T-1)} \in [\delta_1, \delta_2]$. This is so because the transition from excited to refractory is unconditional, and therefore if a cell is in resting state at time-step T, the cell was necesarily excited at time-step $T - 1$. Such interpretation bring more light to the emergence of minimal mobile localizations, e.g., in the rule $R(2222)$. Under the approach of memory advocated here: $\#(1)_i^{(T)} = \#(s_i^{(T-1)} = 1)$ and $\#(2)_i^{(T)} = \#(s_i^{(T-1)} = 2)$.

Rules in the hexagonal tessellation

The *beehive* rule is a totalistic two-dimensional CA rule with three states implemented in the hexagonal tessellation [9, 432, 26] [4].

The *beehive* rule exhibits mobile localized patterns -gliders- which dom-

---

[4]The *beehive* rule assigns the following outputs to each of the 28 possible frequencies of the three states (2, 1, and 0): (0,0,6)→0 (0,1,5)→1 (0,2,4)→2 (0,3,3)→1 (0,4,2)→2 (0,5,1)→0 (0,6,0)→0 (1,0,5)→0 (1,1,4)→2 (1,2,3)→2 (1,3,2)→2 (1,4,1)→1 (1,5,0)→1 (2,0,4)→0 (2,1,3)→0 (2,2,2)→2 (2,3,1)→2 (2,4,0)→0 (3,0,3)→0 (3,1,2)→2 (3,2,1)→2 (3,3,0)→0 (4,0,2)→0 (4,1,1)→0 (4,2,0)→2 (5,0,1)→2 (5,1,0)→0 (6,0,0)→0 .

inate the lattice at the concluding phase of development. Figure 6.7 shows
an example starting from a single $\sigma = 1$ active cell, which produces six
gliders at $T = 9$.

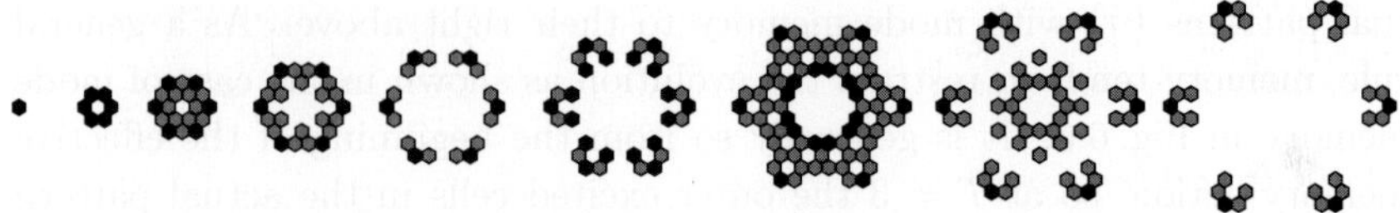

Fig. 6.7   The beehive cellular automaton from a single $\sigma = 1$ cell, up to $T = 9$.

Figure 6.8 shows the effect on the beehive rule of α memories with two
values of the memory factor α: the small value 0.3 and the maximum
$\alpha = 1.0$, the case of full memory in which one: $m_i^{(T)} = \dfrac{1}{T} \sum_{t=1}^{T} \sigma_i^{(t)}$. In the
latter case, memory affects the evolution as early as after $T = 3$: The outer
$\sigma = 1$ cells at $T = 3$, with state history 001, are featured as dead, and the
$\sigma = 2$ cells at $T = 3$, with state histories 002 or 012, are featured as $\sigma = 1$
cells. Consequently, the pattern of the ahistoric and full memory models
diverge from $T = 4$ in Fig. 6.8, leading to a series of patterns which do
not progress far away from the initial seed in the full memory model. The
rounding mechanism [6.1] is rather biased toward the state 1, which tends
to generate $s = 1$ trait states, as occurs in the full memory model in Fig. 6.8.
With memory of level $\alpha = 0.3$, only the six outer $\sigma = 2$ cells at $T = 3$,
with history 002, are not featured by their last state but as $s = 1$. But this
fact determines also a dramatic change in the evolution, characterized by
structured *beehive*-like patterns, unlike the ahistoric model.

Figure 6.9 shows the effect on the beehive rule of two $\delta = t^c$ memory
weights. As expected, the evolution under $\delta = t$ resembles that of full
memory in Fig. 6.8, whereas that of $\delta = t^2$ resembles that with the low
$\alpha = 0.3$ memory factor.

In Fig. 6.10, cells are featured by the parity of the last three state values:
$s_i^{(T)} = (\sigma_i^{(T-2)} + \sigma_i^{(T-1)} + \sigma_i^{(T)})$ *mod* 2 in the case of rule 273 (0101010) and
by the three-state binary number (0101020) rule 276. In the latter case,
the evolution produces the null configuration (all cells dead) at $T = 9$, but
this does not mean extinction: at $T = 10$ a new pattern appears (the same
that at $T = 3$). This odd *cataleptic* phenomenon is not feasible with no
memory.

Last but not least, in Fig. 6.11 the memory rule is the *beehive* rule acting

Fig. 6.8　The beehive rule from a single $\sigma=1$ cell with of α-memory.

Fig. 6.9　Effect of δ-memory on the beehive rule from a single $\sigma=1$ cell.

after $T = 6$. We expected *gliders*, but they do not appear in Fig. 6.11, not even starting at random.

Among the hexagonal three-state two-dimensional cellular automata, the *spiral* rule $[6, 433]$[5] is noteworthy. It is shown in Fig. 6.12 starting

---

[5] At variance with the beehive rule, the spiral rule considers self-interaction, with transitions from frequencies (n_2,n_1,n_0): $(7,0,0)\to 0$ $(6,0,1)\to 0$ $(6,1,0)\to 0$ $(5,0,2)\to 0$ $(5,1,1)\to 0$ $(5,2,0)\to 2$ $(4,0,3)\to 0$ $(4,1,2)\to 0$ $(4,2,1)\to 2$ $(4,3,0)\to 1$ $(3,0,4)\to 0$ $(3,1,4)\to 2$ $(3,2,2)\to 2$

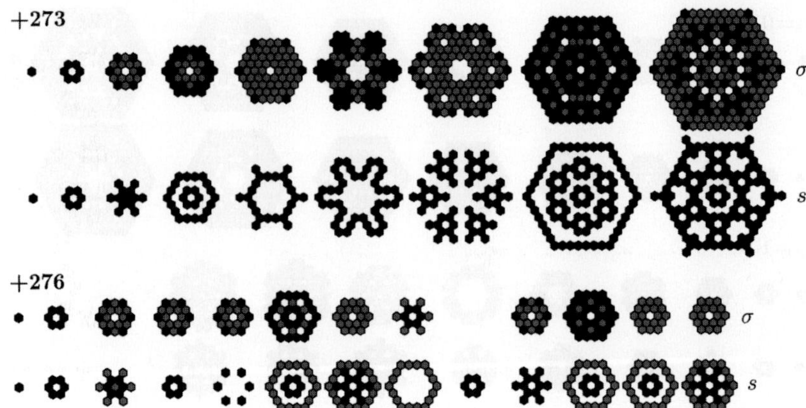

Fig. 6.10 The evolving patterns of the *beehive* rule, with two parity rules of the last three states memory, starting from a single σ=1 cell.

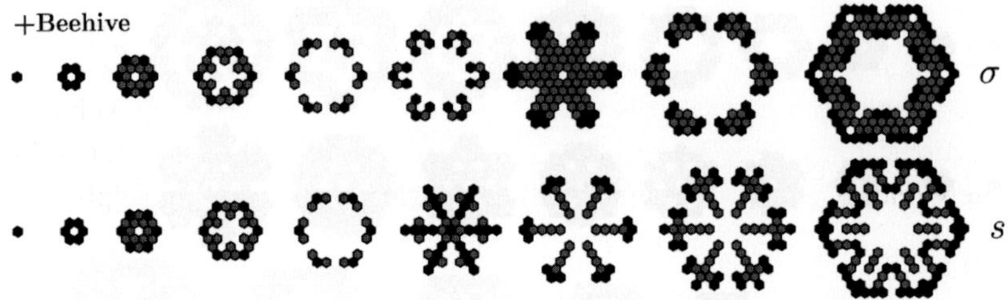

Fig. 6.11 The *beehive* rule with the *beehive* as memory rule, from a single σ=1 cell.

from a single σ=1 cell, which produces six *gliders* at $T = 5$. Figure 6.13 shows the evolving patterns of the spiral rule with spiral memory from T=7 when starting from a single σ=1 cell

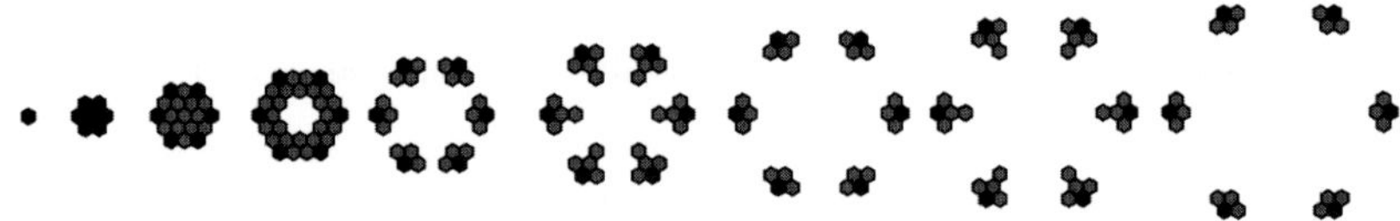

Fig. 6.12 The evolving patterns of the *spiral* starting from a single σ=1 cell.

(3,3,1)→1 (3,4,0)→2 (2,0,5)→0 (2,1,4)→0 (2,2,3)→2 (2,3,2)→1 (2,4,1)→2 (2,5,0)→2
(1,0,6)→0 (1,1,5)→2 (1,2,4)→2 (1,3,3)→1 (1,4,2)→2 (1,5,1)→2 (1,6,0)→2 (0,0,7)→0
(0,1,6)→1 (0,2,5)→2 (0,3,4)→1 (0,4,3)→2 (0,5,2)→2 (0,6,1)→2 (0,7,0)→2.

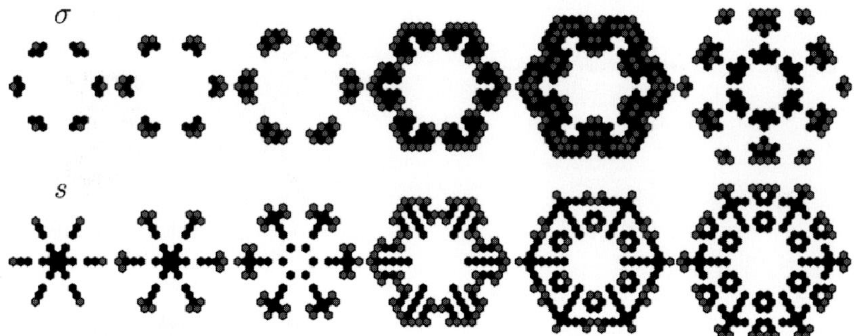

Fig. 6.13   The *spiral* rule with the *spiral* as memory rule, starting from a single $\sigma=1$ cell, shown from $T=7$.

The spiral rule has been found to exhibit both stationary and mobile localizations (eaters and gliders), and generators of mobile localizations (glider-guns). A remarkable feature of this automaton is the existence of spiral glider-guns, a discrete analog of a spiral wave that splits into localized wave-fragments (gliders) at some distance from the spiral tip. It is demonstrated that the rich spatio-temporal dynamics of interacting traveling localizations and their generators can be used to implement computation, namely manipulation with signals, binary logical operations, multiple-value operations, and finite-state machines.

Chapter 7

Reversible dynamics

7.1 Characterization

Reversible systems are of interest since they preserve information and energy and allow unambiguous backtracking [146]. They are studied in computer science in order to design computers which would consume less energy [388]. Reversibility is also an important issue in fundamental physics [151, 272, 393, 400]. Geraldt 't Hooft, in a speculative paper [398], suggests that a suitably defined deterministic, local reversible CA might provide a viable formalism for constructing field theories on a Planck scale. Svozil [384] also asks for changes in the underlying assumptions of current field theories in order to make their discretization appear more CA-like. Reversible quantum CA with memory are considered in [98]. Applications of reversible CA with memory in cryptography are being intensively scrutinized [61, 62, 136, 216, 270, 279].

The second-order in time CA implementation based on the subtraction modulo the number of states (denoted $\ominus$): $\sigma_i^{(T+1)} = \phi\left(\sigma_j^{(T)} \in \mathcal{N}_i\right) \ominus \sigma_i^{(T-1)}$, readily reverses as: $\sigma_i^{(T-1)} = \phi\left(\sigma_j^{(T)} \in \mathcal{N}_i\right) \ominus \sigma_i^{(T+1)}$. To preserve the reversible feature, memory has to be endowed only in the pivotal component of the rule transition, so: $\sigma_i^{(T-1)} = \phi\left(s_j^{(T)} \in \mathcal{N}_i\right) \ominus \sigma_i^{(T+1)}$.

For reversing from T it is necessary to know not only $\sigma_i^{(T)}$ and $\sigma_i^{(T+1)}$ but also $\omega_i^{(T)}$ to be compared to $\Omega(T)$, to obtain:

$$
s_i^{(T)} = \begin{cases}
0 & \text{if } 2\omega_i^{(T)} < \Omega(T) \\
\sigma_i^{(T+1)} & \text{if } 2\omega_i^{(T)} = \Omega(T) \\
1 & \text{if } 2\omega_i^{(T)} > \Omega(T)
\end{cases} .
$$

Then to progress in the reversing, to obtain $s_i^{(T-1)} = round\left(\dfrac{\omega_i^{(T-1)}}{\Omega(T-1)}\right)$,

it is necessary to calculate $\omega_i^{(T-1)} = \dfrac{1}{\alpha}\left(\omega_i^{(T)} - \sigma_i^{(T)}\right)$. But in order to avoid the division by the memory factor (recall that operations with real numbers are not exact in computer arithmetic), it is preferable to work with

$\gamma_i^{(T-1)} = \omega_i^{(T)} - \sigma_i^{(T)}$, and to compare these values to $\Gamma(T-1) = \displaystyle\sum_{t=1}^{T-1} \alpha^{T-t}$.

This leads to: $s_i^{(T-1)} = \begin{cases} 0 & \text{if } 2\gamma_i^{(T-1)} < \Gamma(T-1) \\ \sigma_i^{(T)} & \text{if } 2\gamma_i^{(T-1)} = \Gamma(T-1) \\ 1 & \text{if } 2\gamma_i^{(T-1)} > \Gamma(T-1) \end{cases}$. Continuing in the

reversing process: $\gamma_i^{(T-2)} = \gamma_i^{(T-1)} - \alpha\sigma_i^{(T-1)}$ and $\Gamma(T-2) = \displaystyle\sum_{t=1}^{T-2} \alpha^{T-t}$.

In general: $\gamma_i^{(T-\tau)} = \gamma_i^{(T-\tau+1)} - \alpha^{\tau-1}\sigma_i^{(T-\tau+1)}$ and $\Gamma(T-\tau) = \displaystyle\sum_{t=1}^{T-\tau} \alpha^{T-t}$,

giving: $s_i^{(T-\tau)} = \begin{cases} 0 & \text{if } 2\gamma_i^{(T-\tau)} < \Gamma(T-\tau) \\ \sigma_i^{(T-\tau+1)} & \text{if } 2\gamma_i^{(T-\tau)} = \Gamma(T-\tau) \\ 1 & \text{if } 2\gamma_i^{(T-\tau)} > \Gamma(T-\tau) \end{cases}$.

In the three-state scenario, the assignments generalize as:

$$s_i^{(T)} = \begin{cases} 0 & \text{if } 2\omega_i^{(T)} < \Omega(T) \\ \sigma_i^{(T+1)} & \text{if } 2\omega_i^{(T)} = \Omega(T) \\ 1 & \text{if } \Omega(T) < 2\omega_i^{(T)} < 3\Omega(T) \\ \sigma_i^{(T+1)} & \text{if } 2\omega_i^{(T)} = 3\Omega(T) \\ 2 & \text{if } 2\omega_i^{(T)} > 3\Omega(T) \end{cases} ,$$

$$s_i^{(T-1)} = \begin{cases} 0 & \text{if } 2\gamma_i^{(T-1)} < \Gamma(T-1) \\ \sigma_i^{(T)} & \text{if } 2\gamma_i^{(T-1)} = \Gamma(T-1) \\ 1 & \text{if } \Gamma(T-1) < 2\gamma_i^{(T-1)} < 3\Gamma(T-1) \\ \sigma_i^{(T)} & \text{if } 2\gamma_i^{(T-1)} = 3\Gamma(T-1) \\ 2 & \text{if } 2\gamma_i^{(T-1)} > 3\Gamma(T-1) \end{cases} ,$$

$$
s_i^{(T-\tau)} = \begin{cases}
0 & \text{if } 2\gamma_i^{(T-\tau)} < \Gamma(T-\tau) \\
\sigma_i^{(T-\tau+1)} & \text{if } 2\gamma_i^{(T-\tau)} = \Gamma(T-\tau) \\
1 & \text{if } \Gamma(T-\tau) < 2\gamma_i^{(T-\tau)} > 3\Gamma(T-\tau) \\
\sigma_i^{(T-\tau+1)} & \text{if } 2\gamma_i^{(T-\tau)} = 3\Gamma(T-\tau) \\
2 & \text{if } 2\gamma_i^{(T-\tau)} > 3\Gamma(T-\tau)
\end{cases} \cdot
$$

7.2 Reversible rules with memory

In the following reversible simulations, the initial pattern at $T = 0$ is the same as that at $T = 1$. This is so both starting with a single live cell and at random.

Two-dimensional reversible CA

Figure 7.1 shows the effect of memory on the reversible parity rule starting from a single site live cell, the scenario of Fig. 2.1 with the reversible qualification. As noted for Fig. 2.1, *(i)* the simulations corresponding to $\alpha = 0.6$ or below shows the ahistoric pattern at $T = 4$, whereas memory leads to a pattern different from $\alpha = 0.7$, and *(ii)* the pattern at $T = 5$ for $\alpha = 0.54$ and $\alpha = 0.55$ differ.

Figure 7.2 (a reversible analogue to Fig. 2.2) shows the effect of minimal memory, and, again as in the case of Fig. 2.2, *(i)* the configuration of the patterns is notably altered, *(ii)* the speed of diffusion of the area affected are notably reduced, even by minimal memory, *(iii)* high levels of memory tend to freeze the dynamics from the early time-steps. A study of the effect of memory on reversible two-dimensional CA starting from a single active cell is made in [29].

Figures 7.3 and 7.4 show the effect of α memory on the parity rule in the hexagonal and triangular tessellations starting as in Figs. 2.4 and 2.5. In the hexagonal scenario with full memory, when starting as in Fig. 7.3 a period four oscillator appears at $T=5$.

Figure 7.5 (a reversible analog to Figs. 6.7 and 6.8) shows the effect of two δ memory type implementations on the beehive rule.

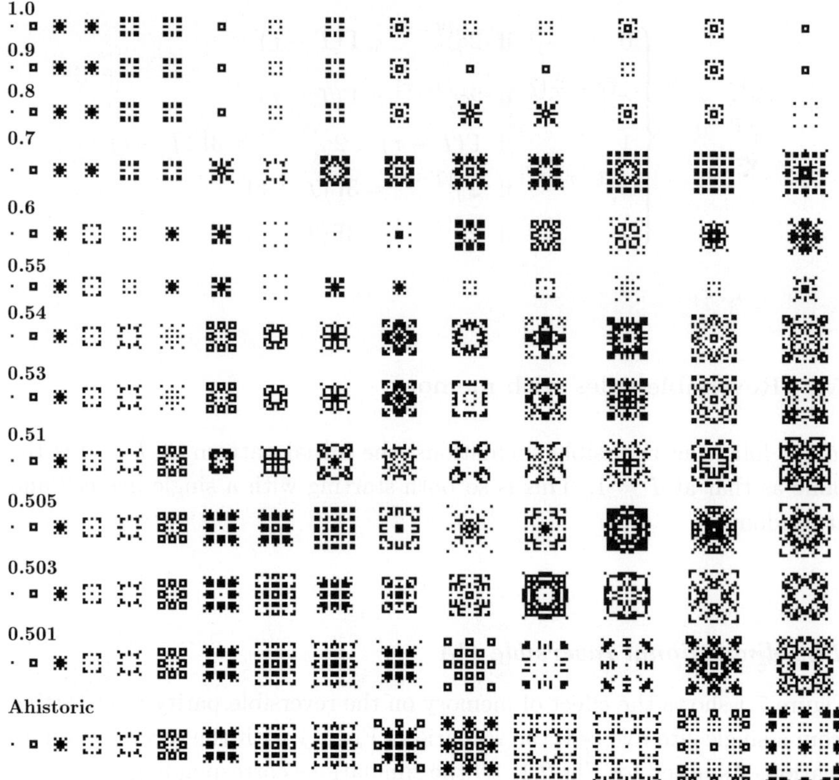

Fig. 7.1   The reversible parity rule with memory.

One-dimensional reversible CA

Figure 7.6 shows the evolving patterns of reversible elementary rules starting from a single site seed with α-memory. Evolution is shown up to 63 time steps The confinement of the disruption generated by a single cell when memory is active becomes very clear. When full history is considered, the evolution dynamics tend to generate oscillators. Only rule 94 dies out, and only in the fully historic model. Some rules show unexpected similar evolving patterns: rules 126 and 254 for example.

Figure 7.7 adopts the same scenario as Fig. 7.6 but with $\delta = t^c$ memory [50]. The patterns in both figures are reminiscent, with an apparent restraining effect of memory. Again, oscillators are frequent when full memory is implemented ($c = 0$) and for some rules with higher c values, $c = 1$

Fig. 7.2   The reversible parity rule with minimal memory: $\alpha = 0.501$.

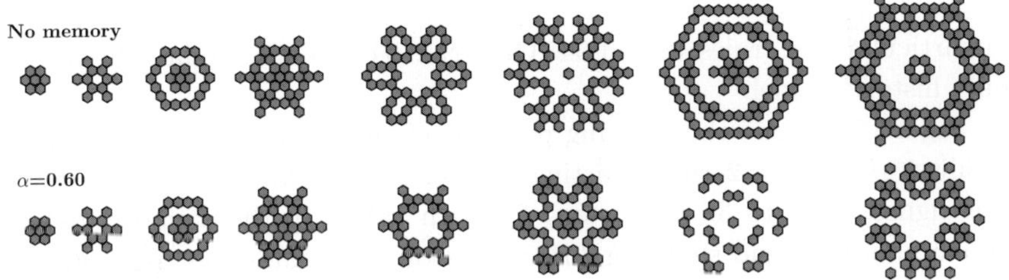

Fig. 7.3   The parity reversible rule in the hexagonal tessellation.

Fig. 7.4   The parity reversible rule in the triangular tessellation.

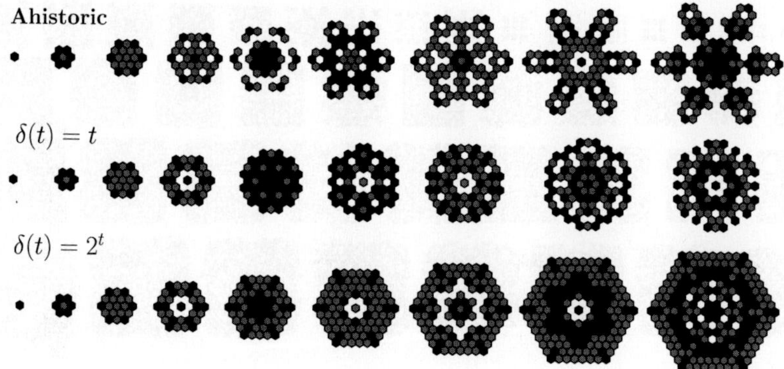

Fig. 7.5   The beehive reversible rule with δ-memory.

in particular, e.g., rules 18 and 54. Only rule 94 dies out, again only in the fully historic model.

Figure 7.8 shows the evolving patterns of the reversible formulation of totalistic $k = r = 2$ rules starting from a single live cell. Evolution in this figure is up to 36 time steps for high values of the memory factor and up to 217 time steps for low values of α. Again, the confinement of the disruption generated by a single live cell becomes very clear. No extinction has been found in this scenario. Configuration oscillators are frequent when full history is considered and in some rules for smaller α values, such as rule 46 for $\alpha = 0.9$ and rules 18 and 50 for $\alpha \leq 0.7$. Memory operates on the rules in Fig. 7.8 in a rather foreseeable manner when starting at random ([47] includes all these patterns). Figure A.3 shows two paradigmatic examples, that of rules 42 and 52 starting from the same random initial configuration as in the irreversible configuration. No extinction has been found starting from a disordered configuration in the reversible implementation. Most of the *chaotic* rules, the Class I rules 36 and 54 and the *complex* rules 20 and 52 present a chaotic behavior facing the change in the initial center state in their reversible formulation. Memory constrains the growth in the damaged region. This effect is, as a rule, gradual, and for most rules the error inhibition effect is significant only at high α values; but in some rules, e.g., rule 18, the depletion on the spread of the error is notable already in $\alpha = 0.6$. The case of rule 12 for $\alpha = 0.6$ is atypical: its damaged region in $\alpha = 0.6$ is broader than that of the ahistoric model. Historic memory has no significant effect on the reversible formulation of most class I and II rules and on the Class III rules

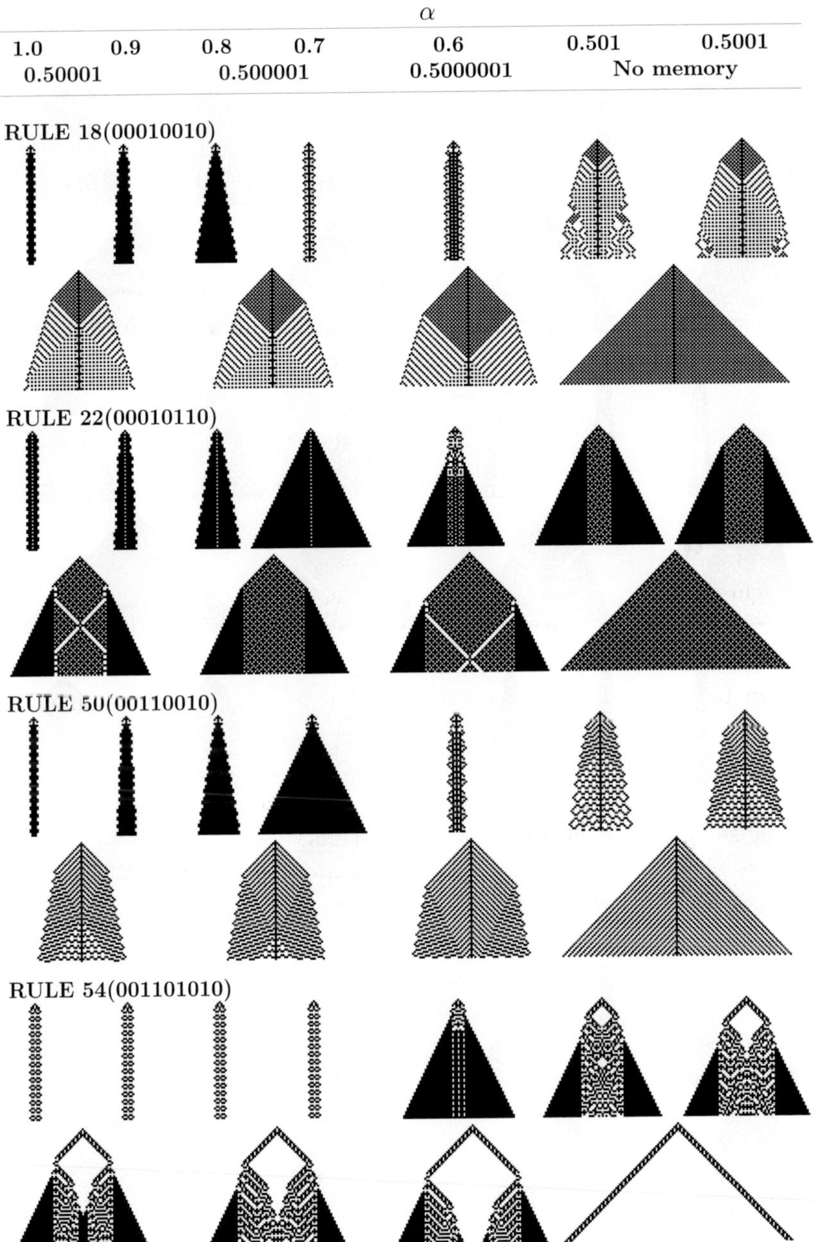

Fig. 7.6 Reversible elementary rules starting from a single seed.

RULE 90(01011010)

RULE 94(01011110)

RULE 122(01111010)

RULE 126(01111110)

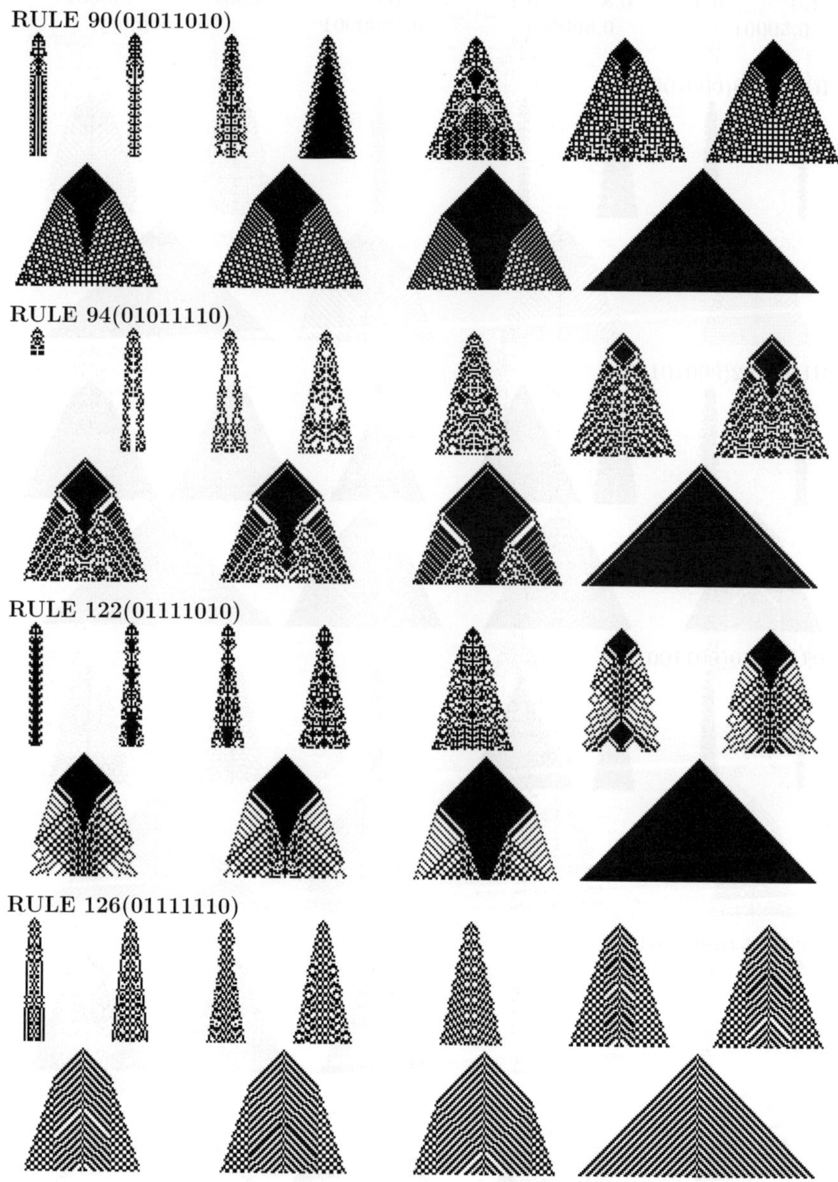

Fig. 7.6   (*continued*)

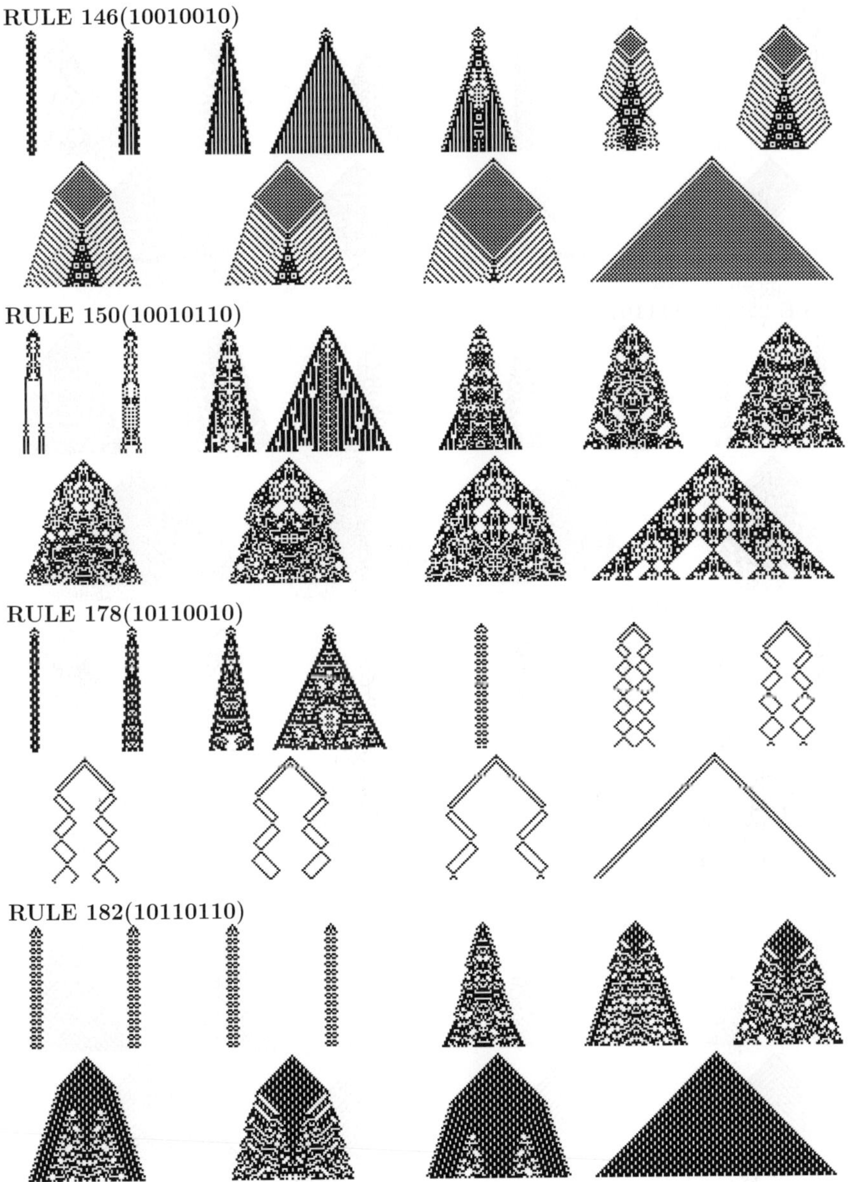

Fig. 7.6 *(continued)*

Discrete Systems with Memory

RULE 218(11011010)

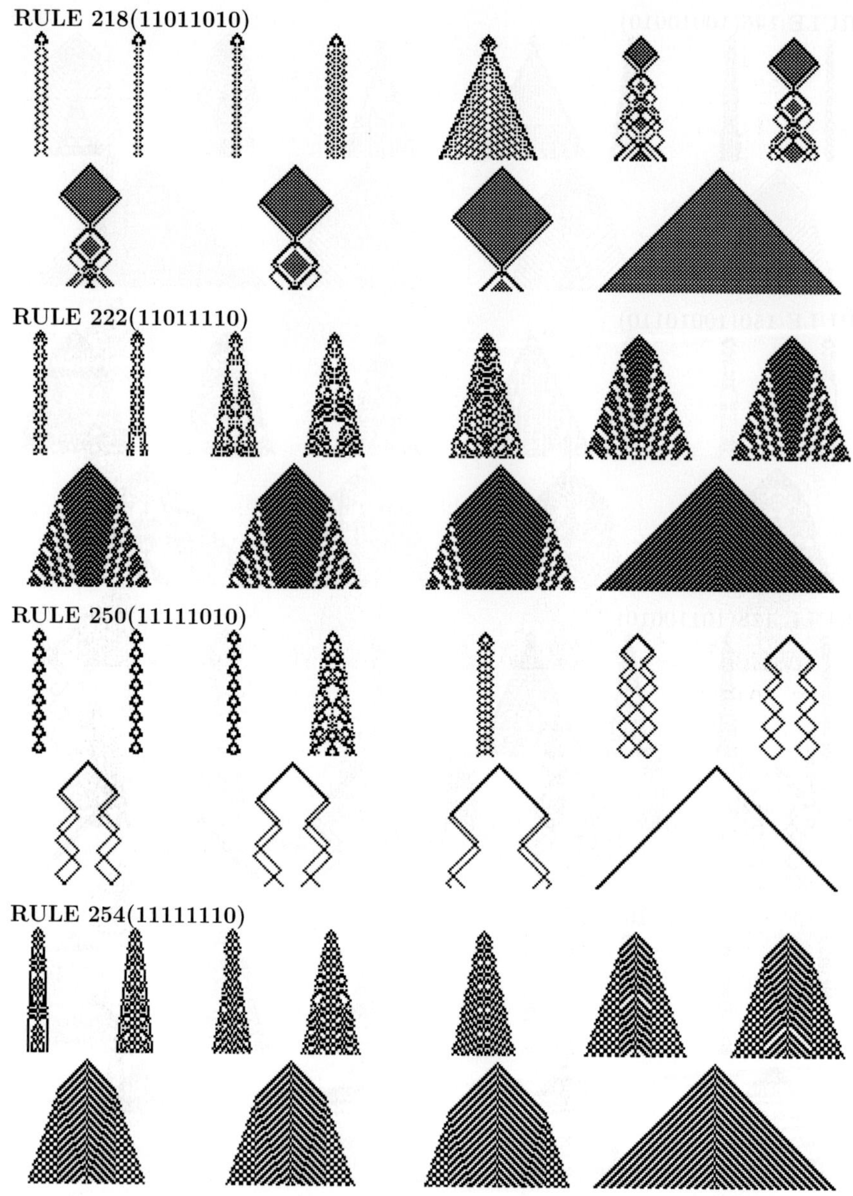

RULE 222(11011110)

RULE 250(11111010)

RULE 254(11111110)

Fig. 7.6 *(continued)*

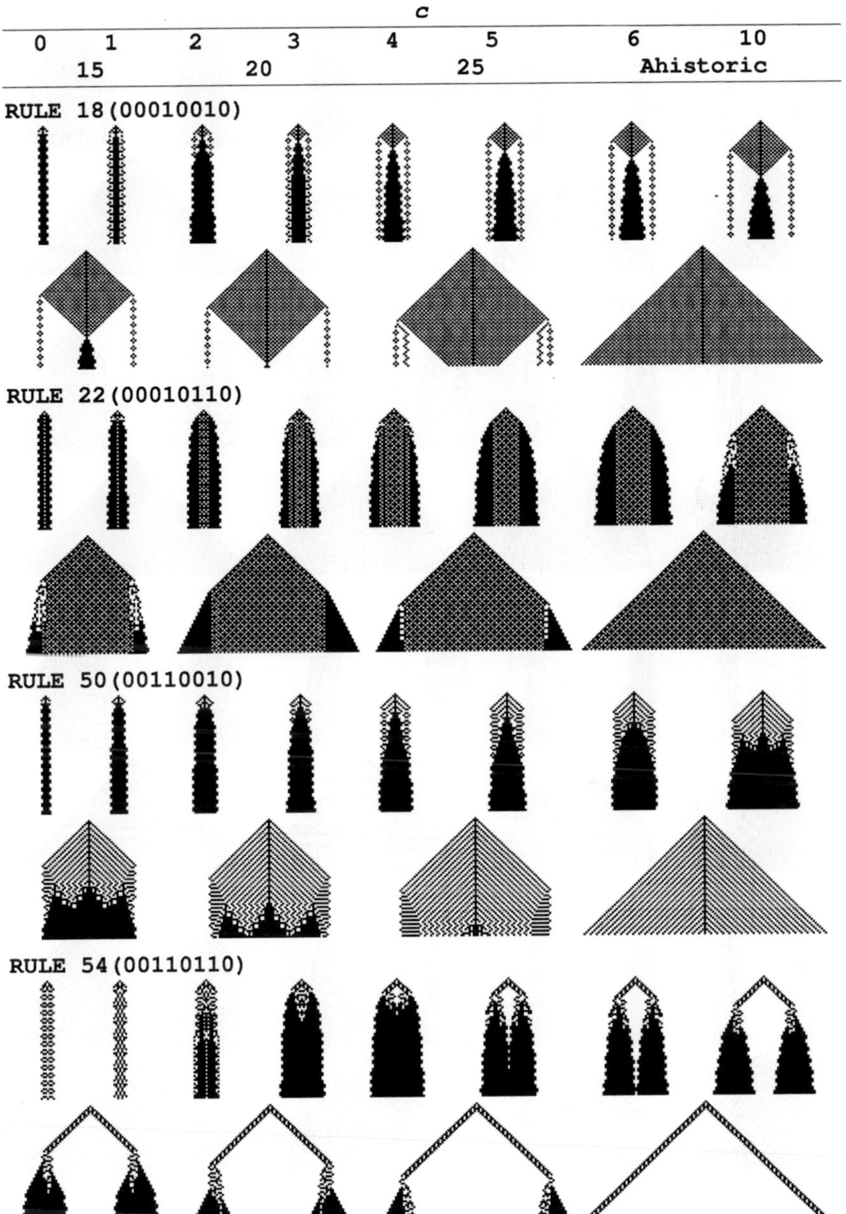

Fig. 7.7 Reversible elementary rules with $\delta = t^c$ memory.

RULE 90 (01011010)

RULE 94 (01011110)

RULE 122 (01111010)

RULE 126 (01111110)

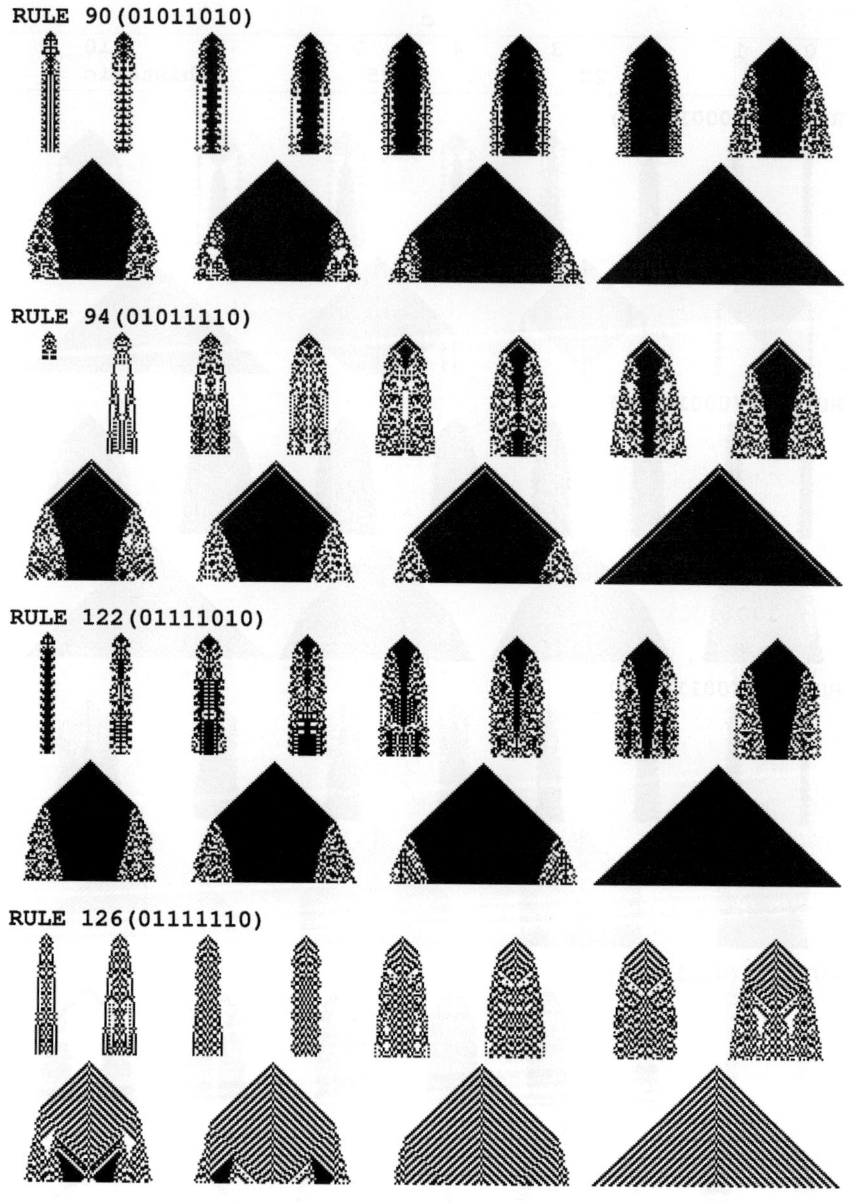

Fig. 7.7   (*continued*)

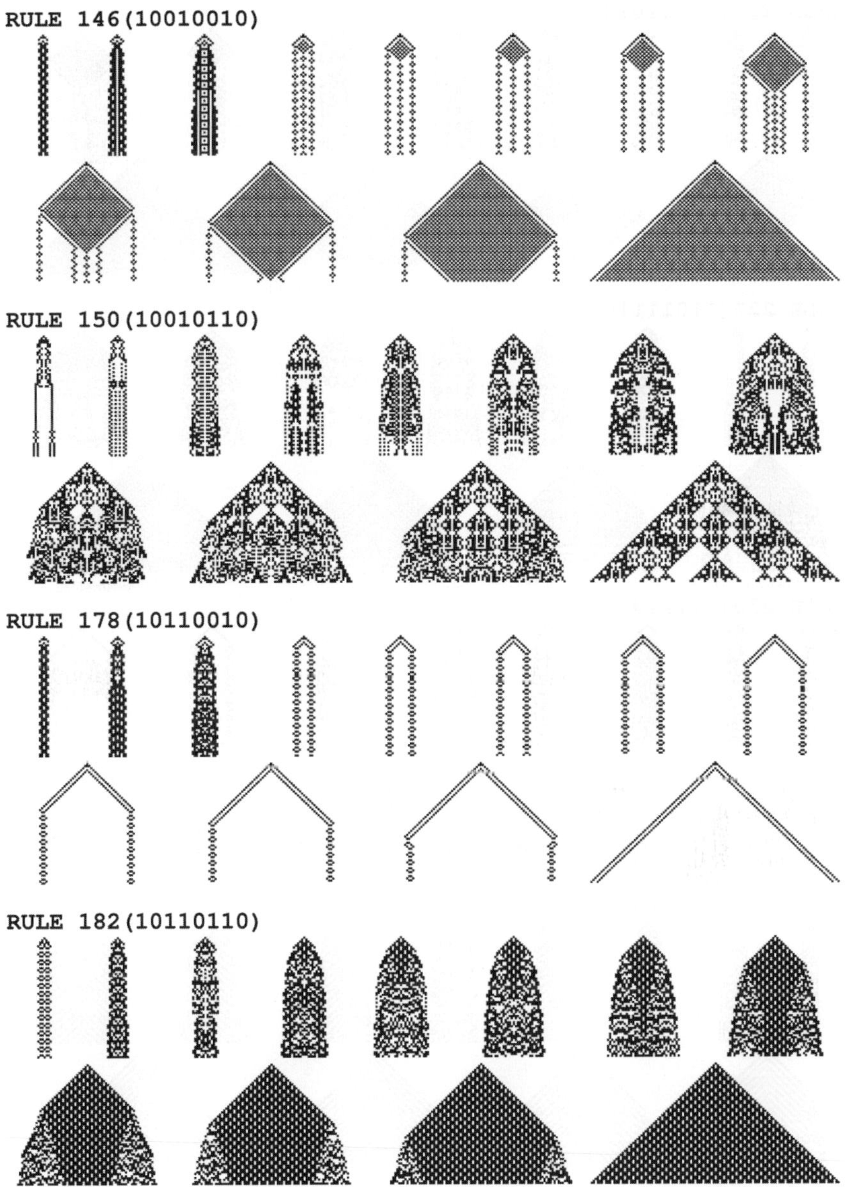

Fig. 7.7 (*continued*)

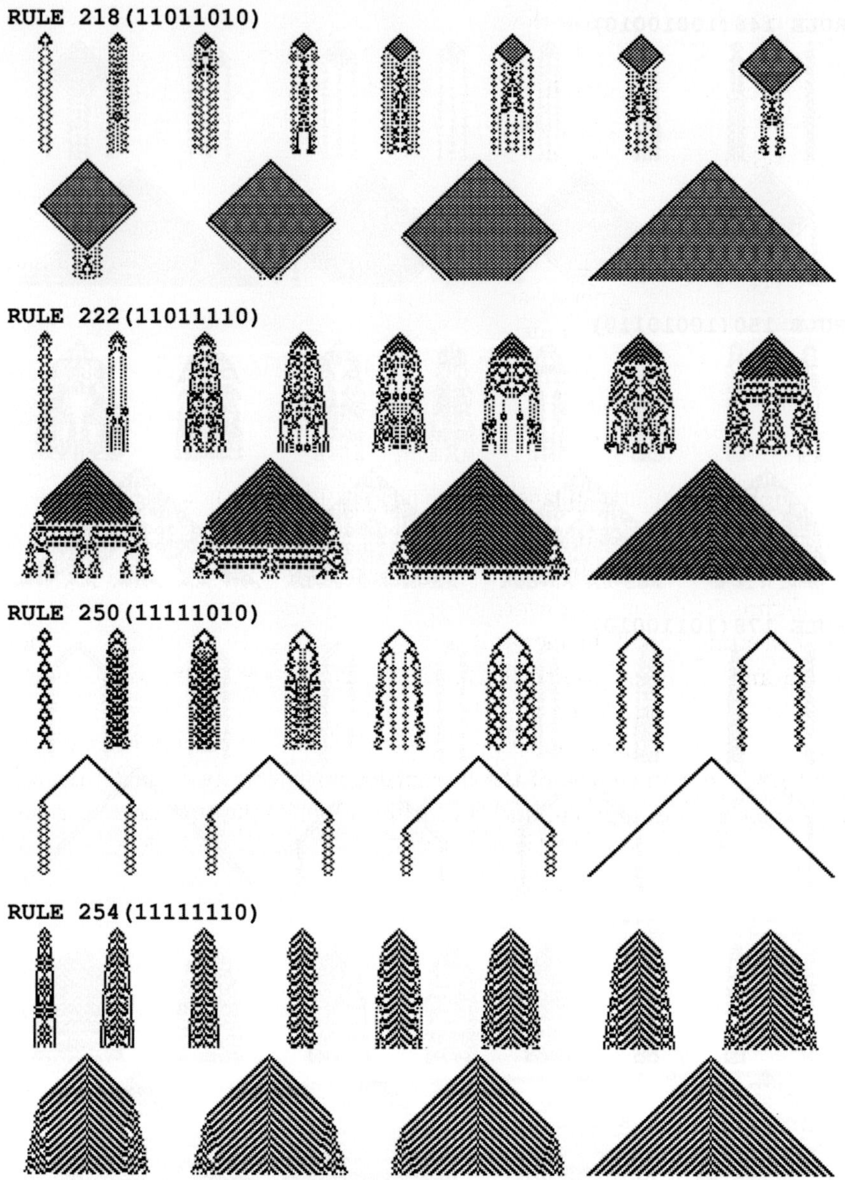

Fig. 7.7   (*continued*)

2, 6 and 34 , which have the disruption confined already in the ahistoric model.

The remaining rules in this section deal with CA with three states (0 blank, 1 black, 2 gray), starting from a single seed with a single $\sigma = 1$ cell, evolving up to $T = 26$.

Figure 7.9 shows the effect of α memory on the evolving patterns of the reversible quiescent totalistic $k = 3$ parity rules, with the memory factor varying as in Fig. 6.1. In contrast to Fig. 6.1 , here rules 276, 303, 519 and 546 do not die but generate small-size oscillators in the ahistoric reversible model, being extinction unfeasible in the standard reversible formulation. Memory confines the pattern growth of these rules from $\alpha = 0.8$. In the reversible formulation of Fig. 7.9, rules 273 and 516 are not confined to the two-state scenario. The patterns of these rules and that of rules 300 and 543 are restricted by memory from $\alpha = 0.6$. Full memory tends to generate oscillators in the reversible scenario (as shown in Fig. 7.9); the *punctuated equilibrium*-like behavior is here much less frequent than in the irreversible formulation.

Figure 7.10 shows the evolving patterns in the scenario of Fig. 7.9, but implementing $\delta = t^c$ memory, c=0,1,2,3,4,5 (in Fig.6 of [48] time reaches T=62 and c applies also to 6, 10, 15, 20, 25, closer to ahistoric model). In contrast to Fig. 6.2, no rule dies out in Fig. 7.10, not even in the ahistoric model. Configuration oscillators are always found in the full memory model. The confinement of the disruption generated by a single live cell due to memory is apparent and gradual in rules 273, 300, 516 and 543 . But contrary to expectation, rules 276, 303, 519, and 546 present wider patterns at the low $c > 0$ values in Fig. 7.10 than those of the ahistoric patterns.

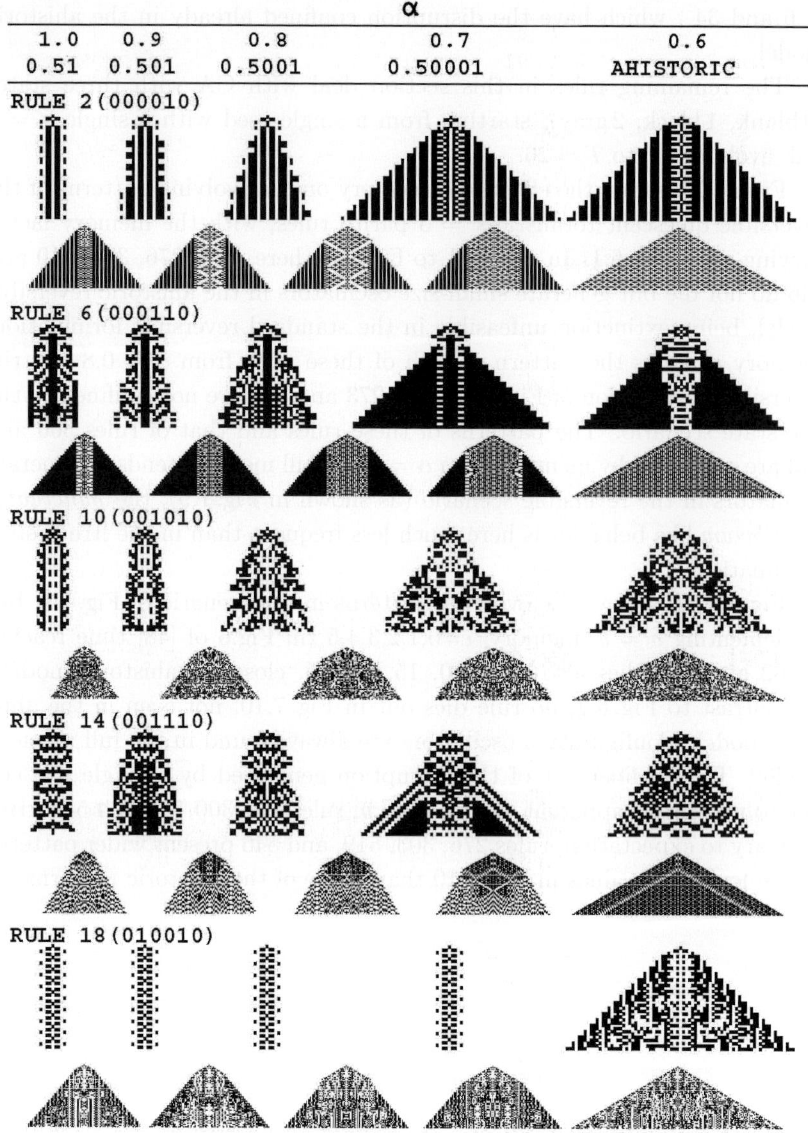

Fig. 7.8   Reversible, totalistic, $k=r=2$ quiescent rules from a single seed. The patterns corresponding to higher α values, presented in the first row of patterns, are zoomed compared to those corresponding low α values (lower rows).

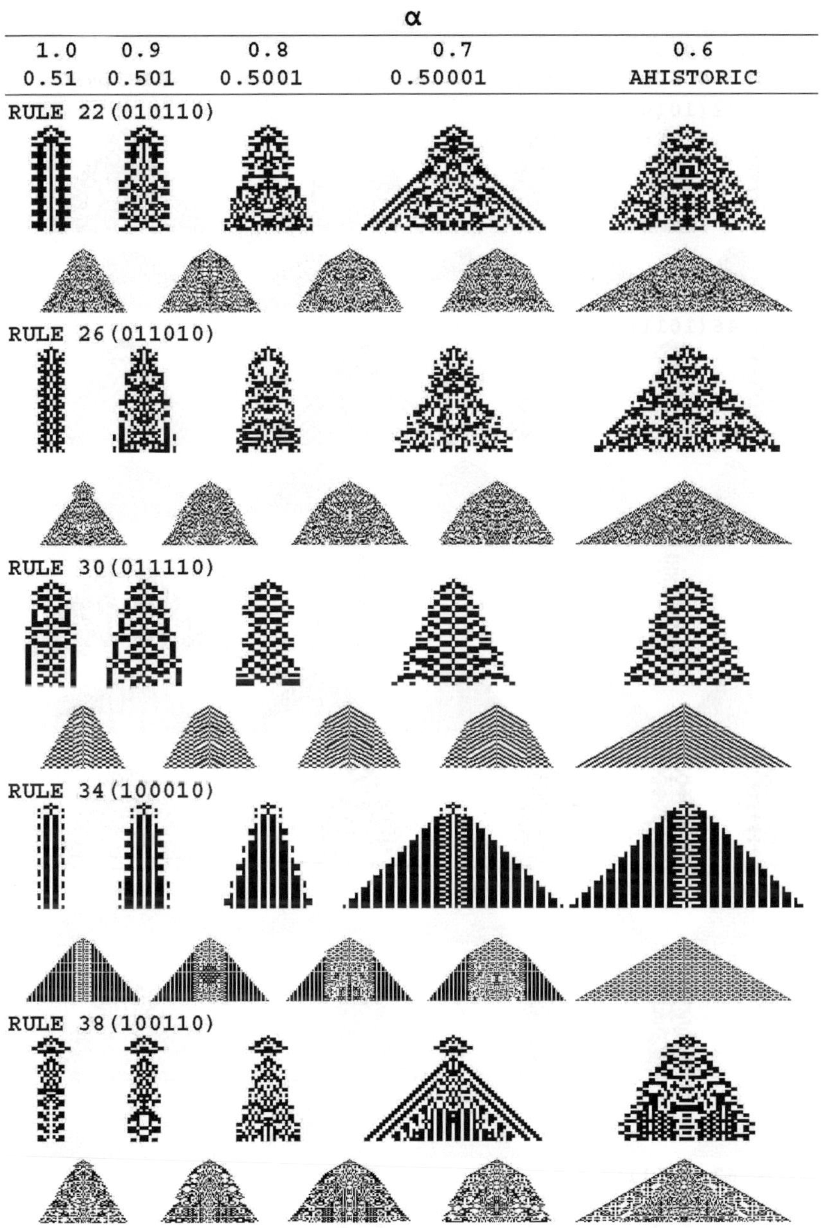

Fig. 7.8 (*continued*)

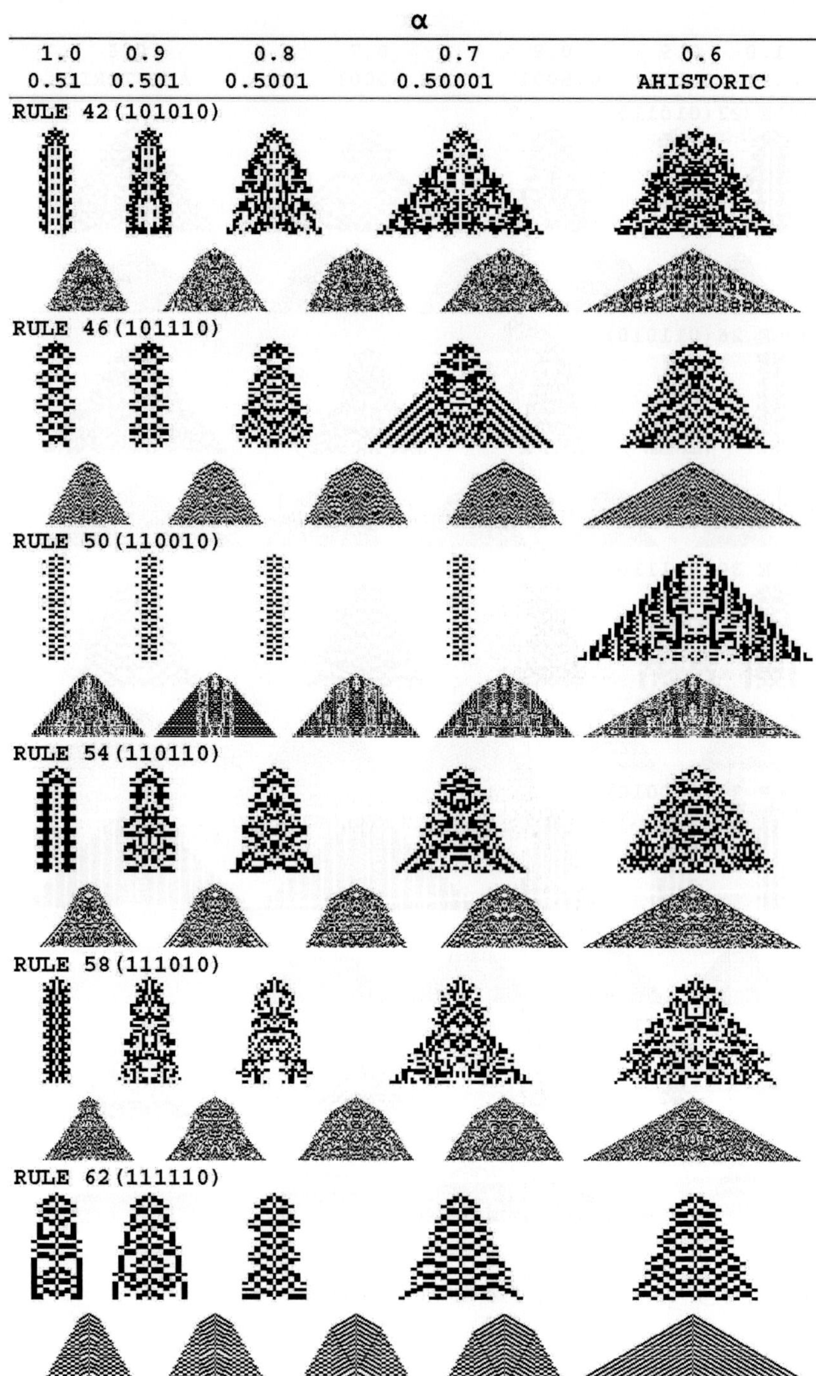

Fig. 7.8 (*continued*)

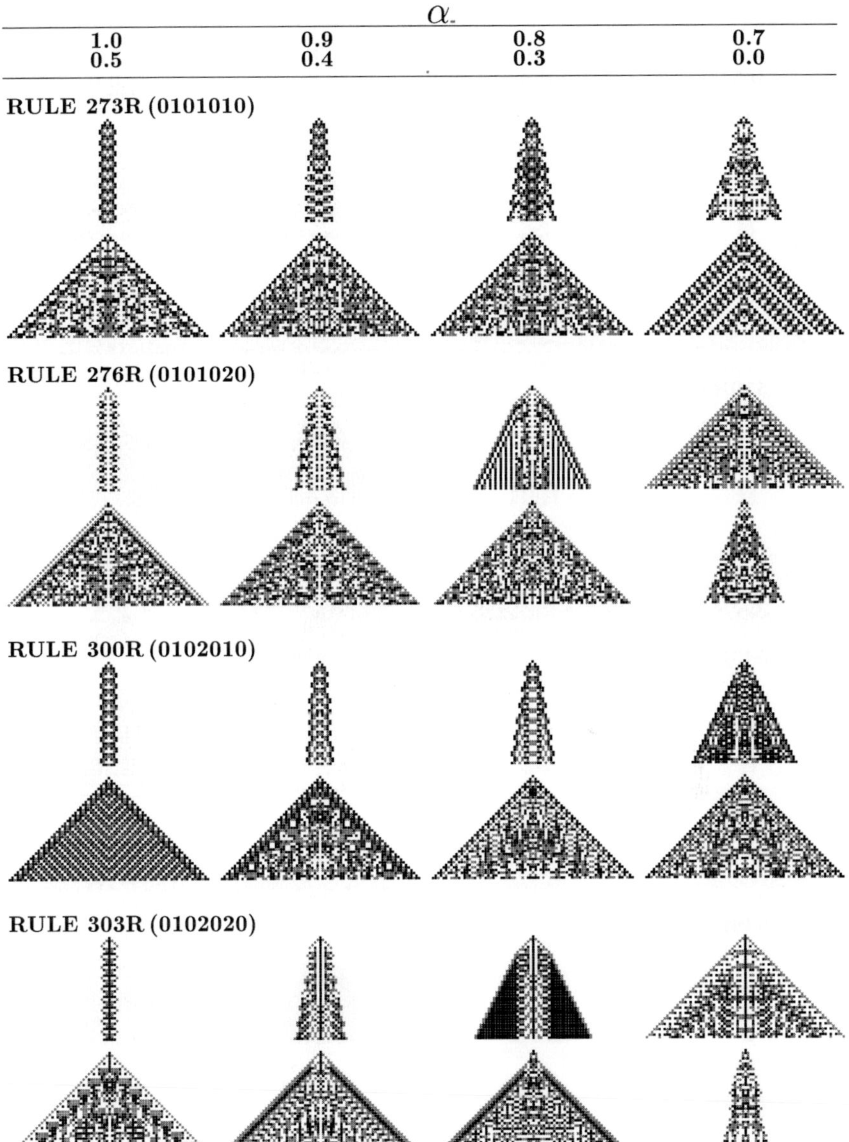

Fig. 7.9 Evolving patterns of the reversible *parity*, $k = 3, r = 1$ rules starting from a single site seed with $\sigma = 1$ at both $T = 0$ and $T = 1$.

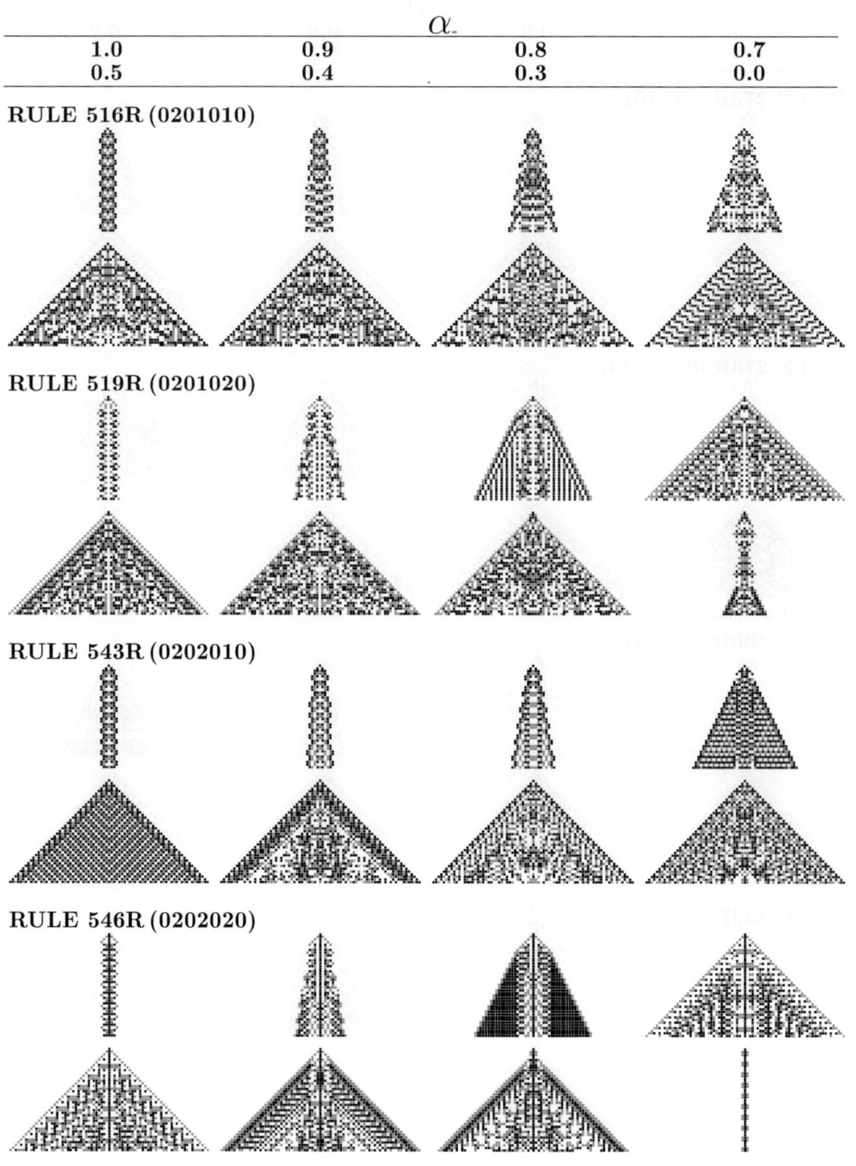

Fig. 7.9   (*continued*)

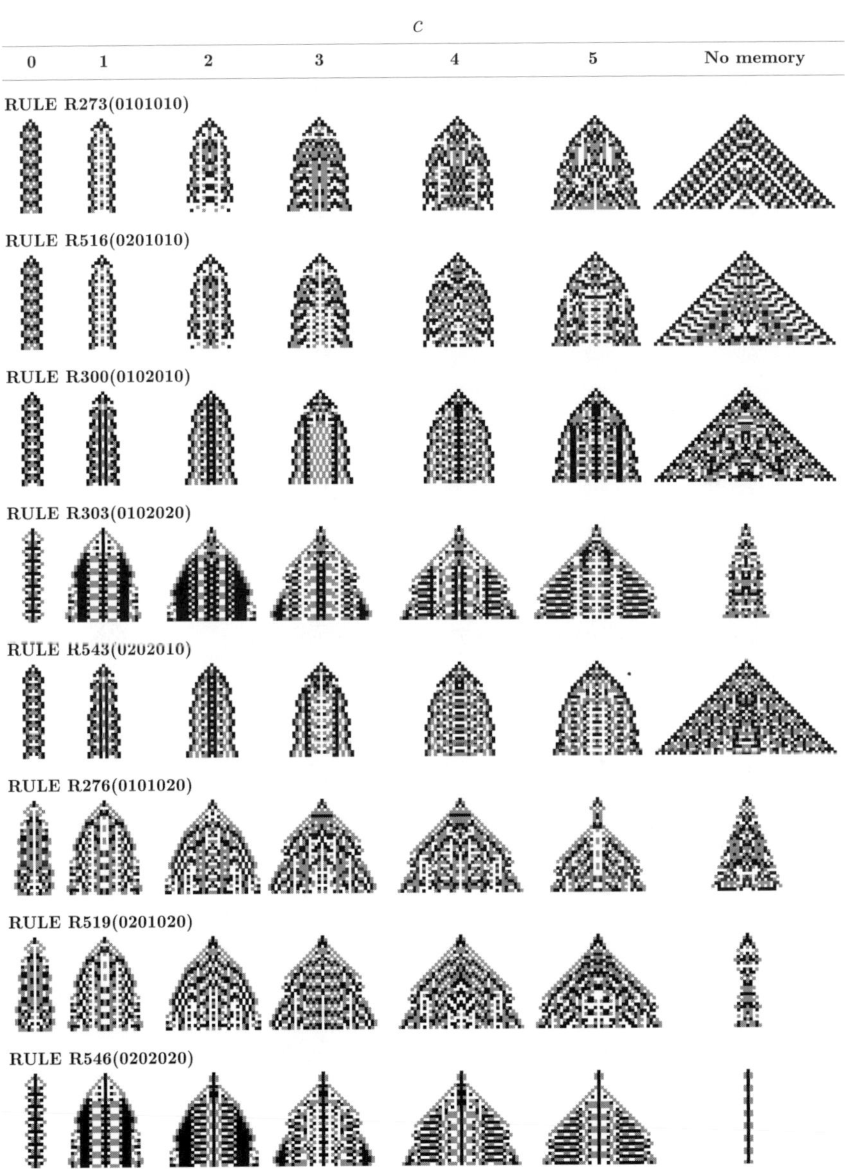

Fig. 7.10 Evolving patterns of the reversible formulation of the $k = 3$ parity rules with $\delta = t^c$ memory in the initial scenario of Fig. 6.2.

Chapter 8

Block cellular automata

8.1 Characterization

A partitioned (or block) CA is a CA with a partitioning scheme such that
the set of cells are partitioned in some periodic way : Every cell belongs to
exactly one block, and any two blocks are connected by a lattice translation.
The update rule of a partitioned CA takes as input an entire block of cells
and outputs the updated state of the entire block. The rule is then applied
alternatingly to the even and to the odd translations.

In the so-called Margolus neighborhood [392], blocks are formed by 2×2
squares of cells in two-dimensional lattices, or simply couples of adjacent
cells in one-dimensional registers. Four block cellular automata rules in the
one-dimensional context are given in Table 8.1 .

Table 8.1   One-dimensional block cellular automata.

Rule I	■■ → ■■	■□ → ■□	□■ → □□	□□ → □■
Rule II	■■ → ■■	■□ → □■	□■ → □■	□□ → □□
Rule III	■■ → ■■	■□ → □■	□■ → ■□	□□ → □□
Rule IV	■■ → □□	■□ → ■□	□■ → □■	□□ → ■■

The far left column of patterns of Fig. 8.1 shows the spatio-temporal
patterns of these elementary one-dimensional block CA starting from a
single full block and starting at random. The evolution in the memoryless
model of the two initially adjacent cells under Rule III in Fig. 8.1, may be
interpreted as of a form of particle collision.

Figure 8.1 also shows the effect of embedding memory of the last three
states and that of unlimited trailing memory.

Rules I, III, and IV are reversible, and rules II and III conserve the

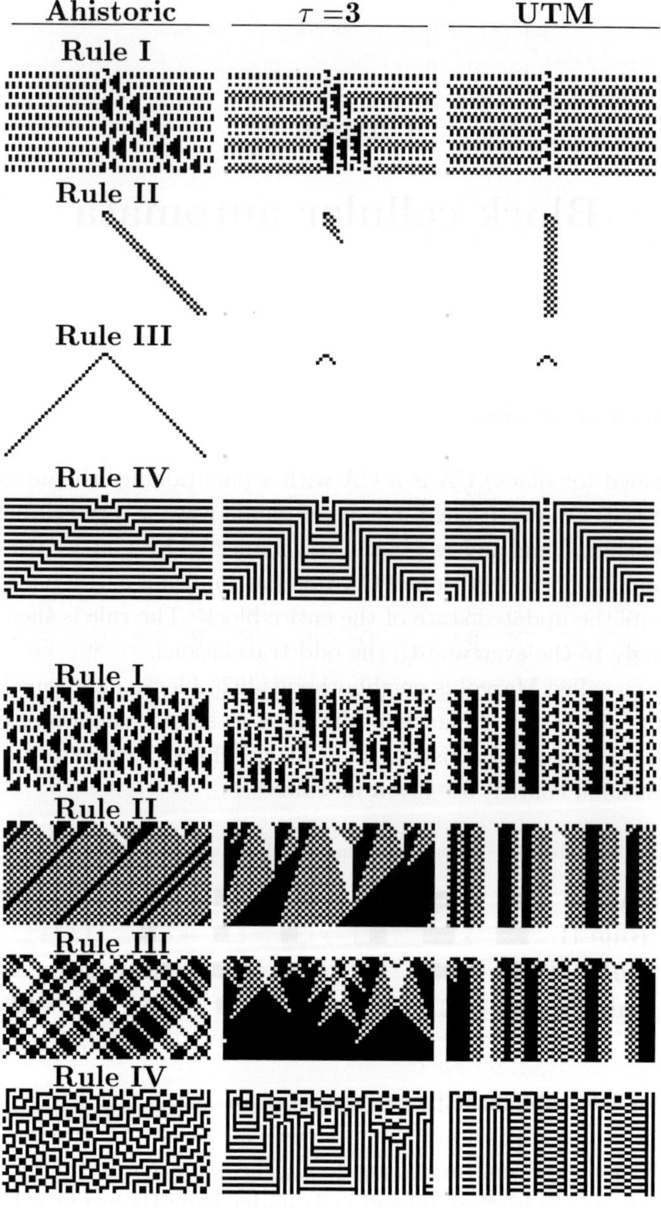

Fig. 8.1   Elementary one-dimensional block CA starting from a single full block (upper), and at random (below).

number of active sites [87, 88], i.e., they keep unaltered the initial density. When implementing the majority memory, these rules lose the number conserving property [154], so that the evolution of the density of active cells (ρ_t) is that shown in Fig. 8.2.

Table 8.2 defines some of the most important two-dimensional block cellular automata, with the lower row showing the evolution of the upper row by rule. Particular attention will be paid here regarding the density classification task to the HPP rule [183, 184], a seminal rule in gas computer modeling. In fact Rule III is the one-dimensional simple version of the HPP rule.

Table 8.2   Two-dimensional block cellular automata. The lower rows show the evolution of the upper row by rule.

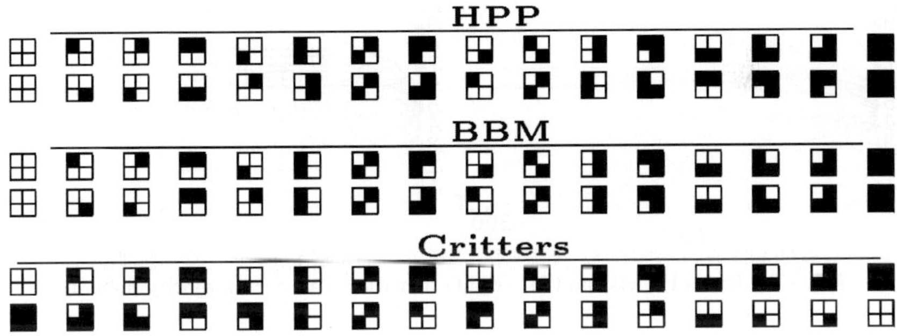

8.2   Density classification task

The density classification task (DCT), also referred as the majority problem, is the problem of finding cellular automaton rules that accurately determine if an initial configuration (IC) has more or less 1s than 0s, i.e., if the initial density (ρ_0) is over or under 0.5. Ideally a correct solution of the DCT must eventually set all cells to zero if $\rho_0 < 0.5$, and must eventually set all cells to one if $\rho_0 > 0.5$. The desired eventual state is unspecified if $\rho_0 = 0.5$.

While solving the DCT is a trivial task for any computational system with central control, this is not the case for fully distributed systems, with local processing, as cellular automata. Solving the DCT has attracted attention in the specialized literature, due to the hope that good results achieved by a specific technique for DCT may be useful regarding other

CA based problems. This is an important test case in measuring the computational power of cellular automaton systems.

Although the DCT was proposed in the seventies, only in 1994 was the solution to the problem, as formulated, proved to be impossible [239]. Thus, research efforts shifted to looking for the best rule dealing with the DCT, i.e., the rule for which the fraction of the possible starting configurations that are correctly classified is the highest.

The DCT may be solved if one relaxes the definition by which the automaton is said to have recognized the initial density [100, 153]. But here we will keep unaltered the specified *natural* criterion: $\rho_* = 0$ if $\rho_0 < 0.5$, $\rho_* = 1$ if $\rho_0 > 0.5$.

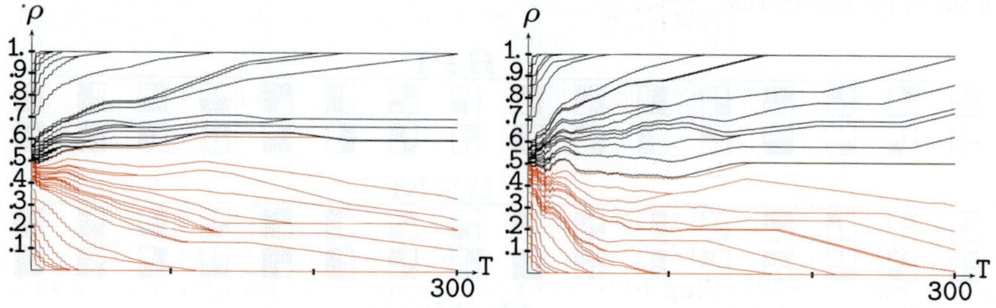

Fig. 8.2   Rules II (left) and III (right) densities with $\tau=3$ majority memory.

One dimension

The dynamics of the density under rules II and III with memory in Fig. 8.2, shows as an overall rule that the patterns tend to drift to fixed configurations either of all 1s or of all 0s depending upon whether or not they lie within the initial configuration $\rho_0 > 0.5$. So far, it is foreseeable that rules II and III, endowed with $\tau=3$ majority memory in cells, will produce good results in classifying density by leading to steady configurations that readily indicate the correct classification. It becomes apparent from Fig. 8.2, that these rules with memory very soon relax to the correct fixed point if the initial density is either $\rho \geq 0.6$ or $\rho \leq 0.4$.

Rule II is equivalent to the elementary CA rule 184, whose properties regarding density classification are studied in Chapter 11. Thus we will focus here on rule III. Two examples of this rule with $\tau=3$ majority memory starting from low and high initial densities are given in Fig. 8.3, showing how this rule with memory readily relaxes in both cases to the fixed point that correctly classifies the initial configuration.

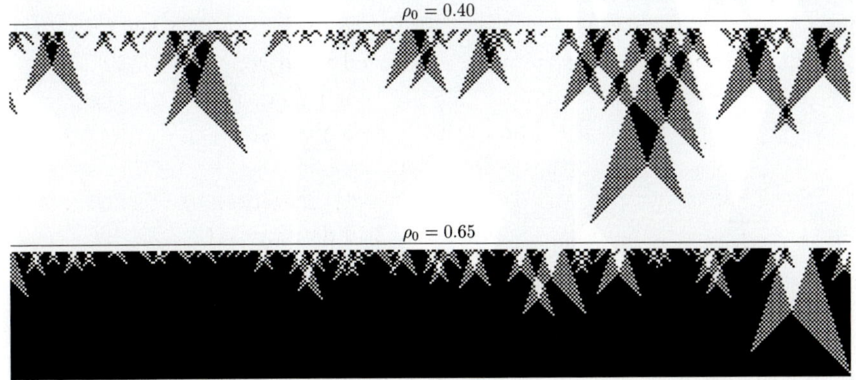

Fig. 8.3 Rule III block CA with $\tau=3$ majority memory, starting at random.

The drift of density is much slower in the oversampled $[0.4, 0.6]$ interval of Fig. 8.2, but the steady-configurations are always reached. This is so even for the particular cases that appear stabilized up to T=300 in the right frame of Fig. 8.2, corresponding to initial actual densities 0.4975 and 0.5025 but coincident after T=150, whose further evolution is shown in Fig. 8.4. Thus only the configuration with initial density 0.4975 in Fig. 8.4 is misclassified by $\tau=3$-Rule III. The spatio-temporal pattern starting from the correctly classified ρ_0=0.5025 is shown in Fig. 8.5, in which the goal of every cell at 1 is achieved unusually late, at T=1530.

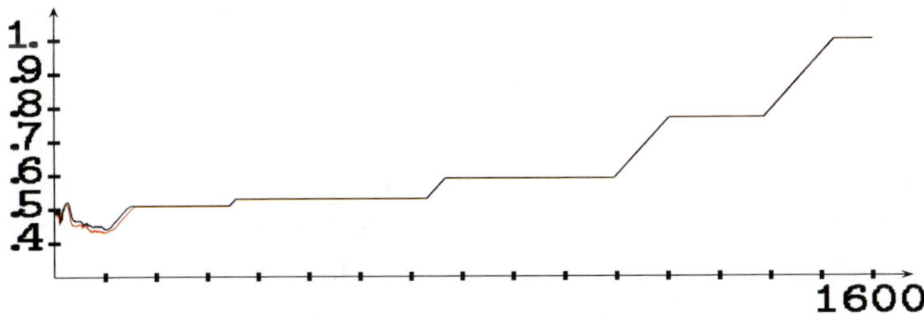

Fig. 8.4 Further evolution of the instances that appear stable in the right frame of Fig. 8.2.

When Rule III with $\tau=3$ majority memory, run up to T=1600, was applied to the simulation of 10 series of 1000 uniformly distributed initial densities in the $[0.0, 1.0]$ interval, reported in Chapter 11, only 66 densi-

Fig. 8.5   Spatio-temporal patterns under Rule III with τ=3 majority memory from a configuration with $\rho = 0.5025$. Evolution up to T=1530.

ties in the $[0, 48, 0.52]$ interval were incorrectly classified, thus, remarkably overcoming the performance of both $\tau = 3$ rule 184 and GKL rules, that misclassified 114 and 160 densities respectively.

As already stated, both GKL and rules with τ=3 mode memory have in common the presence of the mode operation, which seems to be the origin of excellent results in the density task. In fact the majority rule has been implemented in regard to the DCT not only in the perfectly structured CA context, but also in the so called CA on graphs, e.g., in the *small-world* networks in [411], and in the noisy communication between units and asynchronous updating contexts addressed in [293]. This study adopts the majority rule (still with radius three) as a better alternative to the GKL, which shows a poor DCT performance in the afore mentioned non-idealized structure of interactions. The authors support their choice of the majority rule in [293] as -a plausible heuristic to reaching the consensus-, because -it is reasonable to hypothesize that in real-world systems the units make their decisions by using simple heuristics that are robust against errors and do not depend on the precise structure of interactions-.

To test the performance of rules in more challenging scenarios, initial configurations are to be generated *binomially* distributed, i.e., with every cell in the register given a state value equiproblably. In 10^4 binomially generated IC, discarding the 419 IC in which the number of zeros is exactly equal to that of ones, i.e., the unclassifiable $\rho_0 =0.5$, resulted in an efficiency in correct classification (referred to as ϵ_τ) of ϵ_3=81.002 . Increasing the length of the trailing memory up to τ=4, seems to slightly increase the capacity to discriminate density, ϵ_4=81.192 % in our simulation. Beyond this trailing length, inertial effects, characteristic of the majority memory, oppose the drift to all 1s or all 0s intended in the density classification: ϵ_5=78.061 %, ϵ_6=78.238 %, ϵ_7=75.952,%, ϵ_8=75.775 . It seems that τ=3 as length of the majority memory is a good (and simple) choice regarding the density task.

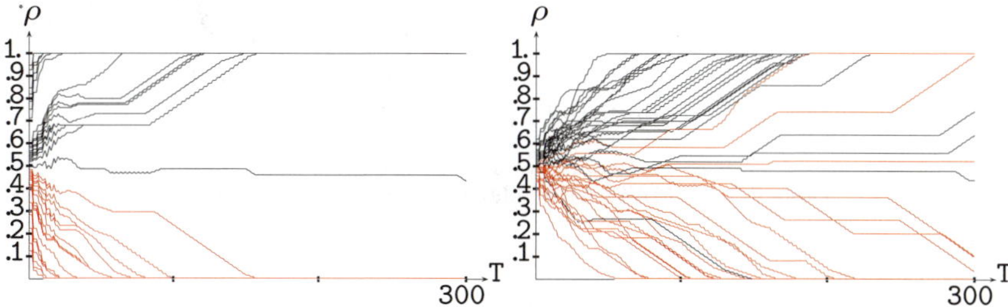

Fig. 8.6   Evolution of the density in rule III with τ=3 majority memory in a n=150 register.   Left: Uniform simulation of initial density, Right: Binomial simulation of initial cell states.

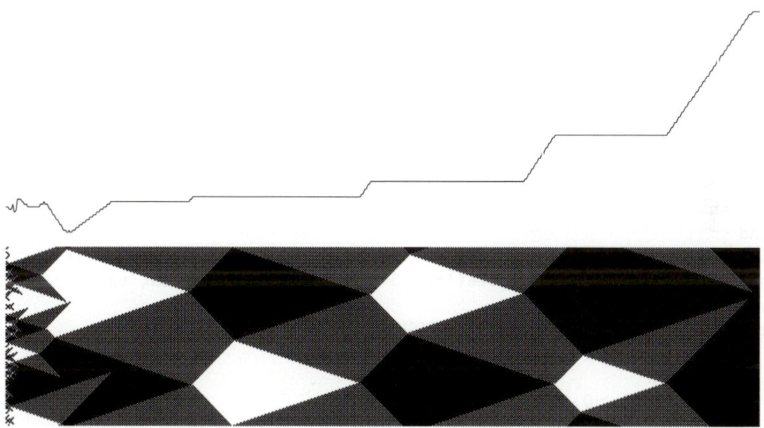

Fig. 8.7   Wrongly classified density ρ_0=0.4933. Patterns and density evolution up to T=650 in a register of size n=150.

In order to compare the block rules with memory with other rules effectively solving the density task, the remaining result reported in this section concerning the one-dimensional scenario apply for registers of size n=149 and n=150. Thus, in Fig. 8.6, with n=150 the τ=3 Rule III proves to be faster in its drift to a steady configuration compared to the n=400 scenario, mainly indicating the correct density classification. Thus only one instance is misclassified in the left frame of Fig. 8.6, with $\rho_0{=}0.5067$, whereas in the binomially simulated IC in the right frame of Fig. 8.6, only three instances are incorrectly classified, one of which is shown in Fig. 8.7.

Figure 8.8 refers to the odd-size n=149. No instance is misclassified in the left frame of Fig. 8.8, and only five instances are incorrectly classified in the binomially IC simulated right frame.

Table 8.3 shows a MATLAB code for Rule III starting at random. The ahistoric evolution, as well as majority memory of lengths $\tau{=}3$ and $\tau{=}4$, are implemented in the code listed in Table 8.3.

Table 8.3   A MATLAB® block cellular automata code.

```
function paca();T=200;N=149;right=[2:N 1];
for tau=2:4
 [SIGMA,SIGMA0] =init(T,N,SIGMA0,tau);
 for t=1:T
    S=SIGMA;ST(t,:)=SIGMA;
    if(tau>2&t>2);SX=ST(t-2,:)+ST(t,:)+ST(t-1,:);
     if(t==3|tau==3);S=(SX>1);end
     if(t >3&tau==4);SX=SX+ST(t-3,:);
       for i=1:N
          if(SX(i)>2)S(i)=1;end;if(SX(i)<2)S(i)=0;end
       end
     end
    end
    ji=1+mod(t+1,2);
    jf=N;if(mod(N,2)==1&ji==1)jf=N-2;end
    for j=ji:2:jf,2;bc=right(j);
      SIGMA(j)=S(bc);SIGMA(bc)=S(j);
    end
 end
 subplot(1,3,tau-1);imagesc(ST);axis('square');colormap(gray);axis('off');
 title('HPP');if(tau==3)title('tau=3');end;if(tau==4)title('tau=4');end
end
function[SIGMA,SIGMA0]=init(T,N,SIGMA0,tau);
  if(tau==2);SIGMA(1:N)=0
    for i=1:N
      x=rand;if(x>0.50);SIGMA(i)=1;end;
    end;
    SIGMA0=SIGMA;
  else;SIGMA=SIGMA0;
  end
```

The results on efficiency in a simulation of one million of binomially generated initial configurations in registers of sizes 150 and 149 (similar sizes to that in [428]), evolving up to T=2000 are shown in Tables 8.4, and 8.5. Table 8.4 concerns majority memory and 8.5 refers to α-memory. In

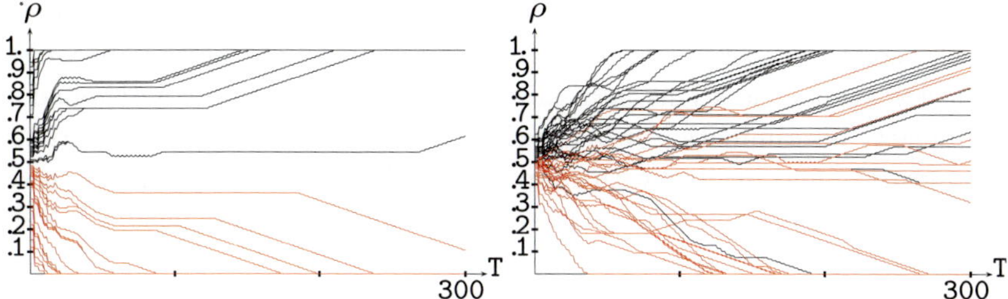

Fig. 8.8 Density evolution in the scenarios of Fig. 8.6, but in a n=149 register.

the n=150 context of Tables 8.4 and 8.5 , $65215 \simeq 6.5\,\%$ configurations were unclassifiable ones with $\rho_0=0.5$.

Table 8.4 Percentage of correctly classified densities in a simulation of one million of binomially generated initial configurations. Rule III with majority memory of length τ in one-dimensional registers of size n .

	$\tau=3$	$\tau=4$	$\tau=5$	$\tau=6$
n=150	85.049	85.641	82.530	82.839
n=149	81.562	82.026	79.152	79.346

Table 8.5 Percentage of correctly classified densities and average convergence time of classified configurations in the same IC as in Table 8.4 under Rule III with α-memory.

	$\alpha=0.51$	$\alpha=0.55$	$\alpha=0.60$	$\alpha=0.65$	$\alpha=0.70$
n=150	88.350 174	86.358 178	86.118 190	84.397 197	83.496 205
n=149	84.049 283	82.632 287	82.459 289	80.511 319	79.779 328

The figures in Tables 8.4 and 8.5 are notable compared to those reported in the literature, nevertheless there are surpassed for the best ones, those reported in [428], which reach $\epsilon_3=88.9$. It turns out that the efficiency in the simulations in registers of even size is higher than in those in the n=149 size registers (in which the GKL rule reaches $\epsilon=81.6\,\%$), and that $\tau=4$ is a slightly better choice than $\tau=3$. As noted above, increasing the trailing length of memory increases the inertial effects that oppose the density discrimination, so that the figures of efficiency in Table 8.4 , are lower for $\tau=5,6$. The small $\alpha=0.51$ emerges as the best choice in Table 8.5 .

A low fraction of the initial configurations turn out to be *unclassified* due to the lack of convergence to any of the extreme configurations. These unclassified configurations are counted as *misclassified* to compute the ef-

ficiency in the density discrimination. An example is given in Fig. 8.9, in
which case the density gets fixed to 0.5 as soon as the pattern becomes
perfectly symmetric. This is usually the kind of dynamics that impedes
the density discrimination, which explains why unclassified configurations
are mostly found dealing with even size registers. In the **n**=150 context
of Table 8.4, the number of unclassified configurations were 6950 $\simeq$ 0.7 %,
4218, 6625 and 4322 from τ=3 up to τ=6. With **n**=149, the number of
unclassified configurations were lower: 1756, 1130, 0 and 3. Nevertheless, misclassified IC, the important numerical component of the not correctly classified configurations, are notably less frequent in the even size
lattice simulations. An example of misclassification with **n**=149 is given in
Fig. 8.10. Both Figs. 8.9 and 8.10 refer to the τ-majority memory context.

Fig. 8.9   Unclassified configuration. Evolution up to T=650. n=149.

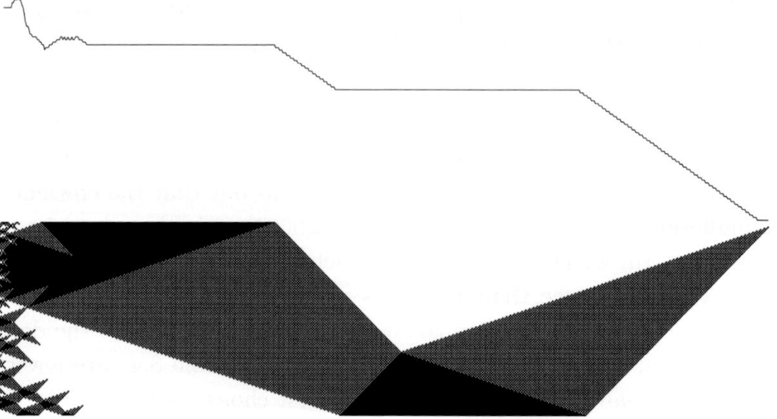

Fig. 8.10   Wrongly classified density ρ_0=0.5370. Evolution up to T=600. n=149.

Two dimensions

Figure 8.11 refers to the evolution of density in two-dimensional lattices [340] evolving under the HPP rule with $\tau=3$ majority memory. The classification under this rule becomes remarkably straightforward, especially if attention is paid to the fact that evolution is shown up to T=100, not up to T=300, as in the equivalent figure in the one-dimensional scenario, i.e., Fig. 8.6. Figure 8.12 shows two examples of the evolution in the first four time-steps, starting from high ($\simeq 90\,\%$) and low ($\simeq 10\,\%$) densities. As soon as memory becomes operative, i.e., at T=4, the patterns are dramatically skewed to follow their correct classifier steady-configuration.

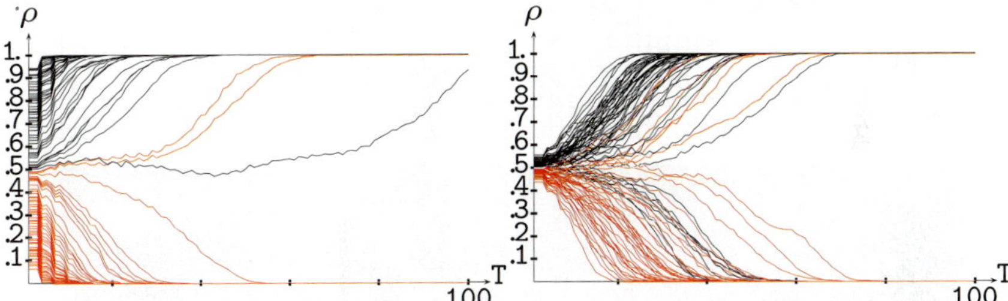

Fig. 8.11   Evolution of the density in the HPP rule with $\tau=3$ majority memory in a 22×22 lattice.

The results on efficiency in a simulation of 10^5 binomially generated initial configurations in 22×22, and 21×21 size lattices (similar sizes to that in [428]), as well as in the bigger lattices of sizes 32×32, and 31×31, are shown in Table 8.6. The automata in Table 8.6 evolve up to T=200 in n=22 lattices, thus one-tenth the time-steps of that in Table 8.4, and up to T=500 in n=32,31 side size lattices. In the n=22 context of Table 8.6, $3593 \simeq 3.6\,\%$ configurations were unclassifiable with $\rho_0=0.5$. The number of IC unclassified due to the lack of convergence to any of the extreme configurations is much lower in the two-dimensional context compared to the one-dimensional, under $0.1\,\%$. The figures for Table 8.6 were 3, 372, 11 and 97 for n=22 (from $\tau=3$ up to $\tau=6$), and 127, 427, 669 and 708 for n=21. The unclassified IC are those that are transformed in stable configurations, often close to the extreme steady configurations as those in Fig. 8.13, generated from $\rho_0=0.523$ and $\rho_0=0.496$, in the $\tau=4$, n=22 scenario.

The efficiencies in the odd-side size lattice Table 8.6 overcomes that of the best rule reported in [316]: $\epsilon \simeq 70.0\,\%$, and that corresponding to $\tau=4$ also surpasses the highest efficiency reported in [428]: $\epsilon=83.27$. In the

Table 8.6   Percentage of correctly classified densities and average convergence time of classified configurations. HPP rule with majority memory of length τ in two-dimensional lattices of size n×n. Simulation of 10^5 binomially generated initial configurations.

	$\tau=3$		$\tau=4$		$\tau=5$		$\tau=6$	
n=22	88.648	43	92.767	42	87.368	52	89.779	48
n=21	84.235	83	87.877	84	82.821	99	84.811	96
n=32	87.171	66	91.131	63	85.798	79	87.744	74
n=31	82.646	130	85.076	130	81.157	157	83.811	152

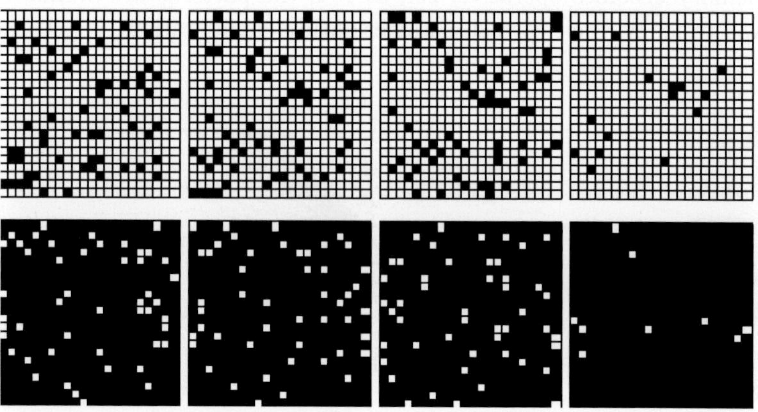

Fig. 8.12   Patterns up to T=4 in the HPP rule with $\tau=3$ majority memory in a 22×22 lattice.

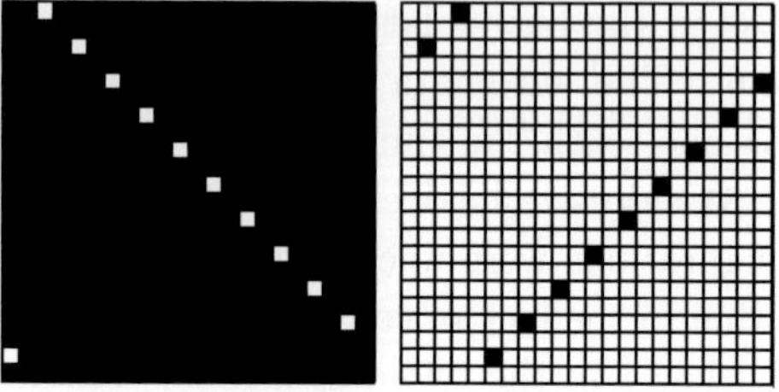

Fig. 8.13   Stable configurations in the HPP rule with $\tau=4$ majority memory in a 22×22 lattice.

odd-size side lattice, the one-dimensional rule III has been applied to the two odd border sides at every time-step. Without this updating of border cells, the efficiency in the density discrimination declines dramatically. The progress made by the HPP rule with memory in the even-size side lattice, in which case the block partitioning covers the whole lattice every time-step, is highly remarkable. The maximum efficiency in Table 8.6 is reached, as in the one-dimensional context, when tracking memory of the four last time-steps. Further comparison to other DCT rules, particularly those reported in [217], will be made in the next section.

Let us stress here that the convergence to the steady configurations is very fast. This is so even also when dealing with binomially generated IC, as already seen graphically with $\tau=3$ in Fig. 8.11 and quantified in Table 8.6. Thus in the odd-side size lattices not only is the efficiency lower than in the even-size, but the convergence to the steady configurations takes longer, approximately double the number of time-steps.

As perhaps foreseeable, as the size of the lattices increases from 21-22 to 31-32, the performance of the rules decreases, and the convergence takes longer. In this comparison, the difference in the convergence time seems notable, but the decay in the classification performance is not too high, lower than 2 %.

Table 8.7 shows the results on DCT efficiency of the HPP rule with α-memory acting on the same IC as in Table 8.6 up to n=32.

Table 8.7   Percentage of correctly classified densities and average convergence time of classified configurations of the same binomially generated IC as in Table 8.6 and 10^4 simulations in the 41×41 and 42×42 lattices. HPP rule with α-memory.

	$\alpha=0.55$	$\alpha=0.6$	$\alpha=0.65$	$\alpha=0.7$	$\alpha=0.8$
n=22	91.652 44	93.869 41	91.959 40	90.766 44	88.544 59
n=21	89.108 94	87.866 88	86.934 90	85.400 93	83.589 138
n=32	92.205 61	91.903 63	90.276 62	88.658 69	86.532 96
n=31	86.612 143	85.357 136	84.848 143	83.772 146	80.591 215
n=42	91.519 82	90.294 87	89.141 87	86.916 96	85.068 136
n=41	84.200 195	84.260 185	83.180 198	81.690 200	80.560 315

Higher α values qualitatively correspond to higher τ lengths of trailing majority memory, whereas $\alpha=0.55$, close to the minimum effective $\alpha = 0.5$, is somewhat similar to the $\tau=3$ majority memory, and $\alpha=0.60$ may be compared to $\tau=4$ majority memory. In any case, α memory appears very effective, in particular with low values of the memory factor, thus incorporating the memory effect but without impeding the drift to the classificaying patterns. The evolution of the density in one hundred IC with binomial simulation of initial cell states, under the HPP rule with

α=0.55 memory in n=31 and n=32 lattices is shown in Fig. 8.14. The three unclassified configurations in the n=32 (left) panel of Fig. 8.14 evolve up to stable (in terms of density) configurations similar to those in Fig. 8.13, thus with density either close to zero or to one.

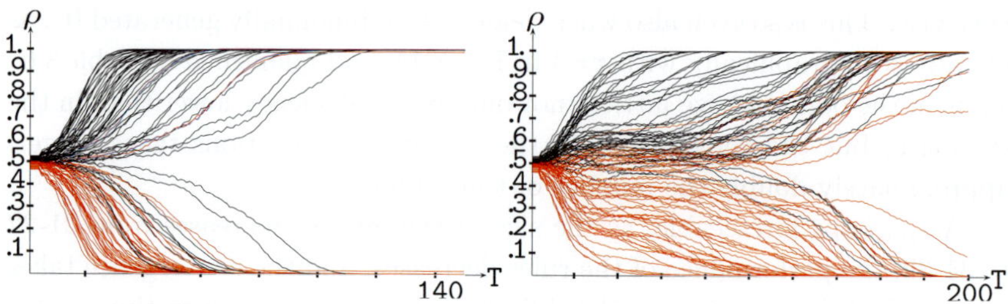

Fig. 8.14   Evolution of the density of one hundred IC with α=0.55 memory in 32×32 (left) and 31×31 (right) lattices.

In parallel with what happens in the τ-majority memory scenario of Table 8.6, the data in Table 8.7 when implementing α-memory, indicate that, i) in the odd-side size lattices the efficiency is lower than in their immediate superior even-size, whereas the convergence to the steady configurations of classified IC takes longer, over double the number of time-steps, ii) as the size of the lattices increases from 21-22 to 31-32, the performance of the rules tend to decrease by less than 2 % (with the exception of $\epsilon_{0.55}$ and n=32 with higher $\epsilon_{0.55}$ than on the n=22 register), whereas the convergence time increases significantly.

The highest efficiencies reported in [217] are those of the so-called 2D GKL, with simulations run up to $2n^2$=881 time steps. For n=21 lattices ϵ=0.873, outperformed by the α=0.55 HPP, run just up to T=300, with ϵ=89.103 in Table 8.7. In the n=31 scenario, 2D GKL achieves ϵ=0.854, which is below the α=0.55 and α=0.60 performances in Table 8.7. Finally, with n=41, 2D GKL achieves ϵ=0.842, comparable to the performances of α=0.55 and α=0.60 in Table 8.7. Incidentally, in the 2D GKL implemented in [217], the one-dimensional GKL rule is applied during one time-step in the horizontal direction and in the next time-step in the vertical direction, an alternating mechanism somehow reminiscent of the characteristic alternating translations of the block CA, which allows for the transmission of information between distant regions of the lattice. The second highest efficiencies reported in [217] for n=21, n=31 and n=41 lattices are ϵ=0.794, ϵ=0.789 and ϵ=0.782 respectively. The HPP rule with the α memory levels in Table 8.7 outperforms these efficiencies. The second efficiencies are achieved in [217] by a majority rule, as might be expected after realizing the importance of the majority rule in the DCT. Nevertheless, it is not a conventional majority rule, but a non-local version with neighbors of every cell picked randomly in its Moore neighborhood of radius $r=10$, another procedure for information transmission.

Although no theoretical attempt is made in this experimental study to explain the good performance of the HPP rule endowed with majority memory in cells, several heuristic reasons may be argued to support it: i) HPP is a number conserving rule, as the reputed DCT-solver elementary rule 184, ii) HPP allows for information transmission as the good classifiers reported in [217] in 2D lattices, and, last but not least, iii) the majority operation seems to be present in most of the good density classifiers, such as the GKL and the non-local majority in structured CA [217], and other generalized majority rules in non-local networks [411] and in noisy environments [293].

Other 2D block cellular automata have been checked regarding the DCT, but showed poor results. Thus for example, in Fig. 8.15, the Billiard Ball Machine (BBM), seems not to be active enough to promote the drift to the extreme steady-configurations, whereas the *Critters* rule (a rule supporting *gliders* [392]), leads to configurations oscillating around 0.5, regardless of the initial density. Both BBM and *Critters* rules are defined in Table 8.2.

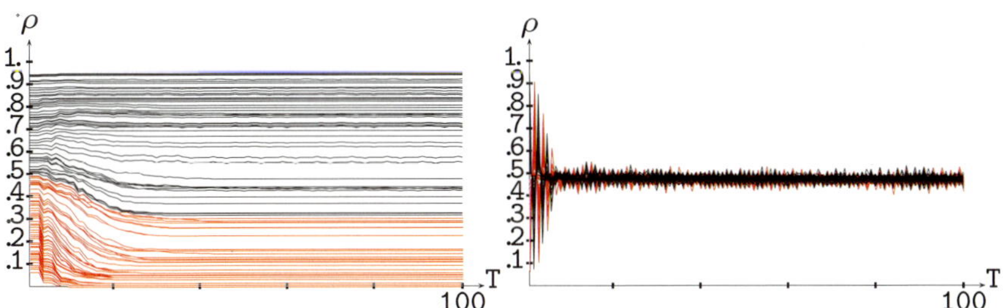

Fig. 8.15   Evolution of the density in the BBM (left) and Critters (right) rules with $\tau=3$ majority memory in a 22×22 lattice.

Evolutionary techniques [217] may help in the search for more efficient, albeit also more sophisticated, partitioned CA with memory density classifiers. Thus, for example non-hybrid Genetic Algorithms, as used in [383] in the context of elementry CA with memory. To obtain high performance solutions, the authors design in [383] a novel representation based on the ternary representation used for Learning Classifier Systems. This representation is found able to produce superior performance to the bit string traditionally used for representing CA.

Chapter 9

Structurally dynamic systems

9.1 Introduction

Structurally dynamic cellular automata (SDCA) were suggested by Ilachinski and Halpern [206, 207]. The essential new feature of this model was that the connections between the cells are allowed to change according to rules similar in nature to the state transition rules associated with the conventional CA. This means that given certain conditions, specified by the *link transition rules*, links between rules may be created and destroyed; the neighborhood of each cell is now dynamic rather than fixed throughout the automaton, so, state and link configurations of an SDCA are *both* dynamic and are continually interacting.

In the Ilachinski and Halpern model, an SDCA consists of a finite set of binary-valued *cells* numbered 1 to N whose connectivity is specified by an $N{\times}N$ connectivity matrix in which λ_{ij} equals 1 if cells i and j are connected; 0 otherwise. So, now: $\mathcal{N}_i^{(T)} = \{j/\lambda_{ij}^{(T)} = 1\}$ and $\sigma_i^{(T+1)} = \phi\left(\{\sigma_j^{(T)}\} \in \mathcal{N}_i^{(T)}\right)$. The geodesic *distance* between two cells i and j, δ_{ij}, is defined as the number of links in the shortest path between i and j. We say that i and j are *direct neighbors* if $\delta_{ij} = 1$, and that i and j are *next-nearest neighbors* if $\delta_{ij} = 2$, so $\mathcal{NN}_i^{(T)} = \{j/\delta_{ij}^{(T)} = 2\}$. There are two types of link transition functions in an SDCA : *couplers* and *decouplers*, the former add new links, the latter remove links. The set of coupler and decoupler determines the link transition rule : $\lambda_{ij}^{(T+1)} = \psi\left(l_{ij}^{(T)}, \sigma_i^{(T)}, \sigma_j^{(T)}\right)$.

Instead of introducing the formalism of the SDCA, we deal here with just one example in which the decoupler rule removes all links connected to cells in which both values are zero ($\lambda_{ij}^{(T)} = 1 \rightarrow \lambda_{ij}^{(T+1)} = 0$ iff $\sigma_i^{(T)} + \sigma_j^{(T)} = 0$) and the coupler rule adds links between all next-nearest neighbor sites in which both values are one ($\lambda_{ij}^{(T)} = 0 \rightarrow \lambda_{ij}^{(T+1)} = 1$ iff $\sigma_i^{(T)} + \sigma_j^{(T)} = 2$

159

and $j \in \mathcal{NN}_i^{(T)}$). All of these totalistic rules are applied at the same time (*in parallel*). The SDCA with these transition rules for connections, together with the *parity* (or *mod* 2) rule for mass values, is considered by Halpern and Caltagirone [190].

Let us consider the case of Fig. 9.1 , in which, again, the initial Euclidean lattice [1] is seeded with a 3×3 block of ones surrounded by zeros. After the first iteration, at time-step $T = 2$, most of the lattice structure has decayed as an effect of the decoupler rule, so that the active value cells and links are confined into a small region. After $T = 6$, the link and value structures become periodic, with a periodicity of two.

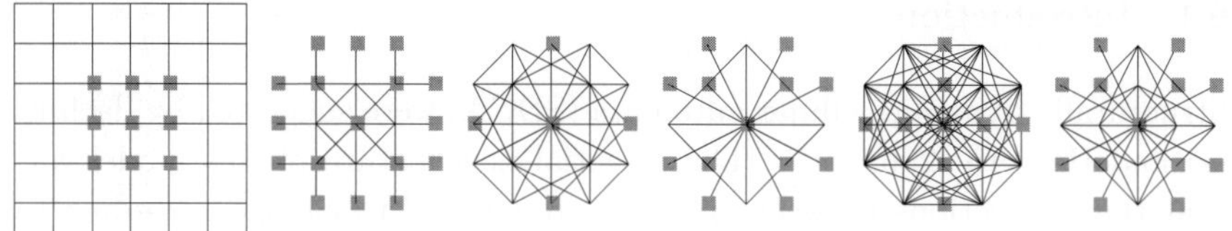

Fig. 9.1   The ahistoric SDCA up to $T = 6$.

The SDCA seems to be particularly appropriate for modeling the human brain function, -updating links between cells imitates variation of synaptic connections between neurons represented by the cells -, in which the relevant role of memory is apparent [2]. Models similar to SDCA have been adopted to build a dynamical network approach to quantum space-time physics [341, 342]. Reversibility is an important issue at such a fundamental physics level. Technical applications of SDCA (and graph theory) may also be traced, so [354] and [345].

Moreover, besides their potential applications, SDCA with memory have an aesthetic and mathematical interest on their own. The study of the effect of memory on CA has been rather neglected and there have been only

---

[1] Von Neumann neighborhood in CA terminology, with next-nearest neighborhood :

[2] Caudle [103] has used conventional 2D CA to explore the hypothesis that ion channels in the membranes of astrocytes (http://en.wikipedia.org/wiki/Astrocite) form a dynamic information storage device. Motion *inertia* is introduced in the cell migration model [67] by - keeping the *memory* of the latest movement of the cell and allowing it to move only towards the three hexagons that are in the forward direction - . Learning and memory, abilities associated with a brain, or, at the very least, neuronal activity, have been observed at cellular level [72, 442].

limited investigations of SDCA since its introduction in the late 1980s [3].
Nevertheless, it seems plausible that further study on SDCA (and Lattice
Gas Automata with dynamical geometry [261]) with memory [4] should turn
out to be profitable.

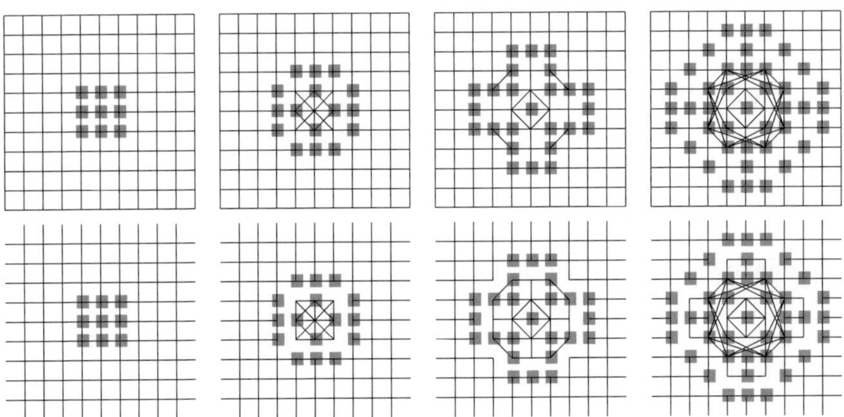

Fig. 9.2 The reversible formulation of the automaton of Fig. 9.1 in a 11 × 11 lattice.
The first row of patterns includes the representation of the toroidal connections of the
border cells; this information is omitted in the second row of patterns.

9.1.1 *Reversible SDCA*

The Fredkin's reversible construction is feasible in the SDCA scenario ex-
tending the $\ominus$ operation also to links: $\lambda_{ij}^{(T+1)} = \psi\left(\lambda_{ij}^{(T)}, \sigma_i^{(T)}, \sigma_j^{(T)}\right) \ominus$
$\lambda_{ij}^{(T-1)}$, together with $\sigma_i^{(T+1)} = \phi\left(\{\sigma_j^{(T)}\} \in \mathcal{N}_i^{(T)}\right) \ominus \sigma_i^{(T-1)}$ [24].

Figure 9.2 shows the evolution of the reversible formulation of the cellular automaton of Fig. 9.1 up to $T = 4$ in a lattice of size 11 × 11 with
periodic boundary conditions. At variance with what happens in the irreversible formulation in Fig. 9.1, the initial Euclidean lattice structure does
not decay at $T = 2$ (nor at posterior time-steps) because of the adding of
the structure at $T = 0$ (at $T - 1$), supposed to be the same that as at $T = 1$.
This adds a new problem into the planar representation of the web of con-

---

[3]To the best of our knowledge, the relevant references on SDCA are [190],[191],[198],
and [266], together with a review chapter in [208] and a section in [7].

[4]Not only in the basic paradigm scenario, but also in SDCA with random but value
dependent rule transitions (which relates SDCA to Kauffman networks [190]), and/or in
SDCA with the extensions considered in [266], such as unidirectional links, also implemented in [342].

nections when, as in our case, periodic boundary conditions are imposed on the edges. This is because the presence of the connections of the border cells, that are linked with their *opposite* ones in the lattice, implies more chance of misunderstanding in the web of connections. In the first row of evolving patterns in Fig. 9.2, the full web of connections is represented. In this case, as an example, the structure of connections of cells around the central zone is easily misunderstood at T=2: these cells seems to remain connected as they began (with a *cross* added), but this is not so. To facilitate the correct understanding of the web of connections, we opted to omit what concerns the border cells, as shown in the second row of patterns in Fig. 9.2. In this representation the links in the central zone, at $T = 2$ for example, are revealed more correctly. But the link representation problem still remains. The central cell, for example, seems to be connected to every cell of its Moore neighborhood at $T = 2$ but this is not so in Fig. 9.2. In fact it is connected only to the *corners* of its Moore neighborhood: the horizontal and vertical segments that cross it connect only its neighbor active cells. Another link representation problem must be taken into account in the reversible implementation, that of auto-connection. Initially every cell is autoconnected (connected with itself) and the link transition rules do not alter this feature. But superposition of patterns leads to the complete disappearance of auto-connections at $T = 2$ in Fig. 9.2. Auto-connections are not represented but of course they affect the mass updating.

We will consider here mostly the Euclidean space, but SDCA in the hexagonal and triangular tessellations are scrutinized in [43] and [25] respectively. Figure 9.3 shows simple examples of the SDCA previously introduced in these non-euclidean tessellations. In the hexagonal context, after T=4 the link and state patterns become a period-two oscillator, whereas in the triangular tessellation at T=5 the state value configuration dies off with only nine links surviving.

Figures 9.4 and 9.5 show the initial patterns of the reversible formulation of the SDCA introduced in Fig. 9.3 .

Figures 9.6 and 9.7 show the patterns starting in Figs. 9.4 and 9.5 at T=13, with and without memory. In both cases, *i)* in the ahistoric case, the web of connections is so dense in the affected area, that it is impossible to discern it, and *ii)* with memory the webs appear dramatically cleared, and the advance of mass is notably restrained, marking again the inertial effect of memory. Incidentally, a likeable *face*, seems to appear in the triangular tessellation with memory if Fig. 9.7.

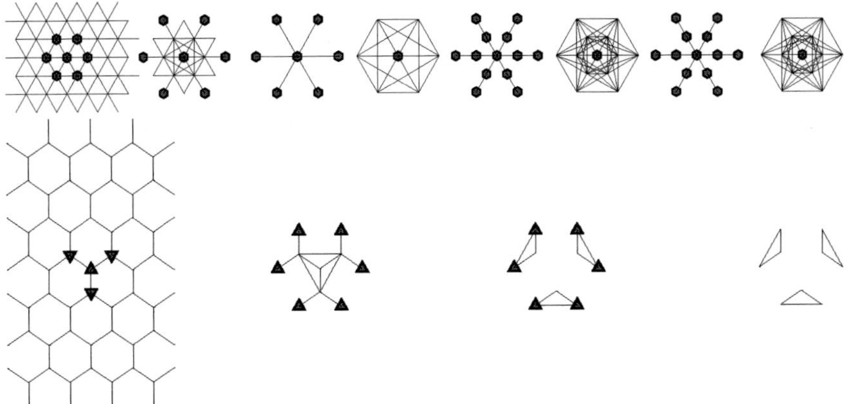

Fig. 9.3   The ahistoric SDCA in the hexagonal and triangular tessellations.

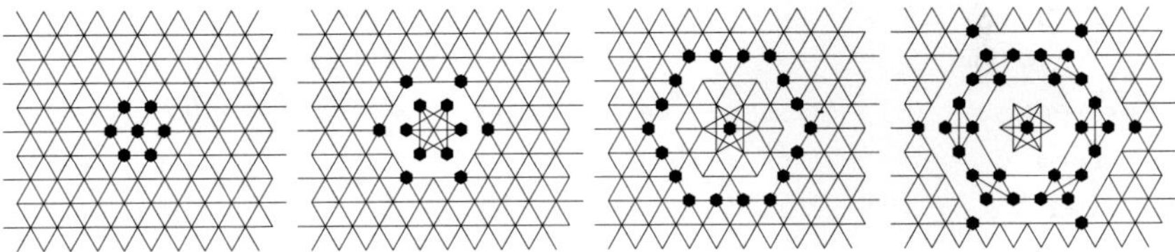

Fig. 9.4   The ahistoric reversible formulation of the cellular automaton in the hexagonal tessellation.

9.2   SDCA with memory

9.2.1   *Two state SDCA with memory*

Memory can be embedded in links in a similar manner as in state values, so the link between any two cells is featured by a mapping of its previous link values: $l_{ij}^{(T)} = l(\lambda_{ij}^{(1)}, \ldots, \lambda_{ij}^{(T)})$. The *distance* between two cells in a historic model (d_{ij}), is defined in terms of the l instead of the λ values, so that i and j are *direct neighbors* if $d_{ij} = 1$, and are *next-nearest neighbors* if $d_{ij} = 2$; $N_i^{(T)} = \{j/d_{ij}^{(T)} = 1\}$, and $NN_i^{(T)} = \{j/d_{ij}^{(T)} = 2\}$. In our approach here, the memory rule for links (l) is the same that of state values (s). Generalizing the approach to embedded memory introduced in section 1.2, the unchanged transition rules (ϕ and ψ) operate now on the featured link and cell state values: $\sigma_i^{(T+1)} = \phi\big(\{s_j^{(T)}\} \in N_i\big)$, $\lambda_{ij}^{(T+1)} = \psi\big(l_{ij}^{(T)}, s_i^{(T)}, s_j^{(T)}\big)$.

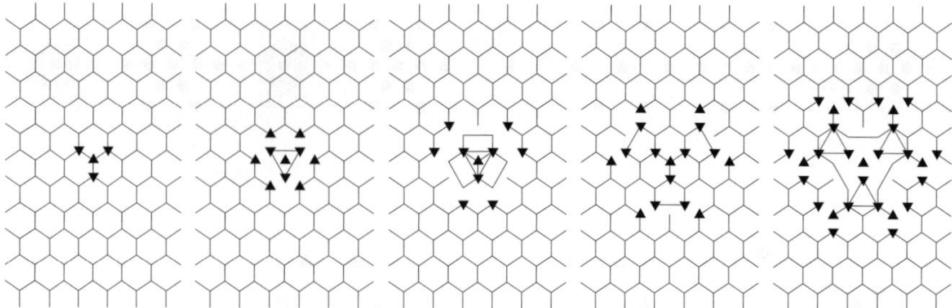

Fig. 9.5   The ahistoric reversible formulation of the cellular automaton in the triangular tessellation.

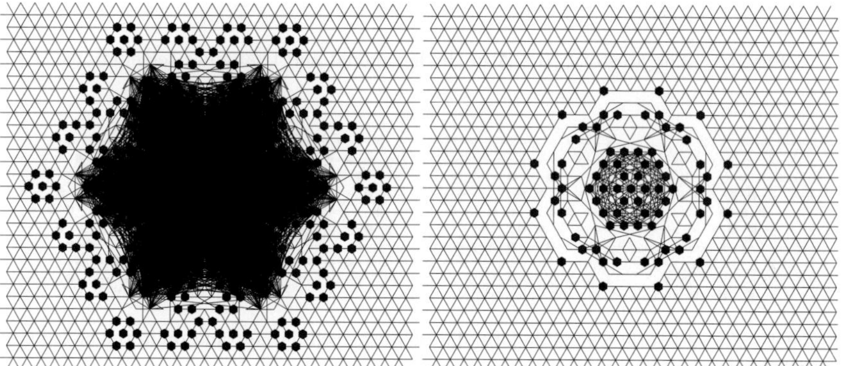

Fig. 9.6   The patterns starting in Fig. 9.4 at $T=13$. Ahistoric (left) and $\alpha=0.6$ memory models.

Figure 9.8 shows the effect of α-memory on the cellular automaton introduced in 9.1 starting as in Fig. 9.1. The last row in Fig. 9.8 shows the evolving patterns of the trait values of cells and links of its immediately above row of actual patterns in the full memory scenario.

Figure 9.9 shows the effect of elementary rules acting as memory [23] on the cellular automaton introduced in section 9.1, starting as in Fig. 9.1.

Reversible SDCA with memory

A generalisation of the Fredkin's reversible construction is feasible in the SDCA scenario endowed with memory as: $\sigma_i^{(T+1)} = \phi(\{s_j^{(T)}\} \in N_i^{(T)}) \ominus \sigma_i^{(T-1)}$, $\lambda_{ij}^{(T+1)} = \psi(l_{ij}^{(T)}, s_i^{(T)}, s_j^{(T)}) \ominus \lambda_{ij}^{(T-1)}$. Now, for reversing from T it is necessary to know not only $\sigma_i^{(T)}$, $l_{ij}^{(T)}$, $\sigma_i^{(T+1)}$, and $l_{ij}^{(T+1)}$, but also $\omega_i^{(T)}$

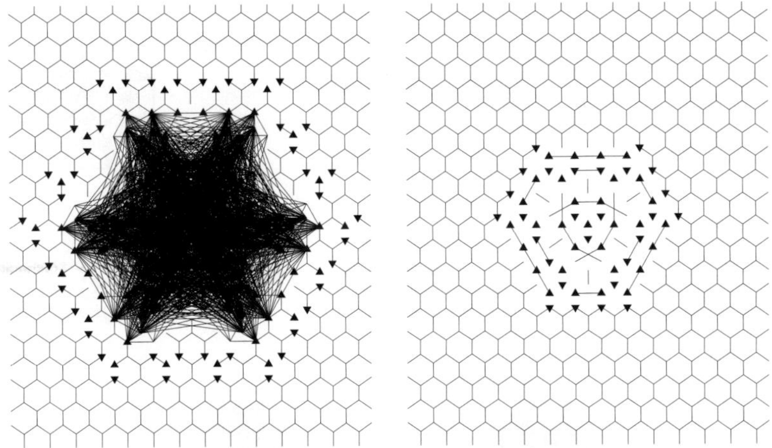

Fig. 9.7   The patterns starting in Fig. 9.5 at $T=13$. Ahistoric (left) and $\alpha=0.7$ memory models.

$$\alpha = 0.6$$

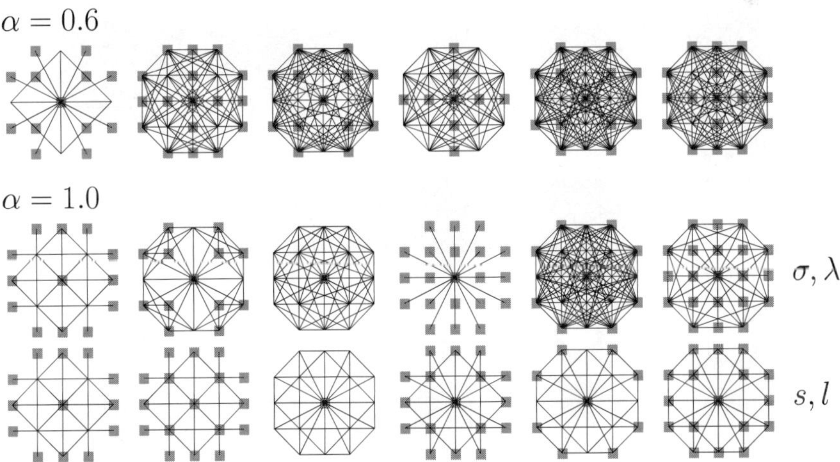

$$\alpha = 1.0$$

σ, λ

s, l

Fig. 9.8   The SD cellular automaton introduced with weighted memory of factor α. Evolution from $T = 4$ up to $T = 9$ starting as in Fig. 9.1.

and $\omega_{ij}^{(T)}$, proceeding for reversing in connections as stated for mass values in section 7.1.

Figure 9.10 shows the initial effect of memory in the initial scenario of Fig. 9.2. The first row of patterns in Fig. 9.10 shows the last patterns (σ and λ), the bottom row shows the weighted mean patterns (s and l). The

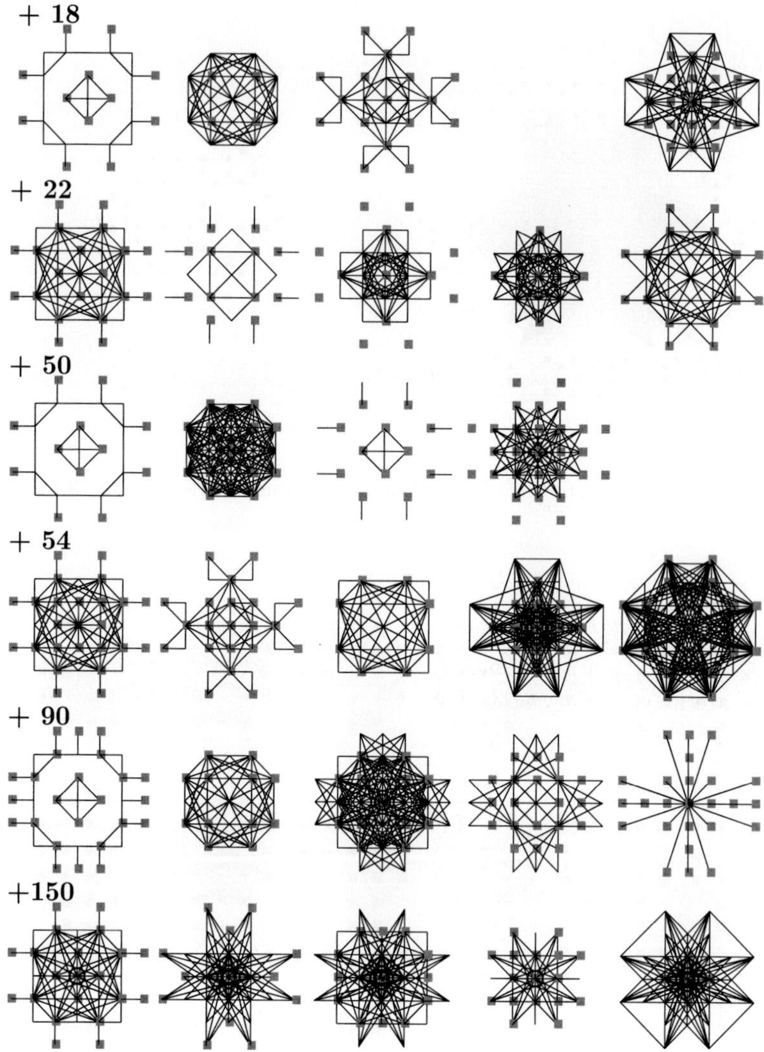

Fig. 9.9   The SD cellular automaton with elementary rules as memory, starting as in Fig. 9.1, from $T = 4$ up to $T = 8$.

patterns of non-null values of the mass ω_i values in the full memory model are shown in between.

Figure 9.11 shows the effect of memory on the evolution of the mass patterns in the initial scenario of Fig. 9.2.

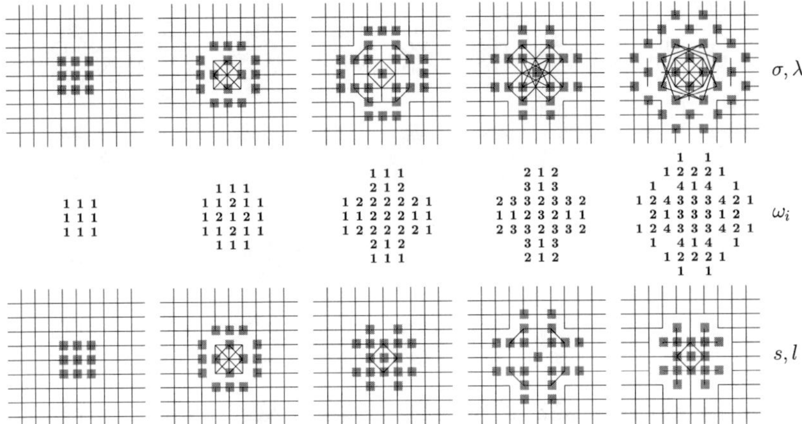

Fig. 9.10 Evolution up to $T = 5$ of the reversible automaton in the initial scenario of Fig. 9.2 with $\alpha \leq 0.7$ memory.

9.2.2 *Three state SDCA*

A plausible wiring dynamics when dealing with excitable CA is that in which the decoupler rule removes all links connected to cells in which both values are at refractory state ($\lambda_{ij}^{(T)} = 1 \to \lambda_{ij}^{(T+1)} = 0$ iff $\sigma_i^{(T)} = 2$ and $\sigma_j^{(T)}=2$) and the coupler rule adds links between all next-nearest neighbors sites in which both values are excited ($\lambda_{ij}^{(T)} = 0 \to \lambda_{ij}^{(T+1)} = 1$ iff $\sigma_i^{(T)} = 1$ and $\sigma_j^{(T)} =1$ and $j \in \mathcal{NN}_i^{(T)}$). In the SDCA treated here, the transition rule for cell states is that of the generalized defensive inhibition rule, excitation interval [1,2] : resting cell is excited if a ratio of excited-and-connected-to-the-cell neighbors to total number of connected neighbors lies in the the interval [1/8,2/8].

The initial scenario of Fig. 9.12 is that of Fig. 6.6 with the wiring network revealed, that of an Euclidean lattice with eight neighbors [5]. No decoupling is verified at the first iteration in Fig. 9.12, but the excited cells generate new connections, most of them lost, together with some of the initial ones, at $T=3$. The excited cells at $T=3$ generate a *crown* of new connections at $T=4$.

Figure 9.13 shows the effect of mode memory in the scenario of Fig. 9.12. The first row of evolving patterns in Fig. 9.13 applies to the actual patterns,

---

[5]Moore neighborhood in CA terminology, in which, the generic cell □ has sixteen next-nearest neighbors :

Fig. 9.11   Mass evolution up to $T = 13$ of the reversible automaton introduced in Fig. 9.2 with increasing α-memory.

the second row shows the evolving patterns of the featuring states of cell and links. Again, memory has an *inertial* effect. Also, in relation to the link dynamics, the initial wiring network tends to be restored as made apparent in the underlying patterns of Fig. 9.13.

But inertial effect does not mean full restraint, as Fig. 9.14 shows. This figure, the ahistoric and mode memory patterns at $T = 20$, again makes apparent the preserving effect of memory.

Memory can be implemented only in the cell state dynamics but not in the link dynamics or vice versa. Figure 9.15 shows the effect of such partial

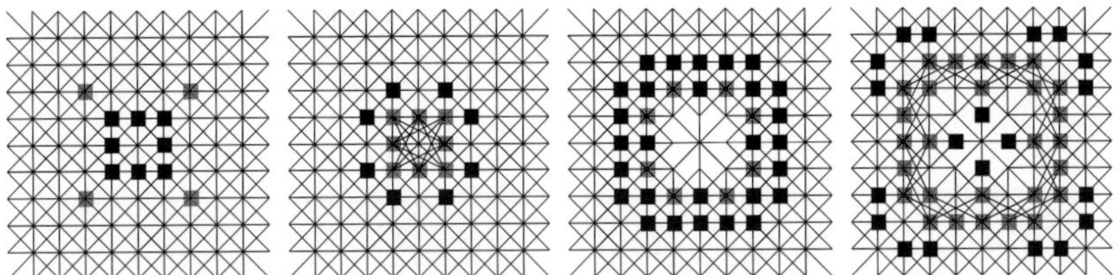

Fig. 9.12   The excitable SDCA described in section 9.2.2 .

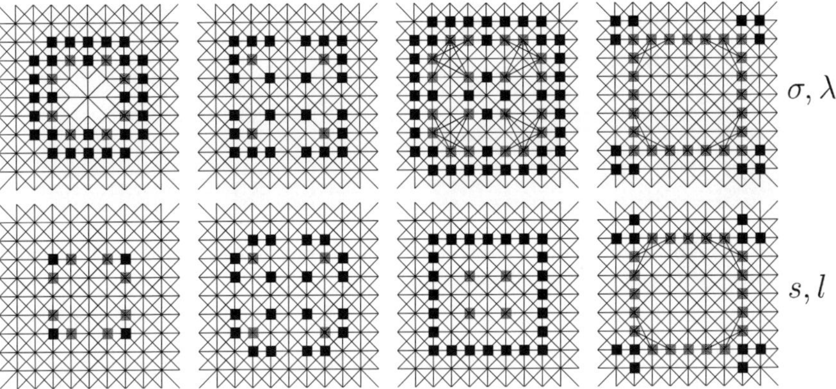

Fig. 9.13   The SD cellular automaton with mode memory. Evolution from $T = 3$ up to $T = 6$ starting as in Fig. 9.12.

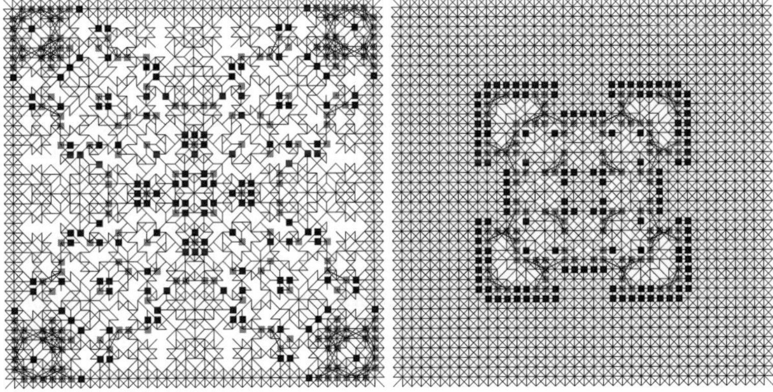

Fig. 9.14   The SD cellular automaton starting as in Fig. 9.12 at T=20, with no memory (left) and mode memory in both cell states and links.

memory implementations starting as in Fig. 9.12. Memory in cells appears much more determinant in preserving the initial features than memory in links.

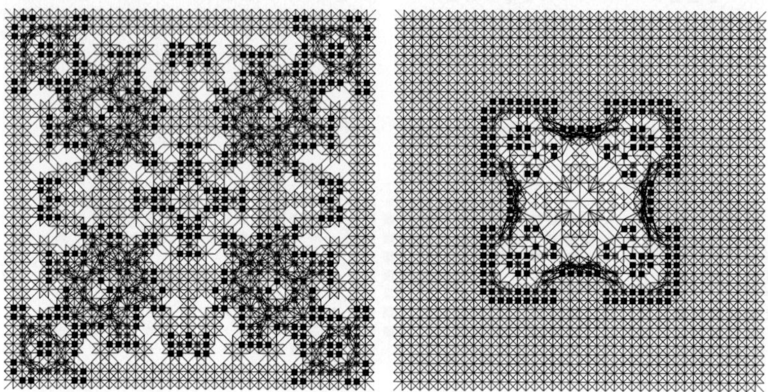

Fig. 9.15 The SD cellular automaton starting as in Fig. 9.12 at T=20, with memory only in links (left) and only in cell states (right).

Figure 9.16 shows the effect of mode memory at $T = 20$ on the SDCA in section 9.2.2 when starting at random with cell state levels in an initially eight-neighbors lattice of size 45×45 with periodic boundary conditions. The wiring of the border cells are not shown in Fig. 9.16 to avoid the situation where the connections of the border cells (linked to their *opposite* ones in the lattice) mask the whole pattern. The ahistoric pattern shows a low density of non-resting cells, together with a irregular tessellation which clearly differs from the initial wiring network, but with a notable absence of links *traversing* the pattern. In contrast, the historic pattern exhibits a higher presence of excited and refractory cells as well as a higher density of links, somehow more reminiscent of the initial structure but much *traversed* by connections created during the evolution with memory.

Figure 9.17 shows the evolution up to $T = 50$ of the excited cell density, the average number of nearest neighbors and next-nearest neighbors per site and the *effective dimension*: the average ratio of the number of next-nearest to nearest neighbors per site [6] (a discrete analogue to the continuous Hausdorff dimension), in the simulations of Fig. 9.16. The evolution in the mode memory model shown in Fig. 9.17 is that of the featured patterns in

---

[6]Starting from an Euclidean lattice with eight neighbors, the effective dimension is 16/8=2.

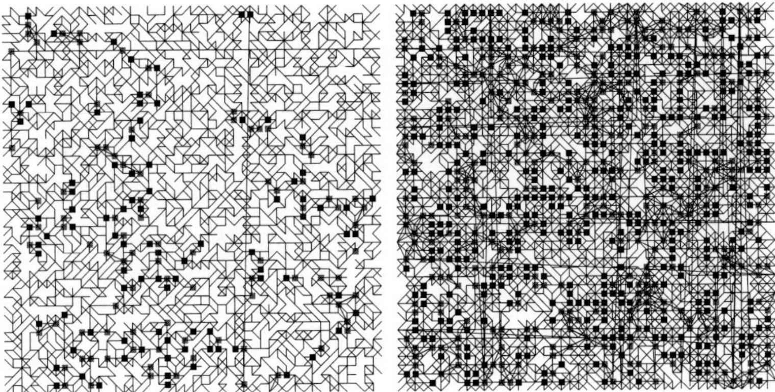

Fig. 9.16   The SD cellular automaton starting at random. Patterns at T=50 with no memory (left), and mode memory.

all the parameters, and also that of the excited cell density and nearest neighbors of the actual patterns, corresponding to the without-dots curves. Again the inertial effect of memory is shown in the evolution of excited cell density and neighbor averages: the ahistoric curves (unjoined dots) stabilize in levels more distant from (lower than) the initial ones than those of the ahistoric. The excited cell density evolution shows a fairly consistent parallelism between the actual (dotted) and featured patterns, whereas the actual and featured average number of nearest neighbors tend to approach each other, after an initial perturbation of the former. The general validity of the form of the curves shown in Fig. 9.17, as well as the aspect of patterns at $T = 20$, has been assessed by running ten more simulations with different random initial configurations.

In the example of Fig. 9.18, the decoupler rule removes all links connected to cells in which neither value is zero ($\lambda_{ij}^{(T)} = 1 \rightarrow \lambda_{ij}^{(T+1)} = 0$ iff $\sigma_i^{(T)} + \sigma_j^{(T)} = 0$) and the coupler rule adds links between all next-nearest neighbor sites in which both values are not dead ($\lambda_{ij}^{(T)} = 0 \rightarrow \lambda_{ij}^{(T+1)} = 1$ iff $\sigma_i^{(T)} > 0$, $\sigma_j^{(T)} > 0$ and $\delta_{ij}^{(T)} = 2$).

In Fig. 9.18 the hexagonal tessellation [7] is initially seeded with a *ring* of $\sigma = 1$ cells. After the first iteration, at time-step $T = 2$, most of the lattice structure has decayed as an effect of the decoupler rule, so that the active value cells and links are confined into a small region. Evolution in Fig. 9.18

---

[7]With next-nearest neighborhood 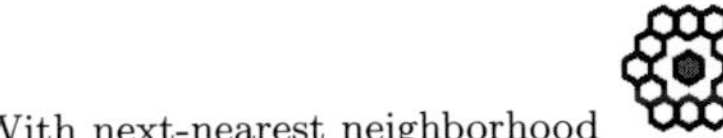 .

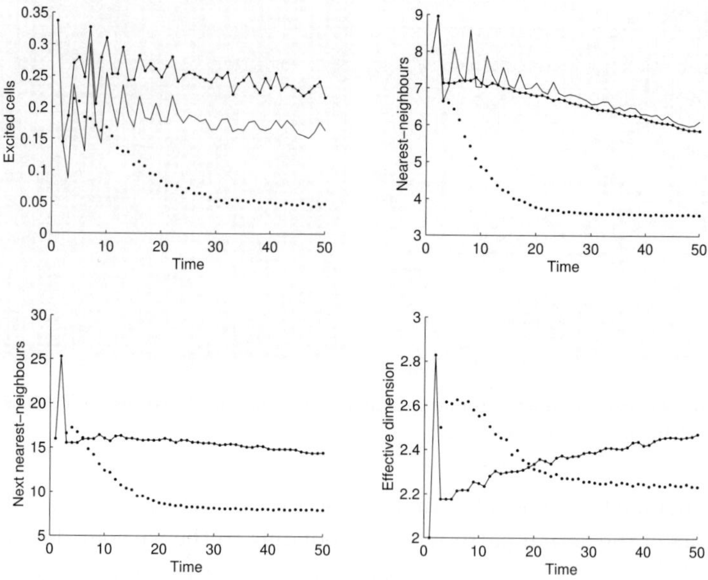

Fig. 9.17   Evolution of excited cell density, average number of nearest neighbors and next-nearest neighbors per site, and effective dimension in the simulation of Fig. 9.16. Ahistoric (unjoined dots) and mode memory (joined dots or unmarked) simulations.

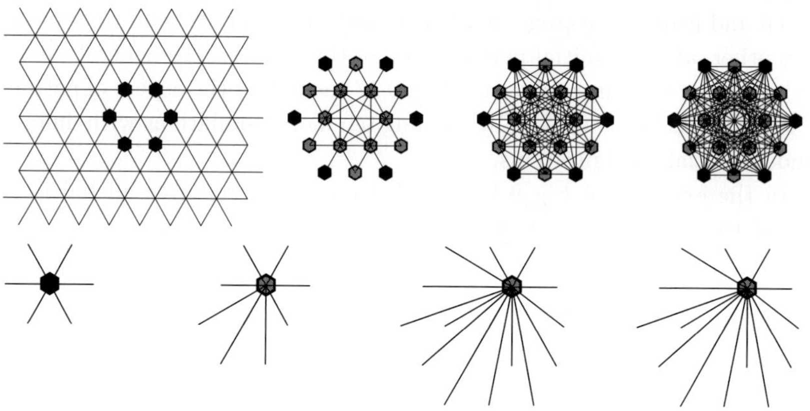

Fig. 9.18   The ahistoric structurally dynamic beehive rule described in text starting from a ring of $\sigma = 1$ cells.

is shown up to $T = 4$, as from this time-step the pattern remains unaltered. By varying the wiring, the frequency of states is not obliged to add up to six. But the *beehive* rule has not been altered, so if the sum of frequencies is over six, the cell value remains unaltered. This explains what happens in Fig. 9.18, where the evolution of a typical non-dead cell at $T = 1$ is given in its lower part.

Ahistoric

Full memory

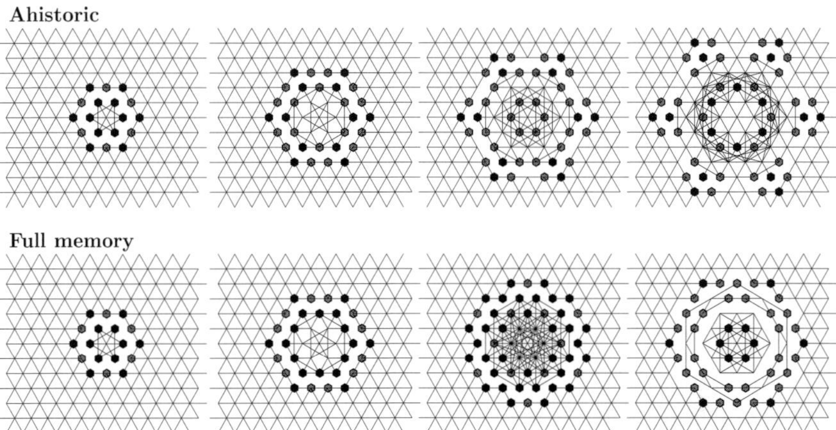

Fig. 9.19   The reversible structurally dynamic beehive rule starting as in Fig. 9.18. Evolution from $T = 2$ up to $T = 5$.

Figure 9.19 shows the evolution of the reversible formulation of the SDCA of Fig. 9.18 up to $T = 5$. At variance with what happens in the irreversible formulation in Fig. 9.18, the initial lattice structure does not decay at $T = 2$ (nor at posterior time-steps) because of the subtraction of the structure at $T = 0$ (at $T - 1$), which is supposed to be the same as that at $T = 1$. Link transition rules do not alter auto-connections, but subtraction of patterns may do so. Thus, for example, in Fig. 9.19 every cell is autoconnected at $T = 0$ and $T = 1$, but the subtraction of these patterns leads to the complete disappearance of auto-connections at $T = 2$. Auto-connections are not represented in figures, but of course they affect the mass updating. To facilitate the visualization, the wiring of border cells is not represented.

Chapter 10

Boolean networks

10.1 Automata on networks

Automata on networks have arbitrary connections, but, as proper CA, the number of inputs per node (K) is constant and they are *homogeneous*, i.e., the same transition rule applies to all nodes [273]. This generalization of the CA paradigm addresses the intermediate class between CA and Boolean Networks (BN, considered in the following section) in which, rules may be different at each site. The dynamics of BN have been extensively studied, but those of CA on networks (also termed *nonlocal* [251] or *disordered* [234, 294] CA) are still far less understood than those of CA and BN.

Random and regular networks are the two topological ends in connectivity types. Both display totally opposite geometric properties. Random networks, also known as the Erdös-Renyi model (ER model), have lower clustering coefficients and shorter average path length between nodes commonly known as *small world* property. On the other hand, regular graphs, have a large average path lengths between nodes and a high clustering coefficients.

In an attempt to build a network with characteristics observed in real networks, large clustering coefficient and a small world property, Watts and Strogatz [412] proposed a model built by randomly rewiring a regular lattice [1]. This Watts and Strogatz (WS) model displays the high clustering coefficient common to regular lattices as well as the small world property seen in the ER model. The WS model interpolates between regular and random networks, taking a single new parameter, the random rewiring degree, i.e., the probability that any node redirects a connection, randomly, to any other. A regular network has no connections rewired, so the connec-

---

[1]Introducing new links between unrelated genes has been proposed as a way to assess how robust and evolvable are genomic networks [78, 211].

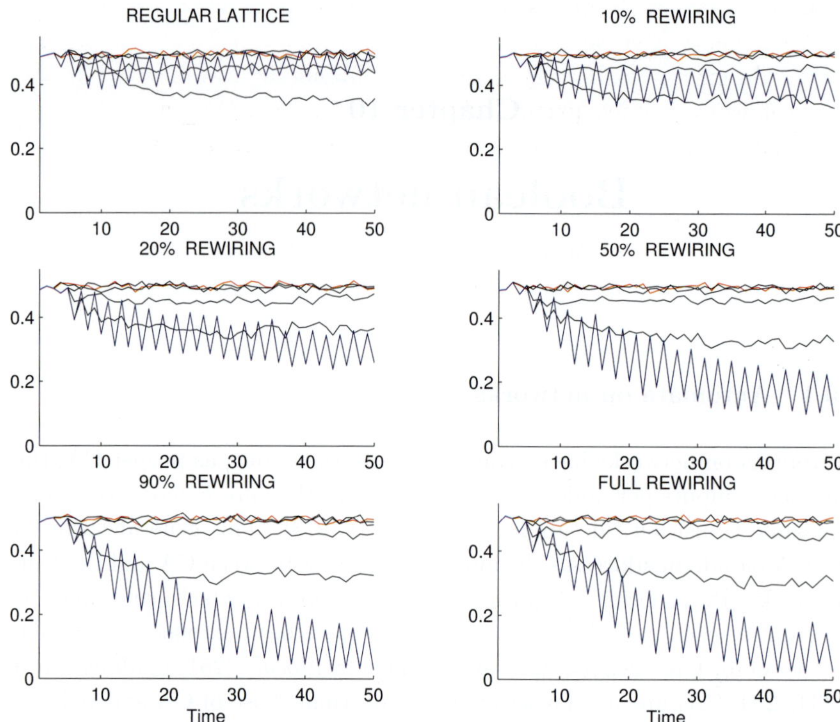

Fig. 10.1   The parity rule with four inputs: effect of memory and topology. Changing rate in a regular graph, graphs with 0.1, 0.2, 0.5, 0.9 and full degree of random rewiring, each considering the ahistoric model (red) and memory models of α levels: 0.6, 0.7, 0.8, 0.9 (black), and full memory (blue).

tions are only between near neighbors, whereas in a fully rewired network all of its connections are randomly rewired.

The underlying lattice structure of the WS model produces a locally clustered network, and the long-range links introduced by the random-ization procedure dramatically reduce the diameter of the network (small world property), even when very few such links are randomly rewired. The small world property has been related to a higher flow in the information transmission. The opinion evolution based on cellular automata rules with memory has been studied in [434].

Figure 10.1 shows the effect of memory and topology on the parity rule with four inputs in a lattice of size 65×65 (with periodic boundary conditions in the case of regular topology), starting at random. This figure shows the relative Hamming distance between two consecutive patterns in the regular graph, in graphs with increasing degrees of rewiring, each considering the *ahistoric* model (red) and memory models of α levels: 0.6, 0.9 and 1.0 (blue). As expected, memory depletes the Hamming distance in relation to the *ahistoric* model, particularly when the degree of rewiring is high. With full memory, quasi-oscillators tends to appear. As a rule, the higher the curve the lower the memory factor α, but in the particular case of the regular lattice (and of the lattice with 10 % of rewiring), the evolution of the distance in the full memory model turns out rather atypical, as it is maintained over some memory models with lower α parameters.

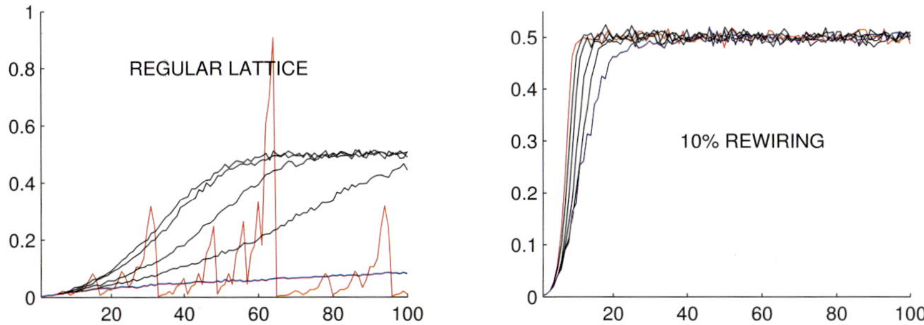

Fig. 10.2   Damage up to $T = 100$ in the parity CA of Fig. 10.1.

Figure 10.2 shows the evolution of the damage spread when reversing the initial state of the 3×3 central cells in the initial scenario of Fig. 10.1. The fraction of cells with state value reversed is plotted in the regular and 10 % of rewiring scenarios. The plots corresponding to higher rates of rewiring are very similar to that of the 10 % case in Fig. 10.2, because of that, they are omitted. Alternatively, Fig. 10.3 shows the evolution of the damage at the first ten time-steps in networks with low degrees of rewiring. Apparently the damage spreads fast once rewiring is present, even to a small degree. That agrees with similar results reported in other contexts, such for example in [244].

Heterogeneous scale free networks

The homogeneous connectivity of regular and random networks is a characteristic that is not realistic. Anti-intuitively, many real networks exhibit high heterogeneity in their connectivity. It is generally assumed cells connectivity in a network, which would fit a bell curve, with most cells having an average of $(\overline{K})$ of links; but this is often not the case where the proba-

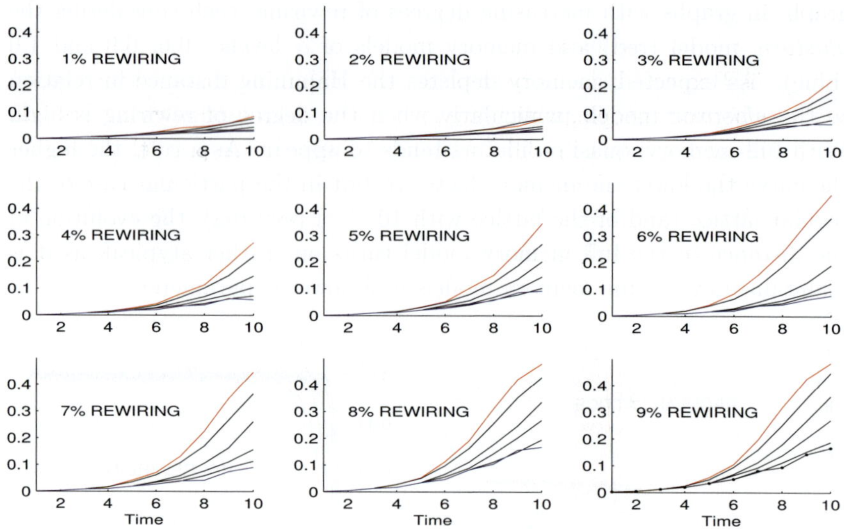

Fig. 10.3   Initial damage in the scenario of Fig. 10.2 with low levels of rewiring.

bility distribution of connectivity ($P(K)$) follows a power law. The power law relationship indicates that there is no such preference, i.e., connectivity is unrelated to scale unless it is characterized by a single value of K. The probability $P(K)$ that an arbitrary cell of the network is connected to exactly K other ones adopt the form $P(K) \sim K^{-\gamma}$, where γ, usually called the *scale free* exponent, may vary from approximately 2.0 to 3.0 in most real systems. Most cells in these networks have low connectivity, but a few (the *hubs*) are highly connected cells, the power law distribution being an indication of that. As their heterogeneity and dynamics are independent of the system size N, these networks are termed *scale-free*.

Real networks grow, and rewiring is hardly the mechanism responsible for the small world. Growth is anything but random; the relationship between new and existing elements in the system is almost always of a nontrivial nature and generates heterogeneous network connectivity in a dynamic growth. Therefore, the small world effect (also featuring scale-free networks) is not the product of probabilistic random rewiring as in the WS model. Rather, the connectivity pattern, in keeping with some manner of the rule -the product of a *tinkering* process [214]-, generates *hubs*, that are the elements in charge of quickly transmitting the flow of information through the network, in which the average path length between cells would be shortened.

An anthropologically motivated interpolation between Barábasi-Albert and Erdös-Renyi models has been proposed [358] in the context of the social relations between people, which seldom follow regular lattice structures and where people prefer to help those who helped them in the past, thus keeping memory of the act. This fact drives the way in which links evolve, an alternative to the *preferential attachment* [73] model in which nodes link to the existing network structure with a probability proportional to the number of nodes previously attached.

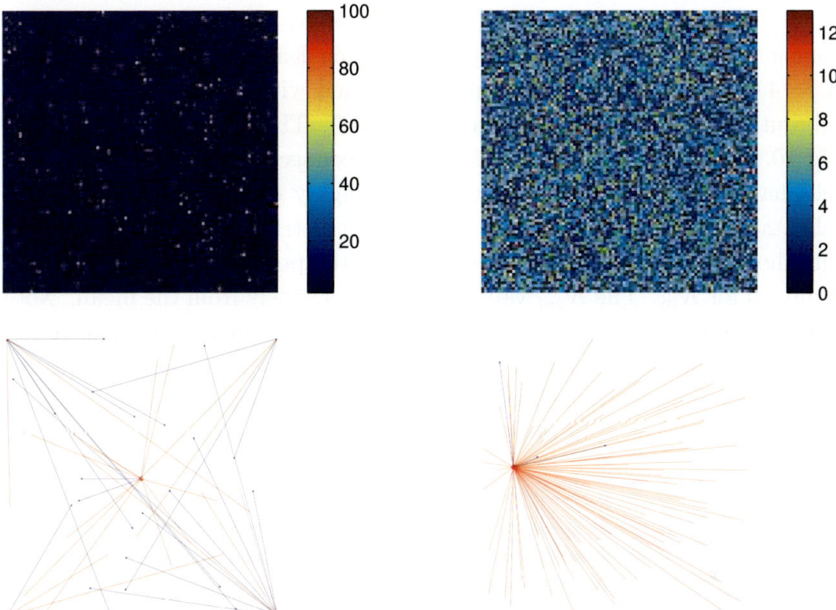

Fig. 10.4   Spatial distribution of K_{in} (upper left) and K_{out} (upper right). Both distributions have a mean of $\overline{K} \simeq 4$. Lower left: connectivity in the corners and central cell; lower right: connectivity of a *hub*. Inputs and outputs are red and blue colored respectively.

Implementation of a scale-free network

With a view to modelling the heterogeneous connectivity present in real scale-free networks, the model proposed here does not use a dynamic growth process, but with a fixed number of nodes N, *in-degree* cell connectivity is defined to be non-homogeneous. The model distinguishes between connections that arrive to a cell (K_{in}) and ones that leave a cell (K_{out}). The network is directed, and the topology of the input connections is typically not the same as the topology of the output connections. To generate the heterogeneity, a simple power law distribution is used to determine the input connections, so the number of inputs of each cell is a random variable that follows the probability distribution $P(K_{in}) \sim K_{in}^{-\gamma}$. When the K_{in} connected cells are chosen randomly from anywhere in the network, the number of K_{out} turns out to be a random variable that follows a Poisson-like distribution whose average is determined by the *scale free* exponent γ^2.

The power-law probability distribution of K_{in} in Fig. 10.4 has an exponent $\gamma = 2.52$ and domain [2,100], which leads to a expected mean $\overline{K} = 4.086$. Simulating the figure (performed with the NAG® G05MZF subroutine) led to an actual mean $\overline{K} = 4.007$. The upper left snapshot in Fig. 10.4 shows the simulation of the input connectivity matrix, with most cells having low values of K_{in} and a few highly connected *hubs* (red-like cells), a pattern characteristic of a power law distribution. The upper right snapshot shows the output connections for the power law distribution determined for K_{in}. The K_{out} values deviate less widely from the mean. Note that K_{in} and K_{out} have different color scale in the figure. The lower left drawing of Fig. 10.4 shows the *in-degree* and *out-degree* (dashed lines) of the corners and central cell of the above network. A *hub* with $K_{in} = 100$ is drawn on the lower right of Fig. 10.4.

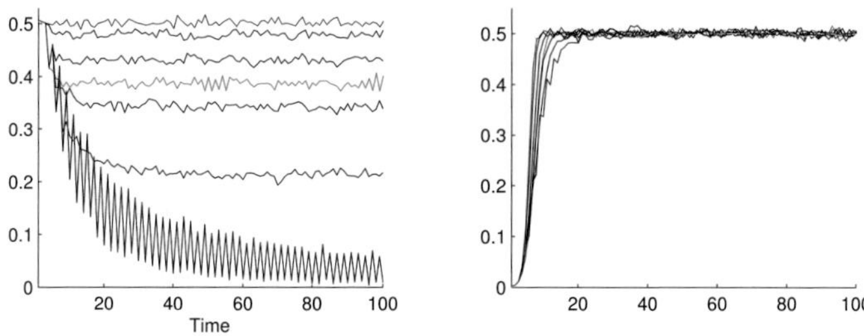

Fig. 10.5   Changing rate (*left*) and damage spreading (*right*) starting at random in the *parity* rule scenario. *Scale free* network with $\overline{K}_{in} \simeq 4$ described in Fig. 10.4. Red, black and blue colors code as in Fig. 10.1. Green color curve corresponds to the $\tau=3$ majority memory model.

---

[2]The case of RBN with nodes with the same number of incoming connections and scale-free outgoing connections has been studied in [364].

Qualitatively speaking, the same effects of memory are found in scale-free networks as in the small-world context when considering the parity rule, as shown in the comparison of Fig. 10.5 to the rewired scenarios in Fig. 10.1 and to Fig. 10.2. In particular, again, memory is not able to control the propagation of damage. The high activity or disorder generated by the *parity* rule could explain this behavior.

Two alternative rule models are implemented to observe the effect of memory in scenarios with a less erratic transition rule: the *threshold* and the *probabilistic* transition rules. In both rule models an activation parameter is computed for every cell, namely its ratio of active cells to the total number of cells connected (K_{in}). In the threshold model if the ratio lies in a given activation interval, the cell becomes (or remains) active. In the probabilistic model the ratio represents the probability of cell activation. Although every cell has its own ratio at every time-step, the transition rule is common to every cell, so these probabilistic automata are not Boolean networks, strictly speaking.

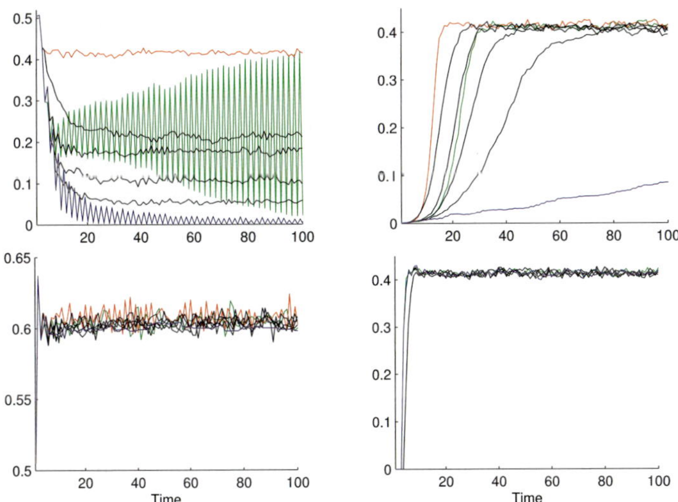

Fig. 10.6   Changing rate (*top left*), damage spreading (*top right*) density (*lower left*) and distance to the ahistoric model (*lower right*) in the scenario of Fig. 10.4 with the [0.25,0.75] threshold transition rule. Color code as in Fig. 10.5.

Figure 10.6 shows the effect of memory and *scale free* topology in the scenario of Fig. 10.4, in the threshold rule model with activation interval [0.25,0.75]. The lower activity of the threshold model with respect to the *parity* rule (in Fig. 10.5) can be observed by comparing the changing rate and damage spreading plateaus reached in the ahistoric model in both scenarios. The percentage of active cells is close to 10 % lower in the threshold model than in the parity model. Approximately the same magnitude of reduction of activity can also be observed in the changing rate in the historic models (left of figure). In the same vein, this decline can be observed by comparing the spread of damage (right plot) between the two rule scenarios in every model with not full memory. In the case of the full memory, the damage spreads slowly, reaching approximately a 10 % of damage at time-step $T = 100$, whereas in the *parity* rule scenario full memory is also unable to control the damage spreading. In the remaining historic models, after a initial containment of the damage propagation, it tends to climb to levels of 40 %, as in the ahistoric model, which is in any event below the 50 % observed in the parity rule in Fig. 10.5. Figure 10.6 shows also the density and the relative Hamming distance between the ahistoric pattern and those historic in the threshold model scenario with the *scale free* topology considered here. Density oscillates around 60 % regardless of the memory level, and, surprisingly, after a very short transition period, the relative distance between the historic scenarios and the ahistoric one hovers around a fairly constant 0.4. An unexpected result appears in the changing rate in the case of $\tau=3$ majority memory, as its corresponding green curve in the top left panel displays an increasing oscillation dynamics that advances to fluctuate between those of the ahistoric and full memory levels. Nevertheless, the damage spreading in this $\tau=3$ memory model, tends, as expected, to be similar to that of with $\alpha=0.8$ memory.

Figure 10.7 shows the effect of memory in the probabilistic transition rule scenario. All the simulations evolve with identical realizations of the stochastic noise, i.e., with equal sequence of random numbers. The upper row of charts corresponds to the initial pattern implemented in the two former simulations, with a initial density $\rho_0 = 0.505$, while the lower row correspond to a initial pattern density of 0.499, just under the 0.5 landmark. In both scenarios, the ahistoric model exhibits fairly consistent density and changing rates dynamics at around 0.5. When account is taken of memory, however, evolution varies dramatically: density tends to one with $\rho_0 = 0.505$, but wanes to nought with $\rho_0 = 0.499$, if $\alpha \in [0.6, 0.9]$ (recall that memory has no remarkable effect on density in the threshold rule model, as shown in Fig. 10.6). Here also, the mode model approaches the $\alpha = 0.8$ dynamics. These features of the dynamics under the probabilistic transition rule shown in Fig. 10.7 were already detected in the homogeneous

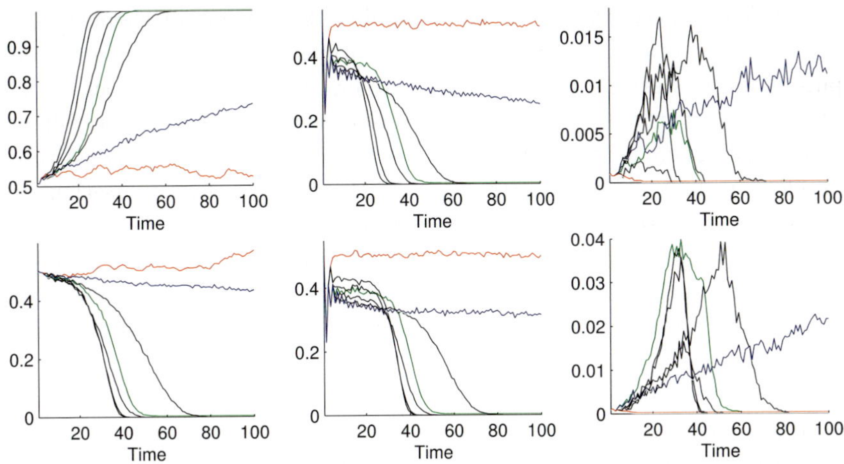

Fig. 10.7 Density (*left*), changing rate (*center*), and damage spreading (*right*) in the probabilistic transition rule scenario. Upper row: initial density $\rho_o=0.505$, lower row $\rho_o=0.499$. Color code as in Fig. 10.5.

$K = 4$ context of Figs. 4.7 and 4.8.

10.2 Boolean networks

In Boolean Networks (BN), instead of what happens in canonical CA, rules may be different at each site: $\sigma_i^{(T+1)} = \phi_i\big(\sigma_j^{(T)} \in \mathcal{N}_i\big)$, that becomes with memory $\sigma_i^{(T+1)} = \phi_i\big(s_j^{(T)} \in \mathcal{N}_i\big)$. Working with *totalistic* rules: $\sigma_i^{(T+1)} = \phi_i\Big(\sum_{j\in\mathcal{N}_i} s_j^{(T)} \Big)$. Boolean networks are also referred to as hybrid, inhomogeneous [401, 369], or non-uniform [370] CA.

In Random Boolean networks (RBN) [222] each cell has K arbitrary connections to other nodes in the network, all update synchronously based upon the current state of those K nodes. Cells may have arbitrary connections. RBN were originally introduced to explore aspects of biological genetic regulatory networks and can be viewed as a generalization of binary cellular automata. Since then they have been used as a tool in a wide range of areas such as self-organization (e.g., [224]), computation (e.g., [144]), robotics (e.g., [336])and artificial creativity (e.g. [129]). Updating in RBN is typically executed in synchrony across the network but asynchronous versions have also been presented (e.g., [193]), leading to a classification of the space of possible forms of RBN [163].

A Boolean network with a finite number N of Boolean elements (or nodes), has a total of 2^N possible configurations. Since the configuration space is finite, starting from any initial configuration, the dynamics of the network will eventually return to previously visited configurations, falling into a cyclic behavior thereafter. The set of configurations that constitute a cycle is usually called an attractor and the number of configurations it contains is the attractor length. The number and the typical length of attractors are important characteristics of RBN. Both magnitudes depend on the automata behavior and this relies essentially on the connection degree K. The attractors typically contain an increasing number of states with increasing K.

Three phases of behaviour are suggested: *i)* ordered when $K=1$, with attractors consisting of one or a few states, *ii)* critical regime around $K_c=2$, where similar states lie on trajectories that tend to neither diverge nor converge and 5-15 % of nodes change state per attractor cycle (see [224] for discussions of this critical regime, e.g., with respect to perturbations), *iii)* chaotic if K is larger than a critical K_c. In this regime the successive configurations are random with respect to the preceding ones and the dynamic is very sensitive to minimal disturbances. In the chaotic regime the number of attractors and their lengths are very large, and the exploration, or evolution probabilities of the system, increase exponentially. On the other hand, if the connection degree K is lower than K_c, the behavior is completely frozen, the automata are insensitive to disturbances, the possibilities to evolve disappear and the number and length of attractors becomes very small compared with the chaotic regime.

But when $K = K_c$ the network behavior changes drastically. The sensitivity to minimal disturbances exists but, the mutations create typically only slight variations in automaton dynamics and only some rare mutations evoke the radical, cascading changes in the automata *programs*. When $K = K_c$, the phase space structure in terms of attractor periods, the number of different attractors and the distribution of the number of configuration that converge into an attractor (basins of attraction), are complex, showing many properties reminiscent of biological networks. This is the behavior at the edge of chaos, at the borderland between chaos and order, when the BN is complex. Kauffman argues that the case $K = K_c$ provides the background conditions for an evolution of genetic cybernetic systems, only in this complex regime have the systems the *ability to evolve* [224–226].

Figure 10.8 shows the dynamics of the changing rate damage spreading of five randomly created (in both rule a state value assignments) $N=1000$

BN, for varying K. Thus, it can be seen how the changing rate increases with K in every frame, and the high propagation of damage in the $K=7$ scenario.

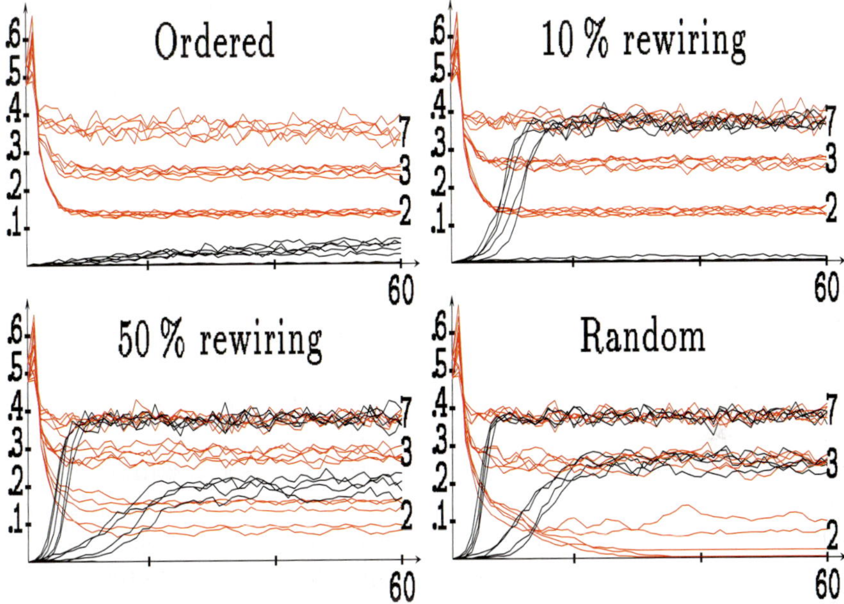

Fig. 10.8 Changing rate (red) and damage spreading from reversion of the state value of a single node (black) in N=1000 Boolean networks with K=2, 3, and 7 (values indicated at right).

As we mentioned before, the key factor in determining the network behavior is the connectivity. Nevertheless, the disorder behavior could be reduced by introducing other parameters. Reduced disorder means returning, or being close, to the system in the ordered regime. The parameter *p-bias* indicates the automaton probability of activation in relation to Boolean functions. When *p-bias*=0.5 the K_c value is 2. The interesting thing is that when *p-bias* is greater than 0.5, the K_c value is greater than 2: the complex behavior is displaced.

The pictures that follows in this section analyze the effect of memory on BN with four inputs ($K = 4$), and rules assigned at random to cells among the pool of 2^5 available rules [3]. In this case the K value will be over the

---

[3]Note that there is no filtering in the selection of rules, not even considering only *quiescent* rules: $\phi(0) = 0$. Simulations considering only quiescent rules render similar

critical K_c since working with totalistic rules, chosen at random without any filtering, fixes the value of *p-bias* to 0.5 ($K_c = 2$).

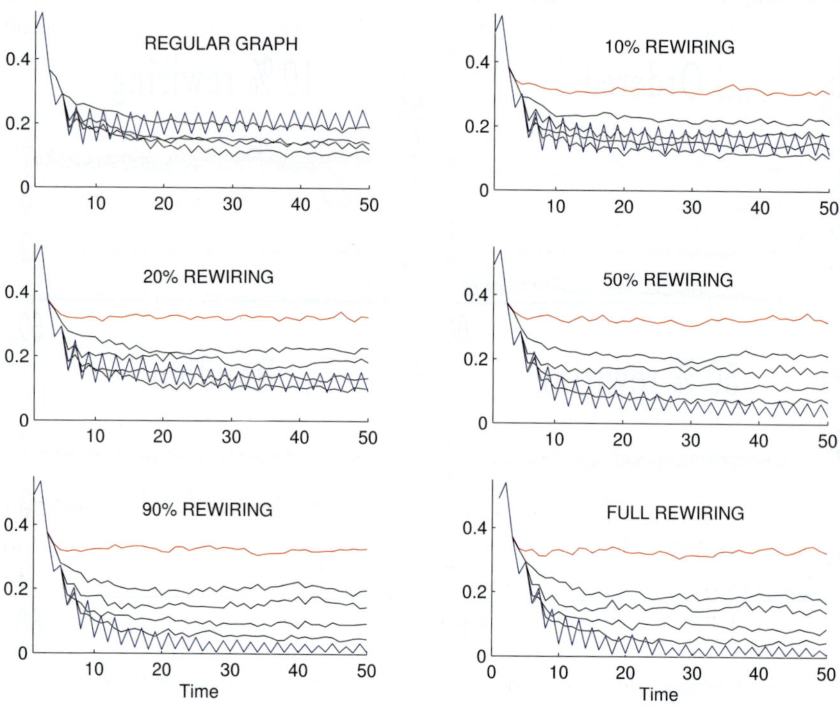

Fig: 10.9   Changing rate in Boolean network with totalistic, $K = 4$ rules in the scenario of Fig. 10.1.

Figure 10.9 shows the changing rate in a BN with totalistic $K = 4$ rules in the scenario of Fig. 10.1. Comparing successive patterns in the dynamics of BN, that evolves in a disordered regime, it is possible to observe the effect of memory as an order parameter. In the *ahistoric* model, the network maintains a fairly high changing rate (around 0.30: slightly under this value in the regular lattice, slightly over it in the rewired ones) this would generate the instability of the system in front of disturbances when evolving in a disordered regime. The interesting thing is that when memory is introduced into the system, memory reduces the Hamming distance compared to the *ahistoric* model. The main features of the effect of memory in Fig. 10.1 are preserved in Fig. 10.9: *i*) the ordering of the *historic* networks tends to be

---

results to that reported here, albeit not fully coincident.

stronger with a high memory factor, *ii*) with full memory, quasi-oscillators appear (it seems that full memory tends to induce *oscillation*), *iii*) in the particular case of the regular graph (and to a lesser extent in the networks with low rewiring), the evolution of the full memory model turns out to be rather atypical, as it is maintained over some of those memory models with lower α parameters.

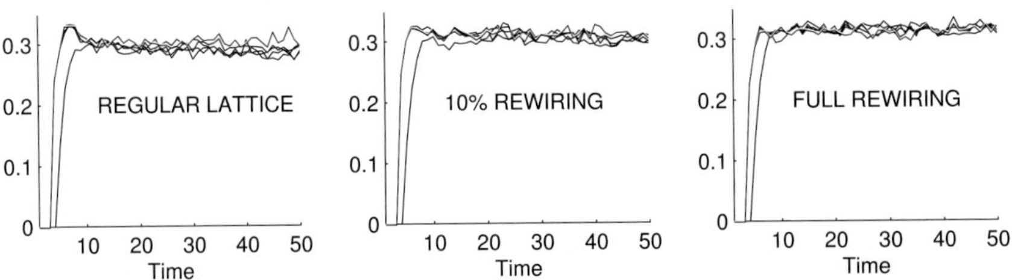

Fig. 10.10   Distance between the ahistoric patterns and those with memory in the scenario of Fig. 10.9.

Figure 10.10 shows the relative Hamming distance between the *ahistoric* patterns and those *historic* in a regular lattice, and lattices with 10 % and full connections rewired, considering memory models of α levels from 0.6 to 1.0 by 0.1 intervals. Surprisingly, the relative Hamming distance between the historic-ahistoric scenarios tends to be fairly constant around 0.3, after a very short transition period.

Figure 10.11 shows the patterns of changes of Fig. 10.9 at $T = 50$. The activity remains in the *ahistoric* model regardless of the topology, but full memory induces a depletion in the activity pattern, null in the case of full rewiring.

Figure 10.12 shows the evolution of the damage when reversing the initial state of the 3×3 central cells [40, 42, 262]. The fraction of cells with state value reversed is plotted in the ahistoric model and with $\alpha = 0.6, 0.7, 0.8, 0.9$ and full memory models. As a rule in every frame, corresponding to increasing rates of random rewiring, the higher the curve the lower the memory factor α.

The *damage vanishing* effect induced by memory does result apparent in the regular scenario of Fig. 10.12, but only full memory controls the damage spreading when the rewiring degree is not high, the dynamics with the remaining α levels tend to the damage propagation that characterizes the *ahistoric* model. Thus, up to a 10 % of connections rewired, full memory notably controls the spreading, but this control capacity tends to disappear

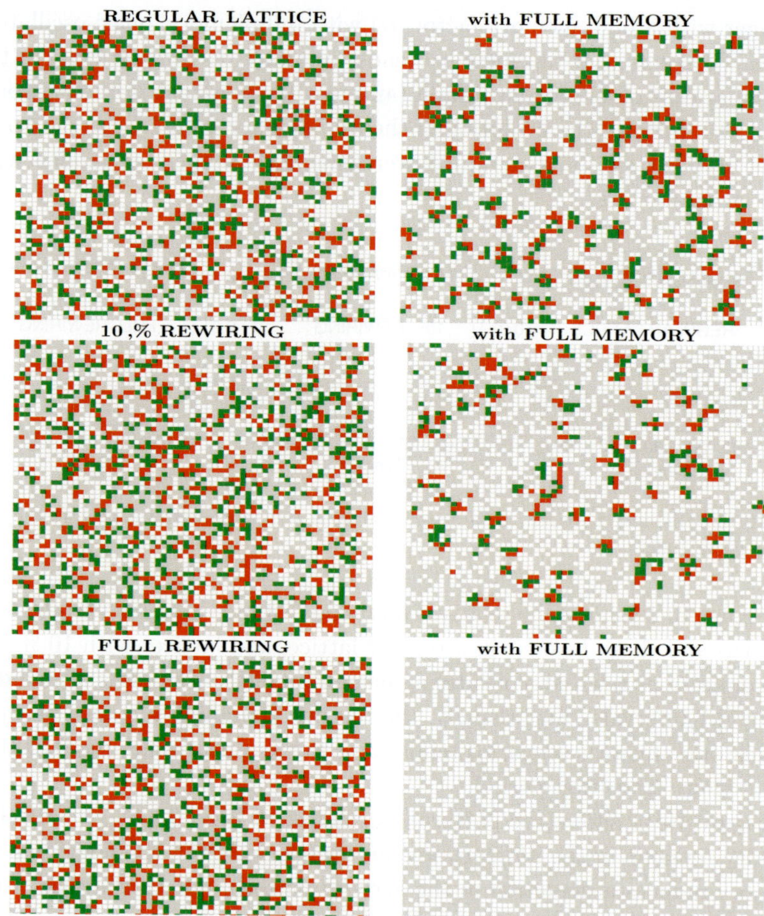

Fig. 10.11   Changing patterns at $T = 50$ in Fig. 10.9. $0 \to 1$ ■, $1 \to 0$ ■.

with higher percentage of rewiring connections. In fact, with rewiring of 50 % or higher, not even full memory seems to be very effective in altering the final rate of damage, which tends to reach a plateau around 30 %[4] regardless of the scenario. The damage control effect of memory turns out minimized when the network topology has a high percentage of connections rewired.

Figure 10.13 shows the initial damage and damage at $T = 3$ in the

____

[4]Notably coincident with the percolation threshold in site percolation in the simple cubic lattice [380], and the critical point for the nearest neighbor Kaufmann model on the square lattice [379] : 0.31 .

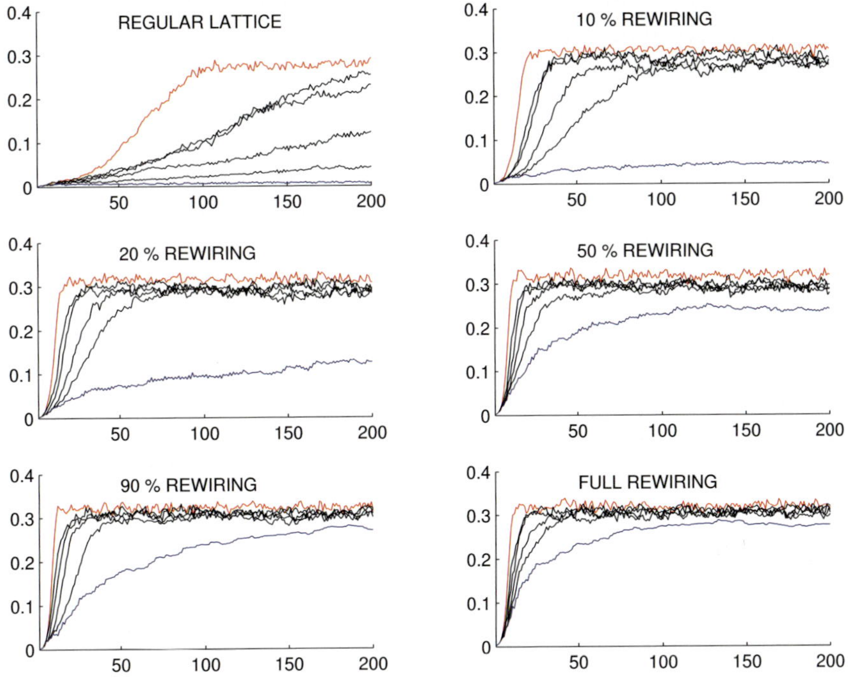

Fig. 10.12 Evolution up to $T = 200$ of the damage when reversing the initial state of the 3×3 central cells in the scenario of Fig. 10.9.

regular and random wiring scenarios of Fig. 10.12, and Fig. C.1 in the Appendix C displays the damage at $T = 200$ when in the regular lattice scenario of Fig. 10.12. In the *ahistoric* model, the damage spreads through the whole lattice. But, when memory is introduced, the damage tends to decrease, notably when the memory degree is high ($\alpha \geq 0.8$).

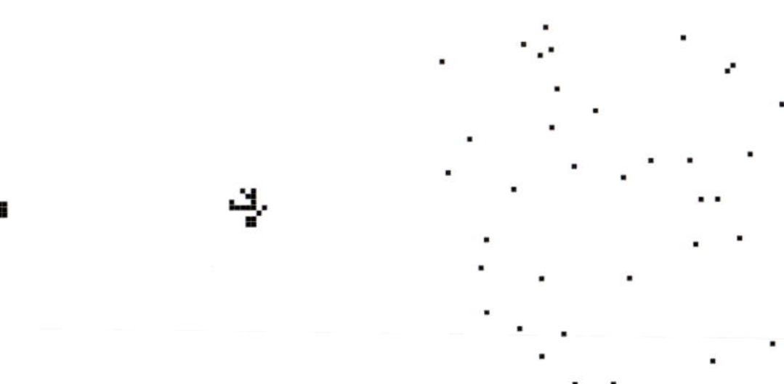

Fig. 10.13 Initial damage and damage at $T = 3$ in the regular (center) and random wiring (right) scenarios of Fig. 10.12.

Figure 10.14 relates to Fig. 10.12 by showing the effect of memory on the damage spreading, but in a bigger lattice: that of size 100×100. In the regular topology scenario, the damage spreads in the *ahistoric* model in a smoother way and its stabilization is obtained later compared to the previous smaller system (65×65). This indicates a size affect in the propagation of memory effect. Nevertheless, when a low degree of random rewiring is introduced (10 %) the system has a behavior similar to the small system in the same rewiring scenario. Again, the small world effect could explain this: with a low random rewiring, the size does not matter so much as in the case of the regular topology.

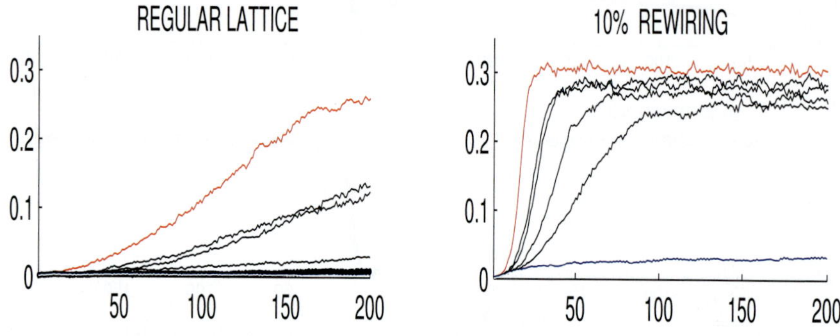

Fig. 10.14   Damage in the scenario of Fig. 10.12 but in a 100×100 lattice.

Figure 10.15 shows the return maps up to $T{=}60$ of six simulations (coded with different colors) of random BN in the ahistoric scenario, with unlimited majority memory and with unlimited parity memory. Unlike in the return maps of Chapter 5, e.g., Fig. 5.1, the (x_T, x_{T+1}) trajectories are joined in Fig. 10.15, which makes its snapshots reminiscent of Jackson Pollock's artwork [102], maybe *linearized*. The restraining effect of majority memory (center column) becomes apparent even with high connectivity. Conversely, the activating effect of parity memory (far right) also acts in the simple $K{=}2$ scenario.

10.3   Automata on proximity graphs

Automata on β-skeletons

In computational geometry and geometric graph theory, a β-skeleton or beta skeleton is an undirected proximity graph defined from a set of points

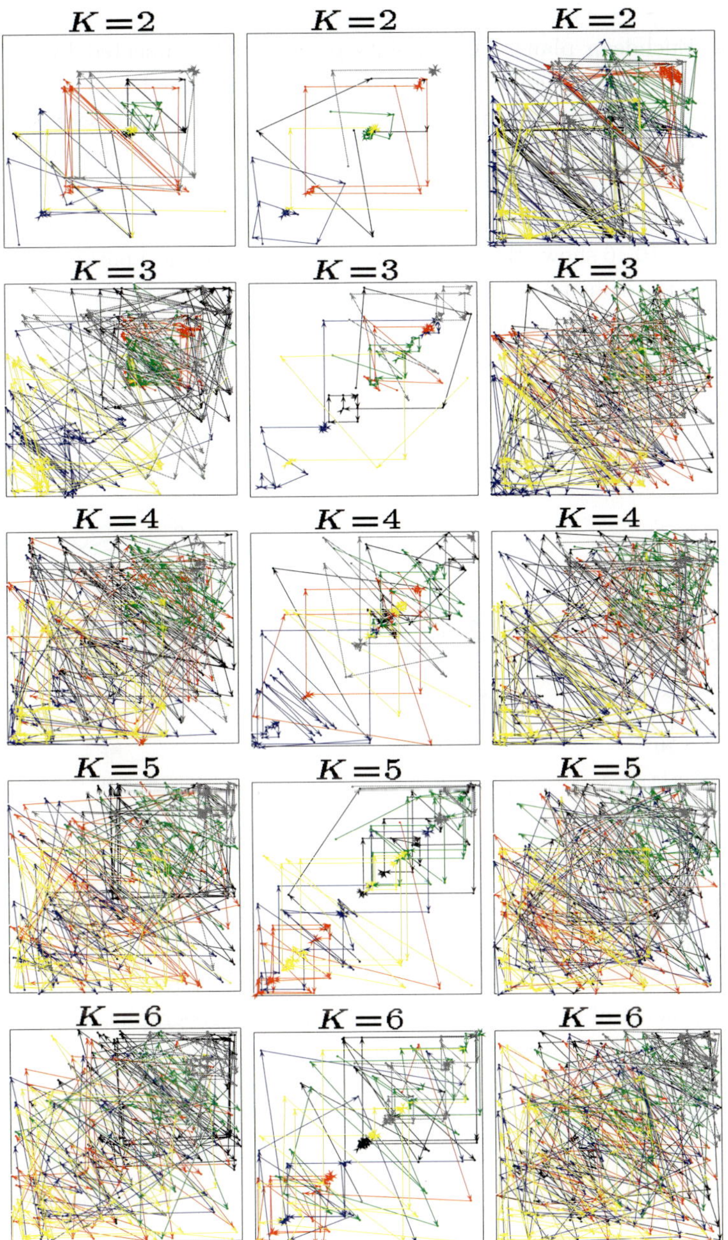

Fig. 10.15 Returns maps in random BN. No memory (left), unlimited majority memory (center), and unlimited parity memory (right).

in the Euclidean plane. Two points p and q are connected by an edge whenever their β neighborhood is empty [231]. In lune-based β-skeletons, the β-neighborhood is defined as: the intersection of two circles of radius $d(p,q)/2\beta$ that pass through p and q [5], if $\beta \in [0,1]$; the intersection of two circles of radius $\beta d(p,q)/2$ centered at the points $(1 - \beta/2)p + (\beta/2)q$ and $(\beta/2)p + (1 - \beta/2)q$, if $\beta \geq 1$.

Figure 10.16 shows three β-skeletons with increasing β parameter value, based on the same ten nodes. The figure shows also the β-neighborhoods of the node labelled 1 (upper-right). In the transition from β=0.9 to β=1.0, the 1-node loses its 1-6 and 1-8 links, whereas from β=1.0 to β=1.5 it loses the 1-5 link.

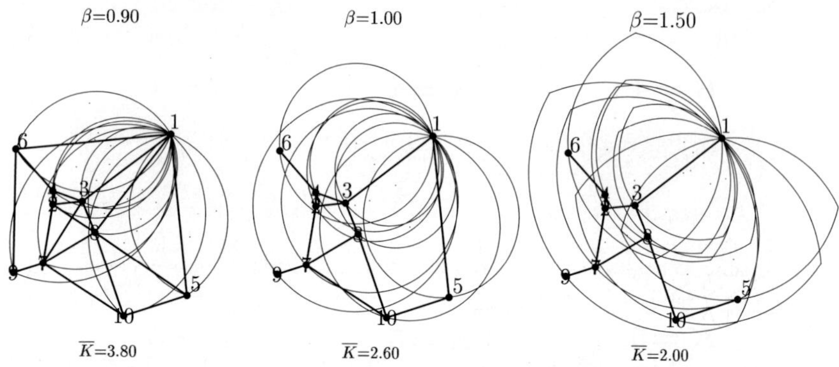

Fig. 10.16   Three β-skeletons on the same ten nodes.

β-Skeletons belongs to a family of proximity graphs, which are monotonously parameterised by the parameter β. The structure of proximity graphs represents a wide range of natural systems and is applied in many fields of science. Few examples include geographical variational analysis [156, 283, 375], evolutionary biology [265], simulation of epidemics [397], study of percolation [84] and magnetic field [378], design of ad hoc wireless networks [252, 376, 352, 300, 407]. Thus developing and analysing computational models of spatially-extended systems on proximity graphs will shed a light onto basic mechanisms of activity propagation on natural systems.

Automata on β-skeletons were originally introduced in [10] and studied in a context of excitation dynamics. In the automata on beta-skeletons

---

[5]Thus, centered in: $x_c = x_m \pm (y_p - y_q)d(p,q)/2$, $y_c = y_m \pm (x_q - x_p)d(p,q)/2$, with $x_c = (x_p + x_q)/2$, $y_c = (y_p + y_q)/2$.

studied here, each node is characterized by an internal state whose value belongs to the $\{0,1\}$ set. The updating of these states is made simultaneously (*à la* cellular automata) according to a common local transition rule involving only the neighborhood of each node .

Figure 10.17 shows the initial evolving patterns of a simulation of the parity rule on $\beta=0.9$, $\beta=1$, and $\beta=2$ skeletons based on the same N=100 nodes distributed at random in a unit square. Black squares denote node state values equal one, gray squares denote zero state values. The effect of endowing nodes with memory of the *majority* of the last three states is shown at $T=4$. The encircled node exemplifies the initial effect of memory.

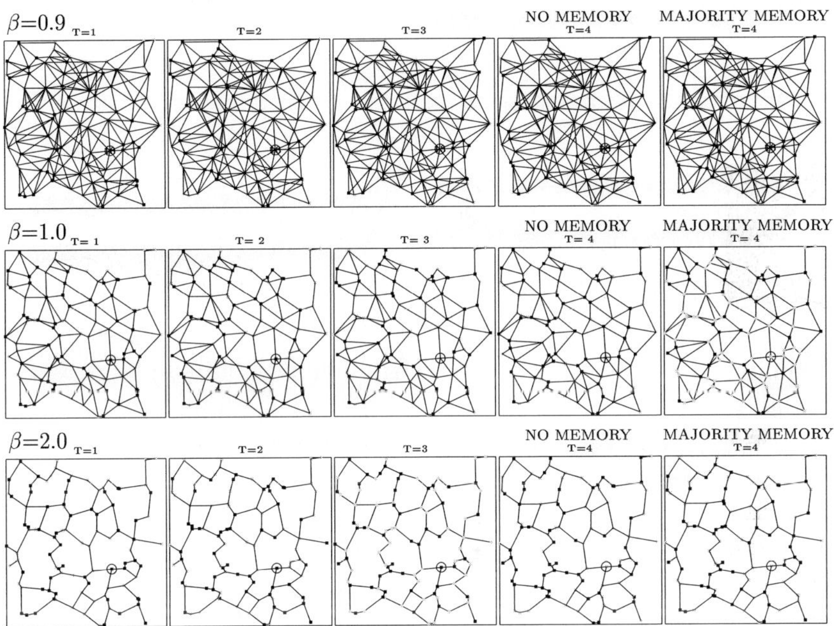

Fig. 10.17   Evolution up to $T=4$ of the parity rule on β-skeletons with one hundred nodes.

Figure 10.18 shows the evolution of the changing rate in eleven different $\beta=1$ simulations based in one thousand nodes distributed at random in a unit square. In the ahistoric simulations the parity rule exhibits a very high level of changing rate, oscillating around 0.5 . Figure 10.18 shows also the effect on the changing rate of endowing nodes with memory of the last τ state values. The inertial effect of memory tends to reduce the changing rate compared to the ahistoric model, particularly when β is odd. With high memory charges, such as $\tau=19$ in the lower left panel, the changing rate tends to vary in the long term in the $[0.1, 0.2]$ interval, after an initial *almost-oscillatory* behaviour which ceases by $T > \tau+1=20$. With unlimited trailing memory (lower right panel), this *oscillatory* pattern is never truncated, so that a rather unexpected quasi-oscillatory behaviour turns out with full memory. With no exception, the proportion of node states having one given state value (*density*), oscillates near to 0.5 regardless of the model considered.

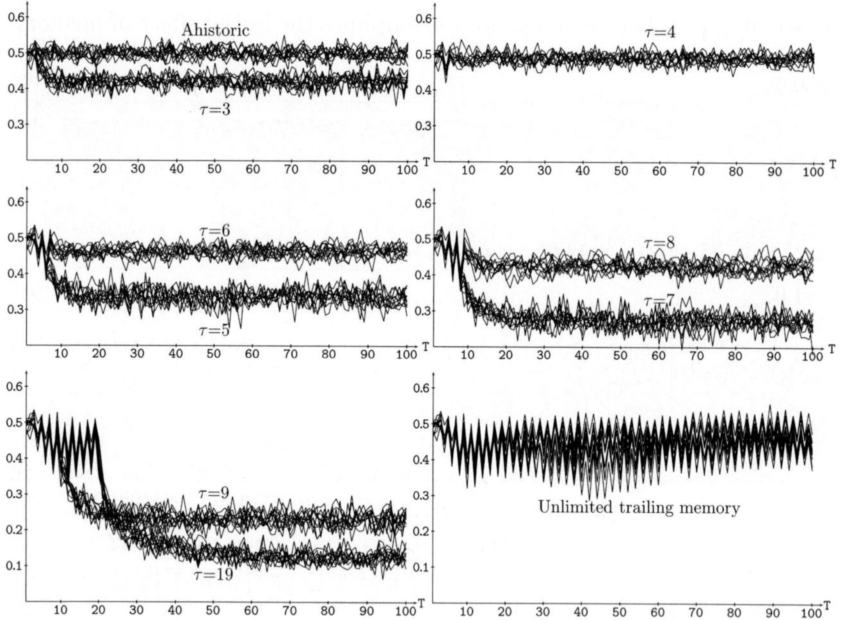

Fig. 10.18   Changing rate in eleven simulations up to $T=100$ of the parity rule on $\beta=1$ skeletons. N=1000.

Figure 10.19 shows the evolution of the changing rate in eleven different $\beta=2$ simulations based in the same nodes and initial states as in Fig. 10.18. In this scenario, even low memory charges, e.g., $\tau=3, 4$, have an apparent effect on the changing rate, and higher ones, e.g., $\tau=9, 19$, led this parameter to very low values. Thus for example with $\tau=19$ the changing rate varies in the $[0,0.1]$ interval. With unlimited trailing memory (lower right panel), the non-truncated *oscillatory* patterns soon reach a very high level near 0.5. $\beta=2$-skeletons, also termed relative neighbourhood graphs, have been implemented in simulation of human-made road networks [409, 410], and in the formal representation of biological transport networks, particularly foraging trails of ants [4].

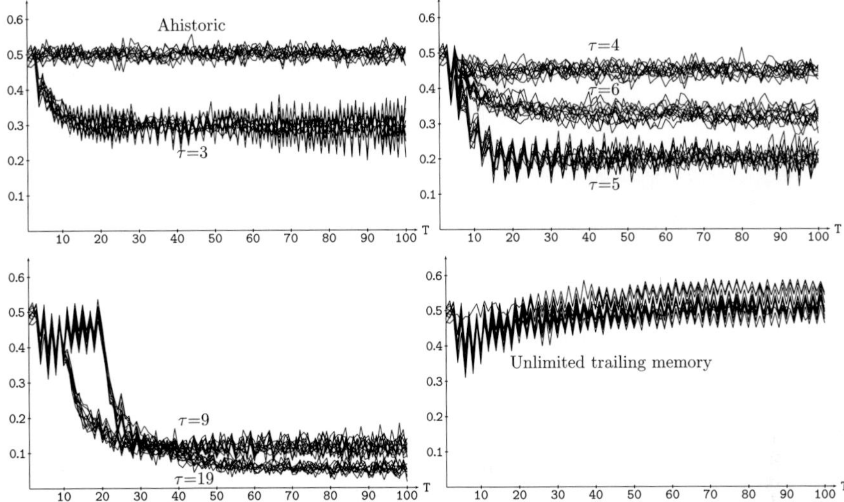

Fig. 10.19   Changing rate in eleven simulations up to $T=100$ of the parity rule on $\beta=2$ skeletons. N=1000.

Figure 10.20 shows the evolution of the changing rate in eleven different $\beta=0.9$ simulations. based in the same nodes and initial states as in Fig. 10.18. In this scenario, with high connectivity, low memory charges, e.g., $\tau=3,4$, have not an apparent effect on the changing rate, so that there are not presented in the figure. Memory charges of $\tau=5$ and $\tau=7$ have a limited effect, whereas $\tau=9,19$ already have an apparent effect. With unlimited trailing memory, the *oscillatory* patterns get a notable amplitude, albeit below 0.5.

Figure 10.21 shows the changing rate with α-memory in parity β-skeletons of one thousand nodes, starting from a single active node. Without memory, the changing rate progresses very fast regardless of the β value. On the contrary, full memory, i.e., $\alpha=1.0$, dramatically restrains the changing rate. In the $\beta=2.0$ scenario, thus with low connectivity, even the smallest memory charges, e.g., $\alpha=0.6$, lead the changing rate to extinction. In the more connected networks, low memories such as $\alpha=0.6$ and $\alpha=0.7$ only delay the increase of the changing rate, that reaches 0.5 in the figure in both the $\beta=1.0$ and $\beta=0.9$ scenarios. With $\alpha=0.8$ memory, the 0.5 level is not reached up to $T=100$ in the $\beta=1.0$ simulation, but it is

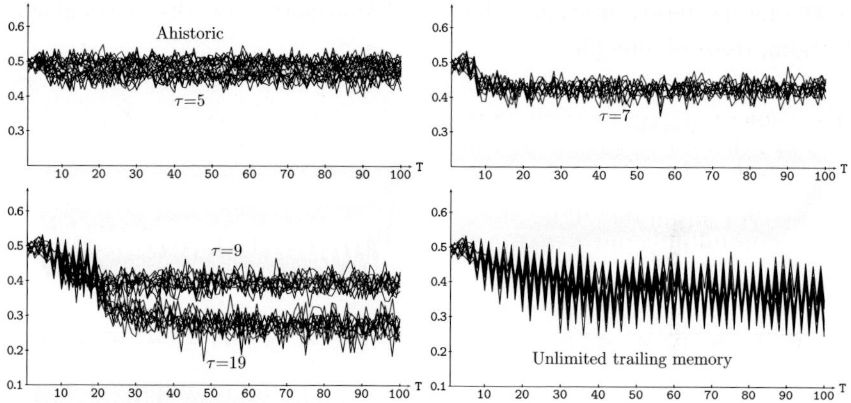

Fig. 10.20   Changing rate in eleven simulations up to T=100 of the parity rule on β=2 skeletons. N=1000.

reached in the β=0.9 simulation. Finally, with α=0.9 the changing rate is notably restrained in both scenarios, though it seems not stabilized as it is with full memory. Thus, in the simple context of Fig. 10.21, full memory exerts its expected effect, i.e., maximum dynamic inhibition. Further study of damage spreading on β-skeletons may be found in [32].

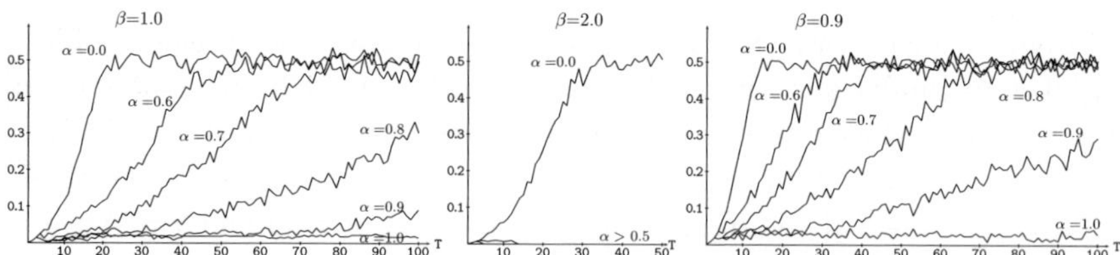

Fig. 10.21   Changing rate in parity β-skeletons with α-memory from a single active node. N=1000.

Automata on Delaunay triangulations

A Delaunay triangulation for a set $\mathbf{P}$ of **n** points in the plane is a triangulation DT($\mathbf{P}$) such that no point in $\mathbf{P}$ is inside the circumcircle of any triangle in DT($\mathbf{P}$). The Delaunay triangulation DT($\mathbf{P}$) corresponds to the dual graph of the Voronoi diagram for $\mathbf{P}$, thus the latter is obtained joining the centers of the cimcurcircles in DT($\mathbf{P}$) defining neighboring triangles [80]. Table 10.1 shows a simple Delaunay triangulation on four points an its corresponding Voronoi diagram. The number of vertices in the Voronoi diagram of a set of **n** points is at most 2n-5, and the number of edges is at most 3n-6. Both upper limits are reached in the Voronoi example of Table 10.1: 2×4-5=3 points, 3×4-6=6 edges. A Voronoi-like partition of lattice in cellular automata has been proposed in [8].

Table 10.1   A Delanaunay triangulation and its corresponding Voronoi diagram with four nodes.

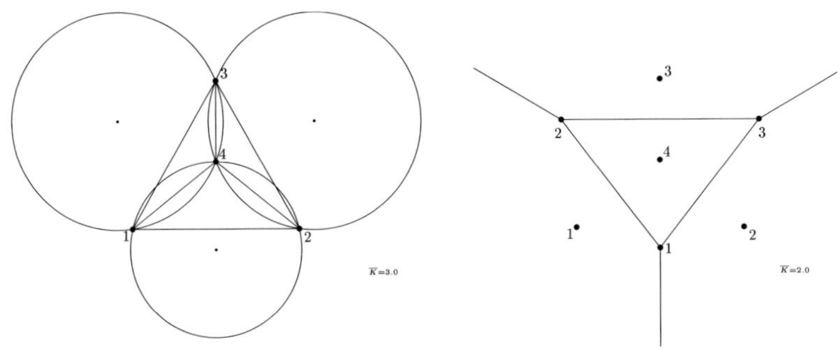

Figure 10.22 shows a Delaunay triangulation in the same nodes of Fig. 10.16 and the effect of memory at T=4 in the parity automaton.

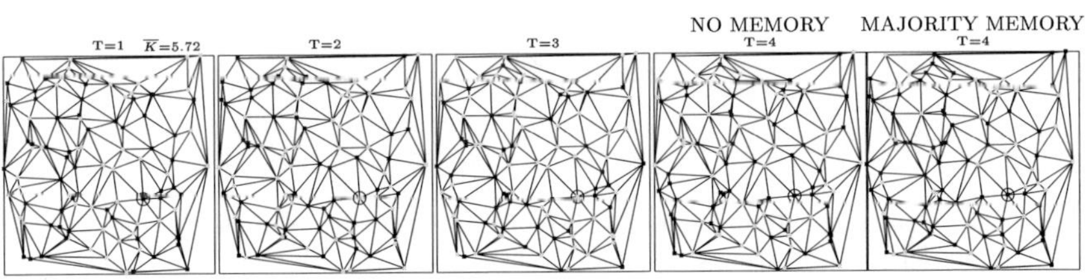

Fig. 10.22   A Delaunay triangulation in the same nodes of Fig. 10.16.

Figure 10.23 shows the changing rate in ten simulations of the parity rule on Delaunay triangulations based the same sets of one thousand points of Fig. 10.18. The mean connectivity in the triangulations considered in the figure is fairly high: $\overline{K}$=6.91, a value similar to that achieved with β=0.9 skeletons in the previous simulations (in fact the DT in Fig. 10.22 and the α=0.9 in Fig. 10.16 are somehow alike). Consequently, the high activity of the parity rule is hardly restrained with memory. So for example, the changing rate with τ=19 majority memory oscillates near 0.3.

Accordingly with the high connectivity of the DT just mentioned, α-memory turns out rather inefficient in avoiding the spread of activity start-

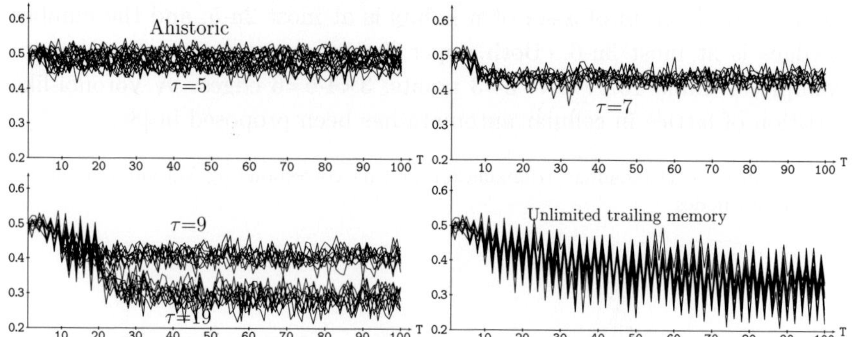

Fig. 10.23   Changing rate in ten simulations up to $T=100$ of the parity rule on Delaunay triangulations on sets of one thousand nodes.

ing from a single active node under the parity rule as shown in the left panel of Fig. 10.24. The ineffectiveness of α-memory is also apparent starting at random, as shown in the right panel of Fig. 10.24.

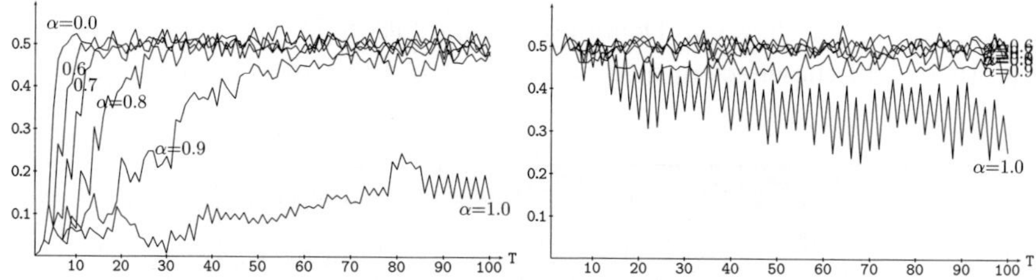

Fig. 10.24   Changing rate in the parity DT of Fig. 10.23 with α-memory from a single active node (left) and starting at random (right).

Apollonian networks

Connecting the centers of touching spheres in a three-dimensional Apollonian gasket by edges given a graph known as the Apollonian network. This process is illustrated in Table 10.2 for the case of the planar Apollonian gasket up to the fourth generation (the order-two network has the connectivity of the Fano plane). The construction of a fourth circle tangent to three given tangent circles is facilitated by the Descartes formulas:

$$k_4 = k_1 + k_2 + k_3 \pm 2\sqrt{k_1 k_2 + k_1 k_3 + k_2 k_3}\,,\ z_4 = (z_1 k_1 + z_2 k_2 + z_3 k_3 + z_4 k_4 \pm$$
$$\sqrt{z_1 z_2 k_1 k_2 + z_2 z_3 k_2 k_3 + z_1 z_3 k_1 k_3})/k_4\,,$$

where k stands for curvature and z for center location given as a complex number. The plus sign in the $\pm$

symbol in the the formulas applies to the internally tangent circle, whereas the minus applies to the externally tangent circle.

Table 10.2 The Apollonian network up to the fourth generation.

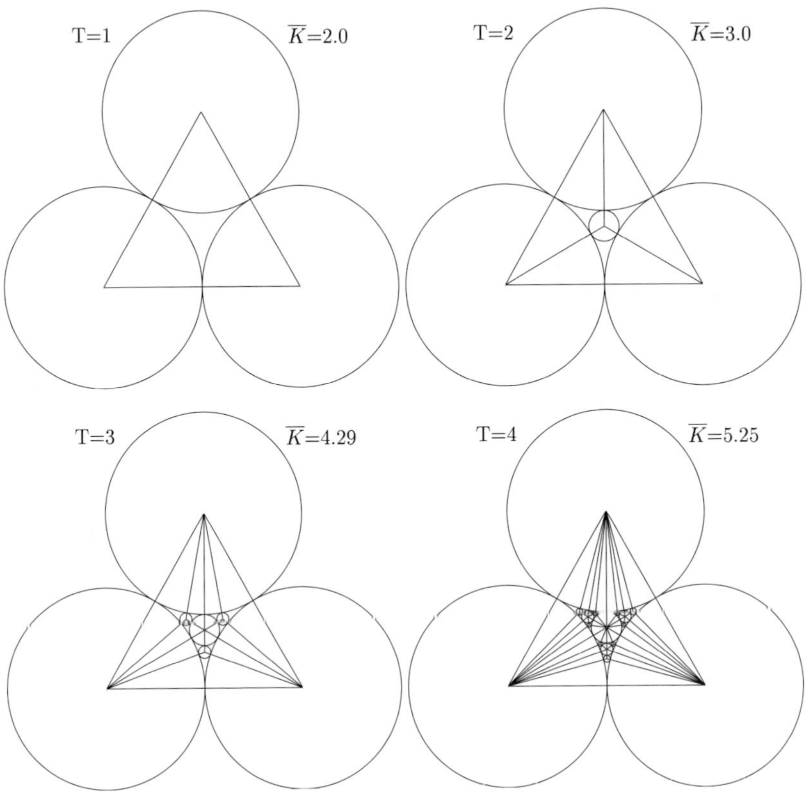

This network turns out to have some very special properties: They are simultaneously scale-free, display small-world effects [63, 255] can be embedded in an Euclidean lattice, and show space filling [196] as well as matching graph properties. These networks describe force chains in granular packings [66], fragmented porous media, hierarchical road systems, and area-covering electrical supply networks. Apollonian networks share many features of neuronal systems, and have been used to study the brain.

Figure 10.25 shows the density dynamics of the parity rule in the fourth generation Apollonian automata, thus with only sixteen nodes. Both the ahistoric and the effect of minimal memory induced by τ=3 majority mem-

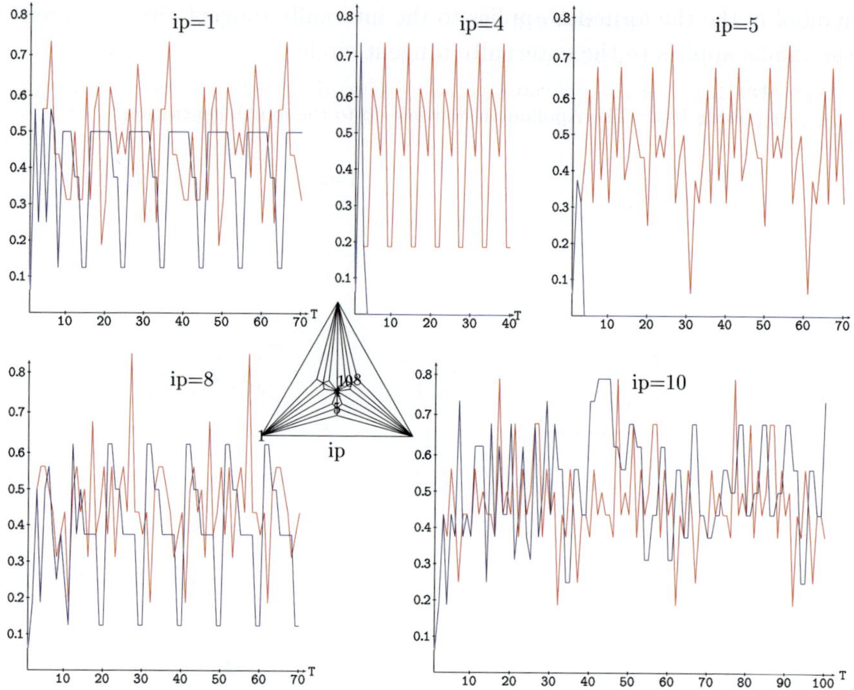

Fig. 10.25   Density in fourth generation parity Apollonian automata. Ahistoric (red) and with $\tau=3$ majority memory (blue) dynamics.

ory are considered in the figure. In the scenario of Fig. 10.25 only a single node (ip) is initally active. The five frames in the figure correspond to the five classes in which the points may be classified: 1 a point of the first generation, 4 the new point appearing at the second generation, 5 a point of the third generation, 8 and 10 point classes appearing at the fourth generation. The density in the ahistoric model varies very erratically, but as is to be expected in the context of a network with a small number of nodes, periodic behaviour is reached fairly soon. This is particularly true starting from the second generation (ip=4) active point. The effect of $\tau=3$ majority memory turns out very different depending on the inial active node. Thus, when starting either with the second generation (3) point or a third generation (5) point, memory quickly induces the all-zero state configuration, whereas in the remaining scenarios the density keeps varying: in an apparent periodic way when starting from a first generation (1) point or one point of one of the two fourth generation classes, and in a fairly erratic way when starting from one active point of the alternative fourth generation class, that labelled 10.

Chapter 11

Coupled layers

Many systems of interest may be viewed as consisting of multiple networks, each with their own internal structure and *coupling* structure to their external world partners. Examples include economic markets, social networks, ecologies, and within organisms such as neuron-immune networks.

An easy way of coupling two networks of the same size is that of connecting their homologous cells. This is so in the one-dimensional networks of the subsequent figures.

11.1 Coupled cellular automata

Thus, the parity rule in Fig. 11.1 remains active on cells with three inputs: their nearest neighbors in their own layer and their homologous cell in the partner layer. Noting σ and $[\sigma]$ the current state values and s and $[s]$ the trait states in left and right layers respectively, in the formulation of the parity rule in Fig. 11.1 it is: $\sigma_i^{(T+1)} = \sigma_{i-1}^{(T)} \oplus [\sigma_i^{(T)}] \oplus \sigma_{i+1}^{(T)}$, $[\sigma_i^{(T+1)}] = [\sigma_{i-1}^{(T)}] \oplus \sigma_i^{(T)} \oplus [\sigma_{i+1}^{(T)}]$, whereas with memory: $\sigma_i^{(T+1)} = s_{i-1}^{(T)} \oplus [s_i^{(T)}] \oplus s_{i+1}^{(T)}$, $[s_i^{(T+1)}] = \left([s_{i-1}^{(T)}] \oplus \sigma_i^{(T)} \oplus [s_{i+1}^{(T)}]\right)$, with $s_i^{(T)} = mode\left(\sigma_i^{(T)}, \sigma_i^{(T-1)}, \sigma_i^{(T-2)}\right)$, $[s_i^{(T)}] = mode\left([\sigma_i^{(T)}], [\sigma_i^{(T-1)}], [\sigma_i^{(T-2)}]\right)$, and initial assignations $s_i^{(1)} = \sigma_i^{(1)}$, $s_i^{(2)} = \sigma_i^{(2)}$, $[s_i^{(1)}] = [\sigma_i^{(1)}]$, $[s_i^{(2)}] = [\sigma_i^{(2)}]$.

Initially only the central cell of the left layer is active in Fig. 11.1, in which evolution is shown up to $T=30$. The restraining effect of the short-term memory of the majority of the last three states becomes apparent also in the context of coupled automata.

The parity rule acts in Fig. 11.2 on cells with five inputs: nearest and next-nearest neighbors plus the homologous cell in the partner network.

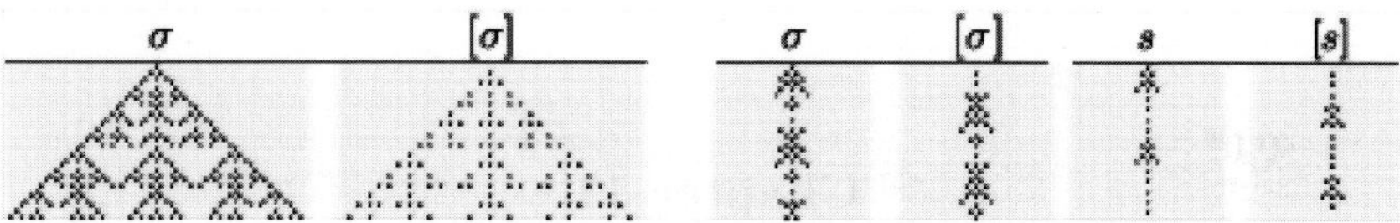

Fig. 11.1   The $K{=}3$ coupled parity rule. Ahistoric (left) and $\tau{=}3$ majority memory (center), with their trait states shown in the far right.

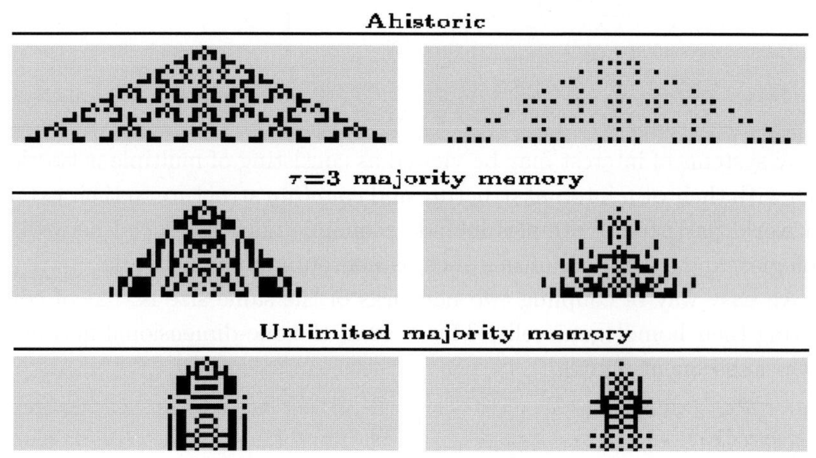

Fig. 11.2   Coupled parity rule with five inputs.

Stochastically coupled CA and Boolean Networks have been implemented to study synchronization of coupled chaotic systems [295, 294, 170, 210]. The reduction induced by memory in the perturbation from a single live cell means a reduction in their differences of the spatio-temporal patterns of coupled CA. That might be an indicator of a higher synchrony in the dynamics of coupled CA with memory in general. But this is not necessarily so, at least no in the case of the *vivid* parity rule, as indicated in Fig. 11.3 by the differences in the patterns of coupled CA starting from the same random initial configurations and periodic boundary conditions imposed on the border cells.

The trait states with unlimited majority memory in Figs. 11.2 and 11.3 are :

$$s_i^{(T)} = mode\left(\sigma_i^{(T)}, \sigma_i^{(T-1)}, \ldots, \sigma_i^{(1)}\right),$$

$$[s_i^{(T)}] = mode\left([\sigma_i^{(T)}], [\sigma_i^{(T-1)}], \ldots, [\sigma_i^{(1)}]\right).$$

As a rule, increasing the length of the majority memory produces an in-

Fig. 11.3   Coupled $K=3$ parity rule starting at random. Ahistoric model, and models with $\tau=3$ and unlimited majority memories. The last column of graphics shows the differences in the patterns of the two coupled layers.

creasing inertial effect, so that in the unlimited trailing majority memory scenario, oscillators or quasi-oscillators, tend to be generated. This well-known effect in the one-layer scenario acts also in the coupled one, as envisaged in Figs. 11.2 and 11.3 .

It can be noted that an extension of the Java based CA system *JCAsim* [152] supports coupled CA simulation [94] .

Perturbations

Two coupled CA with identical transition rules, wiring scheme and initial state configuration will, of course, evolve identically. In the case of the inter-wiring scheme being considered here, i.e., connection of homologous cells, the evolution of two $K=2+1$ coupled networks is that of a $K=3$ standard one-layer CA with self-interaction.

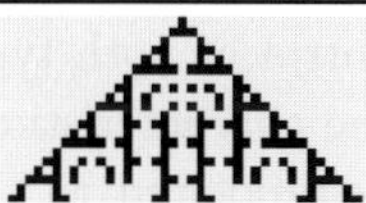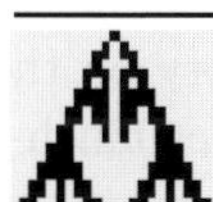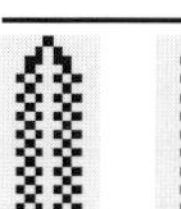

Fig. 11.4   Coupled $K=3$ parity rule with the central cell of the left network without connection with its homologous cell. Ahistoric (left), $\tau=3$ (center) and unlimited trailing memory (right) models. Both central cells are initially at state 1.

In Figs. 11.4 and 11.5 both networks are initiated with an active central cell. But the ahistoric and $\tau=3$ memory patterns in these figures are not those corresponding to the one-layer context because there is a *default* in the coupling: the central cell of the left network does not receive the input from its homologous cell in the right one. Because of that it is not active at $T=2$ in the left layer. This simple default in wiring very much alters the patterns compared to that of the one-layer scenario, i.e., those of rule 150 in Fig. 2.6 and Fig. 3.5 in the $K=3$ example of Fig. 11.4, and those of rule 42 in Fig. 3.7 in the case of Fig. 11.5

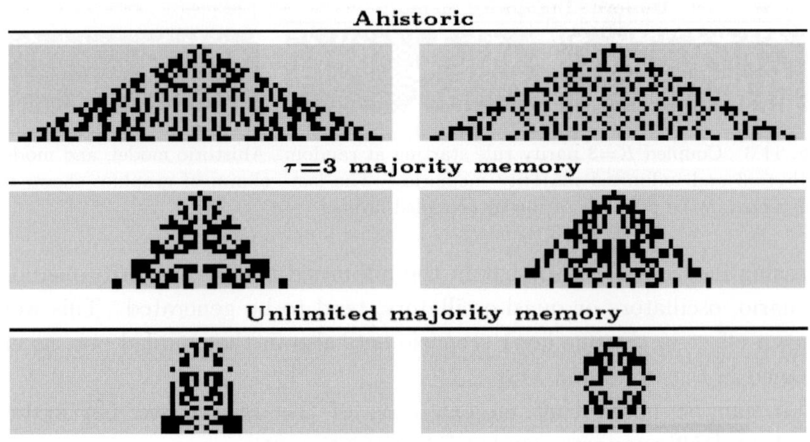

Fig. 11.5   The coupled $K=5$ parity rule in the scenario of Fig. 11.4 .

Figures 11.6 and 11.7 show the evolving patterns in coupled twin layers of size 122 up to $T=20$, from the same random initial configuration. The common spatio-temporal pattern is shown in Figs. 11.6 and 11.7 with the active cells in gray tone. Darker cells show the *damage* (rather perturbation, [379]) induced by two different causes: *i)* the just described lack of inter-connection of the central cell of left network in Fig. 11.6, and *ii)* the reversion of the state value of this cell. Qualitatively speaking, the restraining effect of memory on the control of both types of perturbation spreading is similar. In fact much of the characteristics that are observed starting from a single seed may be extrapolated to the general damage spreading scenario: starting from a single site is a limit case of damage in fully quiescent automata.

Coupled automata with elementary rules

Elementary rules operate on nearest neighbors:
$$\sigma_i^{(T+1)} = \phi_i\left(\sigma_{i-1}^{(T)}, \sigma_i^{(T)}, \sigma_{i+1}^{(T)}\right).$$ Thus, coupled elementary rules evolve

Fig. 11.6   Effect of the lack of inter-connection of the central cell of the left layer. $K=3$ parity rule acting on twin layers.

Fig. 11.7   Effect of reversing the state value of the central cell of the left layer in the scenario of Fig. 11.6.

as :

$$\sigma_i^{(T+1)} = \phi\big(\sigma_{i-1}^{(T)}, [\sigma_i^{(T)}], \sigma_{i+1}^{(T)}\big) \quad , \quad [\sigma_i^{(T+1)}] = \phi\big([\sigma_{i-1}^{(T)}], \sigma_i^{(T)}, [\sigma_{i+1}^{(T)}]\big) \ .$$

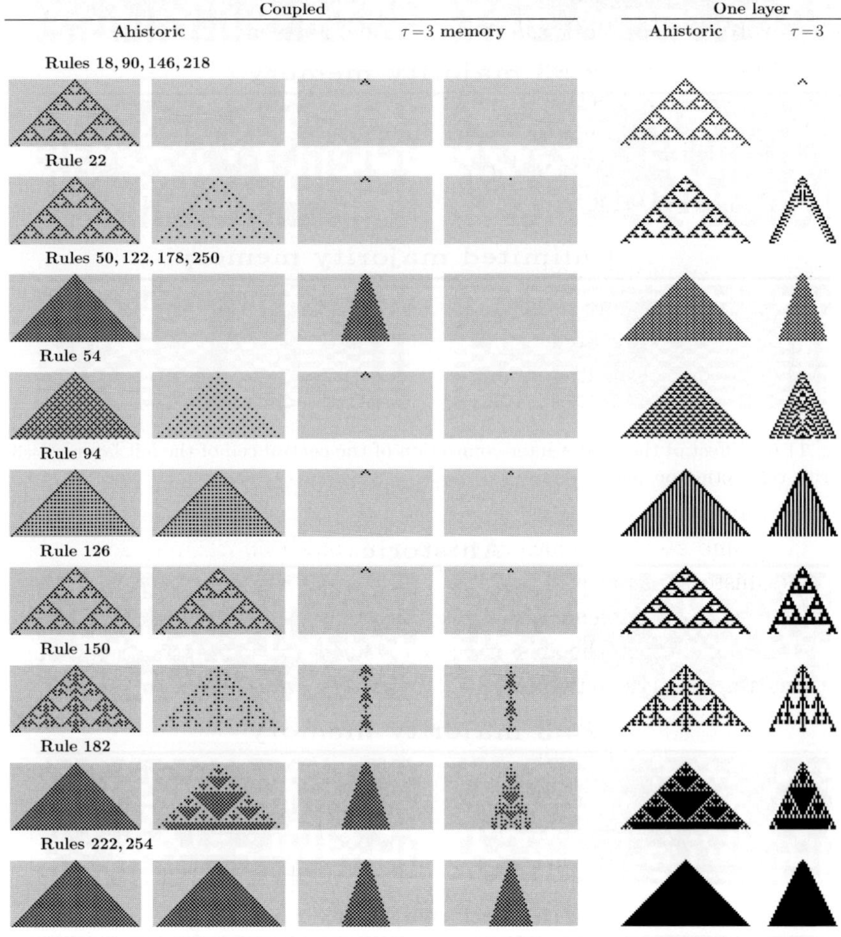

Fig. 11.8   Coupled and one-layer elementary legal CA starting from a single active cell in the left lattice. Ahistoric and mode of the last three state memory models.

The far right column of patterns in Fig. 11.8 summarizes the effect of $\tau=3$ majority memory in the legal rules significantly affected by memory when starting from a single site seed. The overall effect of memory is that of restraining the expansion of the spatio-temporal patterns, with particular incidence in rules 18, 90, 146, and 218 which extinguish as soon as $T=4$.

Automata with transition rules with no self-interaction in their one-layer formulation, i.e., $\sigma_i^{(T+1)} = \phi\left(\sigma_{i-1}^{(T)}, \sigma_{i+1}^{(T)}\right)$, are unaltered if coupled in the simple way considered here. Examples are the left and right shift rules 170 and 240 $\left(\sigma_i^{(T+1)} = \sigma_{i+1}^{(T)}, \ \sigma_i^{(T+1)} = \sigma_{i-1}^{(T)}\right)$ and rule 90: $\sigma_i^{(T+1)} = \sigma_{i-1}^{(T)} \oplus \sigma_{i+1}^{(T)}$, which evolves in Fig. 11.8 *isolated* as in the one-layer scenario.

Rules with $\beta_6 = \beta_8 = 0$, i.e., those which do not generate active cells when the neighbors are not active (regardless of the state of the cell itself) in the one-layer scenario, do not *transmit* activity to an initially fully inactive layer when coupled. Among the 64 elementary rules with $\beta_6 = \beta_8 = 0$, are the legal rules 18,90,146,218, 50,122,178,250 reported in Fig. 11.8, and the *majority* rule 232. Also neighbor-quiescent are the asymmetric shift rules 170 and 240 (equivalent under the reflection transformation), and rules 226 and 184 (equivalent under the negative transformation).

Figure 11.8 shows under the heading -coupled- the evolving patterns of coupled legal elementary rules significantly affected by memory when starting from a single seed in one layer (the left one in Fig. 11.8). In this context, memory kills, as in the ahistoric model, the evolution of the group of rules $18, 90, 146$, and 218, and also that of rules $22, 54, 94$, and 126. Memory notably narrows the evolution of the parity rule 150 (as advanced in Fig. 11.1), and also very much alters that of rules 182, 222, and 254.

The ahistoric patterns of rules 22, 54, 94, 126, 150, 182, 222, and 254 become split in two complementary ones in their coupled evolution in Fig. 11.8 This splitting effect in the ahistoric model affects every quiescent rule when starting from a single active cell, so that the mere superposition (OR operation without the need of the XOR) of the patterns in the coupled formulation generates the ahistoric ones.

The split of the one-layer pattern, starting from a single cell when coupling, is not a particular property of elementary rules, but of the simple way of coupling adopted here. Thus for example it holds with the parity rule with five inputs as shown in Fig. 11.9.

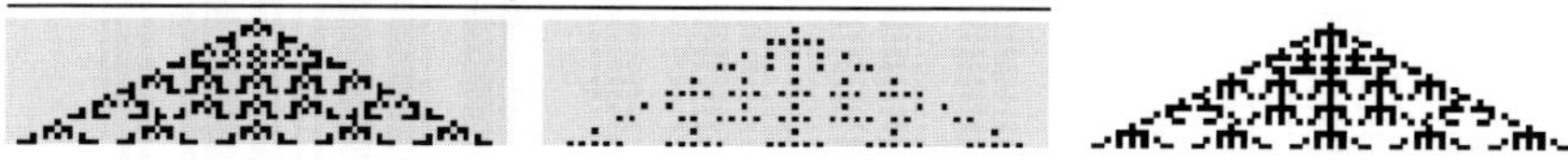

Fig. 11.9   The parity rule with five inputs. Coupled (line above left) and one-layer (right) ahistoric models.

Starting at random from a sole layer, only the linear rules 60 (00111100), $\sigma_i^{(T+1)} = \sigma_i^{(T)} \oplus \sigma_{i-1}^{(T)}$, and its reflected rule 102 (01100110), $\sigma_i^{(T+1)} = \sigma_i^{(T)} \oplus$

$\sigma_{i+1}^{(T)}$, together with rules 150 and 204, retain this superposition property in the ahistoric context.

In Fig. 11.10 only the left layer starts with live cells, thus $\sigma_i^{(2)}=\sigma_{i-1}^{(1)} \oplus \sigma_{i+1}^{(1)}$, and $[\sigma_i^{(2)}]=\sigma_i^{(1)}$, so that the initial configuration is transmitted to the right layer after the first iteration, then, at $T=3$, $[\sigma_i^{(3)}]=\sigma_{i-1}^{(1)} \oplus (\sigma_{i-1}^{(1)} \oplus \sigma_{i+1}^{(1)}) \oplus \sigma_{i+1}^{(1)} = 0$, and consequently, $[\sigma_i^{(4)}]=\sigma_i^{(3)}$. The left layer is at $T=3$, $\sigma_i^{(3)}=(\sigma_{i-2}^{(1)} \oplus \sigma_i^{(1)}) \oplus \sigma_i^{(1)} \oplus \sigma_i^{(1)} \oplus \sigma_{i+2}^{(1)}), =\sigma_{i-2}^{(1)} \oplus \sigma_i^{(1)} \oplus \sigma_i^{(1)} \oplus \sigma_{i+2}^{(1)}$, i.e., the same state value as in the one-layer configuration. The ulterior dynamic generalizes these initial results, so that the initially empty layer is empty at odd time-steps and copies that of the partner layer at even time-steps, whereas the configuration of the initially active layer at even time-steps is the same as in the one-layer model. Qualitatively, the same dynamic is followed in rules 60 and 102, as seen in Fig. 11.10.

Fig. 11.10 Coupled elementary legal CA starting at random in the left lattice (overlined left) which generate via superposition the one-layer evolution (right).

Unlike in Fig. 11.8, starting at random as in Fig. 11.10 to get the one-layer patterns from those in the coupled formulation, addition modulo two is necessary with the linear rules 60, 102 and 150.

With $\tau=3$ majority memory, no rule, except the identity rule 204, $\sigma_i^{(T+1)} = \sigma_i^{(T)}$, verifies this superposition property. Thus, as an example, the patterns of the coupled partner layers in Fig. 11.11 do not add the one-layer ones shown also in the figure. The identity rule 204 is unaffected by memory, so that in the coupled formulation the initial pattern appears at odd time steps in the left layer and at the even ones in the right layer, generating by direct superposition the one-layer lattice as shown in Fig. 11.10.

Rule 60

Rule 102

Rule 150

Fig. 11.11   Coupled elementary CA rules starting at random in the left lattice (overlined left), and one-layer (right) evolution with $\tau=3$ majority memory.

Linear rules are *additive*, i.e., any initial pattern can be decomposed into the superposition of patterns from a single site seed. Each of these configurations can be evolved independently and the results superposed (module two) to obtain the final complete pattern. The upper group of spatio-temporal patterns in Fig. C.3 shows an example.

Linear rules remain linear when cells are endowed with linear memory rules. Thus, endowing the parity rule of the three last states as memory in cells upon the coupled elementary linear rule 150 yields:

$$\sigma_i^{(T+1)} = s_{i-1}^{(T)} \oplus [s_i^{(T)}] \oplus s_{i+1}^{(T)} \quad , \quad [\sigma_i^{(T+1)}] = [s_{i-1}^{(T)}] \oplus s_i^{(T)} \oplus [s_{i+1}^{(T)}] ,$$

$$\text{with, } s_{i+1}^{(T)} = \sigma_{i+1}^{(T)} \oplus \sigma_{i+1}^{(T-1)} \oplus \sigma_{i+1}^{(T-2)} , \ s_i^{(T)} = \sigma_i^{(T)} \oplus \sigma_i^{(T-1)} \oplus \sigma_i^{(T-2)} , \ s_{i-1}^{(T)} =$$

$\sigma_{i-1}^{(T)} \oplus \sigma_{i-1}^{(T-1)} \oplus \sigma_{i-1}^{(T-2)}$, and, $[s_{i+1}^{(T)}] = [\sigma_{i+1}^{(T)}] \oplus [\sigma_{i+1}^{(T-1)}] \oplus [\sigma_{i+1}^{(T-2)}]$, $[s_i^{(T)}] = [\sigma_i^{(T)}] \oplus [\sigma_i^{(T-1)}] \oplus [\sigma_i^{(T-2)}]$, $[s_{i-1}^{(T)}] = [\sigma_{i-1}^{(T)}] \oplus [\sigma_{i-1}^{(T-1)}] \oplus [\sigma_{i-1}^{(T-2)}]$.

Consequently, linear rules remain additive when endowed with linear memory rules. Thus in the lower group of spatio-temporal patterns in Fig. C.3, the evolution patterns starting from the left layer (upper row), and from the right one (central row), if added modulo two generate the evolving pattern starting from both layers active (bottom row).

Number conserving rules

The number conserving rules 184 and 226 when coupled to an empty layer evolve as a left and right shift respectively [1], thus still conserving the total number of live cells. Starting from different random configurations in both layers, both rules 184 and 226 lose their characteristic number conserving property, in respect of the individual layers, but not considering the two layers as a whole, albeit presenting very different spatio-temporal patterns, as shown in Fig. 11.12 and Fig. C.4 for both rules in a lattice of size 400 up to T=100.

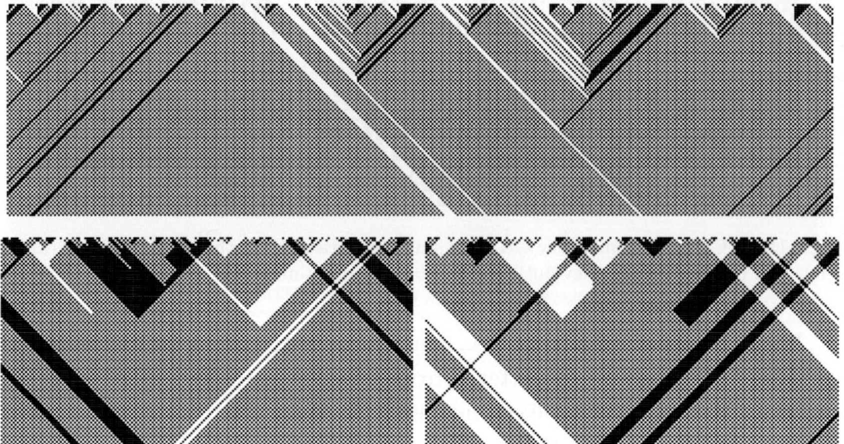

Fig. 11.12   Elementary rule 184 starting at random. One layer (upper) and coupled half (lower) layers.

---

[1]   101 100 001 000 , equivalent to $\sigma_i^{(T+1)}=\sigma_{i-1}^{(T)}$, and   101 100 001 000 , equivalent
       1   1   0   0   1   0   1   0
to $\sigma_i^{(T+1)}=\sigma_{i+1}^{(T)}$.

When implementing majority memory, rules 184 and 226 lose the number-conserving property, so that the pattern drifts to a fixed point of all 1s or all 0s depending upon whether within the configuration the number of 1s was superior to that of 0s or the contrary. The variation in the number of live cells in Fig. 11.13 and Fig. C.5 is low because the initial density is fairly close to the watershed 0.5, so that only by $T=1000$, the pattern fully blackens. The cases of low and high initial densities in Figs. 11.14 show how the rule 184 with $\tau=3$ majority memory readily relaxes in both cases (as in the low density example of rule 226 in Fig. C.6) to a fixed point of all 0s or all 1s, correctly classifying the initial configuration. This is not so when coupling the two halves of the one-layer scenario due to the presence of permanent solitons of the alternative state value. The relaxation of the criterion by which the rules 184 and 226 recognizes majority density [154] must bounce back when these rules are coupled.

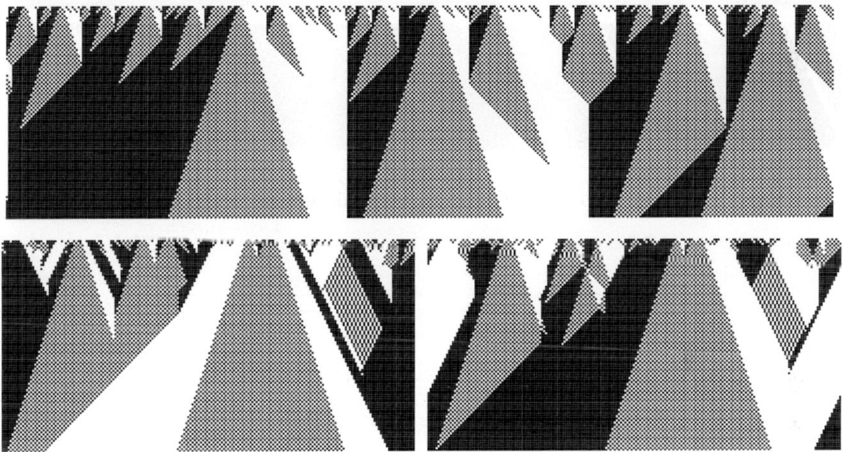

Fig. 11.13   Elementary rule 184 in the scenario of Fig. 11.12 with $\tau=3$ mode memory.

Eventually number-conserving rules

Eventually number-conserving (ENC) CA reach limit sets, after a number of time steps of order of the automaton size, having a constant number of active sizes [85]. Coupled rules 184 and 226 with $\tau=3$ mode memory are reminiscent of the ENC density behavior, as shown in Fig. 11.15 regarding rule 184.

Some ENC rules *emulate* proper number-conserving rules, albeit this is

$$\rho_0 = 0.25$$

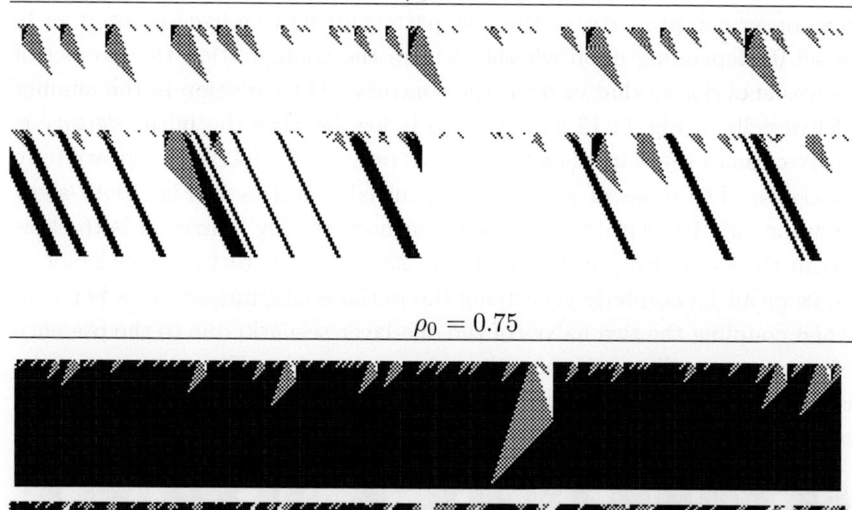

$$\rho_0 = 0.75$$

Fig. 11.14   The scenario of Fig. 11.13 from initial densities 0.25 and 0.75.

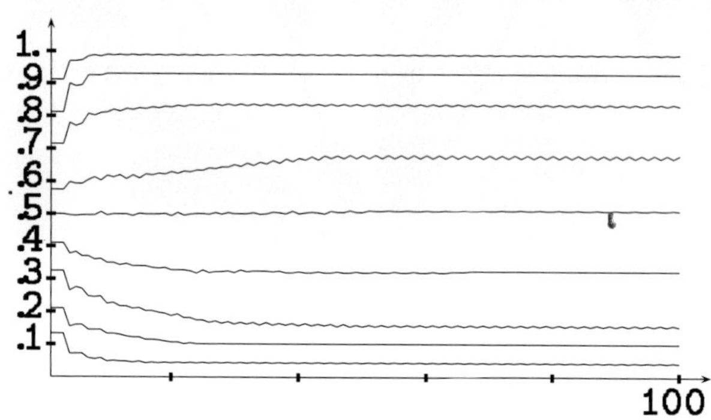

Fig. 11.15   Evolution of density with the coupled rule 184 with $\tau=3$ mode memory.

not a necessary condition. Among the ENC which emulate NC stand the number-nonincreasing rule 176, which emulates both rules 184 and 240, and the non-monotone rule 99 which emulates rule 226, shown in Fig. C.8. When coupled, these rules also evolve as ENC rules, reaching a plateau after a fairly short transition period, as shown in Fig. 11.16. The numbers reached in the steady regime in the one-layer and coupled contexts are not exactly the same, albeit they use to be fairly close.

None of these rules with memory seem to solve the density task as rules 184 and 226 do, because the implementation of memory seems not to discriminate among the different levels of initial density. Thus, the evolution of density is altered in a similar way regardless of ρ_o, either depleting it as happens with rule 176 in Fig. 11.16, or *ii*) leaving it fairly unaltered as in the case of rule 99 in Fig. 11.16 [2]. This is so regardless of the dramatic change that may suffer the spatio-temporal patterns, as shown in Fig. C.8. The dynamics under rule 99 has been also checked by endowing cells with memory of the parity of their three last states. Even in this scenario rule 99 shows a notable resilience to abandon the tendency to stabilize the density close to 0.5. The effect of memory on ENC rules that do not emulate NC rules, e.g. the non-monotone rules 74 and 88 is similar to that just described.

Asymmetric scenarios

Figure 11.17 shows an example of a coupled asymmetric connectivity scenario in which one of the layers has cells with 6+1 inputs, whereas the other one has only 2+1 inputs. It is highly remarkable how the highly connected networks drives the dynamic, *activating* that of the layers with low connectivity.

Figure 11.18 shows the effect of memory in an asymmetric connectivity scenario in which the cells in the automata lying in the left column of the figures have five inputs, whereas those in the right side have three inputs. As before, the scenario is that of nearest (and next-nearest) neighbors and the homologous cell in the partner network. The expected restraining effect of memory is appreciated in this mixed scenario.

Figures 11.19 and 11.20 correspond to an asymmetric memory scenario as only one of the layers (the right one in Figs. 11.19 and 11.20) has cells endowed with memory, whereas the cells of their partner layer are featured by its last state. Both coupled layers have their central cell active initially.

In both scenarios, short-term (τ=3) and unlimited majority memory, memory soon kills the dynamic in the layer with memory in the group of elementary rules 18,90,146,218, and exerts the characteristic restraining

---

[2] Under rule 99, the density departing from the highest values plummets initially in such a way that its change is hidden in the y-axis. Conversely, the variation from the smallest initial densities is hidden in the y-axis due to its initial dramatic increase.

Discrete Systems with Memory

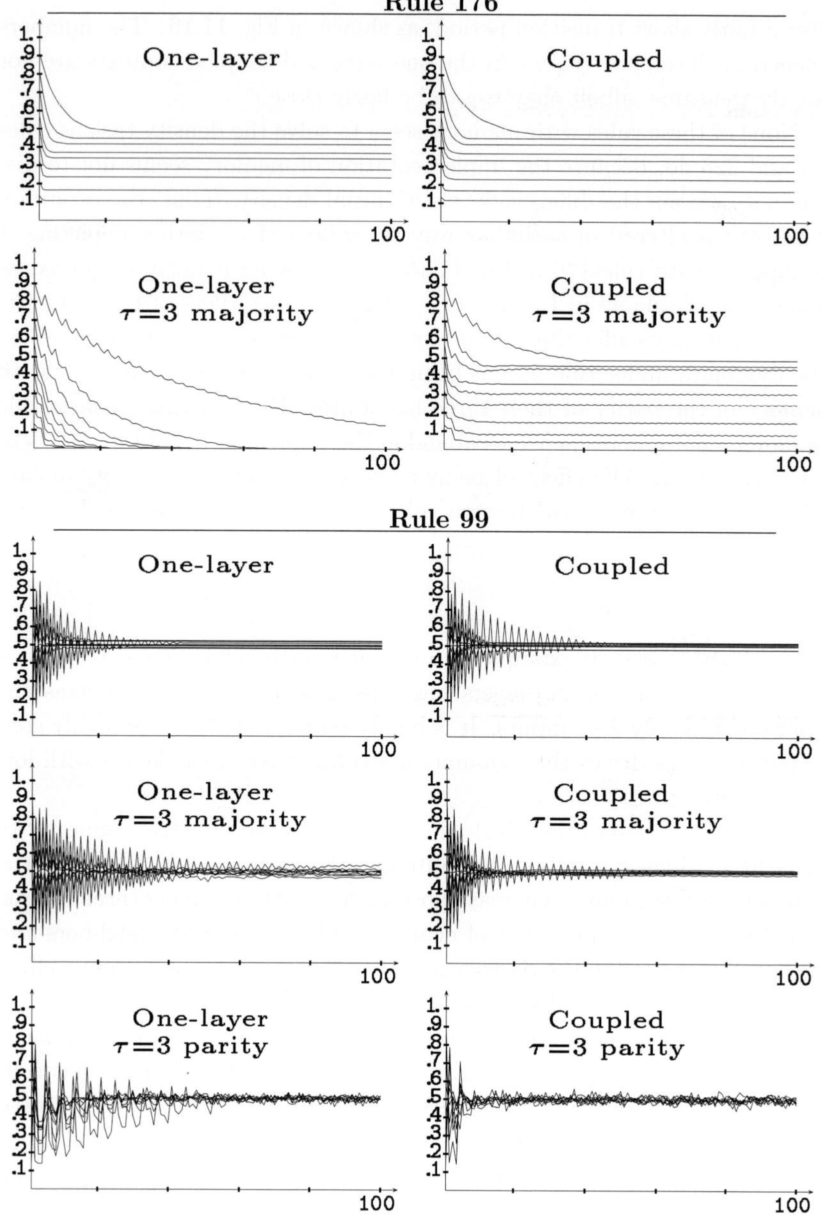

Fig. 11.16   Evolution of density under rules 176 and 99.

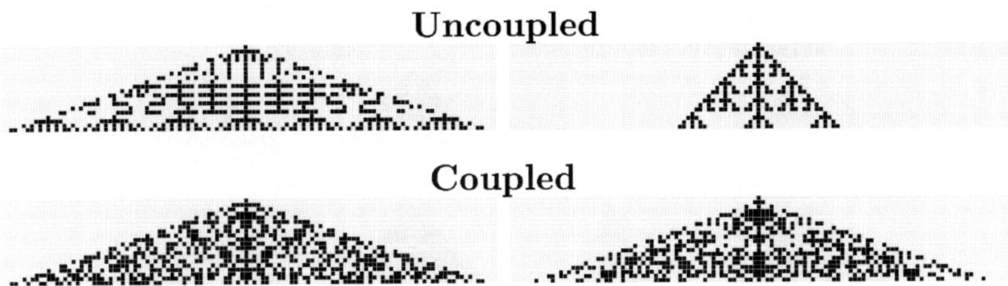

Fig. 11.17   The parity rule on uncoupled (top: K=7 left, K=3 right) and coupled (bottom: K=6+1 left, K=2+1) layers.

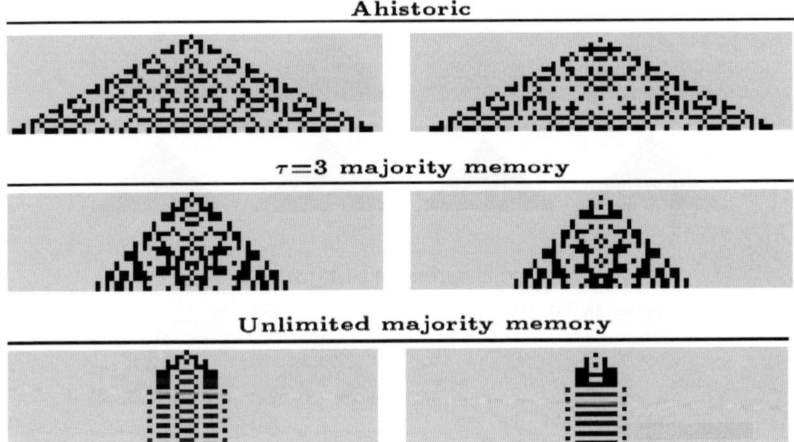

Fig. 11.18   Coupled CA with homologous cells connected and five and three inputs per cell in the left and right disposed networks respectively. Parity rule.

effect in the case of rules 50,122,178, and 250. But the remaining elementary rules, rule 150 in particular, do not evolve in the expected restrained form. This is also the case of the $K=5$ parity rule in Fig. 11.20, as layers with memory do not evolve in a restrained form, so that the ahistoric networks seem to drive the process. This evolution is rather unexpected, particularly in the case of unlimited majority memory, a memory implementation very much tending to *freeze* any dynamical system.

Figure C.9 shows the evolving patterns of every one of the 32 legal rules on the left layer coupled with rule 150 on the right one. The figure shows the spatio-temporal patterns in the ahistoric and parity of the last three state memory models, when starting from a single active cell in the left layer. In some cases, rule 150 lives *alone* in this figure, despite that its

Right layer with $\tau=3$ majority memory

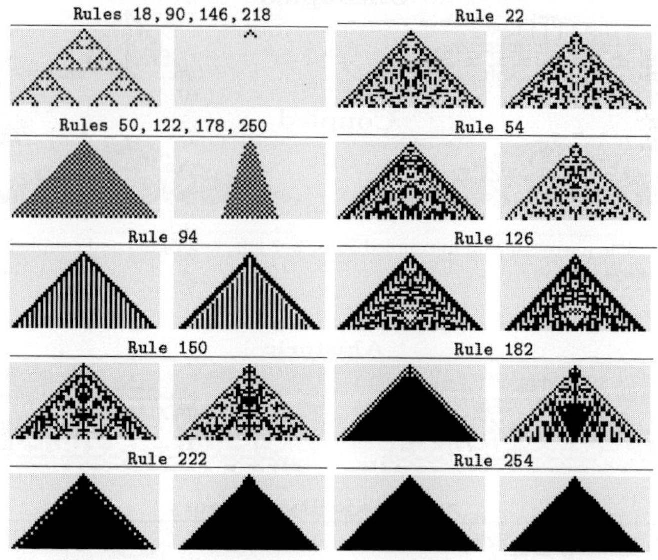

Right layer with unlimited majority memory

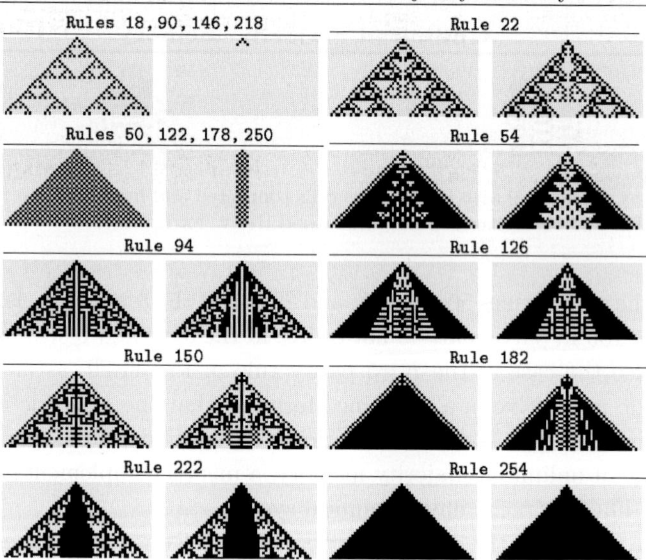

Fig. 11.19 Coupled legal rules with only one layer with memory.

Right layer with $\tau=3$ majority memory

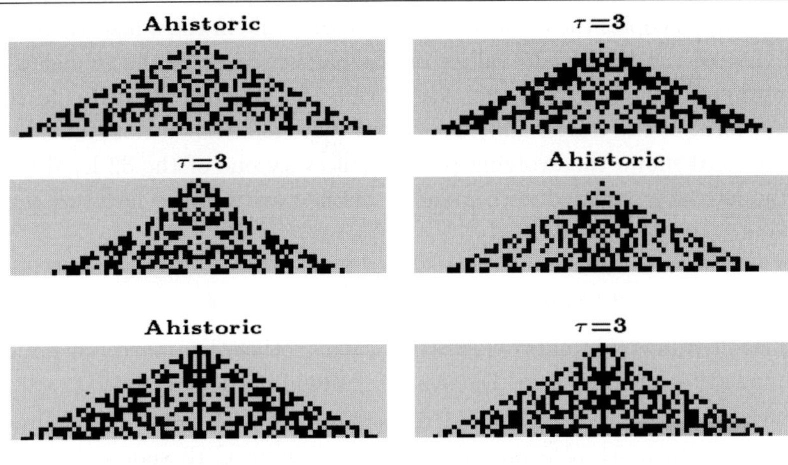

Right layer with unlimited majority memory

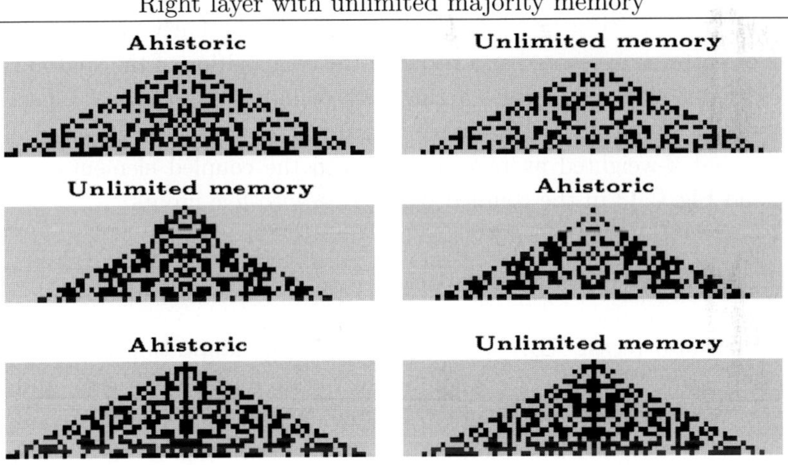

Fig. 11.20   Coupled $K=5$ parity rule with only one layer with majority memory.

pattern rule does not progress at all. The paradigmatic case is that of the association of rule 0 with rule 150, in which case, the pattern under Rule 150 in the ahistoric context is that of the conventional rule 90, i.e., the parity rule with no self-interaction, starting at $T=2$. Rule 150 coupled with rule 0 does not evolve exactly as rule 90 with $\tau=3$ parity memory due to the distortion induced by being $s_c^{(2)}=s_c^{(3)}=1$ in the central cell of the left layer under the coupled rule 0 as an effect of the parity rule acting as memory, instead of being $s_c^{(2)}=0$ as in the one-layer model. The identity rule 204 $\sigma_i^{(T+1)}=\sigma_i^{(T)}$, copies the evolution of its partner rule in the a historic model and that of the trait state values of the partner layer in the model with memory, $\sigma_i^{(T+1)}=[s_i^{(T)}]$. In the case of rule 150, associated with rule 204, the patterns of the trait s and the current σ are alike albeit not coincident. Figure C.10 shows the evolving patterns of every one of the 32 legal rules on the left layer coupled with memory of the parity of the last two time-steps: $s_i^{(T)} = \sigma_i^{(T)} \oplus \sigma_i^{(T-1)}$, $[s_i^{(T)}] = [\sigma_i^{(T)}] \oplus [\sigma_i^{(T-1)}]$.

Other memories in coupled CA

Any Boolean function of previous state values, other than parity or majority, may act as memory rule. Elementary rules in particular, so that: $s_i^{(T)} = \phi(\sigma_i^{(T)}, \sigma_i^{(T-1)}, \sigma_i^{(T-2)})$. Fig. C.11 reports the effect of quiescent asymmetric rules as memory also on rule 150, and Figure C.12 shows the effect of legal rules as memory on rule 150. Initially $s_i^{(1)} = \sigma_i^{(1)}, s_i^{(2)} = \sigma_i^{(2)}$, $[s_i^{(1)}] = [\sigma_i^{(1)}], [s_i^{(2)}] = [\sigma_i^{(2)}]$ in these figures. In the fairly endless patchwork of patterns generated as a result of the composition of the spatial rule 150 with the elementary rules acting with memory, the *original* aspect of the rule 150 is not always traceable. Figure 11.21 shows the effect of variable α and β-weighted average memories in the coupled elementary rule 150, and Fig. C.13 in the coupled parity rule with five inputs.

11.2   Coupled Boolean networks

Kauffman also recast RBN as a model of fitness adaptation for evolving species — the *NK* model — whereby each node represents a gene and its fitness contribution is based upon K other genes, each possible configuration being assigned a random value. It is shown how K effects fitness landscape topology. Later, Kauffman made the point that species do not evolve independently of their ecological partners and presented a coevolutionary model of fitness adaptation [227]. Here each node/gene is coupled to K others locally and to C within each of the S other species with which it interacts — the *NKCS* model. Each species within such models experiences constant changes in the shape of its fitness landscape due to the evolutionary advances of its coupled neighbours until mutual equilibria are reached. The *NKCS* model has been used to explore a number of aspects of natural coevolution, such as symbiosis [e.g., [96]] and collective behaviour [e.g., [95]], along with coevolutionary adaptation in general [e.g., [71]] and

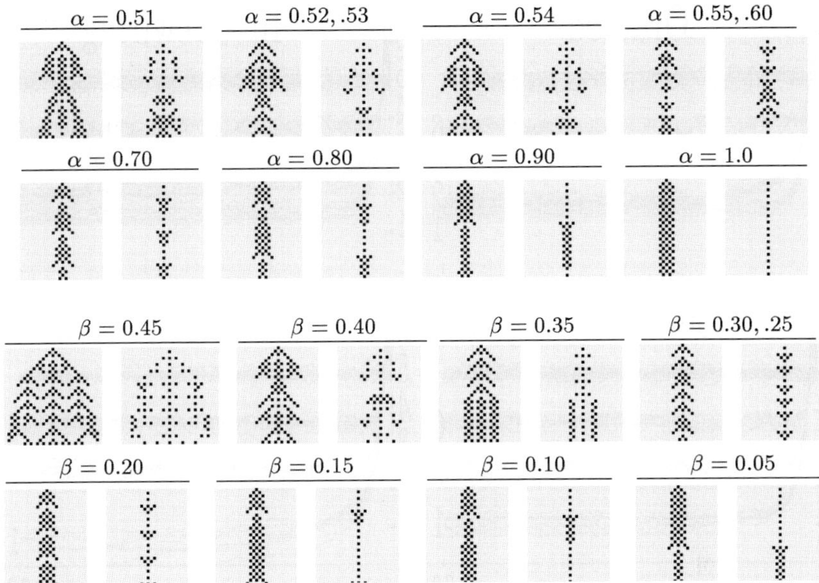

Fig. 11.21   Effect of α and β memories on the coupled $K{=}3$ parity rule.

coevolutionary computation [e.g., [97]]. A similar extension can be made to the RBN framework such that $S{+}1$ networks exist, wherein each node is connected both to K randomly chosen nodes within its own network and C randomly chosen nodes in each of the S other networks. However, unlike the radical change from the NK model to the $NKCS$ model, the effects of adding C connections appears roughly equivalent to increasing K in a single network. That is, given that all connections are randomly assigned and all updates are synchronous, any of the $S{+}1$ networks simply behave as if they have $K + (S \times C)$ connections within a standard RBN. It should be noted that the global network is now of size $(S{+}1)N$. Morelli and Zanette [294] used probabilistic connections between two RBN, wherein they explored synchronization between the networks with K=3 (see Villani et al. [402] for a related study).

In the ordered scenario of Fig. 11.22, every node is connected to its homologous partner in a partner layer (i.e., $S{=}C{=}1$). Thus, noting σ and $[\sigma]$ the state values in both layers, it is: $\sigma_i^{(T+1)} = \phi_i\big([\sigma_i^{(T)}] + \sum_{j \in \mathcal{N}_i} \sigma_j^{(T)}\big)$,

$[\sigma_i^{(T+1)}] = \phi_i\big(\sigma_i^{(T)} + \sum_{j \in \mathcal{N}_i} [\sigma_j^{(T)}]\big)$. Intra-neighbourhoods $\mathcal{N}_i$ of the generic

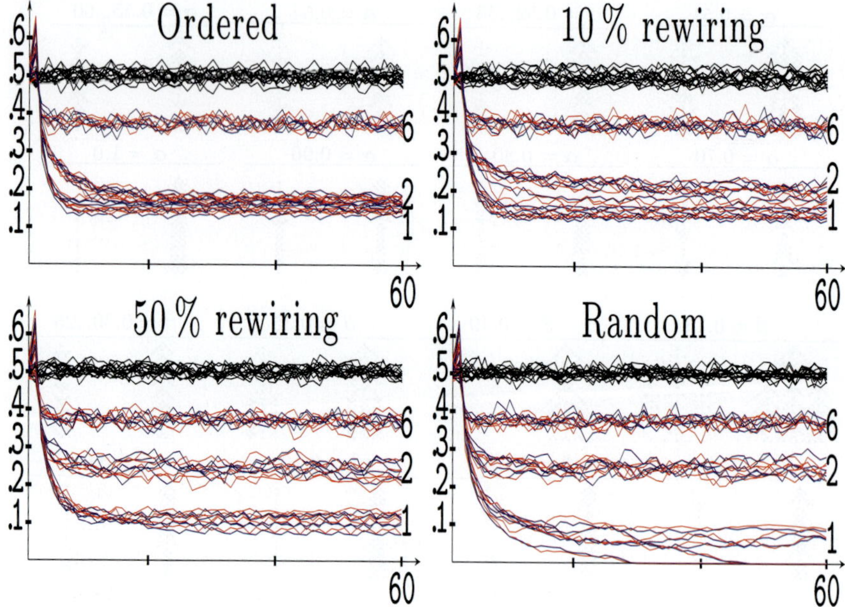

Fig. 11.22   Coupled 1000K11, K=1, 2 and 6 BN. Changing rates in both layers (red, blue) and distance between both layers (black). Connectivities K are indicated at right.

node i are: $i-1$ when K=2, $i-1$ and $i+1$ when K=2, and $i-3, i-2, i-1, i+1, i+2, i+3$ when K=7. Periodic boundary conditions are imposed on the border nodes.

The general aspect of the changing rate patterns in Fig. 11.22 corresponds to that in Fig. 10.8, albeit the *fusion* of the K=1 and 2 changing rates in the ordered coupled layer scenario denotes a notable difference. This *fusion* is lost as the rewiring increases, so that in the random context the evolution of the changing rates is similar in both Fig. 10.8 and Fig. 11.22. No indication of synchrony appears in Fig. 11.22, as the distance between layers remains around 0.5 in every case. The main features in Fig. 10.8 regarding damage are preserved in Fig. 11.23. The perturbation induced by the reversion of a single node state value in one layer (the only kind of damage studied here), with damage shown in red in Fig. 11.23, is transmitted to the partner layer (blue), so that the damaging rates in both layers are fairly *entangled*.

Figures 11.24 and 11.25 deal with cells endowed with memory. Thus,

$$\sigma_i^{(T+1)} = \phi_i\big([s_i^{(T)}] + \sum_{j \in \mathcal{N}_i} s_j^{(T)}\big), \quad [\sigma_i^{(T+1)}] = \phi_i\big(s_i^{(T)} + \sum_{j \in \mathcal{N}_i} [s_j^{(T)}]\big).$$

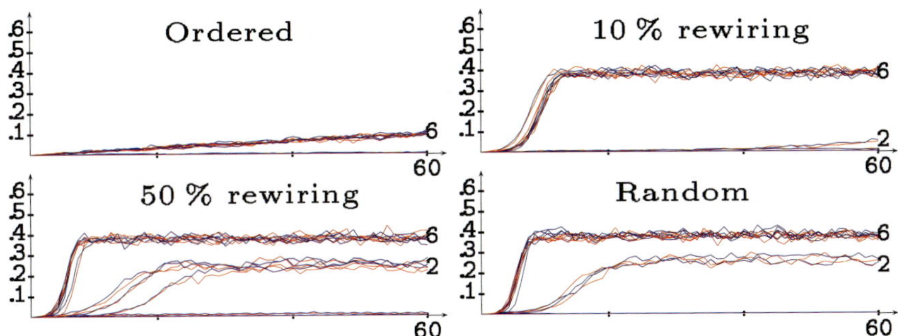

Fig. 11.23 Damage spreading from reversion of the state value of a single node in the *red* layer in the scenario of Fig. 11.22.

Figure 11.24 shows how, using symmetrical K networks as in Fig. 11.22, a similar general inertia effect of memory as reported in the one-layer scenario can be seen when coupling, so that the implementation of majority memory tends to produce a decrease in the rates of change compared to their equivalent non-memory networks. Majority memory induces a kind of oscillation-like effect on the evolution of the changing rate, that appears over-expressed when keeping memory of the last three states, the nodes have three inputs, and the degree of randomization is notable, i.e. in the two lower graphs of Fig. 11.24.

Figure 11.25, to be compared to Fig. 11.23, indicates that, as a rule, memory is not effective in the control of damage spreading in networks with high number of inputs. Thus in the case of seven inputs in the figures, memory of length three does not impede the damage reaching the 0.4 plateau in networks with rewiring. Similar effect is obtained with unlimited majority memory, in which case memory only delays the advance of damage if the degree of rewiring is low [36]. In networks with lower connectivity, the control of damage when endowing memory, particularly unlimited, is more notable, albeit memory of only the last three states just delays the advance of damage in networks with rewiring (lower frames in Fig. 11.25). The study of the effect of parity memory of length three on coupled networks with the same connectivity as well as that of majority memory on coupled $\overline{K}=7$ scale-free networks has been also addressed in [36].

Non-Symmetrical K

After Morelli and Zanette [294], Hung et al. [205] examined two coupled disordered cellular automata with probabilistic connectivity of varying K

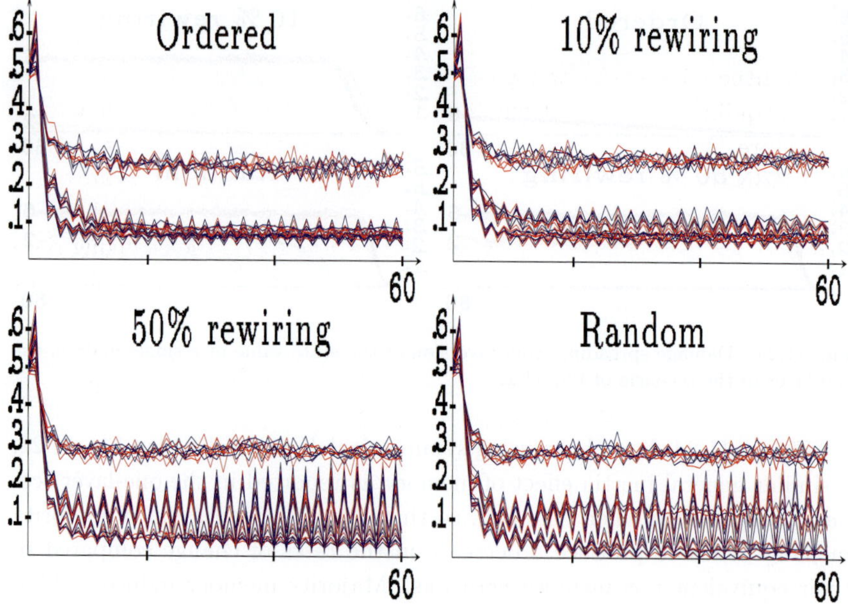

Fig. 11.24   Changing rate in coupled 1000K11, $K=1$, 2, 6 BN with $\tau=3$ majority memory.

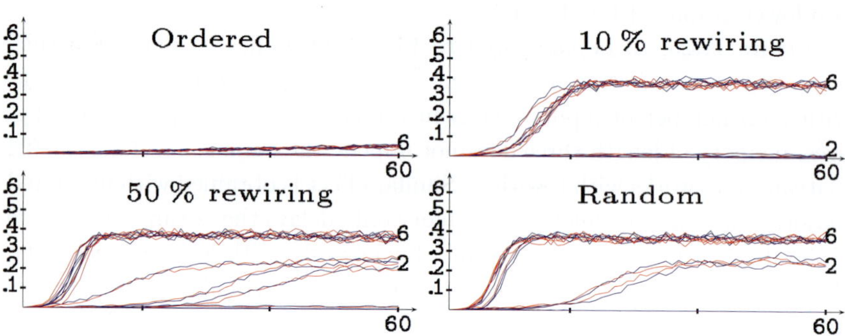

Fig. 11.25   Damage spreading from reversion of the state value of a single node in the *red* layer in the scenario of Fig. 11.24 .

using elementary rule 22. They report a change in the synchronization dynamic when the internal coupling K is different in each network. Indeed, in the broader view, there is no reason to assume that the degree of connectivity within a given network, i.e., its basic internal structure, will be the same as that of its coupled partner network(s). Kauffman and Johnsen [227] also showed changes in behaviour due to heterogeneous values of K for the coevolving populations of the *NKCS* model. In particular, they describe how low K fitness increases for high C when partnered with a high K species before equilibrium but that only low K fitness increases under low C.

Figure 11.26 shows the evolution of the changing rates of $K=2$ and 6, $C=1$ RBN coupled with a $K=C=1$ RBN. As a rule, the changing rate of the $K=C=1$ RBN (blue lines) is smaller than that of their partners, with higher connectivity. But as might be expected from Fig. 11.17, the changing rates of the $K=C=1$ RBN is held by their more connected partner automata, so that, even in the fully random scenario, their changing rates do not decay. Figure 11.27 shows how damage scarcely progresses in the $K=2$ scenario of Fig. 11.26, only two of the five simulations show damage spreading with 50 % or full rewiring, but advances as usual if $K=6$.

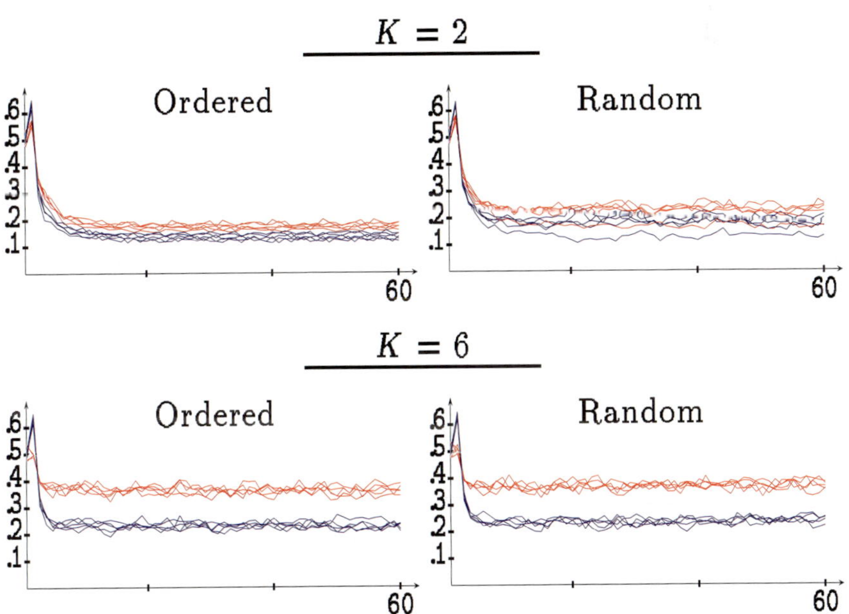

Fig. 11.26   Changing rates in coupled 1000-111 (blue), with 1000-K11, K=2, 6 (red) BN.

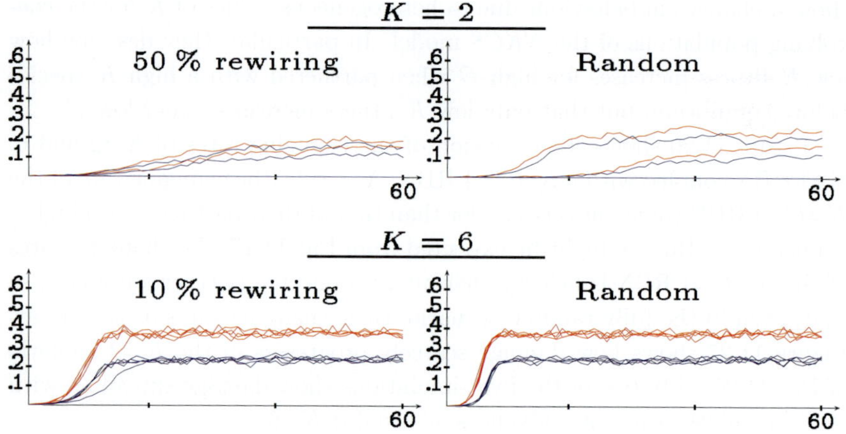

Fig. 11.27   Damage spreading from reversion of the state value of a single node in the *red* layer in the scenario of Fig. 11.26.

Non-Symmetrical memory

Asymmetric memory may now be taken into account, so that different degrees or types of memory may be implemented in every RBN layer. Thus,

$$\sigma_i^{(T+1)} = \phi_i\big([s_i^{(T)}] + \sum_{j \in \mathcal{N}_i} \sigma_j^{(T)}\big), \quad [\sigma_i^{(T+1)}] = \phi_i\big(\sigma_i^{(T)} + \sum_{j \in \mathcal{N}_i} [s_j^{(T)}]\big).$$

The incorporation of memory in only one layer does not significantly restrain the dynamics, which seems to be supported by the ahistoric evolving pattern. The changing rate in coupled RBN with asymmetric memory of this kind shown in Fig. 11.28 is slower in the layer with nodes endowed with memory, slightly slower for low K, appreciably slower if $K{=}7$. As reported previously in other contexts, Fig. 11.29 indicates that memory seems not to be effective in the control of damage spreading out of the ordered lattice scenario. The effect of unlimited trailing memory is similar to that of limited memory, albeit with lower levels of both changing rate and damage spreading, particularly in the $K{=}1$ and 2 simulations.

Coupled RBN with different evolution rates

In all of the aforementioned examples of multiple-network scenarios it is likely that in many cases some networks will be changing at different rates relative to each other: technological networks may change faster than ecological ones, for example. Therefore a new parameter R can be introduced which specifies the number of update cycles a given network undertakes

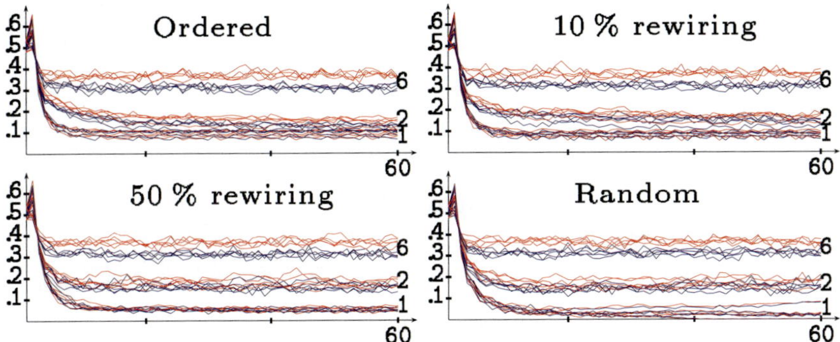

Fig. 11.28   Changing rate in coupled 1000K11, K=1, 2, 6 BN with τ=3 majority memory endowed only to the automata with changing rate shown in blue color.

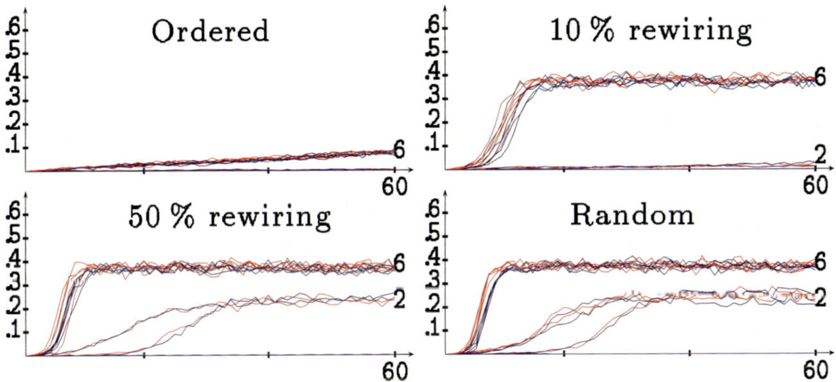

Fig. 11.29   Damage spreading from reversion of the state value of a single node in the *red* layer in the scenario of Fig. 11.28 .

before another performs one update cycle. Thus the updates of individual networks become deterministically asynchronous, somewhat akin to deterministic asynchronous single network RBN [163].

Figure 11.30 shows an example based on the K=7 parity CA rule starting from a single site live cell. The slower dynamics of the automata with R=3 in the uncoupled scenario becomes activated when coupled with the conventional faster automata, a result somewhat similar to that in Fig. 11.17 .

The lack of updating time synchrony generates the *serrated* evolution of changing rate of the network with faster evolution shown in Fig. 11.31 (in which the changing rate appears as constant in the red curves during

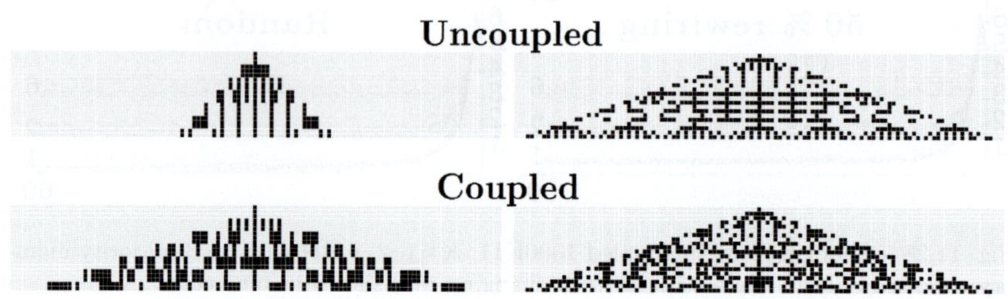

Fig. 11.30   The parity rule on uncoupled layers (top: K=7) and C=1 coupled (bottom: K=6) with R=3 delay in the left automata.

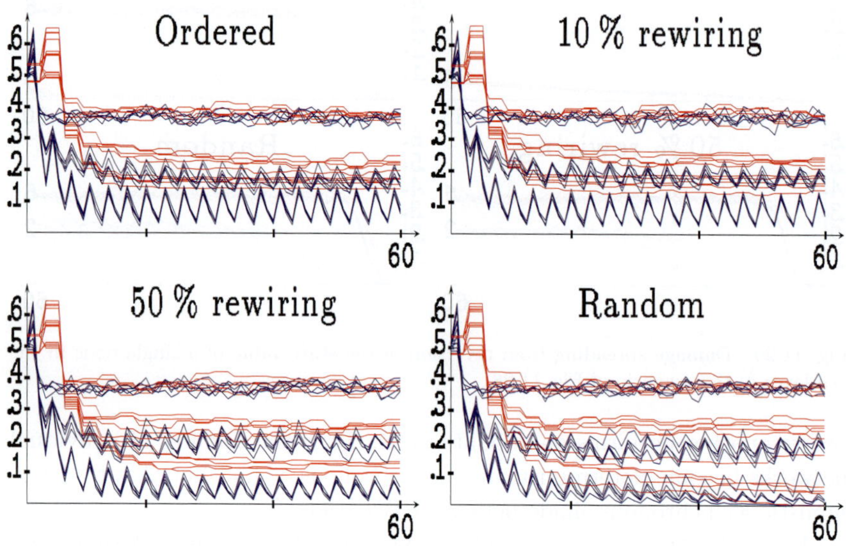

Fig. 11.31   Changing rate in coupled 1000-K11, $K=1$, 2, 6 BN with $R=3$ in the layer shown in red.

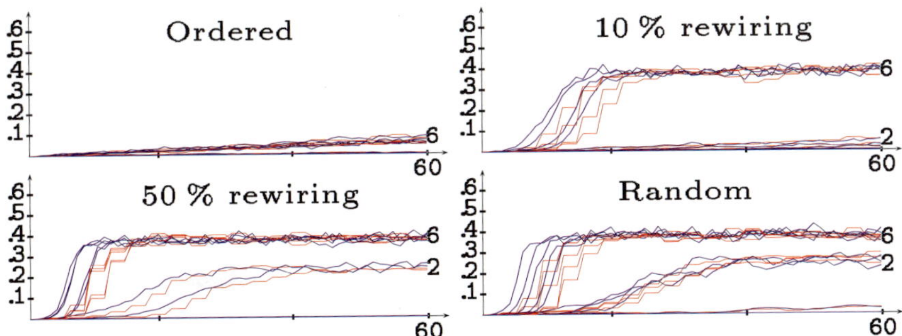

Fig. 11.32   Damage spreading from reversion of the state value of a single node in the *red* layer in the scenario of Fig. 11.31.

their inactive time-steps to facilitate the visualization), a particular feature that does not appear in the evolution of the damage in Fig. 11.32 .

Chapter 12

Continuous state variable

The mechanism of implementation of memory adopted in this book, keeping the transition rule unaltered but applying it to a function of previous states, can be adopted in any spatialized dynamical system. Some of these scenarios are analyzed in this chapter.

12.1   Continuous-valued automata

Historic memory can be embedded in *continuous-valued* CA (or *Coupled Map Lattices*, CML [221]), in which the state variable varies in $\mathbb{R}$, just by considering m instead of σ in the application of the updating rule: $\sigma_i^{(T+1)} = \varphi(m_j^{(T)} \in \mathcal{N}_i^{(T)})$ where φ is a continuous function.

In the very simple case, the new state level of a cell can be the average of its own state level and those of its neighbors. A elementary CA of this kind would be with memory: $\sigma_i^{(T+1)} = \frac{1}{3}(m_{i-1}^{(T)} + m_i^{(T)} + m_{i+1}^{(T)})$. A simple example up to $T = 10$ from a single site live cell is given in Table 12.1, in which historic formulation m is the mean value of the two last states.

The dynamics of moving agents has been modeled in [334] according to probabilistic rules involving the barycenter of the perceived agents and the mean value of the last previous spatial orientations of the direction vector. The two possibilities of movements (the barycenter rule and the memory rule) have different effects: the barycenter rule favors clustering, whereas the memory rule tends to produce the opposite dispersing the effect. Nevertheless, these two effects are not fully orthogonal, and gives rise to interesting non-trivial behaviors.

Fuzzy CA are a sort of continuous CA with states ranging in the real [0,1] interval, obtained by means of *fuzzification* of Boolean CA rules given in

231

232   *Discrete Systems with Memory*

Table 12.1   Evolving patterns of the CML: $\sigma_i^{(T+1)} = \left(m_{i-1}^{(T)} + m_i^{(T)} + m_{i+1}^{(T)}\right)/3$.

Ahistoric	**Historic**
.01 .01	
.01 .01 .02 .02	.01 .01
.01 .02 .03 .04 .04 .04	.01 .01 .02 .02 .03
.04 .05 .06 .07 .07 .08 .08	6.01 .03 .04 .05 .06 .06 .07
.11 .11 .12 .12 .12 .12 .12 .12	.06 .11 .11 .12 .12 .12 .12 .12
.33 .22 .22 .20 .19 .17 .16 .15 .15	.33 .28 .22 .22 .21 .20 .19 .18 .17
1.0 .33 .33 .26 .23 .21 .19 .18 .17 .16	1.0 .33 .33 .31 .27 .25 .23 .21 .20 .19
.33 .22 .22 .20 .19 .17 .16 .15 .15	.33 .28 .22 .22 .21 .20 .19 .18 .17
.11 .11 .12 .12 .12 .12 .12 .12	.06 .11 .11 .12 .12 .12 .12 .12
.04 .05 .06 .07 .07 .08 .08	.01 .03 .04 .05 .06 .06 .07
.01 .02 .03 .04 .04 .04	.01 .01 .02 .02 .03
.01 .01 .02 .02	.01 .01
.01 .01	

their disjunctive normal form, by replacing: $a \vee b$ by $\min(1, a + b), a \wedge b$ by ab, and $\neg a$ by $1 - a$. Thus for example, the elementary rule 90 : $\sigma_i^{(T+1)} = \left(\sigma_{i-1}^{(T)} \wedge (\neg \sigma_{i+1}^{(T)})\right) \vee \left((\neg \sigma_{i-1}^{(T)}) \wedge \sigma_{i+1}^{(T)}\right)$, after fuzzification yields, $\sigma_i^{(T+1)} = \sigma_{i-1}^{(T)} + \sigma_{i+1}^{(T)} - 2\sigma_{i-1}^{(T)}\sigma_{i+1}^{(T)}$; thus incorporating memory: $\sigma_i^{(T+1)} = m_{i-1}^{(T)} + m_{i+1}^{(T)} - 2m_{i-1}^{(T)}m_{i+1}^{(T)}$. An illustration of the effect of memory on the fuzzy rule 90 is given in Table 12.2, with m being the mean value of the two last states (an example of the effect of α-memory is given in [52]).

Table 12.2   Evolving patterns of the historic formulation of the fuzzified rule 90 up to $T = 11$ from a single crisp cell.

```
                                1.0
                            1.0       1.0
                        .50 .50 .50 .50 .50
                    .25 .75 .38 .38 .38 .75 .25
                .13 .63 .45 .52 .49 .52 .45 .63 .13
            .06 .44 .59 .49 .51 .49 .51 .49 .59 .44 .06
        .03 .28 .60 .49 .50 .50 .50 .50 .49 .60 .28 .03
    .02 .17 .52 .53 .50 .50 .50 .50 .50 .50 .53 .52 .17 .02
.01 .10 .40 .55 .50 .50 .50 .50 .50 .50 .50 .50 .55 .40 .10 .01
.06 .29 .53 .51 .50 .50 .50 .50 .50 .50 .50 .50 .51 .53 .29 .06
.03 .20 .47 .52 .50 .50 .50 .50 .50 .50 .50 .50 .50 .52 .47 .20 .03
```

Quantum cellular automata (QCA) are a generalization of (classical) CA, and in particular of reversible CA (an overview of QCA is given in [417]). The earliest one-dimensional quantum CA model was introduced by Grössing and Zeilinger [178], who were the first to coin the term and formalize a QCA. In their approach, every the cell i at time-step T is assigned a complex number representing a quantum mechanical probability amplitude (thus whose absolute square lies between 0 and 1), i.e, the coefficient $c_i^{(T)}$ of the state vector $|\psi_T\rangle$. The updating $|\psi_{T+1}\rangle = U|\psi_T\rangle$, is given by a band-diagonal unitary operator $(UU^+ = I)$. The band-diagonality of U corresponds to a local interaction. After normalization based on $N(T)$, it is: $\sigma_i^{(T+1)} = \dfrac{1}{\sqrt{N(T)}}\left(i\delta\sigma_{i-1}^{(T)} + \sigma_i^{(T)} + i\delta^*\sigma_{i+1}^{(T)}\right)$ [155]. This would become with memory: $\sigma_i^{(T+1)} = \dfrac{1}{\sqrt{N(T)}}\left(i\delta m_{i-1}^{(T)} + m_i^{(T)} + i\delta^* m_{i+1}^{(T)}\right)$. In the historic component in Table 12.3, m is the mean value of the two last states. The Grössing-Zeilinger approach really concerns what one would call today a quantum random walk [304].

Table 12.3 Evolving patterns of the probabilities $|\sigma_i|^2$ of the quantum CA described in text with $\rho = 20$ up to $T = 11$ from a single $\sigma = 1$ cell.

Ahistoric

```
                       .01         .03
              .01      .04         .06
          .05      .10     .12 .01 .14
      .17      .23         .24 .01 .23 .01
  .50      .45 .01 .39 .01 .34 .01 .31
1.0    .66     .51 .01 .42 .01 .36 .02
  .50      .45 .01 .39 .01 .34 .01 .31
      .17      .23         .24 .01 .23 .01
          .05      .10     .12 .01 .14
              .01      .04         .06
                       .01         .03
```

Historic

```
                  .01 .01         .03 .03
          .01 .05 .02 .02 .10 .04 .02
      .10 .14         .14 .20 .01 .14 .22
  .50 .22 .07 .43 .18 .04 .38 .18 .03
1.0    .36 .56 .03 .30 .45 .02 .23 .39
  .50 .22 .07 .43 .18 .04 .38 .18 .03
      .10 .14         .14 .20 .01 .14 .22
          .01 .05 .02 .02 .10 .04 .02
                  .01 .01         .03 .03
```

Loop Quantum Gravity (LQG) is a proposed theory which attempts to reconcile the theories of quantum mechanics and general relativity. LQG preserves many of the main features of general relativity, while at the same time employing quantization of both space and time at the Planck scale, much in the spirit of CA [373]. Loop quantum cosmology predicts that, in simple models, the big bang singularity of classical general relativity is replaced by a quantum bounce. Because of the extreme physical conditions near the bounce, a natural question is whether the universe can retain, after the bounce, its memory about the previous epoch. More precisely, whether the universe retain various properties of the state after evolving unitarily through the bounce or it suffer from some cosmic amnesia. The latter conclusion is achieved by Lee Smolin [374].

12.2 Finite difference equations

12.2.1 *One-dimensional maps*

Memory can be embedded in conventional, non-spatialized finite difference equations (often referred as *discrete dynamical systems*) : $x_{T+1} = f(x_T)$ by means of $x_{T+1} = f(m_T)$, with m_T being an average value of the past states.

Thus, perhaps the canonical example, the logistic map $x_{T+1} = \lambda x_T (1 - x_T)$, becomes with memory $x_{T+1} = \lambda m_T (1 - m_T)$. All the simulations in the logistic map are attracted to $-\infty$ for λ values beyond 4.0 in the conventional model. Because of that, Figs .12.1 and 12.9 show the effect of memory on the logistic map with $\lambda = 4$, so with a fixed point : $x = 0.75$.

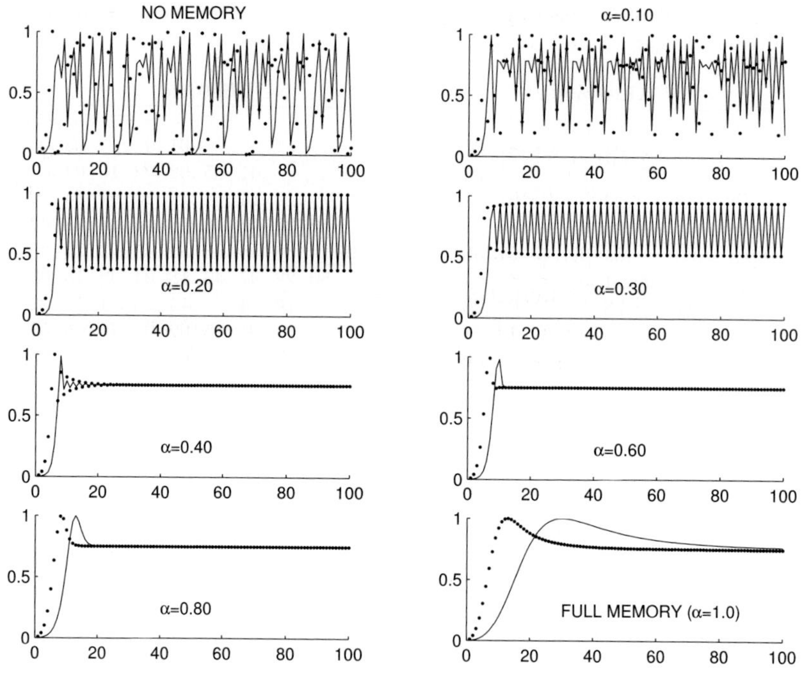

Fig. 12.1   The effect of unlimited trailing memory on the logistic map : $x_{n+1} = m_n + 4m_n(1 - m_n)$. A simulation starting at $x_1 = 0.0001$ is shown with unmarked values joined; that starting at $x_1 = 0.001$ with unjoined dots.

In a general form, at time-step T it is : $\quad m_T = \displaystyle\sum_{t=1}^{T} p_T^{(t)} x_t, \quad$ with

the probabilistic-like normalization condition: $\sum_{t=1}^{T} p_T^{(t)}=1$, $p_T^{(t)} > 0$
[12, 166, 268].

Figure 12.1 shows the effect of average memory:

$$m_T = \frac{x_T + \sum_{t=1}^{T-1} \alpha^{T-t} x_t}{1 + \sum_{t=1}^{T-1} \alpha^{T-t}} \equiv \frac{\omega_T}{\Omega(T)} \ .$$

A low level of memory ($\alpha = 0.1$) smooths the characteristic chaotic dynamics of the standard ahistoric formulation; higher α values generate periodic behavior, and from $\alpha = 0.4$ the evolution leads to the fixed point. It moves from chaos to order by varying not the parameter of the model ($\lambda = 4.0$), but the degree of memory incorporated.

Other approaches to the study of the long-term memory effect in the logistic map either compute the new state as a weighted average of previous states, such as in [377], or alter only one of the arguments of the application, e.g., the linear term in [147], in the latter case, in such a way that the map preserves the fixed point (λ-1)/λ, regardless of the charge of memory implemented (provided that λ lies in the $[0, 4]$ interval). In the study of the periodically perturbed logistic equation made in [175]: $x_{T+1} = (\lambda + (-1)^T \epsilon) x_T (1 - x_T)$, the perturbed parameter in also maintained below 4.0.

Figure 12.2 shows the bifurcation diagram of the logistic map, in its conventional formulation and with α-memory. The Fig. 12.2 shows the last fifty state values in simulations of one thousand iterations starting from $x_1 = 0.10$. At variance with what happens in the ahistoric formulation, with memory it is possible to consider values of the parameter beyond $\lambda = 4.0$.

The inertial effect that α-memory exerts, causes an overall *delay* in the aspect of the bifurcation diagram of Fig. 12.2. Thus, for example, memory delays the λ value at which the first bifurcation arises. These parameters, and the corresponding x state values at which the first bifurcation arises, are given in Table 12.4.

Table 12.4   Parameters of the first bifurcation in the logistic map with α-memory.

	α						
	0.0	0.1	0.2	0.3	0.4	0.5	0.6
λ_1	2.997	3.219	3.491	3.853	4.321	4.992	5.981
x_1	0.666	0.689	0.714	0.740	0.768	0.799	0.833

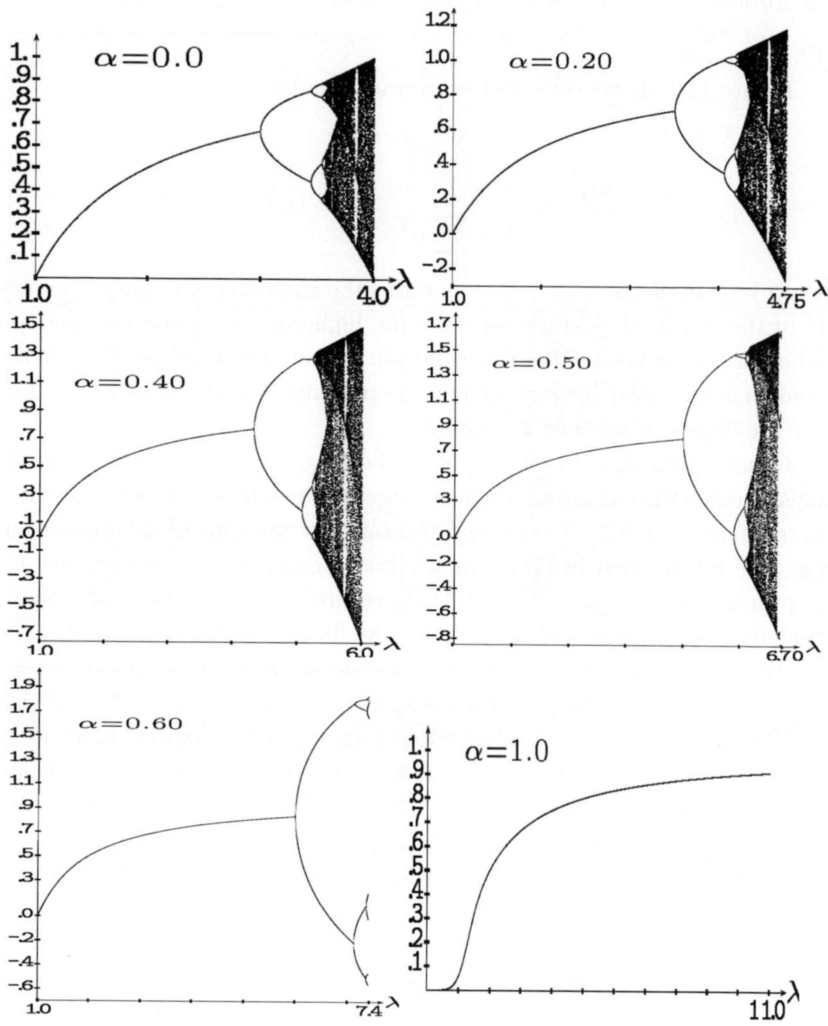

Fig. 12.2   Bifurcation diagram of the logistic map with α-memory.

In the same vein, not too high levels of memory just *delays* the emergence of chaos to higher values of the parameter. Thus, in the $\alpha = 0.2$ and 0.4 panels of Fig. 12.2, the chaotic behaviour appears beyond the ahistoric $\lambda = 3.570$ [287]. Windows of stability remain in the chaotic regions of the $\alpha = 0.2$ and 0.4 simulations, as is better appreciated in their zoomed diagrams in Fig. 12.3. The general aspect of the ahistoric bifurcation diagram seems preserved when implementing low memory, though extending the range of state values, that may be negative and greater than 1.0. Incidentally, the bifurcation diagrams in the chaotic regions with α in the $[0.2, 0.4]$ interval are reminiscent to that of the ahistoric logistic map with the parameter λ in the $[-2, -1]$ interval. Thus, for example, with $\lambda = -2$, the state values vary in the $[-0.5, 1.5]$ interval as shown in the top-left frame of Fig. D.1 of Appendix D.

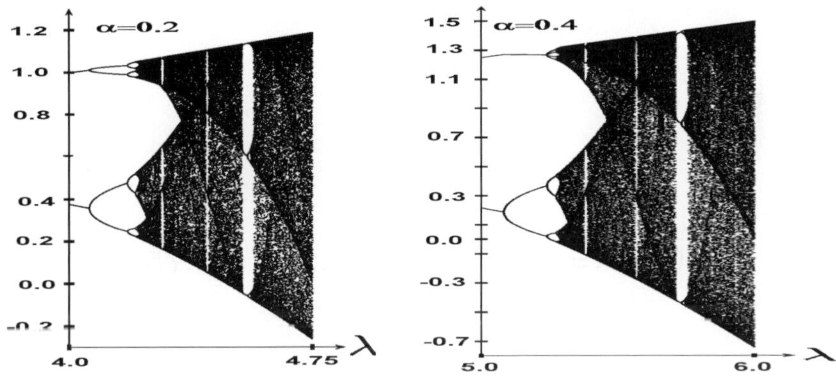

Fig. 12.3   Zoom of the bifurcation diagram of the logistic map with $\alpha = 0.2$ and $\alpha = 0.4$ memory.

Further increasing the memory charge, e.g., $\alpha = 0.5$, shortens the λ-interval in which the chaotic regime emerges. By $\alpha = 0.6$ the studied system shows only three bifurcations before diverging. With full memory (the so-called Cesáro iteration process), the system converges to the fixed point of the logistic map, i.e., $x^* = (\lambda - 1)/\lambda$, up to $\lambda \simeq 11.0$, then it diverges.

The effect of memory when $\lambda < 0$ is shown in the bifurcation diagram of Fig. D.1 of Appendix D. In parallel with what happens in the $\lambda > 0$ scenario, memory allows for values of λ lower than $\lambda = -2$, the minimum attainable in the conventional memoryless scenario. Low values of α, delay and amplify the chaotic region, as shown in Fig. D.1 for $\alpha = 0.2$, 0.4 and 0.5. Higher memory charges only enable a narrow region with bifurcations before divergence, only one in the $\alpha = 0.7$ example in the figure. Full memory permits the map to reach $\lambda \simeq -8.7$, but a zero-level.

Figure 12.4 shows the return maps $((x_T, x_{T+1}))$ of the conventional logistic equation and of three logistic equations with α-memory and high λ values. The maps show the first one thousand iterations starting from $x=0.1$, as in Fig. 12.2. The upper diagrams are zoomed compared with lower ones in Fig. 12.2, and the unit square is drawn in every case to facilitate the comparison. The characteristic *parabolic* aspect of the return map of the ahistoric logistic equation becomes *deformed* in the chaotic contexts with memory and $\lambda > 4.0$.

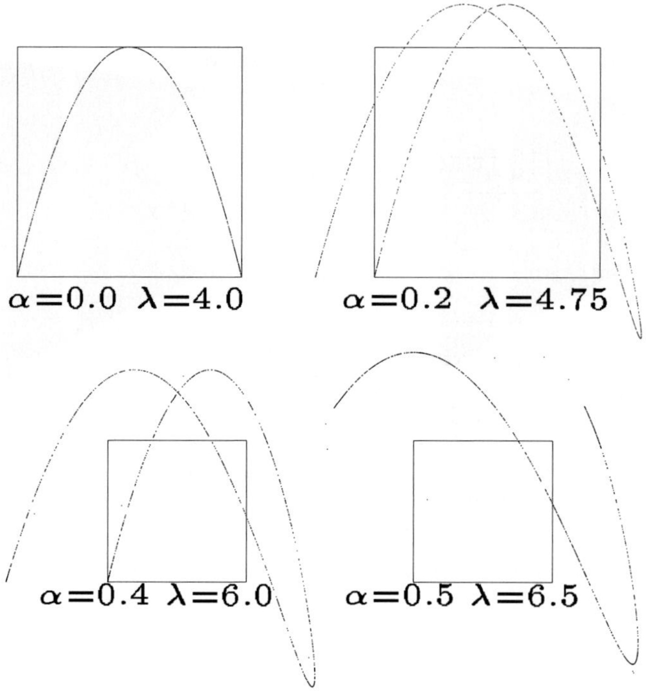

Fig. 12.4   Return maps of the logistic map with α-memory.

The three-dimensional return maps $((x_T, x_{T+1}, x_{T+2}))$ of the simulations with two-dimensional return maps given in Fig. 12.4 are shown in Fig. 12.5. These three-dimensional phase-diagrams reveal the stretching and folding structure of the discrete logistic equation.

Figure 12.6 shows the cobweb diagrams in the thirty first iterations of the simulations in Fig. 12.4. The arrows describe the transitions: $(x_{T-1}, x_T) \rightarrow (m_T, m_T)$, and $(m_T, m_T) \rightarrow (x_T, x_{T+1})$. Again, memory deforms the trait *orthogonal* aspect of the cobweb of the logistic map,

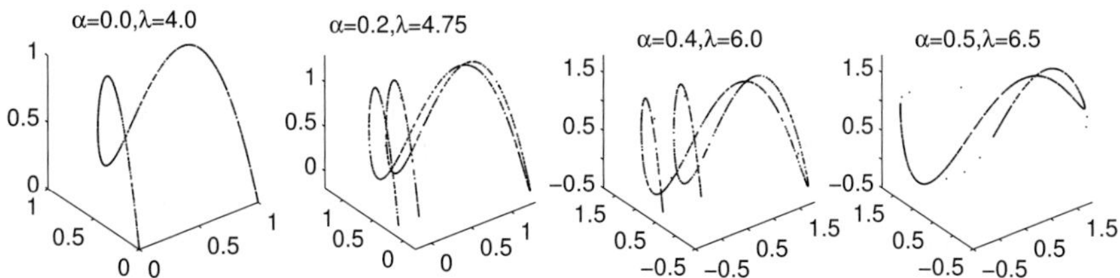

Fig. 12.5   The 3D return map of the simulations with α-memory in Fig. 12.4 .

by, let us say, *stretching* and *shearing* it.

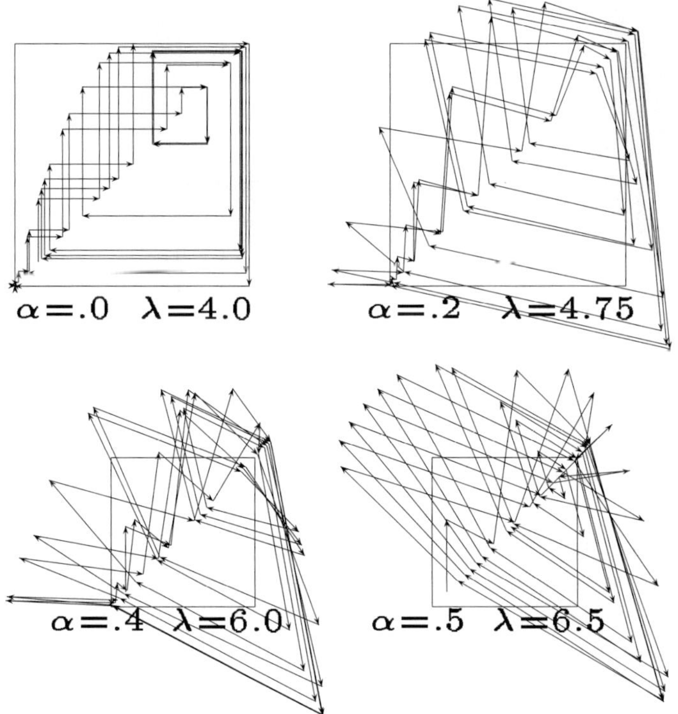

Fig. 12.6   Cobweb of the first thirty iterations in Fig. 12.4.

When taking α as the bifurcation parameter, the diagram obtained is, *prima facie*, similar to the classical diagram with respect to the λ parameter, with the orientation *reversed*. Thus, Fig. 12.7 shows the bifurcation diagram for two given λ values and varying α memory, acting as the bifur-

cation parameter. In both scenarios, increasing memory leads the evolution
from the chaotic context to more predictable asymptotic situations. A diagram like those in Fig. 12.7 is given in [12] for $\lambda = 4.0$, in which case
the variation of the state values remains in the $[0, 1]$ interval, and the diagram converges to the fixed point $x=3/4$ as soon as the parameter α is
not too small, namely $\alpha > 3/7 \simeq 0.33333$. By contrast, in Fig. 12.7 the
interval of variation of x is notably amplified and the α values to which the
convergence to a sole state is achieved are notably increased.

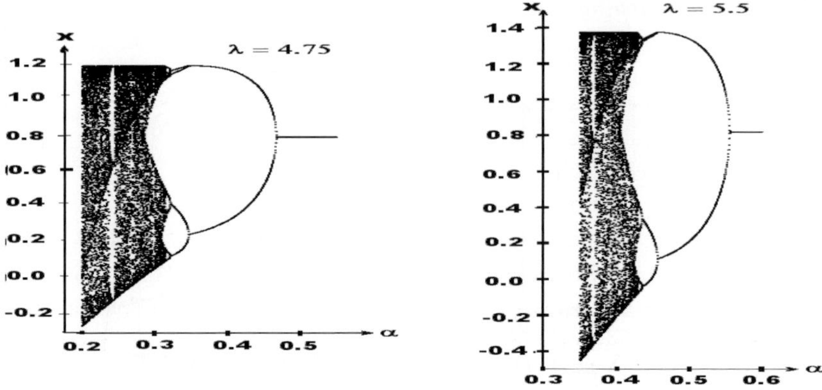

Fig. 12.7   Bifurcation diagram of the logistic map with two fixed λ values and variable
α-memory.

A similar inertial effect of memory has been found in other discrete
chaotic maps. Thus for example, the exponential map (commented in the
seminal paper [287]): $x_{T+1} = x_T \exp(\lambda(1 - x_T))$ [12.2.1], in which when
memory is endowed memory the first bifurcation and the chaotic region
begin beyond their ahistoric values $\lambda = 2.000$ and $\lambda = 2.6924$, as seen in
Fig. 12.8 in the case of $\alpha=0.3$.

Memory of the last two states may be implemented in the form:
$m_T = (1 - \epsilon)x_T + \epsilon x_{T-1}$, $0 \leq \epsilon \leq 1$. This is a more general model
than the α-based: $m_T = \dfrac{x_T + \alpha x_{T-1}}{1 + \alpha}, 0 \leq \alpha \leq 1$, because it allows for a
higher contribution of the past than of the present state. The maximum
attainable in the α memory model is that of $\alpha = 1$, which corresponds
to $\epsilon = 0.5$. Higher levels of the tunable parameter tends to recover the
ahistoric scenario. In the extreme case, if $\epsilon = 1$ it is $m_T = x_{T-1}$, so that
every state of the ahistoric evolution is generated twice. This $\tau=2$ memory

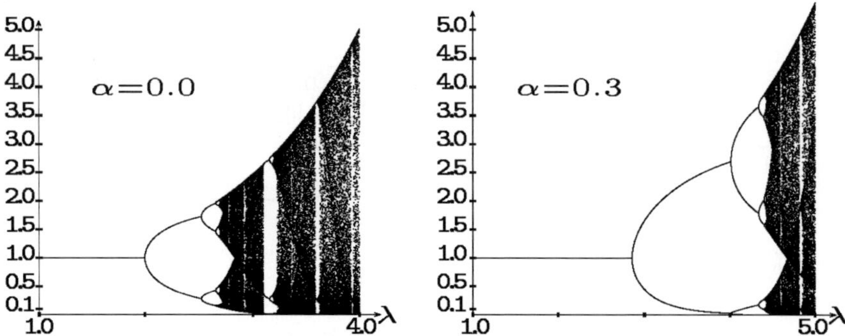

Fig. 12.8   The bifurcation diagram of the exponential map [12.2.1]. Ahistoric and $\alpha{=}0.3$ memory models.

implementation will be often referred as ϵ-memory.

Figure 12.9 shows the effect of such $\tau = 2$ ϵ-memory on the logistic map. Again, a low level of memory of the recent past ($\epsilon = 0.1$) smooths the ahistoric chaotic behavior, higher memory, $\epsilon = 0.2$, generates period-two oscillator, and $\epsilon = 0.3$ leads to the fixed point. Unexpectedly, period-two oscillators of broad amplitude appear again for $\epsilon = 0.5$. Paying more attention to the recent past state than to the last one, i.e., $\epsilon > 0.5$, does not lead to a fixed point nor to periodic behavior but to irregular dynam ics studied in [12] by resorting to the two-dimensional equivalent system:

$$x_{T+1} = f_\lambda((1 - \epsilon)x_T + \epsilon y_T), \quad y_{T+1} = x_T .$$

Figure 12.10 shows some bifurcation diagrams with ϵ-memory in the same scenario of Fig. 12.2: $x_1 = 0.10$, 1000 iterations, last 50 state shown, $\Delta\lambda{=}0.001$. Memory of only two states, enables for $\lambda > 4.0$, as in the case of unlimited trailing memory, albeit, as Fig. 12.10 shows, with only two time-stems memory, the amplification of the range of enabled λ values is notably shorter. Low memory of the recent past as $\epsilon \leq 0.5$ delays the emergence of the bifurcation and the chaos, as is shown in the figure in the case of $\epsilon = 0.5$. This (unweighted) mean memory of the last two states tends to generate a notable discontinuity around $\lambda = 4.0$, which in the (top left) simulation of Fig. 12.10 happens at $\lambda = 3.933$. There is no symmetry in the effect of ϵ-memory: the diagrams for $\epsilon = 0.5 + \delta$ and $\epsilon = 0.5 - \delta$ are dissimilar, as is envisaged in Figure 12.10 and is stressed in Figure 12.14.

Figure 12.11 shows the return maps of the four logistic equations with ϵ-memory of 12.10 and the highest λ values allowed. The maps show the first one thousand iterations starting from $x{=}0.1$, as in Fig. 12.4. The

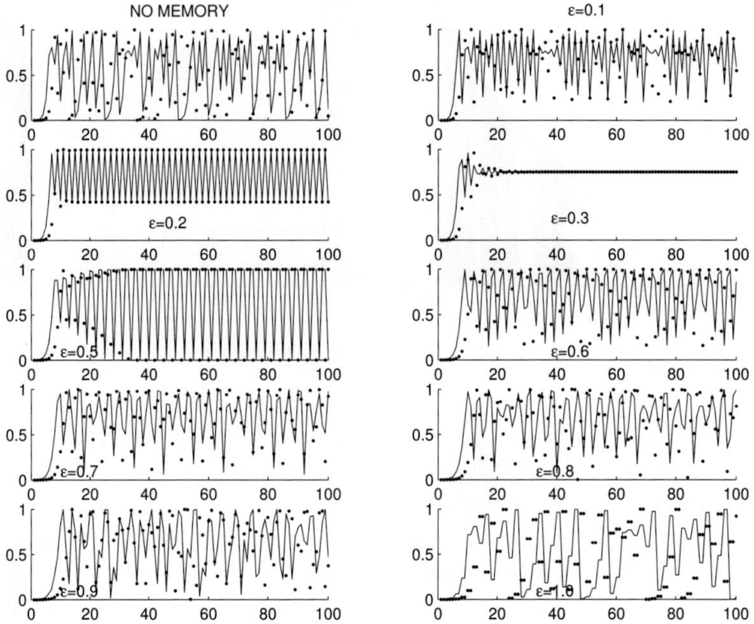

Fig. 12.9   The logistic map with memory of the last two states of type: $m_T = (1 - \epsilon)x_T + \epsilon x_{T-1}$.

unit square is drawn in every panel to provide a measure reference. The scattered aspect of the maps in Fig. 12.11 notably differs from the much more *linear* in Fig. 12.4.

The three-dimensional return maps of the simulations with two-dimensional return maps given in Fig. 12.11 are shown in Fig. 12.12. Again, as in the two-dimensional case, the three-dimensional phase-diagrams in Fig. 12.12 very much differ from that in Fig. 12.5.

Figure 12.13 shows the cobweb diagrams in the thirty first iterations of the simulations in Fig. 12.11. Again, memory deforms the trait *orthogonal* aspect of the cobweb of the logistic map, now in a rather unstructured way, fairly different to that in Fig. 12.6.

Figure 12.14 shows the bifurcation diagram for four given $\lambda > 0$ values and varying ϵ memory. In the $\lambda = 4.0$ scenario, the diagram shows a notable discontinuity at $\epsilon = 0.4573$. Under this value, the $\lambda = 4.0$ figure panel resembles the diagrams in Fig. 12.7, but the for $\epsilon > 0.4573$, it has a fairly different structure. The bifurcation diagram in the $\lambda = 4.20$, 4.25 and

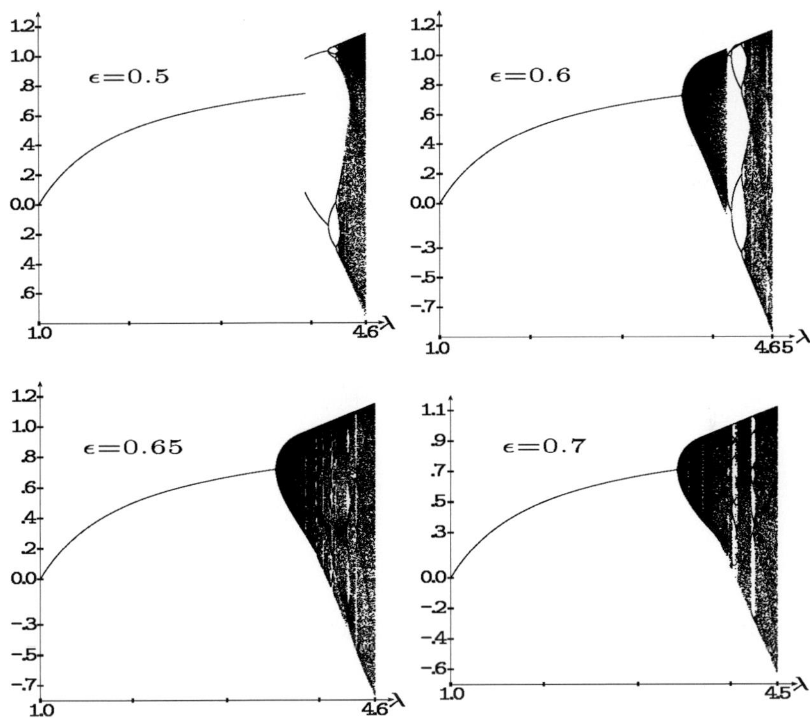

Fig. 12.10   The bifurcation diagram of the logistic map with $\tau=2$ memory.

4.30 scenarios in turn are reminiscent of the corresponding ϵ-sections in the $\lambda = 4.0$ panel, albeit with a broader spectrum of state values allowed. In every case, the discontinuity remains not far beyond $\epsilon =0.4$, and after this, an interval of non-chaotic behaviour, which is amplified for higher λ values, precedes the emergence of chaos with higher ϵ parameter values.

The bifurcation diagram for a set of given $\lambda < 0$ values and varying ϵ memory is shown in Fig. D.3 in Appendix D. Every diagram shows convergence to zero in an interval around $\epsilon=0.35$. This interval is amplified as ϵ increases, covering most of the $[0, 1]$ ϵ interval at $\lambda = -1.25$. Against this, for the smallest λ allowed, e.g., $\lambda = -2.60$, the interval with convergence to zero is very short, and the diagrams show a dramatic chaotic behaviour for large ϵ values. The diagram for the lowest negative λ allowed in the ahistoric context, thus, $\lambda = -2.0$ resembles to that of the highest ahistoric $\lambda = 4.0$ shown in Fig. D.3. Thus, for example, there is a notable

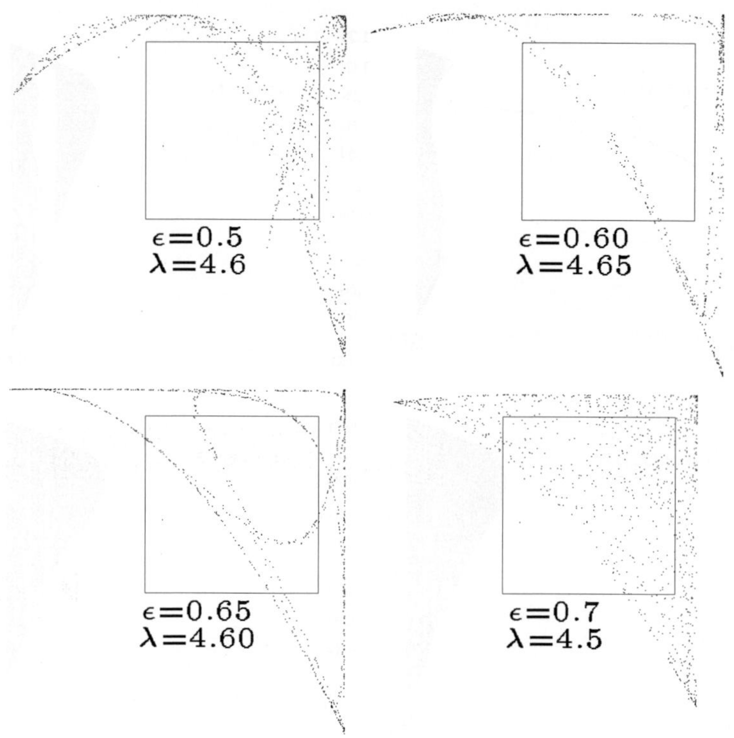

Fig. 12.11   Return maps of the logistic equation with ϵ-memory.

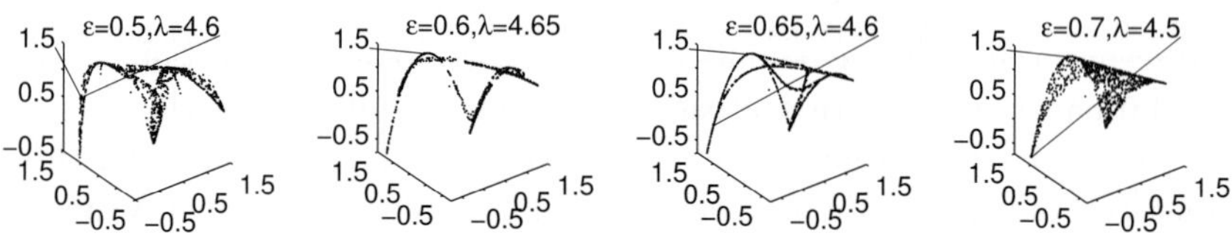

Fig. 12.12   The 3D return map of the simulations with ϵ-memory in Fig. 12.11 .

discontinuity, at $\epsilon = 0.50$ in the negative parameter context.

The term *memory* is often used in the context of discrete dynamic systems as meaning incorporating *delay* terms in the dynamics, as in the simple time-delayed logistic equation: $x_{T+1} = \lambda x_T(1 - x_{T-1})$ [285]. Related somewhat to this approach, partial memory may be implemented as: $x_{T+1} = \lambda(x_T - m_T^2)$, $x_{T+1} = \lambda(m_T - x_T^2)$, $x_{T+1} = \lambda x_T(1 - m_T)$, or $x_{T+1} = \lambda m_T(1 - x_T)$.   We have not found interesting dynamics

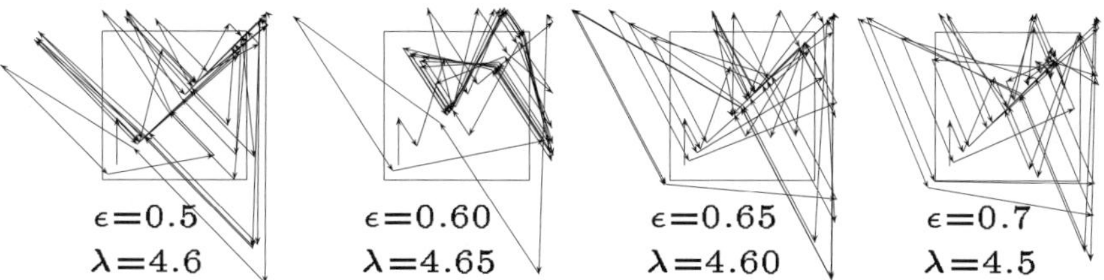

Fig. 12.13   Cobweb of the first thirty iterations in Fig. 12.11.

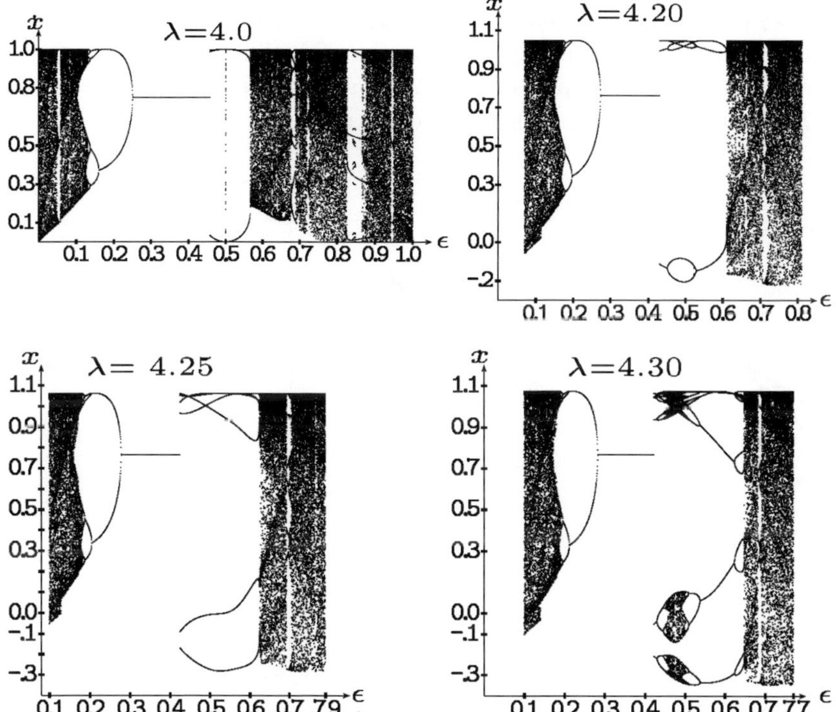

Fig. 12.14   The bifurcation diagram of the logistic map with $\tau=2$ memory at four given λ values.

with this kind of partial memory in the logistic map, but other maps may support them.   More complicated delay approaches are given as: $x_{t+1} = [r_0 - (4 - r_0)tanh(\dfrac{x_t - x_{t-1}}{\epsilon})]x_t(1 - x_t)$ in [130, 305], or by perturbing the quadratic map $x_{t+1} = a - x_t^2$ via reinjection of the *field* $x_{t-\tau}$

with a feedback amplitude b: $x_{t+1} = a - x_t^2 + bx_{t-\tau}$ [158, 159].

12.2.2   *Two-dimensional maps*

It would be interesting to study how memory operates on coupled discrete systems. With logistic maps in particular [185, 221], such as in the maps given by the equations:
$$x_{T+1} = \lambda_x(3y_T + 1)x_T(1 - x_T) \, , \quad y_{T+1} = \lambda_y(3x_T + 1)y_T(1 - y_T) \, .$$
This system exhibits chaotic behaviour when λ_x and λ_y lie in the neighbourhood of the region $[1.032, 1.0843]$. Endowing memory, the system turns out: $x_{T+1} = \lambda_x(3m_T^y + 1)m_T^x(1 - m_T^x), \, y_{T+1} = \lambda_y(3m_T^x + 1)m_T^y(1 - m_T^y)$. The previously mentioned interval becomes amplified with memory, as shown in the two examples of Fig. 12.15, starting from $x_1 = y_1 = 0.1$ up to $T=1000$ and endowed with low levels of memory.

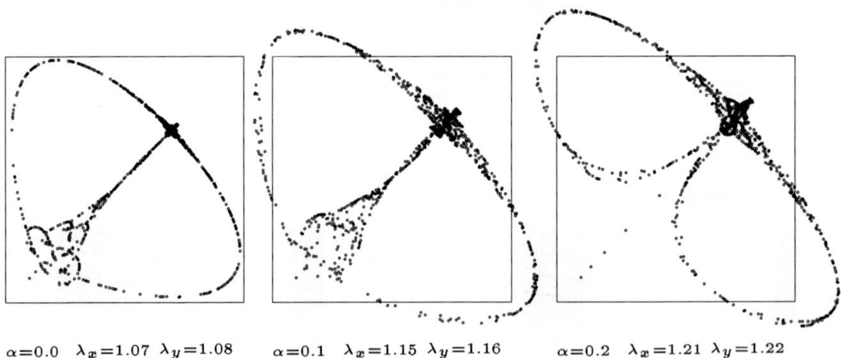

Fig. 12.15   Coupled logistic maps with memory.

In Fig. 12.16, the logistic-like coupled equations are :
$$x_{T+1} = \lambda_x y_T + \lambda_y - x_T^2 \, , \quad y_{T+1} = \lambda_x x_T + \lambda_y - y_T^2 \, .$$
The patterns in Fig. 12.16 display 13000 points starting from $x = 1.0 \times 10^{-7}$, $x = 2.0 \times 10^{-7}$. The particular properties regarding symmetry and self-similarity of this kind of coupled maps are studied in [290]. Symmetry seems fairly well preserved in the simulations with memory in Fig. 12.16. Notice that the simulations with memory and greater λ values are re-scaled, because the range of the point coordinates notably grows in these scenarios.

The quadratic map known as the Henon map [188], :
$$x_{T+1} = 1.0 - \lambda_x x_T + y_T^2 \, , \quad y_{T+1} = \lambda_y x_T \, ,$$
is shown in Fig. 12.17 starting from $x_1=0.1$, $y_1=0.2$. In the ahistoric map it

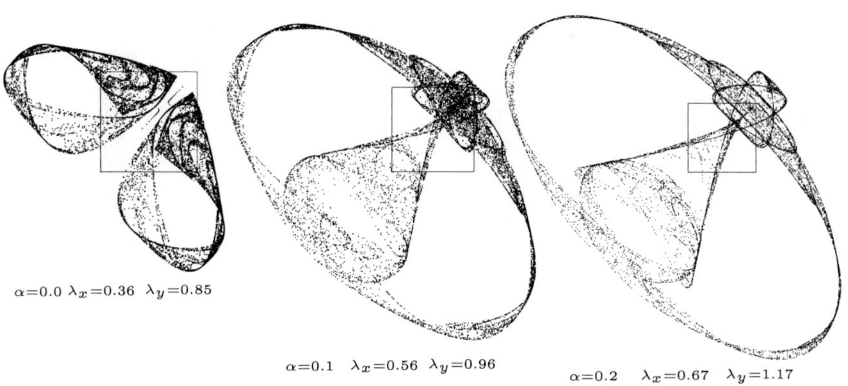

Fig. 12.16   Coupled logistic-like maps with memory.

is λ_x=1.4 and λ_y=0.3, whereas α=0.2 memory allows for a higher λ_x=2.2, with fixed λ_y=0.3 .

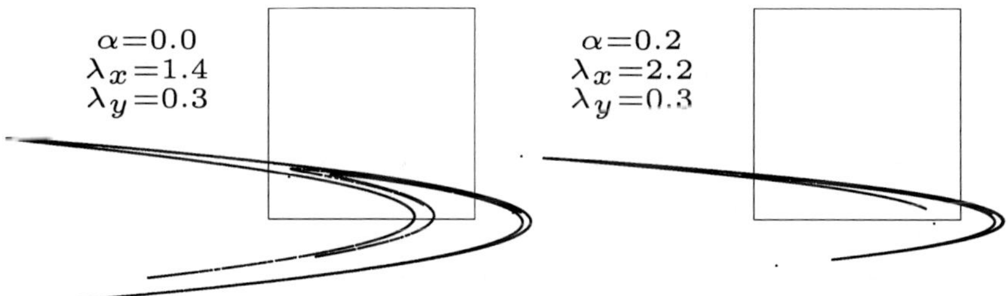

Fig. 12.17   The Henon map. Ahistoric and α=0.2 memory models.

Figure 12.18 shows the bifurcation diagram of the Henon map in the ahistoric and α=0.2 memory models with λ_y=0.3, and $\Delta\lambda_y$=0.001 . Memory allows for values of λ_x up to 2.4 .

The piecewise linear mapping of the plane known as the *Gingerbreadman* map [121, 123] , is defined by:

$$x_{T+1} = 1.0 - y_T + |x_T| , \quad y_{T+1} = x_T .$$

The map is shown in Fig. 12.19 up to T=10000 starting from three different initial points: x_1= -0.1, y_1=0.1, x_1= -0.1, y_1=1.1 (blue marked), and x_1=1.5, y_1=1.5 (red marked) . This area conserving map (framed in the $[-3, 8] \times [-3, 8]$ square) is chaotic in its filled region and stable in the six hexagonal empty regions, as illustrated when starting from (1.5,1.5) in the figure. Memory on the Gingerbreadman map tends to lead the dynamics

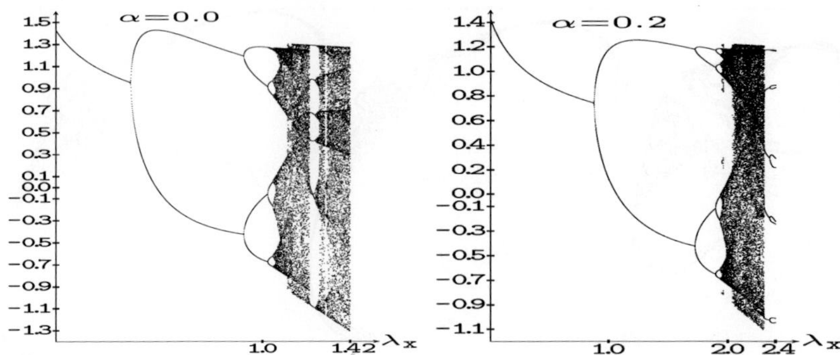

Fig. 12.18   The bifurcation diagram of the λ_x=0.3 Henon map. Ahistoric and α=0.2 memory models.

to the proximity of its fixed point $(1,1)$. To facilitate the visualization of this convergence, the consecutive points in the α=0.1 simulation have been joined. The evolution in the α=0.5 and 1.0 cases appears clear without joining the series of points. The convergence to $(1,1)$ is straitforward in the no full-memory simulations, but when α=1.0, only the trend to $(1,1)$ is marked.

Even very low levels of memory tend to drive the dynamics in the Gingerbreadmap map to $(1,1)$, albeit in fairly tortuous ways. Thus for example, in Fig. 12.20 the three *particles*, not only that starting in $(1.5,1.5)$, seem trapped in central hexagonal region, somehow directed towards $(1,1)$, not describing periodic orbits as $(1.5,1.5)$ does in the ahistoric model.

Even a very little charge of memory dramatically alters the characteristic tessellation produced by the Zaslavsky web map [443]:

$$x_{T+1} = (x_T + K\sin y_T)\cos(2\pi/q) + y_T\sin(2\pi/q),$$
$$y_{T+1} = -(x_T + K\sin y_T)\sin(2\pi/q) + y_T\cos(2\pi/q).$$

Figure D.6 shows an example with K=1.2, q=6.0, starting from $x_1 = 2.86\pi$, $y_1 = 0.75\pi$. The patterns with memory are 8 times zoomed with respect to the ahistoric one.

Coupled maps may be delayed, rather *out of phase*, such as in:

$$x_{T+1} = \lambda_x(3y_T + 1)x_T(1 - x_T), \quad y_{T+1} = \lambda_y(3x_{T+1} + 1)y_T(1 - y_T).$$

This system exhibits chaotic behaviour when λ_x and λ_y both lie in the neighbourhood of the region $[1.032, 1.19]$. Memory may be implemented in maps with delays in various forms. Thus for example, in the simulations with memory in Fig. 12.21, the term *out of phase* remains unaltered:

$$x_{T+1} = \lambda_x(3m_T^y + 1)m_T^x(1 - m_T^x), \; y_{T+1} = \lambda_y(3x_{T+1} + 1)m_T^y(1 - m_T^y).$$

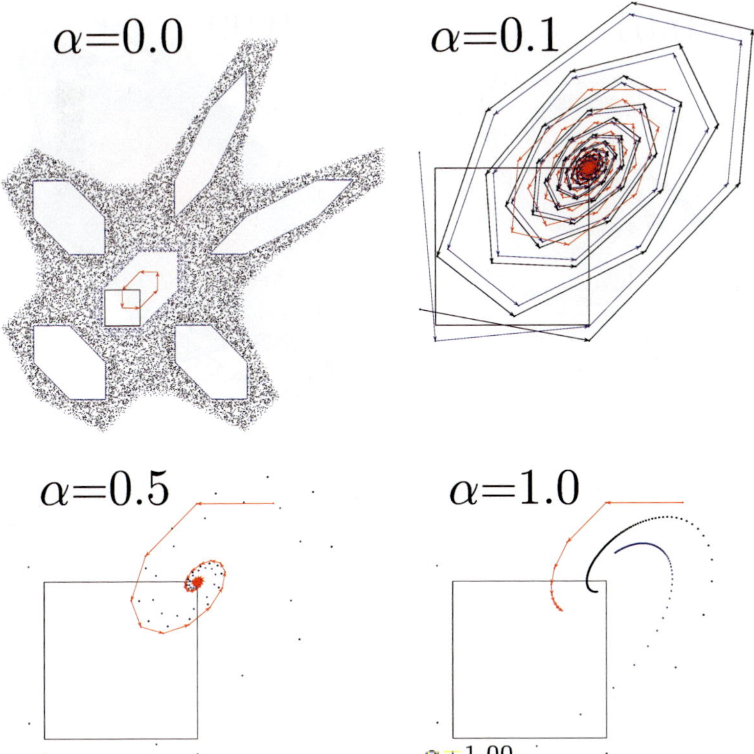

Fig. 12.19   The Gingerbreadman map. Ahistoric and α-memory models. Initial points: (-0.1,0.1), (-0.1,1.1) (blue marked), and (1.5,1.5) (red marked).

Figure 12.21 shows 13000 points starting from $x=0.50$, $y=0.55$, with both equations with the same λ parameter values, and the same low levels of memory of Fig. 12.21, which allow for $\lambda > 1.19$.

We plan to study the effect of memory in the spatialized logistic map [131, 259], starting from its simplest one-dimensional formulation [418]: $x^{(i)}_{T+1} = 1.0 - \frac{\lambda}{2} x^{(i)}_T (x^{(i-1)}_T + x^{(i+1)}_T)$, where i stands for the index of the lattice sites, which becomes with memory: $x^{(i)}_{T+1} = 1.0 - \frac{\lambda}{2} m^{(i)}_T (m^{(i-1)}_T + m^{(i+1)}_T)$.

Networks of logistic maps with auto-feedback memory have been considered in the study of the effect memory on synchronization [274, 280, 296].

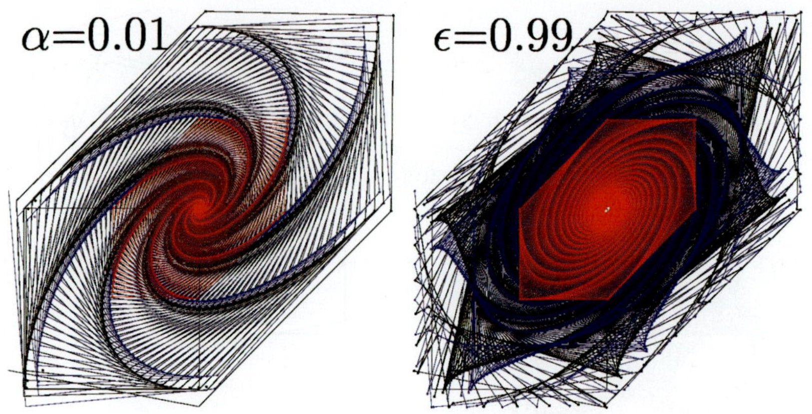

Fig. 12.20   The Gingerbreadman map with very low memory starting as in Fig. 12.19.

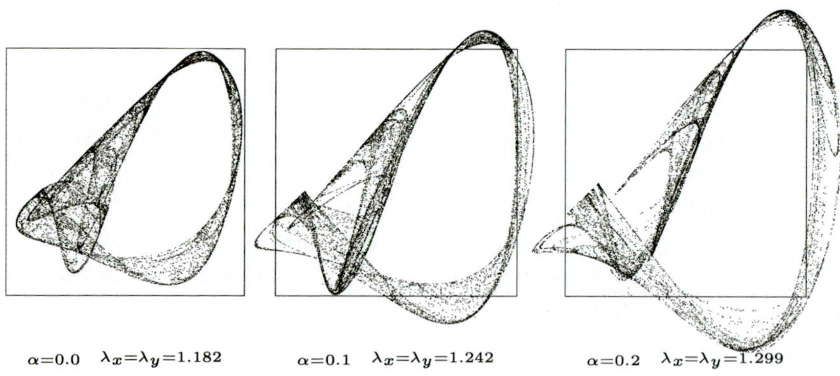

Fig. 12.21   Coupled logistic maps with delay and memory.

12.3   Plane curves

Curlicues

This section deals with a special type of curves, so called *curlicues* [81, 292, 368, 381]:

$$x_{T+1} = x_T + \cos(2\pi f(T)) \,, \quad y_{T+1} = y_T + \sin(2\pi f(T)) \,,$$

a Cartesian version of the complex formulation: $z_{T+1} = z_T + e^{2\pi i f(T)}$.

These polygonal curves advance iteratively by drawing from the generic point (x_T, y_T) a segment of fixed (unit) length in the direction given by the angle $2\pi f(t)$, or describing the dynamics in the complex plane, as a directed graph: $z_1 \to z_2 \to \ldots z_T \to z_{T+1} \to \ldots$ [1].

The generic point z_T is commonly referred to in the specialized literature as the exponential sum $S(T) = \displaystyle\sum_{t=1}^{T} e^{2\pi i f(t)}$. The sum S has been found in two physical contexts, namely the trace of the evolution operator for a quadrapole interaction of a quantum spin in an inhomogeneous field $\left(\mathrm{Tr}(U) = \displaystyle\sum_{n=-L}^{n=L} e^{-iE_n T/\hbar}\,[81,\,238]\right)$ and Fresnel diffraction by a grating with many sharp slits $\left(\psi(x,H) \propto \displaystyle\sum_{n=1}^{n=L} e^{-i\pi(\tau n^2 - \xi n)}\,[81,\,82]\right)$.

Embedding memory in z: $z_{T+1} = \overline{z}_T + e^{2\pi i f(T)}$, or :
$$x_{T+1} = \overline{x}_T + \cos(2\pi f(T))\,, \qquad y_{T+1} = \overline{y}_T + \sin(2\pi f(T))\,.$$
This kind of memory implementation does not alter the directions of the advance, and consequently the structure of the ahistoric pattern remain unchanged. This is shown in Fig. 12.22, run up to $T=5000$ for two $f(T)$ functions. Thus, the first row of figures shows the so-called *Nessie* curve: $f(T) = (log\,T)^4$, whereas the curves below correspond to $f(T) = sin\,(\sqrt{T})$, in which case the curve wanders off to the right, drawing double loops which grow slowly and stay separate from each other. In both types of curlicues in Fig. 12.22, memory does not significantly alter the structure of the ahistoric picture, it just reduces its *size*. In the Nessie scenario producing offspring-like pictures, and in the case of $f(T) = \sin(\sqrt{T})$ reducing both the size and the separation of the loops.

An important kind of curlicues has the form $f(T) = T^k/N$, with its corresponding S sum from $T=0$ up to $T = N - 1$ referred to as a Gauss sum [79]. Figures D.4 (the *bull* up to $T=1002$) and Fig. D.5 (the *bullring* up to $T=1013$) in the Appendix, show how memory in z acts on these curlicues in the *contractive* manner of Fig. 12.22. In the simple $k=2$ case, it is $S(N) = (1 + (-i)^N)\dfrac{1+i}{2}\sqrt{N}$, and the qualitative behaviour depends only on $N\,(mod\,4)$, because $(-i)^N$ depends only on $N\,(mod\,4)$ [247]. The role of $k=2$ Gauss sums in quantum mechanics has been studied in [64]. For $k>2$ [326, 325], or in the more general form $f(T) = aT^b$ (e.g., $b=1/2$ in [249]), the dynamics is far more complicated to analyze.

Unlike the afore-mentioned soft memory effect, a notable alteration of

---

[1] The MIT web addresses *http://scratch.mit.edu/projects/mathjp/xx* provide numerous examples (e.g., *xx= 517068, 512387, 357370, 27830*) of the appealing visualization of the generation of several types of curlicues.

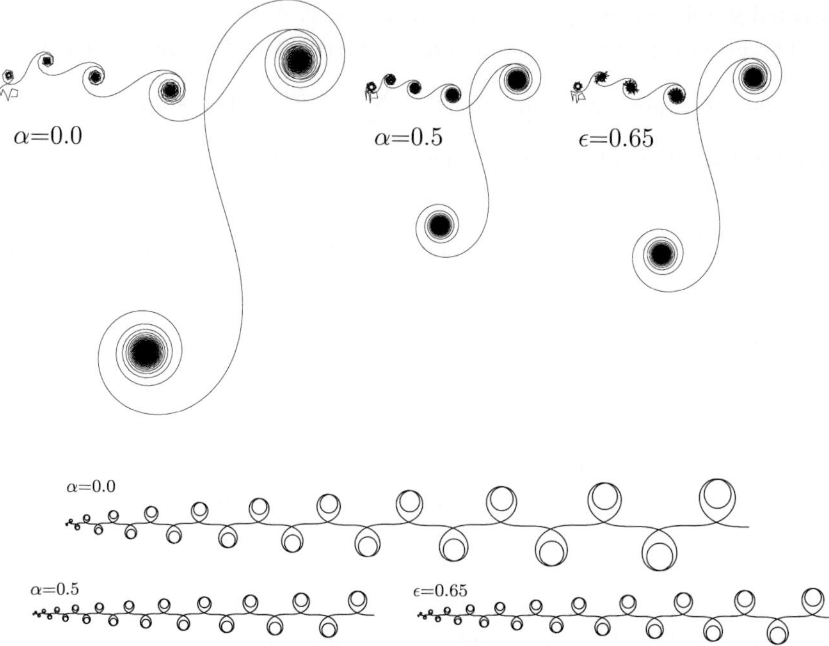

Fig. 12.22   The $f(T) = (log\,T)^4$ (Nessie) and $f(T) = sin\,\sqrt{T}$ curlicues with memory in z: $z_{T+1} = \overline{z}_T + e^{2\pi i f(T)}$.

the dynamics of the curlicues is expected as a result of embedding memory in f: $z_{T+1} = z_T + e^{2\pi i \overline{f}(T)}$, or :

$$x_{T+1} = x_T + \cos(2\pi \overline{f}(T)) , \quad y_{T+1} = y_T + \sin(2\pi \overline{f}(T)) .$$

But before paying attention to curlicues altered by memory in f, let us point out here that some curlicues are fairly unaffected by memory. This is the case of the $f(T) = sin\,\sqrt{T}$ curlicue, which demonstrates in Fig. 12.23 a high robustness in the preservation of the ahistoric structure, being really affected only by full α-memory. The *sin* function appears somehow revealed in such a scenario.

Also remarkable is that, in contrast to the just described minor effect of memory, in some cases memory supports the *emergence* of structures from a plain ahistoric pattern. Two examples with $N\,(mod\,4){=}2$ are given in Fig. 12.24 .

Two examples of the effect of memory in f altering, as expected, the evolving patterns are that on *Nessie* and in the $T^2/321$ curlicue, given in Fig. 12.26 and Fig. 12.28 repectively. The patterns in both figures are re-

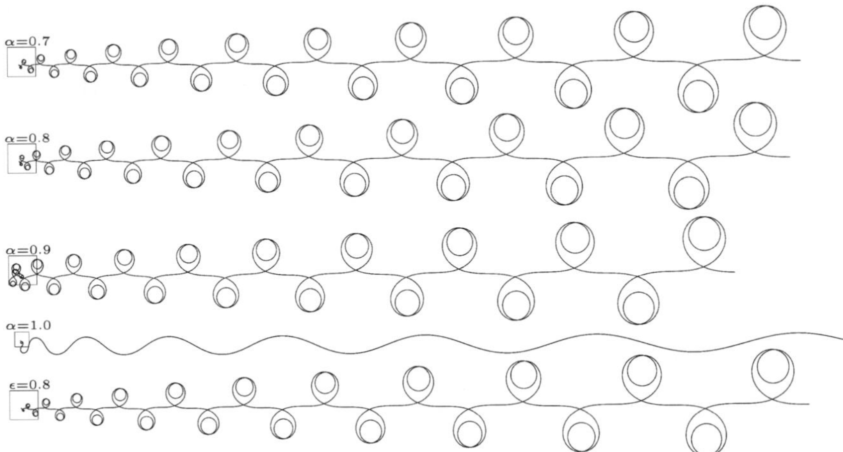

Fig. 12.23   The $f(T) = sin\sqrt{T}$ curlicue with memory in f.

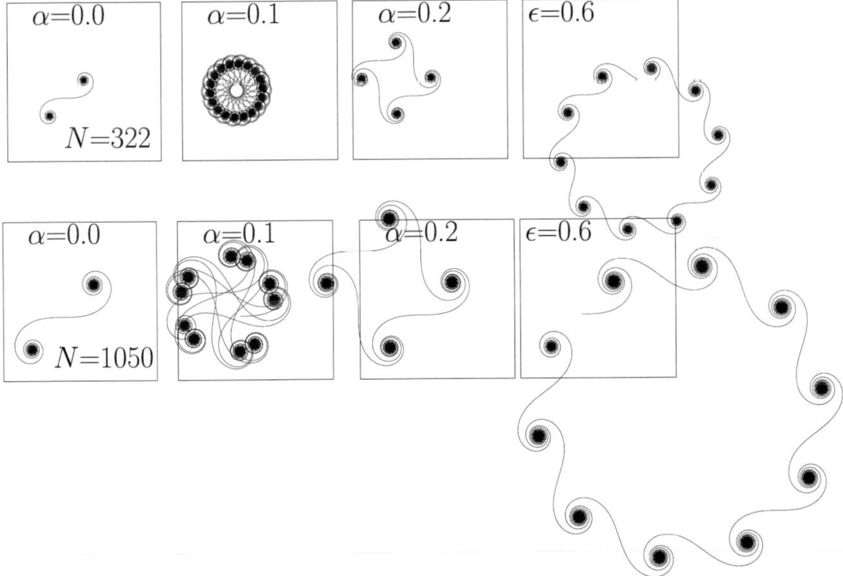

Fig. 12.24   The $f(T) = T^2/N$, N=232, 1050 curlicues with memory in f.

ferred to a common square. In both figures, the similarity of the α=0.5 patterns to the ahistoric one is unexpected. Memory seems to induce some kind of change in the standing way of Nessie, which, except with full memory, appears to keep its *body* elements. The fairly disrupting effect of α=0.6 memory in the $f(T) = (log\,T)^5$ curlicue is given in Fig. D.9 in the Appendix. Memory exerts a more dramatic effect on the $f(T) = T^2/321$ curlicue in Fig. 12.28, either by locking the structure (α=0.1, 0.3, 0.4, 0.7), restraining the evolution to just two concatenated Cornu-like spirals (α=0.2, 1.0), or changing the direction of the advance of the spirals (α=0.5, 0.6, 0.8, 0.9). In the latter case, the patterns produced with α=0.6 and 0.9 resemble the ahistoric one with $f(T) = T^2/323$, whereas that for α=0.8 resembles the ahistoric one with $f(T) = T^2/320$. The single Cornu-like spiral (see Fig. 12.24) produced by the ahistoric curlicues with $f(T) = T^2/N$, N=322, 1050 (recall that the qualitative behaviour in $f = T^2/N$ curlicues depends only on $N \;(mod\; 4)$) relates somehow to the two concatenated ones generated with α=0.2 and α=1.0 in Fig. 12.28.

The sum of the square powers of t is given analytically by: $\Sigma_2(T) =$ $\sum_{t=1}^{T} t^2 = T(T+1)(2T+1)/6$. Thus, in the $f(T) = T^2/N$ simulations with full memory it is: $\overline{f}(T) = (T+1)(2T+1)/6N$. Similarly, $\Sigma_3(T) = \sum_{t=1}^{T} t^3 =$ $T^2(T+1)^2/4$, and $\Sigma_4(T) = \sum_{t=1}^{T} t^4 = T(T+1)(2T+1)(2T^2+3T-1)/30$, thus, with full memory : $\overline{f}(T^3/N) = T(T+1)^2/4N$, and $\overline{f}(T^4/N) = (T+1)(2T+1)(2T^2+3T-1)/30N$.

In the k=1 Gauss scenario, it is $S(N) = \sum_{t=0}^{N-1} e^{2\pi it/N} = 0$. Thus, in the N=9 example of Fig. 12.25, the (0,0) point is reached at time-step T=N-1=8, and the initial (1,0) point is revisited for the first time at T=9, staring a new cycle. Commencing from T=1, the initial point is revisited at T=N=9. It is, $\Sigma_1(T) = \sum_{t=1}^{T} t = T(T+1)/2$, so, in $f(T) = T/N$ simulations with full memory it is: $\overline{f}(T) = (T+1)/2N$. Consequently, full memory generates cycles with T+1=2N points, e.g., the eighteen points in Fig. 12.25. Without full memory, the evolution does not revisit the initial point, as shown in Fig. 12.25 for α=0.8.

Figure D.8 Appendix D shows the effect of τ=2 ϵ-memory in Nessie and the $f(T) = T^2/321$ curlicues. In the latter case, ϵ >0.5 memory produces closed patterns, so that the pair of patterns for ϵ=0.7 and 0.8 are the same, albeit displaced and rotated. The same phenomenon happens with the

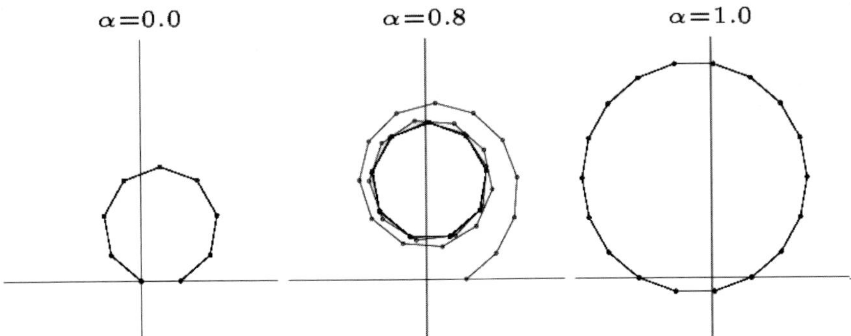

Fig. 12.25   The $f(T) = T/9$ curlicue with memory in f.

pairs $\epsilon=0.6$ (not shown) and 0.8, and $\epsilon=0.65$ and 0.85 (not shown).

As a rule, memory destroys very sophisticated ahistoric patterns as those in Figs. 12.27 and D.7. Nevertheless, for some memory charges, the patterns produced with memory are even more complex than those in the ahistoric formulation. Figures 12.27, and D.7 show some examples. The simulation of Fig. D.7 has been performed in FORTRAN quadruple precision, as double precision was not enough, due to the *explosive* calculus of T^7 involved.

Embedding memory in both z and f: $z_{T+1} = \overline{z}_T + e^{2\pi i \overline{f}(T)}$, or :

$$x_{T+1} = \overline{x}_T + \cos(2\pi \overline{f}(T)) , \quad y_{T+1} = \overline{y}_T + \sin(2\pi \overline{f}(T))$$

produces the patterns generated with only memory in f but contracted, as a result of the memory in z. Figure D.10 in Appendix D shows an example.

Partial memory, i.e., only in the x or y dynamics, produces new patterns. Figures D.11 and D.12 in Appendix D show examples of this. That of Nessie and the $T^2/321$ curlicues with memory endowed only in the f component of y: $x_{T+1} = x_T + \cos(2\pi f(T)) , \quad y_{T+1} = y_T + \sin(2\pi \overline{f}(T))$

Again, the patterns for $\alpha=0.5$ resemble the ahistoric one in both scenarios. Higher levels of memory, particularly full memory, generate very altered *Nessie* patterns. The patterns of the $T^2/321$ with partial memory are notably different to those with memory in the f component of both x and y in Fig. 12.28, although with $\alpha= 0.6$, 0.8 and 0.9 some parallelism regarding the effect of memory can be traced, because in both scenarios the direction of the evolution is deviated. The effect of $\tau=2$ ϵ-memory in Nessie and the $f(T) = T^2/321$ curlicues is shown in Fig. D.13 in the Appendix D. Unlike in Fig. D.8 in the Appendix D, partial ϵ-memory does not produce closed patterns in the $f(T) = T^2/321$ curlicue.

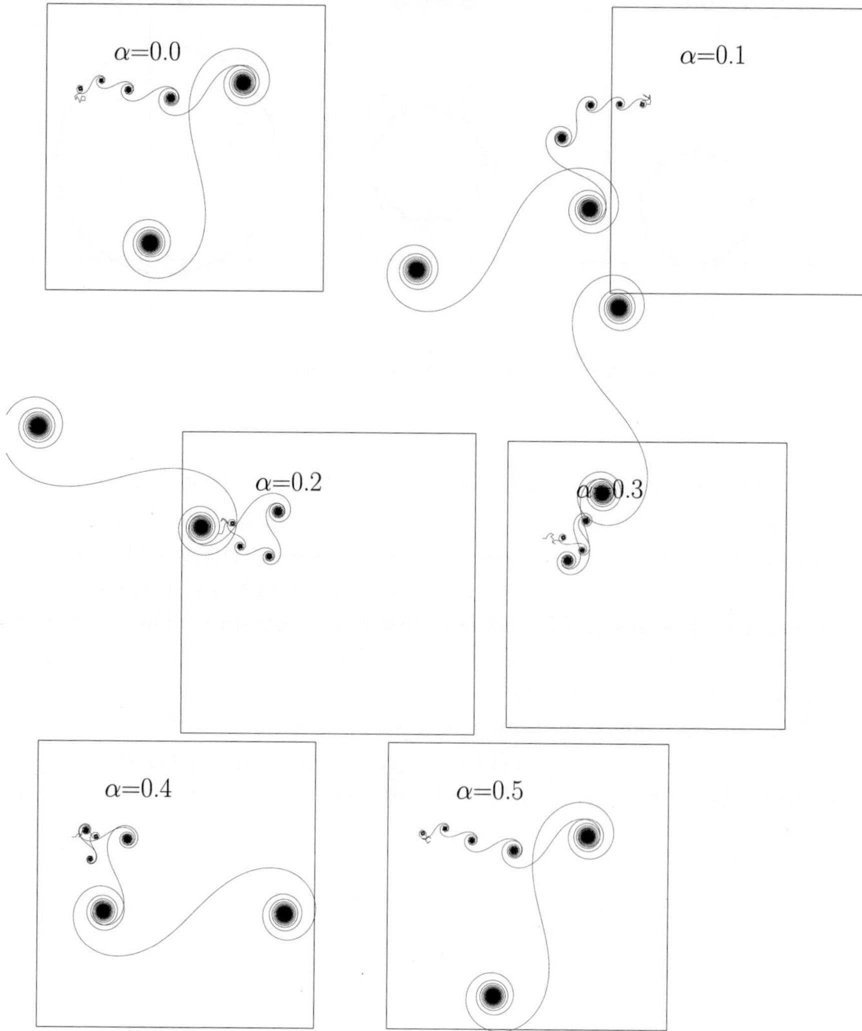

Fig. 12.26　The Nessie curlicue with memory in $f:\ z_{T+1}\ =\ z_T\ +\ e^{2\pi i \overline{f}(T)}$, $f(T) = (log\ T)^4$.

Curlicue fractal

In the so called curlicue fractal [415], after iteratively defining: $\theta(T) = (\theta(T-1)+2\pi\nu)\ (mod\ 2\pi)$, with ν being an irrational number, it is: $f(T) = (f(T-1)+\theta(T-1))\ (mod\ 2\pi)$. Initially, $\theta(0) = f(0) = 0$. Memory in f(T)

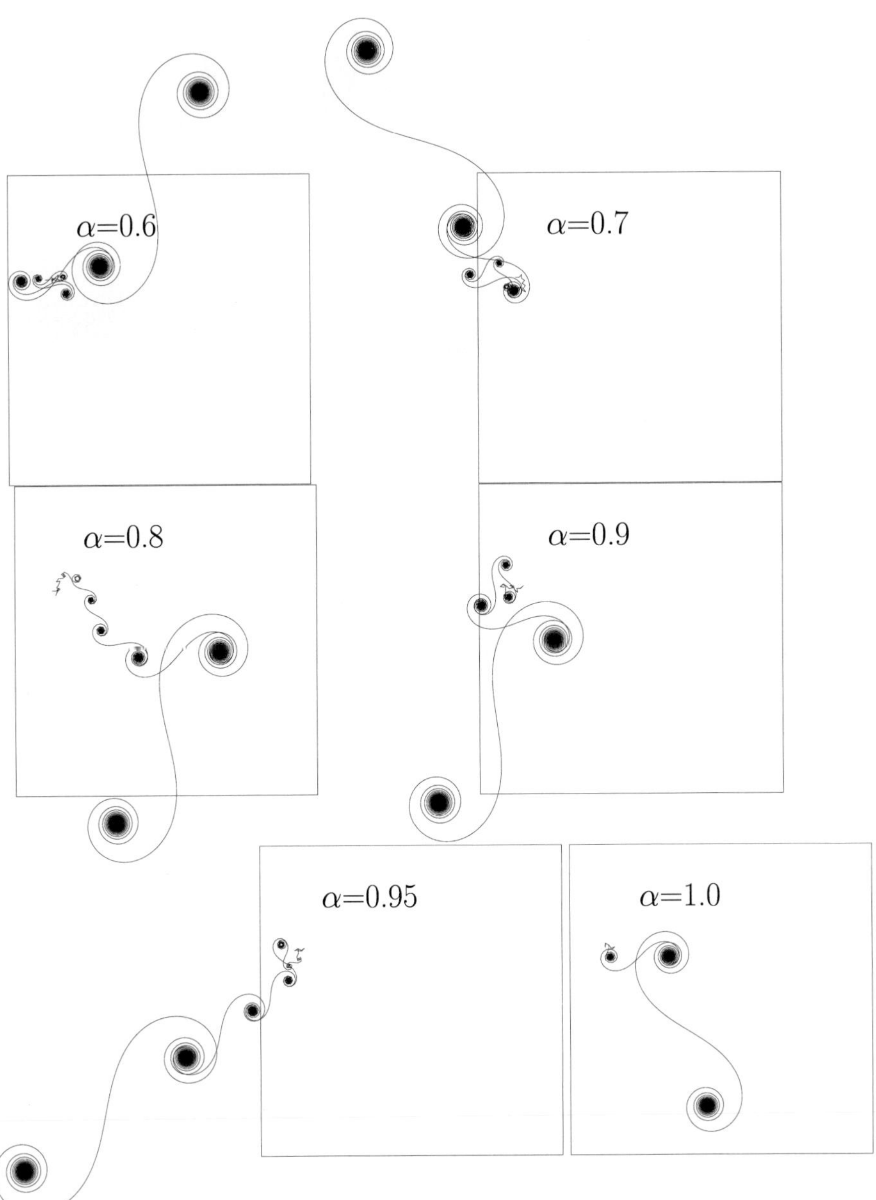

Fig. 12.26   (*continued*)

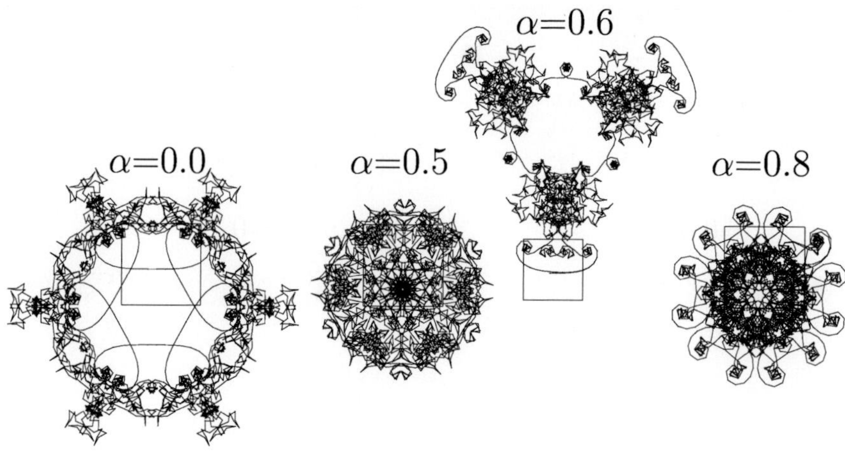

Fig. 12.27   The $f(T) = T^3/2010$ curlicue with memory in f.

soon impedes any organization, producing random-like dynamics. Memory in Fig. 12.29 is implemented on $\theta(T)$, thus: $f(T) = (f(T-1) + \bar{\theta}(T-1))\ (mod\ 2\pi)$. Figure 12.29 shows the patterns corresponding to $\nu = \pi$, with those with $\alpha \leq 0.4$ zoomed. Memory seems to exert a constraining plus rotating effect on the ahistoric pattern. Partial memory in the context of Fig. 12.29 produces patterns much as those in the figure, though *elongated*.

Particular attention has been paid in the theory of curlicues to the study of its *temperature* and *entropy* [122], defined as:

$$\beta = \frac{1}{log\ \dfrac{2l}{2l-h}}, \quad \text{and} \quad S = log\ \frac{2l}{h} + \frac{1/\beta}{e^{1/\beta}-1},$$

with l being its length and h the length the perimeter of its convex hull. The temperature and entropy of a curve are zero only if the curve is a straight line, and increase as the curve becomes more *wiggly*. Thus, these thermodynamic analog parameters appear to be a natural measure of the complexity of a curve.

A general analytical study considering a general $f(T)$ seems too ambitious, thus we will consider first only $f(T) = aT^2$. Even in this scenario, it seems very hard to find a unique rigorous theory so it is probably better to focus on some specific class of a numbers, such for example quadratic irrationals, which have periodic continuous fraction expansion. There are, at least, two different proposed approaches to the analytical study of curlicues: *i)* that by Marklof [271] involving geodesic and horocy-

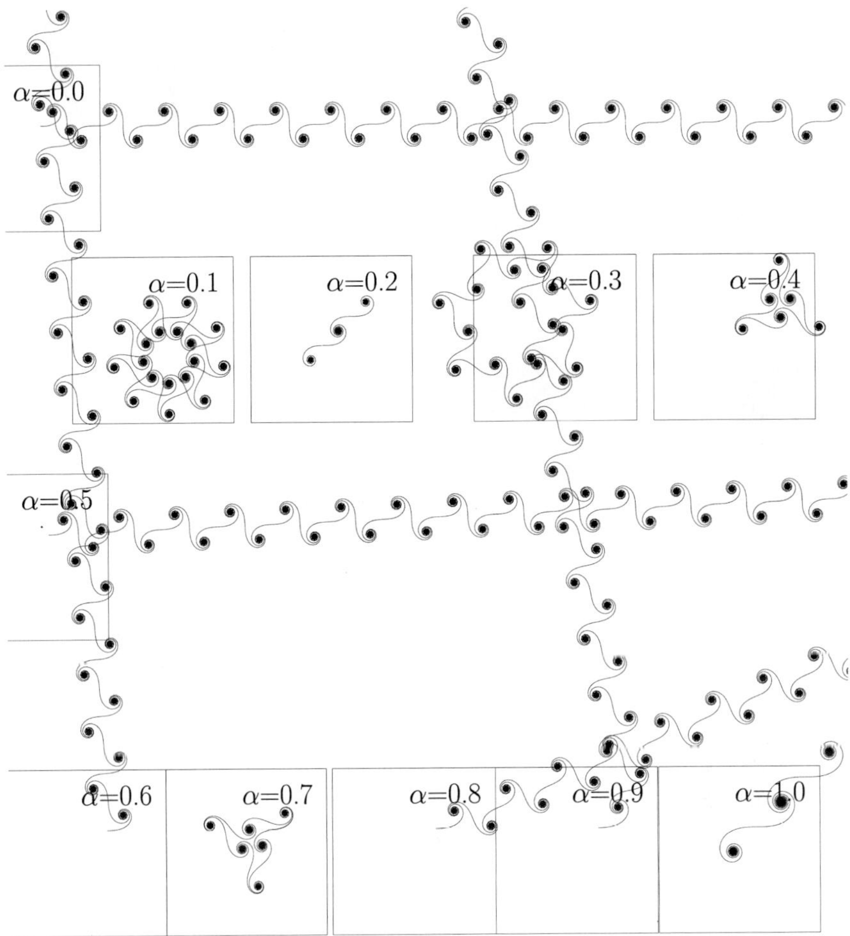

Fig. 12.28   The $f(T) = T^4/321$ curlicue with memory in f: $z_{T+1} = z_T + e^{2\pi i \overline{f}(T)}$.

cle flows on some interesting manifold. This method is quite abstract but incredibly powerful, and *ii)* that by Cellarossi [104], involving continued fractions and some renormalization formula, an apparently simpler idea, but also involving cumbersome technicalities. In a subsequent stage, the work undertaken by Paris [324–327], generalizing the studies [137, 247] on the incomple higher-order Gauss sums, i.e., $S_m(T) = \sum_{t=1}^{T} e^{2\pi i t^m/N}$, will be taken into consideration.

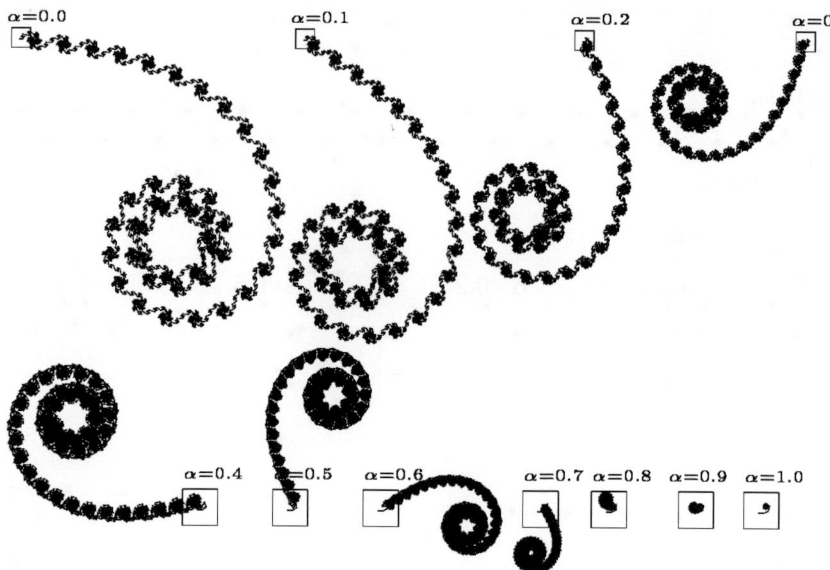

Fig. 12.29   The $\nu = \pi$ curlicue fractal with memory in f.

Rose curves

In mathematics, Rose [189, 243, 260] describes a family of curves, that, up to similarity can all be expressed by a polar equation of form:

$$\rho = \cos(k\theta) \ .$$

Since $\sin(k\theta) = \cos(k\theta - \pi/2) = \cos(k(\theta - \pi/2k))$, the curves given for the polar equations $\rho = \cos(k\theta)$ and $\rho = \sin(k\theta)$ are identical except for a rotation of $\pi/2k$ radians.

If k is an integer, the curve will be rose shaped with k petals if k is odd, and $2k$ petals if k is even. When k is even, the entire graph of the rose will be traced out exactly once when the value of θ changes in any interval of length 2π. When k is odd, this will happen on any interval of length π. If k is rational, the curve is closed and has finite length.

Roses may be discretized as:

$$\rho(\theta_T) = \cos(k\theta_T) \ ,$$

and endowed with memory as:

$$\rho(\theta_T) = \cos(k\bar{\theta}_T) \ .$$

The conversion from polar to Cartesian coordinates turns out,

$$x_T = \cos(k\bar{\theta}_T)\cos(\theta_T) \ , \quad y_T = \cos(k\bar{\theta}_T)\sin(\theta_T) \ .$$

In Fig. 12.30 and Fig. 12.31 it is $\theta_T \equiv T$. In Fig. 12.30, only $\alpha{=}1.0$

memory is implemented, so that $\bar{\theta}_T = T(T+1)/2T = (T+1)/2$. Thus, the curves under the label $\alpha{=}1.0$ in Fig. 12.30 obey the discrete dynamic equation:

$$\rho(T) = cos\left(\frac{k}{2}T + \frac{1}{2}\right), \quad T = 1, 2, \ldots .$$

Consequently, the k-patterns under $\alpha{=}1.0$ in Fig. 12.30 relate to the conventional $k/2$-roses, rotated 0.5 radians, thus, 28.65 degrees, circa one-third of the right angle. Thus for example, the $k{=}1$ circle becomes with full memory the (rotated) $k = 1/2$ conventional pattern, whereas the $k{=}2$ *quadrifolium* becomes a (rotated) $k = 1$ conventional circle. The correspondence from k-$\alpha{=}1.0$ to $k/2$ conventional patterns are also given in the figure for the cases $k = 4 \to k = 2$ and $k = 6 \to k = 3$. The $k{=}3$ and $k{=}5$ *foliums* become roses with six and ten petals, i.e., the conventional roses with $k{=}3/2$ and $k{=}5/2$. When $k = 1/d$, and d is even, the shape of the conventional patterns appear as a series of $d/2$ loops that meet at two small loops at the centre touching (0,0) from the vertical and is symmetrical about the x axis. The examples for $k = 1/2$ and $k = 1/4$ are given in Fig. 12.30. Full memory induces this type of pattern for every $k = 1/d$, because it turns out to be related to conventional $k = 1/2d$ structures. The $k = 1/3$ and $k - 1/5$ examples are given in Fig. 12.30.

A Maurer rose [284, 366] is a plot of a *walk* along a rose in steps of a fixed number of degrees d. Figure D.14 in Appendix D shows the Maurer roses from Fig. 12.30, with $d{=}1$, i.e., the joined points $(\theta_T, \cos(k\bar{\theta}_T))$, $T = 1, 2, \ldots, 200$. The static drawings presented in Fig. D.14 unveil the apparent motion as they are being generated, in some cases with a quite stunning visual effect. The dynamic generation of Maurer roses may be visualized in *http://scratch.mit.edu/projects/mathjp/789739*.

In Fig. 12.31 it is $\alpha < 1$, so: $\bar{\theta}_T = \dfrac{\displaystyle\sum_{t=1}^{T} \alpha^{T-t}\, t}{(\alpha^T - 1)/(\alpha - 1)}$. In general, the sum of t times the t-*th* power of r is given analytically by: $S(T;r) = \displaystyle\sum_{t=1}^{T} tr^t = \dfrac{r - (T+1)r^{(T+1)} + Tr^{(T+2)}}{(r-1)^2}$ [2]. Thus, $w(T;\alpha) = \displaystyle\sum_{t=1}^{T} \alpha^{T-t}\, t$

---

[2] As a particular case of any arithmetico-geometric $(a_t g_t)$ series:

$$S = r + 2r^2 + 3r^3 + \ldots (T-1)r^{T-1} + Tr^T$$
$$rS = r^2 + 2r^3 + 3r^4 + \ldots (T-1)r^T + Tr^{T+1}$$
$$rS - S = -r - r^2 - r^3 - r^4 - \ldots - r^T + Tr^{T+1} = -\frac{(r^{T+1} - r)}{r - 1} + Tr^{T+1}$$

Fig. 12.30   Roses with full memory.

$$S(r-1) = \frac{r - (T+1)r^{T+1} + Tr^{T+2}}{r-1}$$

$$= \alpha^T \sum_{t=1}^{T} t(1/\alpha)^t = \alpha^T \frac{(1/\alpha) - (T+1)(1/\alpha)^{(T+1)} + T(1/\alpha)^{(T+2)}}{((1/\alpha) - 1)^2} =$$

$$\frac{\alpha^{(T+1)} - (T+1)\alpha + T}{(\alpha - 1)^2} \; . \; \text{Finally, } \bar{\theta}_T(\alpha) = \frac{\alpha^{(T+1)} - (T+1)\alpha + T}{(\alpha^T - 1)(\alpha - 1)} \; . \; \text{Non-}$$

high values of the memory factor has little effect on the rose patterns, and even the high values of α in Fig. 12.31, i.e., $\alpha =0.95$, 0.99 do not significantly alter the structure of the conventional roses, that appear lightly left rotated.

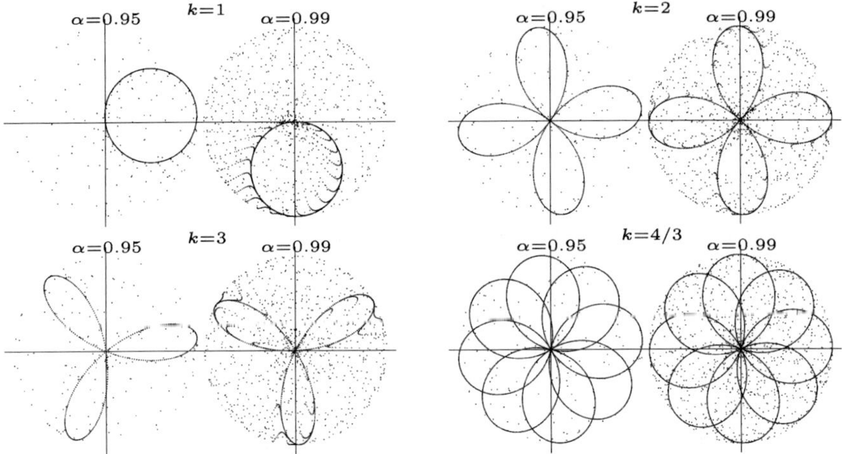

Fig. 12.31   Roses with non-full memory.

The evolution with memory factor near 1.0, e.g., $\alpha=0.999$, as in Fig. 12.32 produces, a sparse distribution of points in the unit circle, regardless of the value of the k parameter.

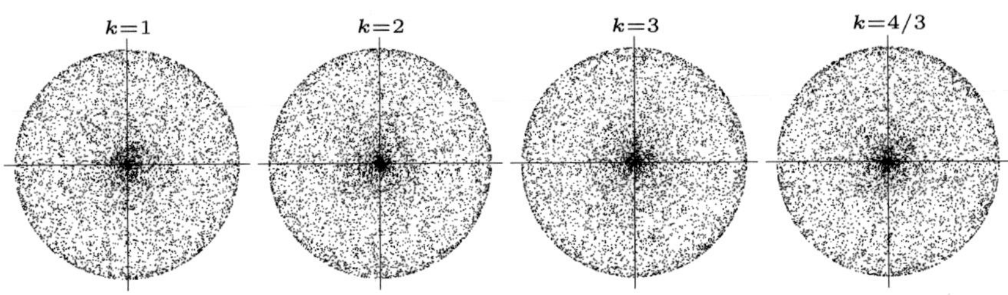

Fig. 12.32   Roses with $\alpha=0.999$ memory.

The transition from the (random) patterns with high memory in Fig. 12.32 to the structured ones with full memory in Fig. 12.30, may be appreciated with the very high α=0.9999999 memory in Fig. 12.33. The rose simulations have been obtained here by means of a FORTRAN code working with quadruple precision.

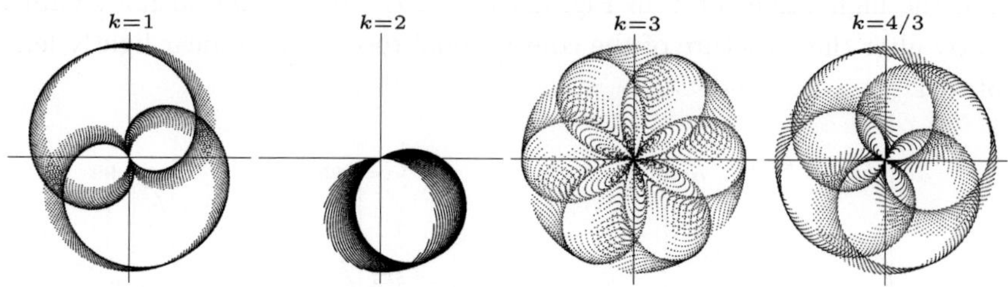

Fig. 12.33   Roses with α=0.9999999 memory.

If k is irrational then the rose curve is not closed and has infinite length. Furthermore, the graph of the rose in this case forms a dense set, i.e., it comes arbitrarily close to every point in the unit disk. The effect of memory in this scenario is not easily recognizable visually.

Roses are related to *epitrochoids*, a kind of *roulette* curves traced by a point attached to a circle of radius b rolling around the outside of a fixed circle of radius a. If the moving circle is rolling around the inside of the fixed one, the curve is called epitrochoid. If h is the distance from the tracing point to the center of the rolling circle, the parametric Cartesian equations for the epitrochoid and hypotrochoid are::

$$x = (a+b)\cos(t)-h\cos((a+b)/b)t)), \quad y = (a+b)\sin(t)-h\sin((a+b)/b)t)),$$

and

$$x = (a-b)\cos(t)+h\cos((a-b)/b)t)), \quad y = (a-b)\sin(t)-h\sin((a-b)/b)t)),$$

respectively.

Particular cases of hypotrochoid are the ellipse with $a = 2b$ and the rose with $a = 2nh/(n+1),\ b = (n-1)h/(n+1)$. In the latter case, $a - b = h$, so (a-b)/b=(n+1)/(n-1). Thus,

$$x = h(\cos(t)+\cos((n+1)/(n-1))t)), \quad y = h(\sin(t)-\sin((n+1)/(n-1))t)),$$

If $h = b$ the curves are called *cycloids*. Thus the hypocicloid has parametric equations:

$$x = (a-b)\cos(t)+b\cos((a-b)/b)t)), \quad y = (a-b)\sin(t)-b\sin((a-b)/b)t)),$$

or,

$$x = b((k-1)\cos(t)+\cos((k-1)t))), \quad y = b((k-1)\sin(t)-\sin((k-1)t))).$$

12.3.0.1 *Other polar curves*

The memory mechanism implemented here in the rose curves may be applied in the same manner to any polar curve. Thus for example, in Fig 12.34:

$$\rho = e^{\sin(\theta)} - 2\cos(4\theta) + sin^5((2\theta - \pi)/12)$$

In which case, the original butterfly [139] transforms with α=1.0 memory in some sort of *bumblebee*.

α=0.0 α=1.0

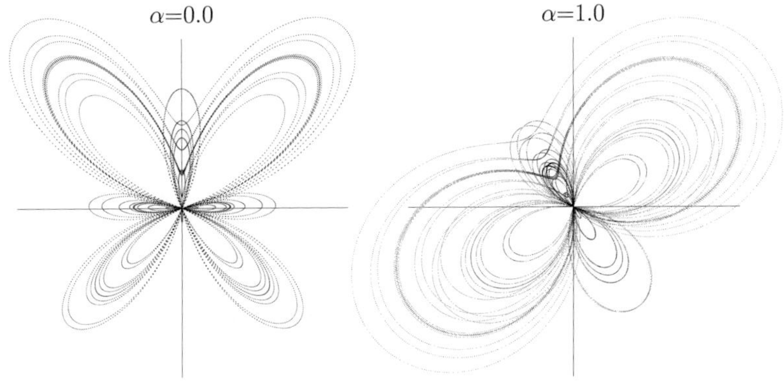

Fig. 12.34   A butterfly with memory.

The last term of the above polar equation defining the butterfly curve, $sin^5((2\theta - \pi)/12)$, is added purely to enhance the aesthetic appeal: the contour of the butterfly (and of the bumblebee) also is genareted if it is suppressing. An alternative way of improving the aestethic appeal is to give the plots some *texture*, as done in [140] by multiplying the base curve by a rapidly varying sinusoidal factor. Thus for example,

$$\rho = \left(e^{\sin(\theta)} - 2\cos(4\theta)\right) sin^4(\lambda\theta),$$

with λ being any large number, chosen $\lambda = 99999999$ in the simulation of Fig. 12.35.

The *texture* produced via the afore mentioned method is very sensitive to the step size $\Delta(\theta)$ [140], which up to now was $\Delta(\theta)$=1. In Fig. 12.36 it is $\Delta(\theta)$=0.0007, thus, θ_T=0.0007 (T-1), in which case the bumblebee appears rotated compared to the previous simulations.

The shape of the bumblebee into which the butterfly transforms with α=1.0 memory is somehow reminiscent of the cycloid of Ceva [438]: $\rho = 1 + 2\cos(2\theta)$, shown in Fig. 12.37. Full memory converts this curve in the

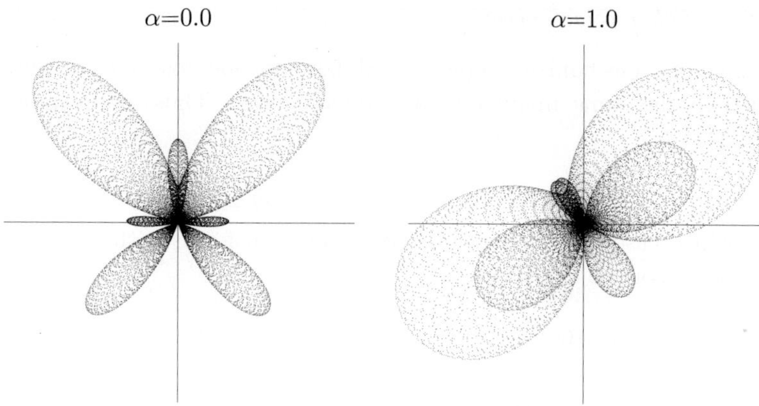

Fig. 12.35 A butterfly with texture and memory.

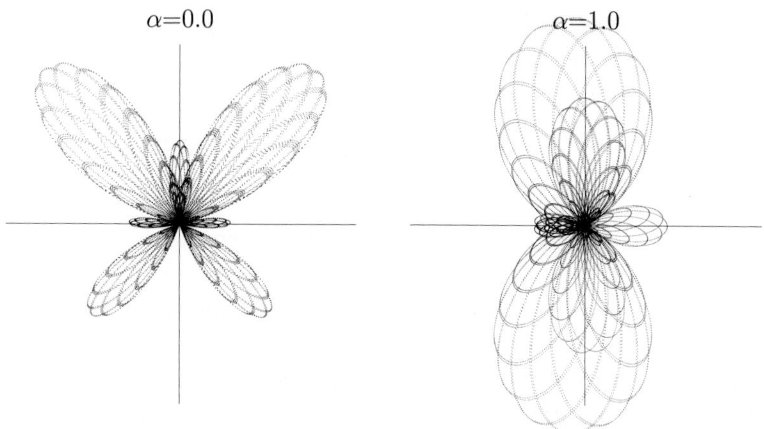

Fig. 12.36 A butterfly with texture and $\Delta(\theta)$=0.0007.

limaçon: $\rho = 1 + 2\cos(\theta + 1)$, also shown in Fig. 12.37. The figure shows how the previously described mechanism of *texturization*, i.e., multiplying the base curve by a rapidly varying sinusoidal factor such as $sin^4(\lambda\theta)$, works in general, not only in the butterfly curve, as well as the sensitivity to the step size of the texture achieved.

Instead of endowing memory in the equation defining the polar curve, memory can be implemented in the conversion from polar to Cartesian coordinates. Thus,

$$\rho(\theta_T) = f(\theta_T) \, ,$$

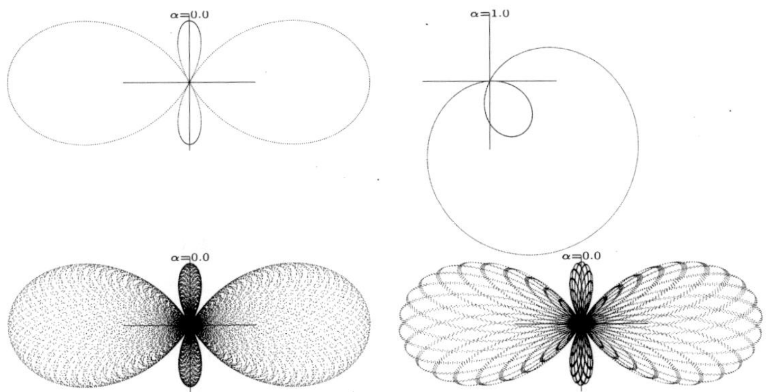

Fig. 12.37   The cycloid of Ceva.

but,
$$x_T = \cos(\overline{\theta}_T)\, f(\theta_T)\,, \quad y_T = \sin(\overline{\theta}_T)\, f(\theta_T)\,.$$
Figure 12.38 shows the disrupting effect of such a memory implementation on the butterfly (Fig. 12.34) and on the cycloid of Ceva (12.37).

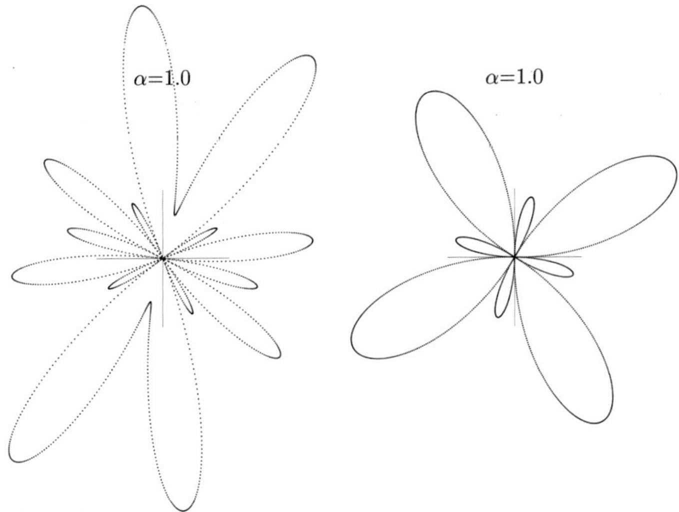

Fig. 12.38   The butterfly (left) and the cycloid of Ceva (right) with full memory in the conversion from polar to Cartesian coordinates.

Figure D.15 in Appendix D shows in the left side of every k-panel, the effect of memory in the conversion to Cartesian coordinates on the roses considered in Figure D.15. No new type of rose seems to appear in this context, which produces roses with non-overlapping $4k$ petals (or propellers with $4k$ blades) for every integer k. That is so from $k{=}1$ and regardless of k being odd or even. Figure D.15 shows also the effect of memory implemented in only one of the coordinate conversions. Thus, either,

$$x_T = \cos(\theta_T)\, f(\theta_T) \,, \quad y_T = \sin(\bar{\theta}_T)\, f(\theta_T) \,,$$

or,

$$x_T = \cos(\bar{\theta}_T)\, f(\theta_T) \,, \quad y_T = \sin(\theta_T)\, f(\theta_T) \,.$$

The patterns produce in this context a different aspect. In the case of k being integer, the aspect of the patterns remain somehow *floral*, but some of them seem to have *thorns*, as asked for in [189].

As happens with memory in ρ, the evolution with high memory charge, up to $\alpha{=}0.999999$ in the coordinate transformation, produces an sparse distribution of points, regardless the value of k. The transition to the structured patterns with full memory may be appreciated with the very high $\alpha{=}0.9999999$ memory factor in the examples of Fig. 12.39. The examples with memory in both coordinates produce some sensation of *movement* in the *propellers*.

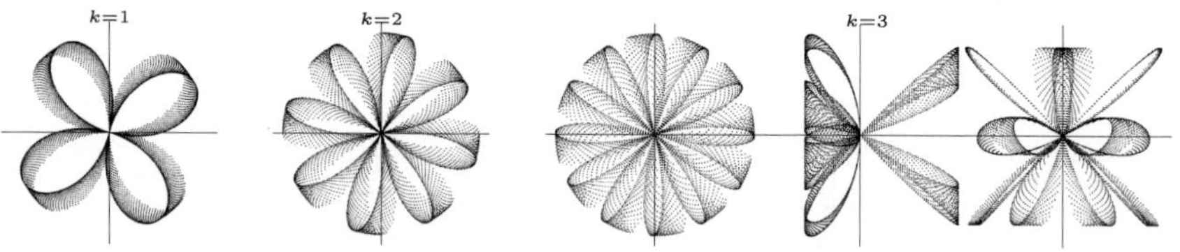

Fig. 12.39   Roses with $\alpha{=}0.9999999$ memory in coordinate transformation.

12.4   Stochastic processes

The probability distribution vector $\mathbf{p}$ evolves in Markovian stochastic processes with probabilty transition matrix $\mathbf{M}$, following the updating rule: $\mathbf{p}'_{T+1} = \mathbf{p}'_T \mathbf{M}$. Memory can be embedded in such systems by means of $\mathbf{p}'_{T+1} = \boldsymbol{\pi}'_T \mathbf{M}$ with $\boldsymbol{\pi}_T$ being a weighted mean of the probability distributions up to T: $\boldsymbol{\pi}_T = \pi(\mathbf{p_1}, \ldots, \mathbf{p_T})$.

Figure 12.40 shows the distribution of probability at $T = 8$ and at $T = 9$ (dotted) in a symmetric $(p_{i-1,i}{=}p_{i,i+1}{=}p{=}0.5)$ random walk starting from $X = 10$ with memory of the last two probability distributions $(\tau = 2)$:

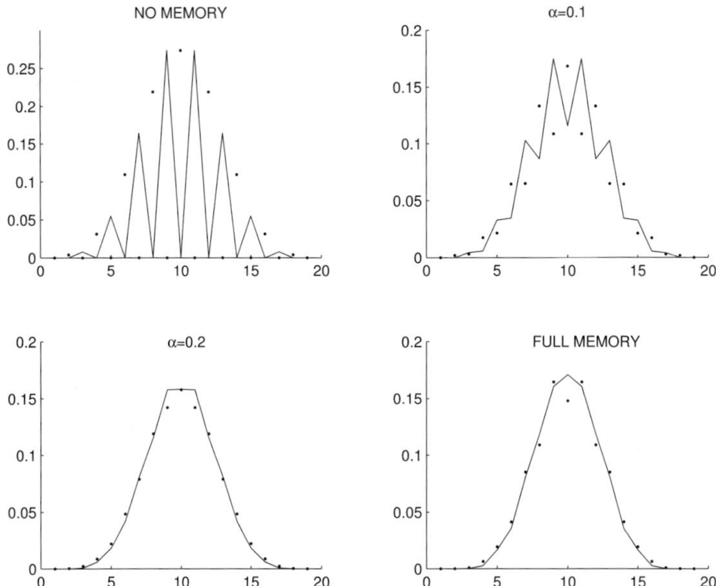

Fig. 12.40 Symmetric random walk with $\tau-2$ memory, starting in $X=10$. Time-steps $T=8$ (unmarked, joined), and $T=9$ (dotted).

$$\pi_T = \frac{\mathbf{p_T} + \alpha \mathbf{p_{T-1}}}{1+\alpha}.$$ A minimal incorporation of memory ($\alpha = 0.1$) alters the characteristic alternating probabilistic distribution in the ahistoric model to a *serrated* form, much smoothed at higher α values as an effect of the mean memory implementation. A similar effect is displayed in [44] when keeping memory of the three last states: $\pi_T = \dfrac{\mathbf{p_T} + \alpha \mathbf{p_{T-1}} + \alpha^2 \mathbf{p_{T-2}}}{1+\alpha+\alpha^2}.$

In the correlated random walk proposed by Bidaux and Boccara [76], the random walker moves at each step either to the right or to the left with probabilities $p/2$, or with probability $q=1\text{-}p$ performs a move which is a stochastic Boolean function of the τ previous steps. In other approaches, the walker remembers the history of prior walk directions, $v_T = \pm$, then it randomly chooses a previous time t and adopts the corresponding v_t with probability p and $-v_t$ with probability $1\text{-}p$. No limit in the trailing memory is established in [362], so that at time step T a instant t from the set $\{1, 2, \ldots, T\}$ is chosen randomly with uniform probability $1/T$. The studies performed in [116] , [199] and [367] consider the case in which the walker can remember only a fraction of the total time elapsed. An analytical description was already given in [362] for the full memory (or *elephant*) walk, whereas the more complicated case of partial memory has been addressed in [236].

High-order Markov Chains [338] and, more generally, Finitarily Makovian processes [297] deal with the generalization of the Markovian paradigm to cope with past states conditioning the probability distribution. The conventional model for τth-order Markov chains has $(k-1)k^\tau$ parameters, where k is the number of states. This large amount of parameters has discouraged the use of the model for $\tau > 1$, even when higher-order dependence is present. Instead of that, the model proposed in [338] involves only one additional parameter for each extra lag. Thus the conditional probability of observing $X_t = j$ given the past, is a convex linear combination from each $X_{t-1}, \ldots, X_{t-\tau}$, much in the way in which the π values are obtained above.

In open quantum systems with non-Markovian dynamics, its asymptotic state strongly depends on the initial conditions, even if the dynamics possess an invariant state. That essentially means memory effects. In particular, the asymptotic state can remember and partially preserve its initial entanglement [77, 117, 258, 439].

Chapter 13

Spatial games

13.1 The prisoner's dilemma

The Prisoner's Dilemma (PD) is a game played by two players (A and B), who may choose either to cooperate (C or 1) or to defect (D or 0). Mutual cooperators each score the *reward* R, mutual defectors score the *punishment* P; D scores the *temptation* $\mathfrak{T}$ against C, who scores S (*sucker's* payoff) in such an encounter. Provided that $\mathfrak{T} > R > P > S$, mutual defection are the only equilibrium strategies (recall that a strategy pair is in Nash equilibrium if each is a best reply to the other). Thus in a single round both players are to be *penalized* instead of both *rewarded*, but cooperation may be rewarded in an iterated (or spatial) formulation. The PD has become a paradigm for animal behavior modeling [385]. The game is simplified (while preserving its essentials) if $P = S = 0$. Choosing $R = 1$, the model will have only one parameter: the *temptation* $\mathfrak{T}=b$.

In the spatial version of the PD dealt with here, each player occupies a site (i,j) in a two-dimensional $N \times N$ lattice. In each generation (T) the payoff of a given individual ($p_{i,j}^{(T)}$) is the sum over all interactions with the eight nearest neighbors and with its own site. In the next generation, an individual cell is assigned the decision ($d_{i,j}^{(T)}$) that received the highest payoff among all the cells of its Moore's neighborhood. In case of a tie, the cell retains its choice.

Nowak and May [312, 313] pioneered the study of this model. They concluded in their original work that spatial structure (or territoriality) can facilitate the survival of cooperators. Thus, the spatialized PD (SPD for short) has proved to be a promising tool to explain how cooperation can hold out against the ever-present threat of exploitation. Further work supported the claim, extending it to models more sophisticated than the simple imitation-of-the-best mechanism adopted by Mowak and May [93].

As a result, on the basis of the PD game it is widely accepted that spatial structure promotes the evolution of cooperation.

This is a task that presents problems in the classic *struggle-for-survival* Darwinian framework, because cooperation is a conundrum, whereas exploitation is not, at least at a first sight. Cooperative entities make a sacrifice: they help others at a cost to themselves. Exploiters, or cheaters, reap the benefits and forego costs. Based on utilitarian principles - be it in the form of evolution by natural selection of the "fittest" type, or the form of "rational" behaviour generating the highest payoff - exploitation should prevail, and cooperation should be rare.

Other spatialized two-person games will come under scrutiny in the last section of this Chapter.

The basic model with memory

In the standard form of the SPD, only the results from the last round are taken into account and the outcomes of previous rounds are neglected, the model just described will be termed *ahistoric*. When dealing with the PD, memory can be embedded not only in choices but also in rewards. Thus, in the *historic* model we consider, following the iteration at time-step T: i) all the payoffs coming from the previous rounds are accumulated,

giving $\pi_{i,j}^{(T)}(p_{i,j}^{(1)}, \ldots, p_{i,j}^{(T)}) = p_{i,j}^{(T)} + \sum_{t=1}^{T-1} \alpha^{T-t} p_{i,j}^{(t)} = p_{i,j}^{(T)} + \alpha \pi_{i,j}^{(T-1)}$, and ii)

players are featured with a summary of past decisions $(\delta_{i,j}^{(T)})$, thus coding cooperation and defection as 1,0 respectively it is:

$$m_{i,j}^{(T)}(d_{i,j}^{(1)}, \ldots, d_{i,j}^{(T)}) = \frac{d_{i,j}^{(T)} + \sum_{t=1}^{T-1} \alpha^{T-t} d_{i,j}^{(t)}}{1 + \sum_{t=1}^{T-1} \alpha^{T-t}} \Rightarrow \delta_{i,j}^{(T)} = round(m_{i,j}^{(T)})$$

Again, in each round or generation T, every player plays with each of the eight neighbors and itself, the decision $\delta^{(T)}$ in the cell of the neighborhood with the highest accumulated payoff $(\pi^{(T)})$ being adopted as the choice to play $(d^{(T+1)})$ in the next round.

This approach to modeling memory has been rather neglected in evolutionary game theory. The usual approach to considering memory in the iterated prisoner's dilemma in particular, consists in designing strategies which determine the next move of a player given the history (genome) of the game. This approach seems at first glance more sophisticated than the somewhat simple-minded one adopted here. But it is to be stressed that, at least concerning the memory mechanism employed, this judgment may be debatable. Thus for example, in the papers by Lindgren and Nordahl [256, 257], and in the probabilistic simulations with genome-type memory in [194], only memory of moves is kept, whereas scores are treated in a Markovian way. In contrast, only payoff memory is taken into account in [11, 335, 365, 372], and the player's *reputation* in [315] is based only in (previous) payoffs. Often memory operates in spatially unstructured populations, such as in the model proposed by Harley [192] on learning developmentally stable strategies, in which animals display most frequently the behaviour which has, up to the present, paid the most. Thus, it seems that no simultaneous proper memory (longer than one time-step) of both payoffs and moves is implemented in the main-stream approaches to memory in the iterated PD. In the same manner, the memory approach to the snowdrift game in [408], or to the minority game in [105, 106] is based on retaining the (last) C/D bits of the past strategy information. Future payoff discounting is implemented in the context of stochastic games [357].

Table 13.1 shows the initial scenario starting from a single defector if $8b > 9 \Leftrightarrow, b > 1.125$, which means that neighbors of the initial defector become defectors at $T = 2$.

Table 13.1   Choices at $T = 1$ and $T = 2$; accumulated payoffs after $T=1$ and $T=2$ starting from a single defector in the SPD. $b > 9/8$

$\mathbf{d}^{(1)} = \delta^{(1)}$	$\mathbf{p}^{(1)} = \pi^{(1)}$	$\mathbf{d}^{(2)} = \delta^{(2)}$	$\pi^{(2)} = \alpha\mathbf{p}^{(1)} + \mathbf{p}^{(2)}$						
		1 1 1 1 1 1 1	$9\alpha+9$	$9\alpha+9$	$9\alpha+9$	$9\alpha+9$	$9\alpha+9$	$9\alpha+9$	$9\alpha+9$
1 1 1 1 1	9 9 9 9 9	1 1 1 1 1 1 1	$9\alpha+9$	$9\alpha+8$	$9\alpha+7$	$9\alpha+6$	$9\alpha+7$	$9\alpha+8$	$9\alpha+9$
1 1 1 1 1	9 8 8 8 9	1 1 0 0 0 1 1	$9\alpha+9$	$9\alpha+9$	$8\alpha+5b$	$8\alpha+3b$	$8\alpha+5b$	$9\alpha+9$	$9\alpha+9$
1 1 0 1 1	9 8 8b 8 9	1 1 0 0 0 1 1	$9\alpha+9$	$9\alpha+9$	$8\alpha+3b$	$8b\alpha$	$8\alpha+3b$	$9\alpha+6$	$9\alpha+9$
1 1 1 1 1	9 8 8 8 9	1 1 0 0 0 1 1	$9\alpha+9$	$9\alpha+9$	$8\alpha+5b$	$8\alpha+3b$	$8\alpha+5b$	$9\alpha+9$	$9\alpha+9$
1 1 1 1 1	9 9 9 9 9	1 1 1 1 1 1 1	$9\alpha+9$	$9\alpha+8$	$9\alpha+7$	$9\alpha+6$	$9\alpha+7$	$9\alpha+8$	$9\alpha+9$
		1 1 1 1 1 1 1	$9\alpha+9$	$9\alpha+9$	$9\alpha+9$	$9\alpha+9$	$9\alpha+9$	$9\alpha+9$	$9\alpha+9$

Nowak and May paid particular attention in their seminal papers to $b = 1.85$, a high but not excessive temptation value which leads to complex dynamics. After $T = 2$, defection can advance to a 5×5 square or be restrained as a 3×3 square, depending on the comparison of $8\alpha+5\times1.85$ (the maximum π value of the recent defectors) with $9\alpha+9$ (the π value of the non affected players). As $8\alpha+5\times1.85 = 9\alpha+9 \rightarrow \alpha = 0.25$, i.e., if $\alpha > 0.25$, defection remains confined to a 3×3 square at $T = 3$. Here we see the paradigmatic effect of memory: it tends to avoid the spread of defection.

NO MEMORY

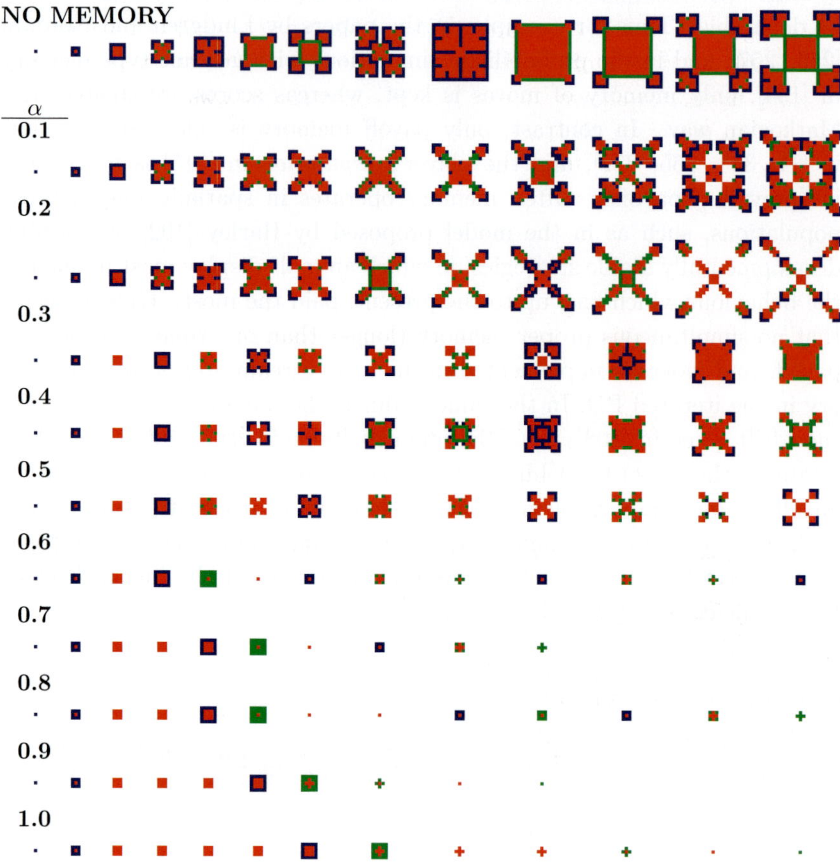

Fig. 13.1   Evolving patterns starting from a single defector up to $T = 13$. Memory factor α. $b = 1.85$. Code color: Blue: Defection (D) after Cooperation (C), Red: D after D, Green: C after D, blank: C after C. This codification of color remains through this PD chapter.

Figure 13.1 shows such a restrictive effect starting with a single defector. Recall that memory of decisions does not operate if $\alpha \leq 0.5$ (section 2.1). Effective memory of decisions (i.e., $\alpha > 0.5$) has a dramatic restrictive effect on the advance of defection in the scenario of Fig. 13.1, e.g., extinction is soon reached with $\alpha \geq 0.7$. With $\alpha = 0.6$, a oscillator of period 23 with number of defectors $\{1,9,5,1,9,5,1,9,5,1,9,5,5,5,5,9,5,5,5,5,21,1,9\}$ is generated at $T = 6$, so only its eight first elements are shown in Fig. 13.1.

Only the payoff memory, i.e., α under 0.5, has a less dramatic effect on the evolving patterns, but the restrictive effect of memory on the defection progress remains (the more moderate the lower the α), always with a remarkable alteration of the defection patterns, even for low values of the memory factor, as shown in Fig. 13.1. Figure 13.2 shows the patterns at $T = 300$ in models with low levels of α when starting from a single defector in a lattice of size 100×100. Table 13.2 shows a MATLAB program rendering Fig. 13.1.

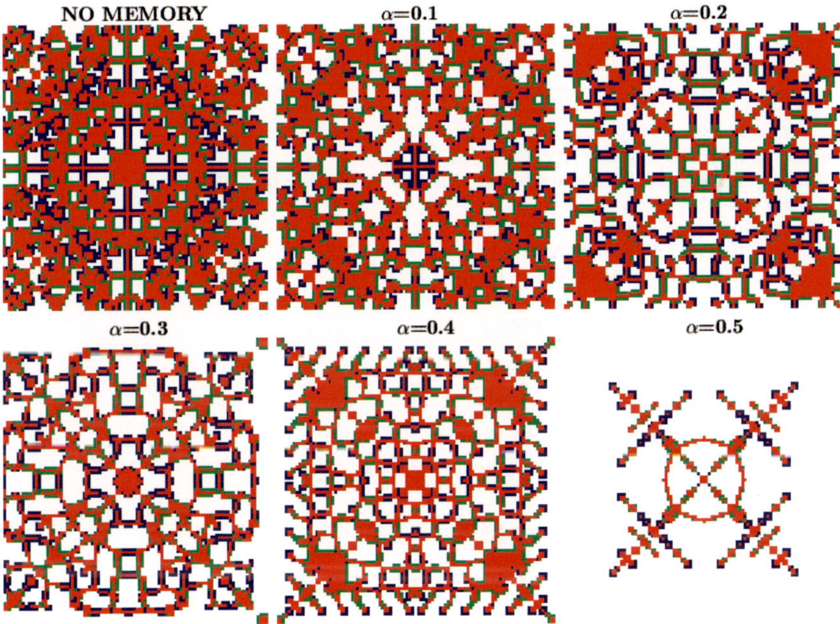

Fig. 13.2   Patterns at T=200 starting from a single defector in a 101×101 lattice.

Figure 13.3 shows the frequency of cooperators (f) up to $T = 100$, starting from a single defector and from a random configuration of defectors $(f = 0.5)$ in a lattice of size 100×100 with periodic boundary conditions when $b = 1.85$. The ahistoric plot (dotted) reveals in both scenarios the convergence of f to 0.318, (*which seems to be the same value regardless of the initial conditions* [312]). Starting from a single defector (a), the plot corresponding to $\alpha = 0.1$ stills shows the features of the ahistoric one but with higher f values, and the frequency of cooperation of the two other models with memory factors under 0.5 reaches plateaus over this value

Table 13.2   A MATLAB® program for the binary prisoner's dilemma.

```
function []=BPD;
    par.b=1.85; par.n=27;par.t=13; par.B=[par.n 1:par.n 1]
    for nal=1:11; [PI,DE,X,par]=ini(nal,par);
        for T=1:par.t
            figure(9);subplot(11,par.t,par.t*(nal-1)+T),imagesc(X,[0,1])
            colormap('bone');axis image;axis('off')
            [P]=PLAY(X,par);
            [PI,DE,DD]=MEMORY(P,PI,T,X,DE,par);
            [X]=DECIDE(PI,DD,T,par);
        end
    end
function [P]=PLAY(X,par);
    for i=1:par.n;for j=1:par.n; x=X(i,j);P(i,j)=0;
    for k=0:2;ik=par.B(i+k);for l=0:2;jl=par.B(j+l);y=X(ik,jl);
            P(i,j)=P(i,j)+(x+(1-x)*par.b)*y;          % Single round Payoffs
    end;end
    end;end
function [PI,DE,DD]=MEMORY(P,PI,T,X,DE,par)
    if(T==1)PI=P;DE=X; 'else;PI=par.alfa*PI+P;DE=par.alfa*DE+X;end
    DD=X;
    for i=1:par.n;for j=1:par.n
            if((DE(i,j))>par.W2(T))DD(i,j)=1;end;
            if((DE(i,j))<par.W2(T))DD(i,j)=0;end;
    end;end
function [X]=DECIDE(PI,DD,T,par);
    for i=1:par.n;for j=1:par.n
            [X(i,j)]=DECISION(PI,DD,T,i,j,par);
    end;end
function [x]=DECISION(PI,DD,T,i,j,par);
    pm=PI(i,j); nc=0;nd=0;delta=0.000001;
    for k=0:2;ik=par.B(i+k);for l=0:2;jl=par.B(j+l);
            px=PI(ik,jl);if(px>pm);pm=px;end;          % max neighbour's payoff
    end;end
    for k=0:2;ik=par.B(i+k);for l=0:2;jl=par.B(j+l);
            px=PI(ik,jl);if(pm-px)¡delta
    if(DD(ik,jl)==1)nc=nc+1;else nd=nd+1;end
    end
    end;end
    x=1;if(nd>nc)x=0;end                     % next choice
function [PI,DE,X,par]=ini(nal,par)
    X=ones(par.n);DE=X;c=(par.n+1)/2;X(c,c)=0;PI=zeros(par.n);
    par.alfa=0.1*(nal-1);w(par.t)=1;
    for tt=1:par.t-1;w(tt)=par.alfa^(par.t-tt);end; sx=0;
    for tt=1:par.t;sx=sx+w(par.t-tt+1);par.W(tt)=sx;end
    par.W2=0.5*par.W
```

sooner and in a smoother way. Values of α over 0.6 lead to the extinction of defection, so to $f = 1$.

Starting at random, right panel in Fig. 13.3, cooperation plummets to nearly extinction after the first round as shown in Figs. E.1, E.2, and E.3 of Appendix E. Hopefully some cluster of cooperators *resists* and become

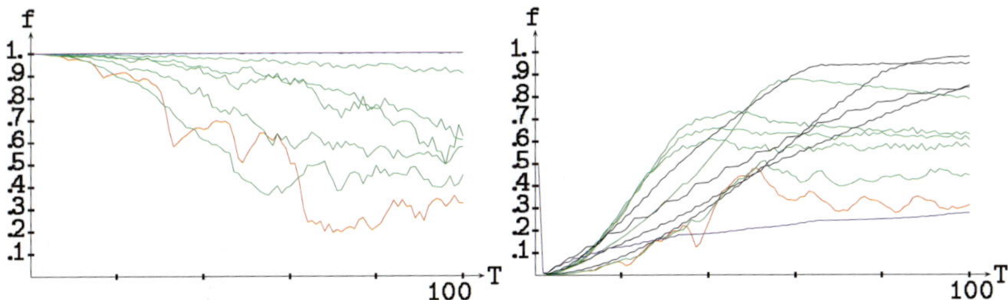

Fig. 13.3   Frequency of Cooperators. (Left) From a single defector, (right) starting at random. Red: ahistoric, Green: $\alpha=0.1$ to 0.5 by 0.1, Black: $\alpha=0.6$ to 0.9 by 0.1, Blue: $\alpha=1.0$.

the seed for a likely slow recovering of cooperation. In the ahistoric simulation of Fig. E.1 just a cluster remains active in recovering cooperation at $T = 4$. The other two small ones, are just *rotators*, rotating in opposite direction in the actual simulation. With memory, cooperation *nucleates* as shown in Fig. E.2, so that cooperation recovers faster and reaches higher levels compared to the ahistoric scenario. That is so, except, again, with full memory, in which case the cooperation recovers slowly, as already detected in Fig. E.3, and reaches levels similar to the ahistoric model (the *magic* 0.318). It seems that some degree of *forgetfulness* helps in supporting cooperation. Figure 13.4 confirms this general dynamic. The patterns at time-steps 100 and 200 in Fig. E.4 remind the initial nucleation (with defection percolating) for small memory factor levels, show only an elongated cluster of defection when $\alpha = 0.7$, and how the initial nucleation turns out revealed again with full memory.

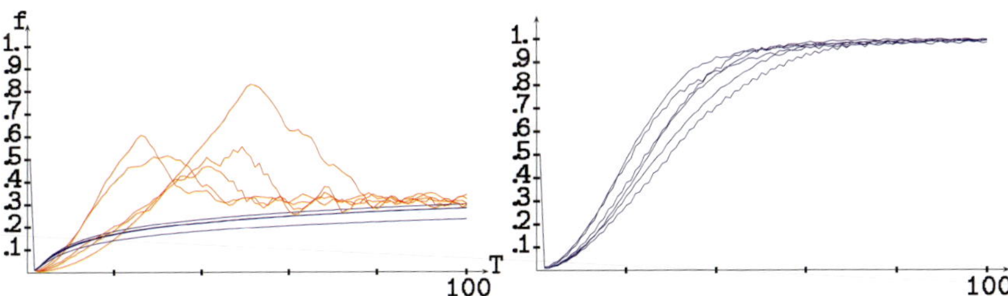

Fig. 13.4   Five alternative simulations. Unlimited trailing memory (left) and memory of the three last time-steps (right).

Similar results are found if memory is limited to the last three iterations [55]: $\pi_{i,j}^{(T)} = \alpha^2 p_{i,j}^{(T-2)} + \alpha p_{i,j}^{(T-1)} + p_{i,j}^{(T)}, \quad m_{i,j}^{(T)} = \left(\alpha^2 d_{i,j}^{(T-2)} + \alpha d_{i,j}^{(T-1)} + \right.$

$d_{i,j}^{(T)})/(\alpha^2 + \alpha + 1)\,, \quad \Rightarrow \quad \delta_{i,j}^{(T)} = round(m_{i,j}^{(T)})$, with initial assignations, $\pi_{i,j}^{(1)} = p_{i,j}^{(1)}\,, \; \delta_{i,j}^{(1)} = d_{i,j}^{(1)}, \; \pi_{i,j}^{(2)} = \alpha\pi_{i,j}^{(1)} + \pi_{i,j}^{(2)}\,, \; \delta_{i,j}^{(2)} = d_{i,j}^{(2)}$. In this scenario, memory of decisions does not operate if $\alpha \leq 0.61805$ (section 2.1), but if $\alpha > 0.61805$, the memory mechanism on decisions becomes that of selecting the *mode* of the last three states. With $\tau = 3$ and full memory the frequency of cooperators also grows monotonically to reach almost full cooperation as proved in Fig. 13.4. Defection persists as scattered unconnected very small oscillators (*D-blinkers*).

Spatial invasion of cooperation

Spatial structure enables cooperators to form clusters such that they are more likely to interact with other cooperators instead of being exploited by defectors. As a result, clusters of a few cooperators may expand (and invade) in a world of defectors [240]. Thus, for example, perfect squares of cooperators, even the simple 2×2 square, or the so called *grower* [309] shown in Fig. E.1, expands in a sea of defectors.

In such a context, memory may inhibit the expansion of cooperation. Thus, in Fig. 13.5 it is shown how the expansion of the *grower* is restrained when historic memory is implemented, even leading to extinction of cooperation with $\alpha=0.6$ as soon as at $T=7$. The *rotator* in Fig. 13.6 remains unaltered with $\alpha=0.1$, and extinguishes at $T=6$ if $\alpha \geq 0.6$, i.e., with effective memory of choices. Surprisingly, in the parameter interval $[0.2, 0.5]$ the cooperation advances from the *rotator*. The *glider* in Fig. 13.7 still glides with $\alpha=0.1$, but extinguishes with $\alpha=0.2$ and $\alpha=0.6$. Unexpectedly, the evolution of the glider with $\alpha >0.6$ resembles that of the grower, leading to the generation of almost perfect squares of decreasing size as the memory factor increases in value.

The long-term fate of these figures very much depends on its topology and on the memory charge implemented. The evolution of their frequency of cooperation (f) in a lattice of size 100×100 in a simulation run up to $T=100$ is shown in Fig. 13.8, and some corresponding patterns at $T=100$ are shown in Fig. E.5. As expected, the curve of the evolution of f from the grower with no memory, i.e., the red curve in the left panel of Fig. 13.8, is much like that starting at random in Fig. 13.3, thus tending to reach $f=0.318$. With $\alpha=0.7$, the grower and the glider virtually invade the space with small clusters of defectors remaining active as shown in Fig. E.5, and in both scenarios, with full memory the cooperation is constrained to the proximity of the site of appearance of the set of cooperators. With $\alpha=0.5$,

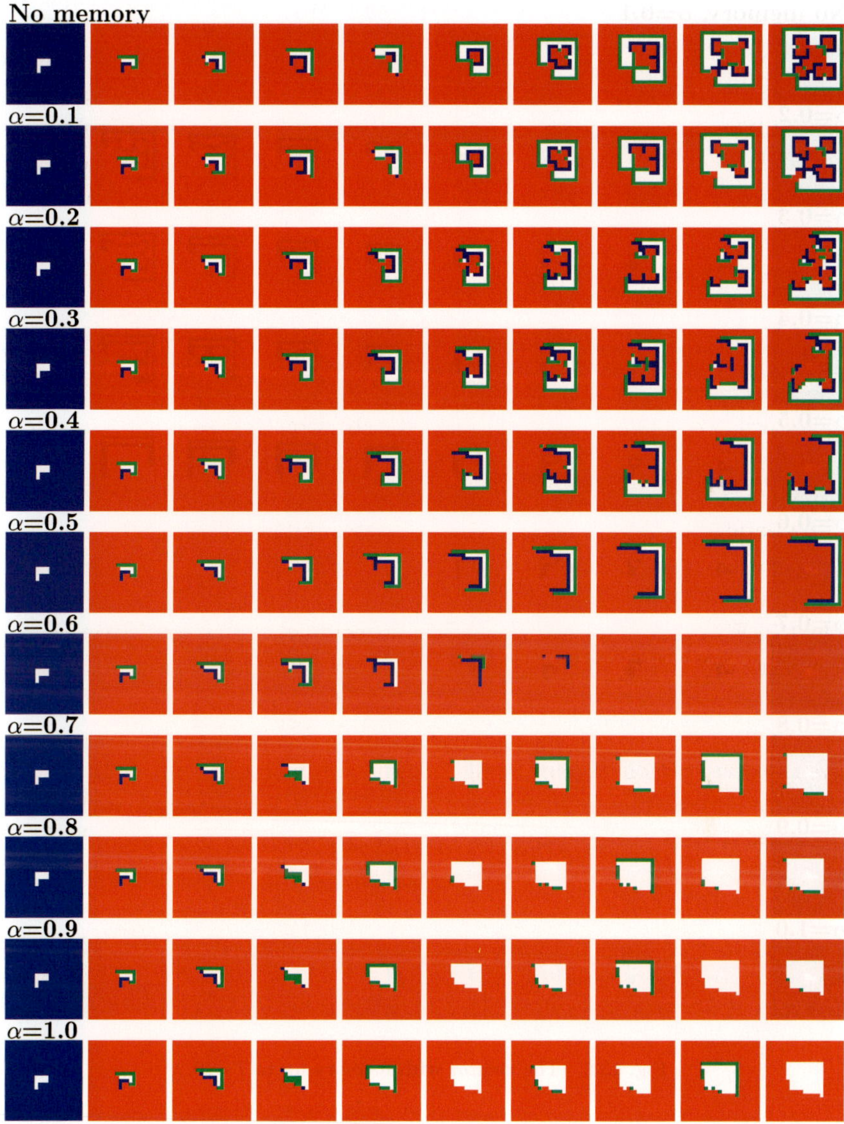

Fig. 13.5   The *grower* with memory up to $T=10$.

the grower and the rotator also tend to spread cooperation up to very high proportions, producing in both scenarios patterns such as that of the rotator shown in Fig. E.5 . In both scenarios, the *three-sides* pattern of cooperation

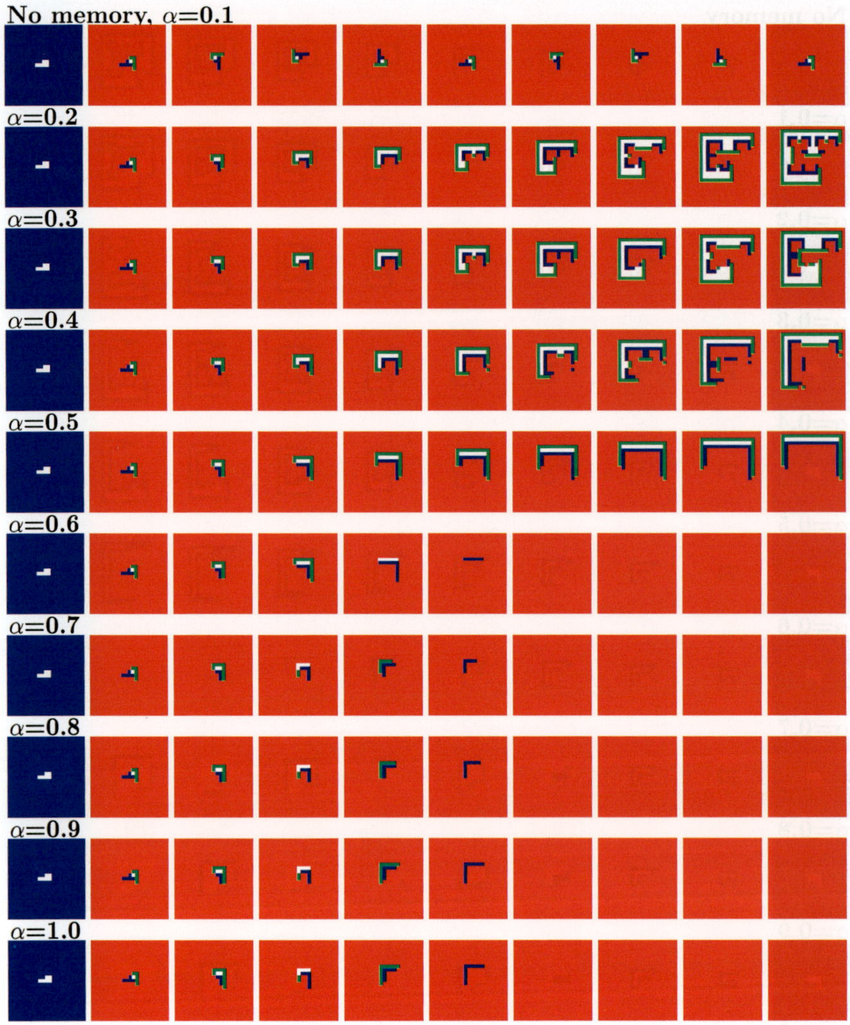

Fig. 13.6   The *rotator* with memory up to T=10.

appearing with $alpha$=0.5 in Figs.13.5 and 13.6 elongates and moves up to the border, where the two 0 opposite *wings* interact generating a kernel of cooperation that enables a fast spread of cooperation. This dynamics is reflected in the sharp increase of f in Fig. 13.8 by $T \simeq 50 = N/2$. The patterns with the rotator and the glider with α=0.3 shown in Fig. E.5 are somewhat similar, showing some striped-like areas of defection.

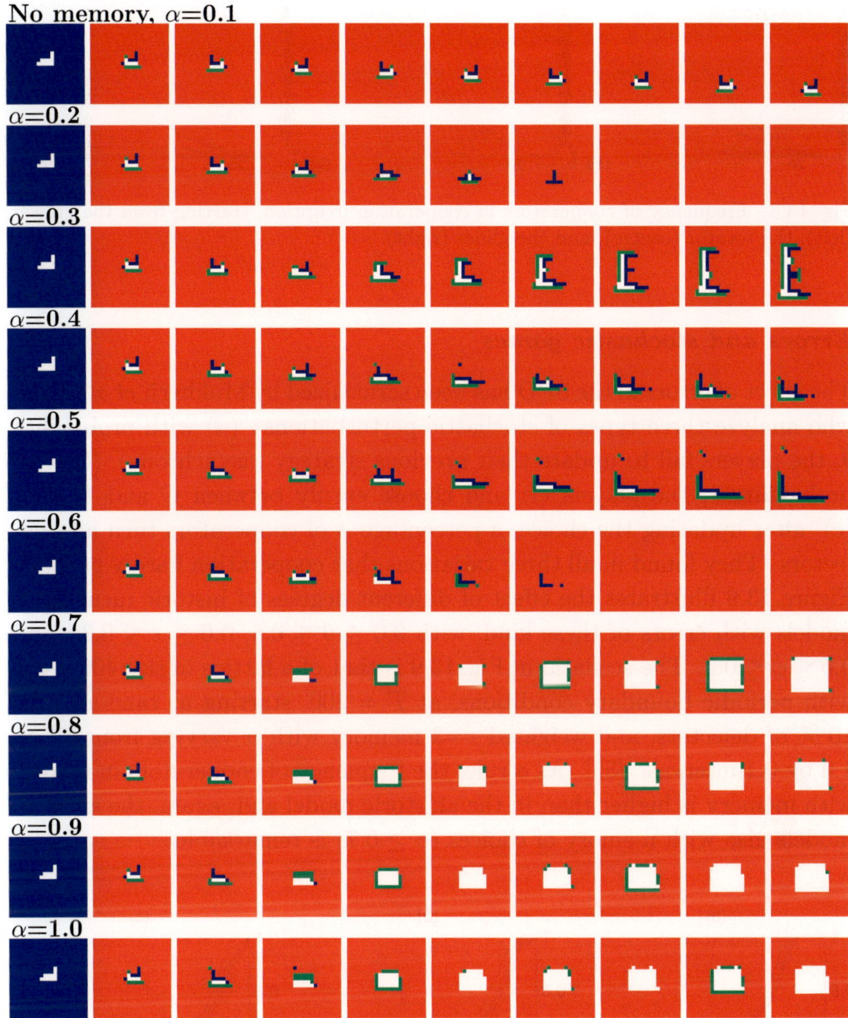

Fig. 13.7   The *glider* with memory up to $T=10$.

Qualitatively speaking, similar conclusions on the effect of memory on the spatialized PD are found for any temptation value in the parameter region $0.8 < b < 2.0$, in which spatial chaos is characteristic in the ahistoric model.

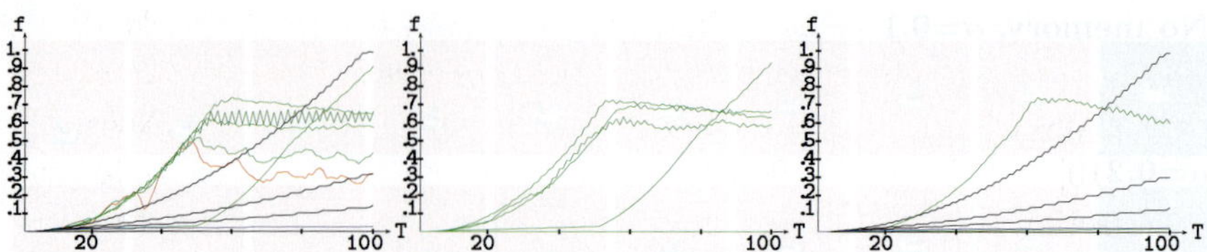

Fig. 13.8   Frequency of cooperators in cooperation irruption, starting from the grower (left), the rotator (centre), and the glider (right).

Errors and stochastic games

The effect of errors in the SPD has been scrutinized by Mukherji *et al.* [299], who analyze three types of stochastic perturbations: (a) with probability θ, the players fail to update their previous strategy (asynchrony), (b) with probability ϵ, the players err and choose evenly between C and D, and (c) after updating the choice, a percentage ϕ of cooperators turn into defectors. They found in all three variations that cooperation hardly persists. Figure 13.9 illustrates the effect of different degrees of historic memory in models with errors of these kind, with $0.0 \leq \theta \leq 0.5$, $0.0 \leq \epsilon \leq 0.3$, and $0.0 \leq \phi \leq 0.2$. The scenario in Fig. 13.9 is that of a lattice of size 400×400 with periodic boundary conditions, at $T = 200$, starting at random with 10 % of defectors, and weighted $\tau{=}3$ memory with α varying from 0.1 to 1.0 by 0.1 intervals [55]. As a rule, the frequency of cooperators in models with memory is higher than in the ahistoric model and, again, the effect of models also with memory of choices ($\alpha \geq 0.7$) is remarkable .

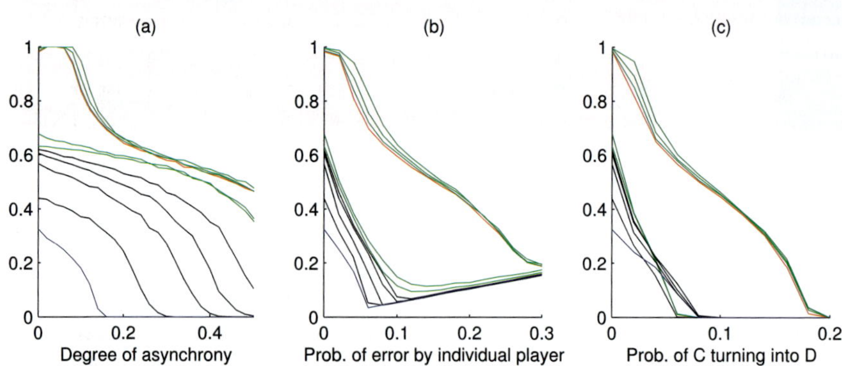

Fig. 13.9   Frequency of cooperators with memory of the last three iterations and errors of type: (a) asynchrony, (b) evenly choosing between C and D, (c) C turning into D .

Particularly important is that asynchrony (a) tends to provoke defection in the ahistoric model. With memory cooperation persists, in a high degree (even with a high degree of asynchrony) if $\alpha \geq 0.7$. Nowak *et al.* [310], allude to the fact that asynchronous updating tends to lead to defection by resorting to temptation values under 1.8, or to *stochastic games*, in which a cell will become a cooperator (defector) with a probability proportional to the total payoff to cooperators (defectors) among its neighbors [311]. Thus, denoting by $\mathcal{N}_i$ the neighborhood of the cell labelled i, the probability that it is occupied by a cooperator at time step T is: $P_i^{(T)} = \left(\sum_{j \in \mathcal{N}_i} {p_j^{(T)}}^m d_j^{(T)} \right) \Big/ \left(\sum_{j \in \mathcal{N}_i} {p_j^{(T)}}^m \right)$. The weighting factor m favors the most successful neighbor: the larger m, the more likely is it that the cell will adopt the strategy of the most successful neighbor. We adopt here $m = 1.0$ (no weighting), thus the probability becomes with memory: $P_i^{(T)} = \left(\sum_{j \in \mathcal{N}_i} \pi_j^{(T)} \delta_j^{(T)} \right) \Big/ \left(\sum_{j \in \mathcal{N}_i} \pi_j^{(T)} \right)$. Memory notably boosts cooperation in this stochastic scenario, dramatically when $\alpha \geq 0.7$ [55, 57].

J.Alonso et al.[15] propose an extension of the basic Nowak-May model, in which every player automatically copies the action of the most successful neighbour if its score p^{msn} is above a minimum threshold p_{min}. On the contrary case, the player has a probability p of adopting the opposite action. It would be interesting to endow the players with memory, and compare $\left(\pi^{(T)} / \omega(T) \right)^{msn}$ to p_{min}.

The effect of memory in disordered networks

In this section we consider the case of networks with irregular wiring obtained by randomly rewiring a regular network.

Figure 13.10 shows the evolution of the cooperation starting from a single defector in lattices with percentages of rewiring from 10 to 100 %, increased by 10 % intervals. In these scenarios the defection spreads *at distance* (see Figs.E.6 and E.7 in the Appendix E), so that the chance of maintaining cooperation is lower than in the ordered lattice context. Thus for example, with only a 10 % of rewiring the cooperation stabilizes in the ahistoric model between 10 % and 20 % (under 0.318), and, as expected, higher percentages of rewiring lead to lower frequencies of cooperation. Memory helps to maintain cooperation in Fig. 13.10 in the scenarios with low rewiring, but at lower levels than in the ordered scenario of Fig. 13.3. With high levels of rewiring, the initial fall of cooperation is deeper and

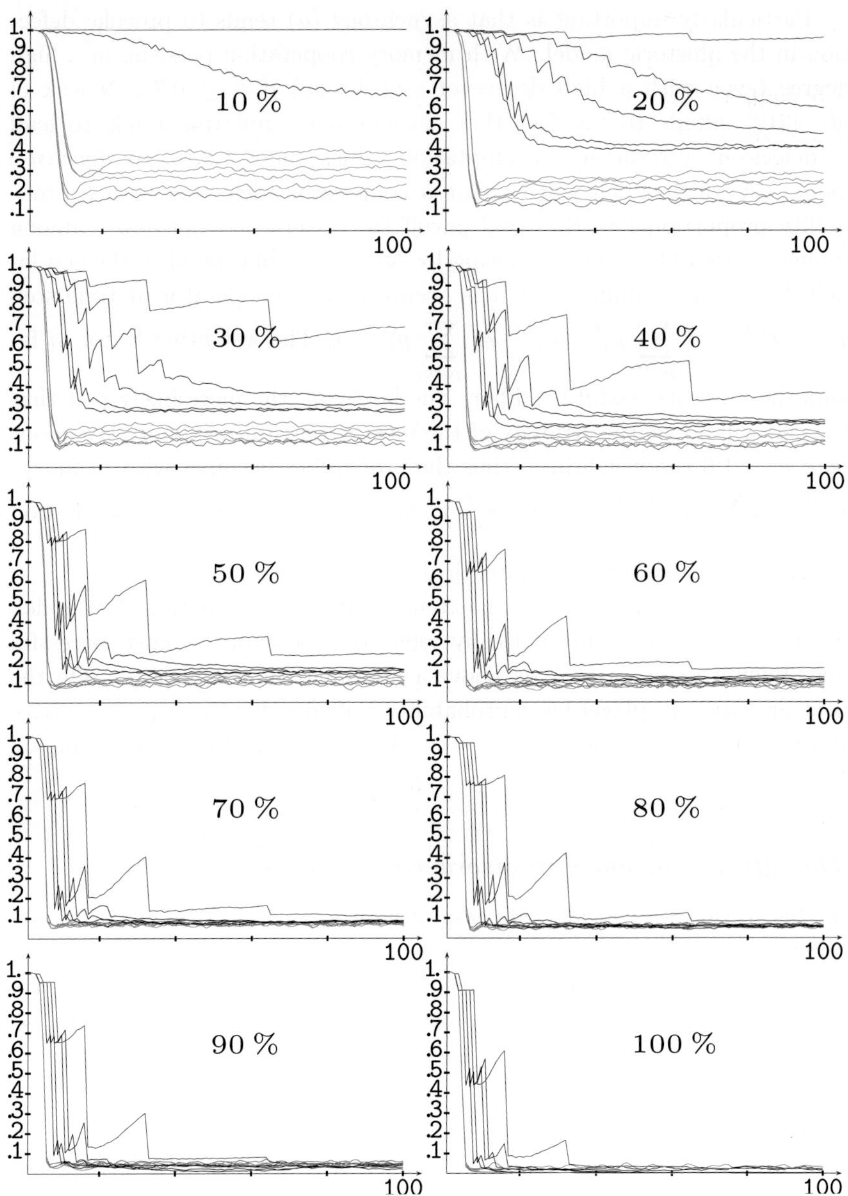

Fig. 13.10   Evolution of the frequency of cooperators starting from a single defector with the percentage of rewiring indicated in each frame.

memory is unable to help in an effective recovering. The effect of full memory is here that foreseen, as it is that of maximum support of cooperation, enhancing the *discontinuities* in the evolution of the frequencies of cooperation that appear in most of the scenarios. In the simulations with rewiring over 20 %, full memory in the early stages appears to induce a persistent increase in the cooperation, but this tendency is interrupted with very sharp decays in the frequency of cooperation.

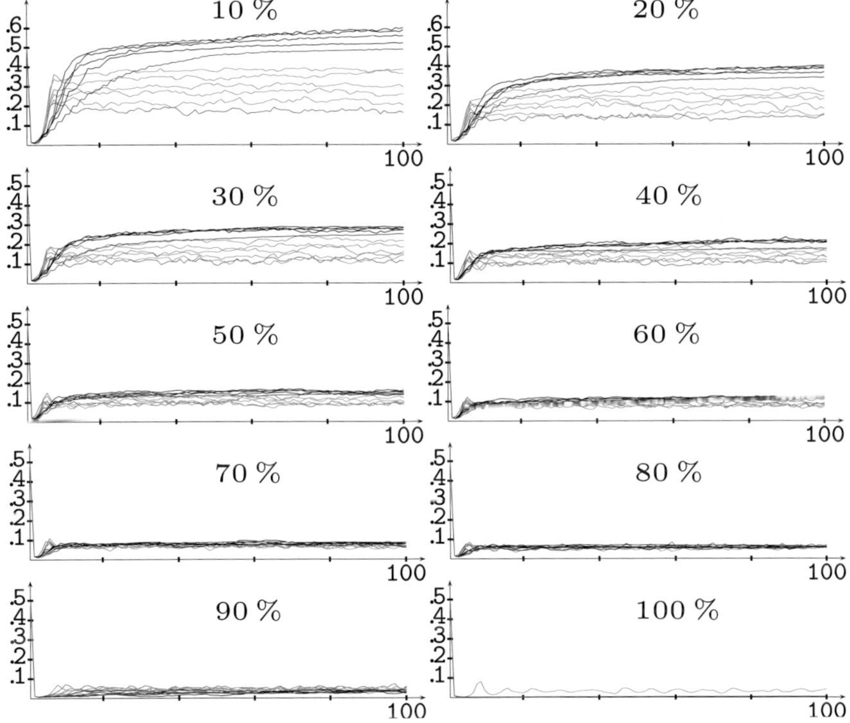

Fig. 13.11   Evolution of cooperation starting at random with the percentage of rewiring indicated in each frame.

Figure 13.11 shows the evolution of cooperation starting at random in lattices with the same percentages of rewiring as in Fig. 13.10. Again, as in the ordered lattice contexts, cooperation plummets initially and recovers later on, but to lower levels as rewiring increases. Full memory keeps the curves under those with also memory of choices (black), but greater than ahistoric (red) and with only payoff memory (green). The nucleation of cooperation phenomenon observed in the ordered lattice context, appears also with low levels of rewiring as shown in Figs.E.8 and E.9 of the Appendix E.

In short, memory notably stimulates cooperation in ordered lattices, but it is unable to boost cooperation as the wiring network becomes highly disordered.

13.2 Degrees of cooperation and strategies

Degrees of cooperation

As a natural extension of the described binary model, the strong 0-1 assumption regarding the move can be relaxed by allowing *degrees* of cooperation in a continuous scenario[1]. Thus in the continuous-valued PD (CVPD for short), denoting x the degree of cooperation of player A and y the degree of cooperation of the player B, a consistent way to specify the pay-off for values of x and y other than zero or one is to simply interpolate between the extreme payoffs of the binary case [149, 269]. Denoting the payoff matrix as $\mathbf{G} = \begin{pmatrix} R & S \\ T & P \end{pmatrix}$, it is:

$$p_A(x,y) = (x, 1-x)\mathbf{G}\begin{pmatrix} y \\ 1-y \end{pmatrix} \ , \ p_B(x,y) = (x, 1-x)\mathbf{G'}\begin{pmatrix} y \\ 1-y \end{pmatrix}.$$

Adopting $\mathbf{G} = \begin{pmatrix} 1 & 0 \\ b & 0 \end{pmatrix}$, it is: $p_A(x,y) = (x + (1-x)b)y$, $p_B(x,y) = p_A(y,x)$.

Incidentally, if x and y will represent probabilities of cooperation, the above expressions of p_A and p_B give the mean payoff of every player.

In the continuous-valued historic formulation, it is not necessary to round-off the m values to obtain the δ ones, simply $\delta \equiv m$, including $\delta_{i,j}^{(2)} = \left(\alpha x_{i,j}^{(1)} + x_{i,j}^{(2)}\right)/(\alpha + 1)$. Table 13.3 illustrates the initial scenario starting from a single (full) defector. Thus, historic memory in the continuous valued scenario of Table 13.3 assigns to the neighbors of the initial defector following their (full) defection at $T{=}2$ the trait degree of cooperation $\delta^{(2)} = \alpha/(1 + \alpha)$, not zero as in the binary model. Consequently, this is the value of x that expands up to a $4{\times}4$ square if $\alpha < 0.25$, and remains confined in the $3{\times}3$ square if $\alpha > 0.25$. Unlike in the binary model, in

---

[1] Discrete levels of cooperation have been also considered in the literature. Thus for example, four levels in [109, 111, 118, 437]. In these articles the levels of cooperation are coded as $\{1, 1/3, -1/3, -1\}$, and the values of the pay-off matrix are: $\mathfrak{T} = 5, R = 4, P = 1, S = 0$. If c_A and c_B are the cooperation levels of players A and B, by linear interpolation of the conventional two-choice model, the payoff of player A is given by: $p_A = 2.5 - 0.5c_A + 2c_B$.

Table 13.3 Weighted mean degrees of cooperation after $T = 2$ and degrees of cooperation at $T = 3$ from a single defector in the CVPD. $b = 1.85$.

$\boldsymbol{\delta}^{(2)}$

1	1	1	1	1
1	$\frac{\alpha}{1+\alpha}$	$\frac{\alpha}{1+\alpha}$	$\frac{\alpha}{1+\alpha}$	1
1	$\frac{\alpha}{1+\alpha}$	0	$\frac{\alpha}{1+\alpha}$	1
1	$\frac{\alpha}{1+\alpha}$	$\frac{\alpha}{1+\alpha}$	$\frac{\alpha}{1+\alpha}$	1
1	1	1	1	1

$\mathbf{x}^{(3)}(\alpha < 0.25)$

1	1	1	1	1	1	1
1	$\frac{\alpha}{1+\alpha}$	$\frac{\alpha}{1+\alpha}$	$\frac{\alpha}{1+\alpha}$	$\frac{\alpha}{1+\alpha}$	$\frac{\alpha}{1+\alpha}$	1
1	$\frac{\alpha}{1+\alpha}$	$\frac{\alpha}{1+\alpha}$	$\frac{\alpha}{1+\alpha}$	$\frac{\alpha}{1+\alpha}$	$\frac{\alpha}{1+\alpha}$	1
1	$\frac{\alpha}{1+\alpha}$	$\frac{\alpha}{1+\alpha}$	$\frac{\alpha}{1+\alpha}$	$\frac{\alpha}{1+\alpha}$	$\frac{\alpha}{1+\alpha}$	1
1	$\frac{\alpha}{1+\alpha}$	$\frac{\alpha}{1+\alpha}$	$\frac{\alpha}{1+\alpha}$	$\frac{\alpha}{1+\alpha}$	$\frac{\alpha}{1+\alpha}$	1
1	$\frac{\alpha}{1+\alpha}$	$\frac{\alpha}{1+\alpha}$	$\frac{\alpha}{1+\alpha}$	$\frac{\alpha}{1+\alpha}$	$\frac{\alpha}{1+\alpha}$	1
1	1	1	1	1	1	1

$\mathbf{x}^{(3)}(\alpha \geq 0.25)$

1	1	1	1	1	1	1
1	1	1	1	1	1	1
1	1	$\frac{\alpha}{1+\alpha}$	$\frac{\alpha}{1+\alpha}$	$\frac{\alpha}{1+\alpha}$	1	1
1	1	$\frac{\alpha}{1+\alpha}$	$\frac{\alpha}{1+\alpha}$	$\frac{\alpha}{1+\alpha}$	1	1
1	1	$\frac{\alpha}{1+\alpha}$	$\frac{\alpha}{1+\alpha}$	$\frac{\alpha}{1+\alpha}$	1	1
1	1	1	1	1	1	1
1	1	1	1	1	1	1

$\mathbf{x}^{(3)}(\alpha = 0.2)$

1	1	1	1	1	1	1
1	1/6	1/6	1/6	1/6	1/6	1
1	1/6	1/6	1/6	1/6	1/6	1
1	1/6	1/6	1/6	1/6	1/6	1
1	1/6	1/6	1/6	1/6	1/6	1
1	1/6	1/6	1/6	1/6	1/6	1
1	1	1	1	1	1	1

$\mathbf{x}^{(3)}(\alpha = 1.0)$

1	1	1	1	1	1	1
1	1	1	1	1	1	1
1	1	0.5	0.5	0.5	1	1
1	1	0.5	0.5	0.5	1	1
1	1	0.5	0.5	0.5	1	1
1	1	1	1	1	1	1
1	1	1	1	1	1	1

which the initial defector never becomes cooperator, the initial defector cooperates at $T=3$ at the degree $\alpha/(1+\alpha)$, the trait degree of cooperation of its neighbors at $T=2$ (in particular that of those in the corners with highest accumulated payoff: $\pi^{(2)} = 8\alpha + 5b > 8b\alpha$, if $b=1.85$).

Figure 13.12 shows the advance of the defection up to $T=13$ in the CVPD starting from a single full defector in a small 11×11 lattice. Memory, even at low levels, avoids the full defection invasion that occurs in the ahistoric model. Table 13.4 shows a MATLAB program rendering Fig. 13.12. Gray tones correspond to degrees of defection in Fig. 13.12 and in the figures of the Appendix regarding the CVPD.

The upper group of patterns in Fig. E.10 in the Appendix E shows the patterns at $T=100$ in the CVPD starting from a single defector in a 101×101 lattice, whereas in Fig. E.11 the patterns when starting at random in a 100×100 lattice are shown. In both cases periodic boundary conditions are imposed on the edges.

In the subsequent figures of this section, a study is made of the evolution in the CVPD of the mean degree of cooperation: $f = \dfrac{1}{N \times N} \sum\limits_{i=1}^{N} \sum\limits_{j=1}^{N} x_{i,j}$.

Figure 13.13 shows the evolution of the mean degree of cooperation in the continuous-valued model, with unlimited trailing memory in a lattice of size 100×100. Memory dramatically constrains the advance of defection

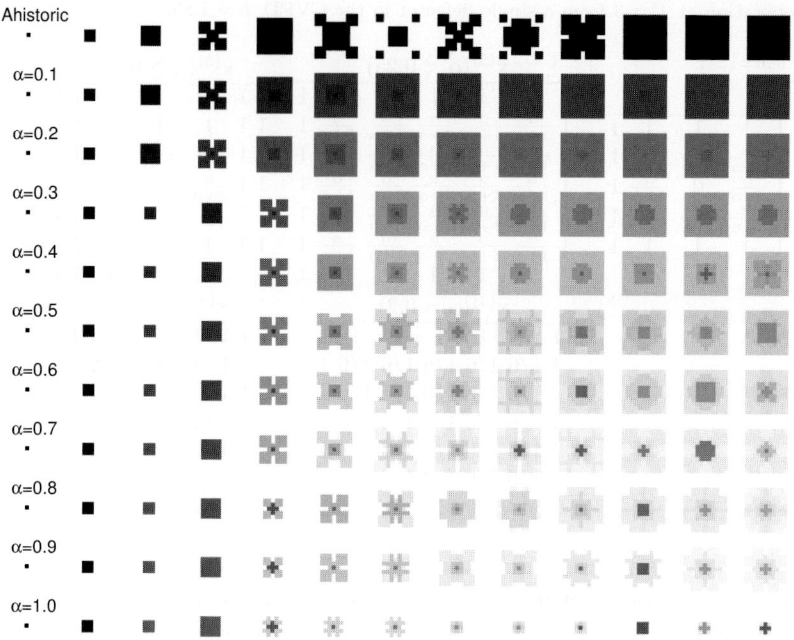

Fig. 13.12   The advance of defection up to $T{=}13$ in the continuous-valued PD starting from a single full defector in a $11{\times}11$ lattice.

in a smooth way when starting from a single full defector, even for the low level $\alpha = 0.1$. The effect appears much more homogeneous compared to the binary model, as memory of decisions is always operative in the continuous-valued model. In parallel with the binary model in Fig. 13.3, full memory notably restrains in Fig. 13.13 the recovering of cooperation after its initial dramatic decrease when starting at random. Figure 13.14 confirms this effect across ten different simulations. In Fig. 13.14, as well as in Fig. 13.4, there is a simulation that does not overcome the initial dramatic fall in cooperation (only a five-cell rotator survives in it).

Figure 13.15 concerns the CVPD with i) memory of the last three iterations [55]: $\pi_{i,j}^{(T)} = \alpha^2 p_{i,j}^{(T-2)} + \alpha p_{i,j}^{(T-1)} + p_{i,j}^{(T)}$, $\delta_{i,j}^{(T)} = \left(\alpha^2 d_{i,j}^{(T-2)} + \alpha d_{i,j}^{(T-1)} + d_{i,j}^{(T)}\right)/\left(\alpha^2 + \alpha + 1\right)$, with initial assignations, $\pi_{i,j}^{(1)} = p_{i,j}^{(1)}$, $\delta_{i,j}^{(1)} = d_{i,j}^{(1)}$, $\pi_{i,j}^{(2)} = \alpha \pi_{i,j}^{(1)} + \pi_{i,j}^{(2)}$, $\delta_{i,j}^{(2)} = \alpha d_{i,j}^{(1)} + d_{i,j}^{(2)}$, and ii) with minimal, $\tau{=}2$ memory: $\pi_{i,j}^{(T)} = \alpha p_{i,j}^{(T-1)} + p_{i,j}^{(T)}$, $\delta_{i,j}^{(T)} = \alpha d_{i,j}^{(T-1)} + d_{i,j}^{(T)})/\left(\alpha + 1\right)$. In both scenarios, short-range memory clearly boosts cooperation, with no decay in

Table 13.4   A MATLAB® program for the CVPD.

```
function []=BPD;
    par.b=1.85; par.n=27;par.t=13; par.B=[par.n 1:par.n 1]
    for nal=1:11; [PI,DE,X,par]=ini(nal,par);
        for T=1:par.t
            figure(9);subplot(11,par.t,par.t*(nal-1)+T),imagesc(X,[0,1])
            colormap('bone');axis image;axis('off')
            [P]=PLAY(X,par);
            [PI,DE,DD]=MEMORY(P,PI,T,X,DE,par);
            [X]=DECIDE(PI,DD,T,par);
        end
    end
function [P]=PLAY(X,par);
    for i=1:par.n;for j=1:par.n; x=X(i,j);P(i,j)=0;
    for k=0:2;ik=par.B(i+k);for l=0:2;jl=par.B(j+l);y=X(ik,jl);
            P(i,j)=P(i,j)+(x+(1-x)*par.b)*y;          % Single round Payoffs
    end;end
    end;end
function [PI,DE,DD]=MEMORY(P,PI,T,X,DE,par)
    if(T==1)PI=P;DE=X; 'else;PI=par.alfa*PI+P;DE=par.alfa*DE+X;end
    DD=X;
    for i=1:par.n;for j=1:par.n
            if((DE(i,j))>par.W2(T))DD(i,j)=1;end;
            if((DE(i,j))<par.W2(T))DD(i,j)=0;end;
    end;end
function [X]=DECIDE(PI,DD,T,par);
    for i=1:par.n;for j=1:par.n
        [X(i,j)]=DECISION(PI,DD,T,i,j,par);
    end;end
function [x]=DECISION(PI,DD,T,i,j,par);
    pm=PI(i,j); nc=0;nd=0;delta=0.000001;
    for k=0:2;ik=par.B(i+k);for l=0:2;jl=par.B(j+l);
            px=PI(ik,jl);if(px>pm);pm=px;end;          % max neighbour's payoff
    end;end
    for k=0:2;ik=par.B(i+k);for l=0:2;jl=par.B(j+l);
            px=PI(ik,jl);if(pm-px)¡delta
    if(DD(ik,jl)==1)nc=nc+1;else nd=nd+1;end
    end
    end;end
    x=1;if(nd>nc)x=0;end               % next choice
function [PI,DE,X,par]=ini(nal,par)
    X=ones(par.n);DE=X;c=(par.n+1)/2;X(c,c)=0;PI=zeros(par.n);
    par.alfa=0.1*(nal-1);w(par.t)=1;
    for tt=1:par.t-1;w(tt)=par.alfa∧(par.t-tt);end; sx=0;
    for tt=1:par.t;sx=sx+w(par.t-tt+1);par.W(tt)=sx;end
    par.W2=0.5*par.W
```

the particular case of full memory as in the case of unlimited trailing memory (albeit, when starting at random, the evolution of cooperation with full memory reaches a plateau that is not the highest one). The lower groups of patterns in Figs.E.10 and E.11 of Appendix E show the patterns at $T=100$ from a single initial defector and starting at random with $\tau=2$ memory.

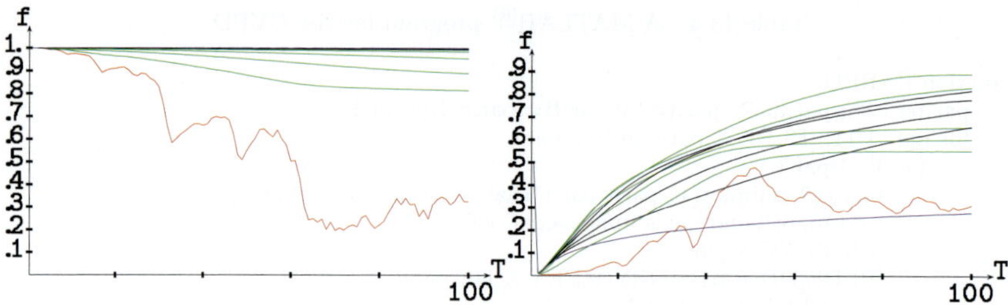

Fig. 13.13   Mean degree of cooperation (f) in the continuous-valued PD.

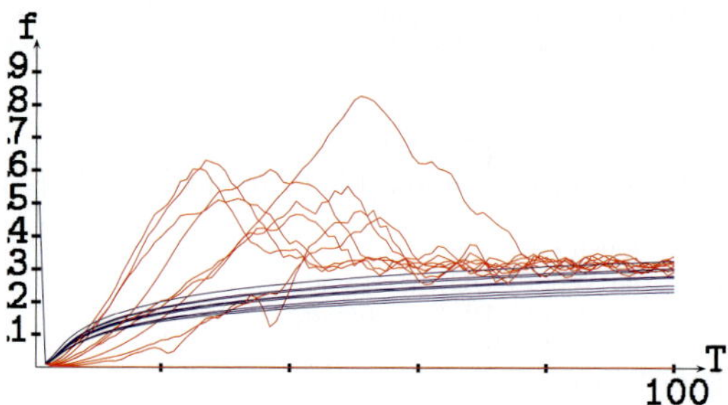

Fig. 13.14   Ten alternative simulations in the CVPD.

Figure 13.16 shows the evolution of the cooperation in the CVPD starting from a single defector in lattices with percentages of rewiring from 10
to 100 %,. Thus for example, with only a 10 % of rewiring the cooperation
stabilizes in the ahistoric model in between 10 % and 20 % (under 0.318),
and, as expected, higher percentages of rewiring lead to lower mean degrees of cooperation. Memory helps to maintain cooperation in Fig. 13.16
in the scenarios with low rewiring, but at lower levels than in the ordered
scenario of Fig. 13.13 . With high levels of rewiring, the initial fall of cooperation is deeper and memory is unable to help in an effective recovering of
cooperation. Increasing levels of memory induce an increasing support of
cooperation in every scenario of Fig. 13.16 .

Figure 13.17 shows the evolution of cooperation in the CVPD starting
at random in lattices with percentages of rewiring varying from 10 to 100%
by 10% intervals. Again, as in the ordered lattice contexts, cooperation
plummets initially and recovers later on, but to lower levels as rewiring
increases. With low levels of rewiring, e.g., 10 and 20 %, full memory shows
again dynamics of the degree of cooperation lower than that corresponding

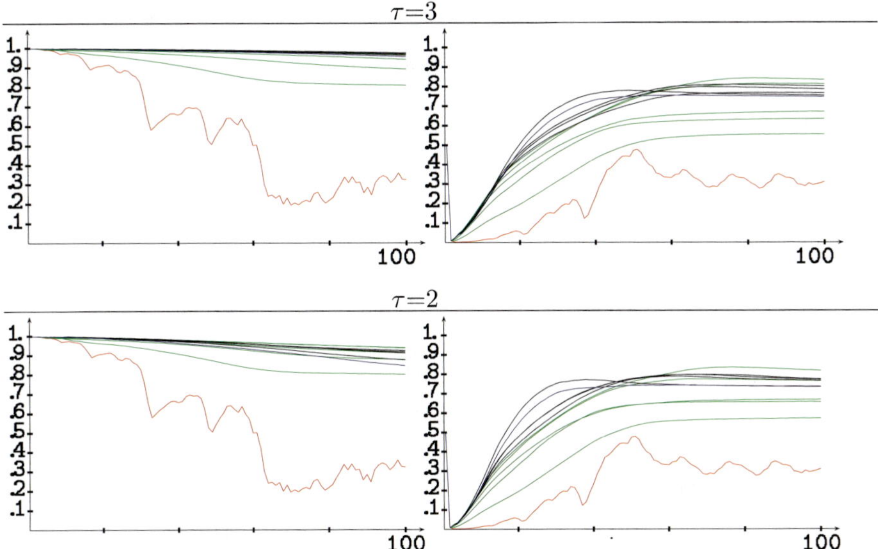

Fig. 13.15   Mean degree of cooperation in the CVPD with short-range memory in the CVPD.

to lower memory factors, but unlike in the ordered lattice scenario notably greater than in the ahistoric (red) model. Masuda and Aihara [282] have analyzed how cooperation is affected when moving from a regular (one-dimensional) ring to a fully random network via rewiring. They extend their study to a wide interval of variation of the temptation parameter b, but do not consider memory. The authors claim in [282] that the small-world topology is the optimal structure in terms of a rapid convergence to a equilibrium state with many cooperators. The rapid convergence to a relatively high degree of cooperation (over the 0.318 rate expected in the ahistoric, fully structured model) achieved with memory in models with low rewiring (up to 20 %), seems to agree somehow with the conclusion reached by Masuda and Aihara. Moreover taking into account that in the perfectly structured lattice scenario, full memory fails in promoting cooperation in the CVPD.

Figure E.12 shows the initial patterns in the CVPD with $\alpha=0.9$ and full memories starting at random in a 100×100 lattice, whereas Fig. E.13 shows the initial patterns in the CVPD with $\alpha=0.9$ memory starting at random in a 100×100 network with a 10% of rewiring. Some kind of nucleation seems to characterize the patterns in Fig. E.12, whereas Fig. E.13. appears

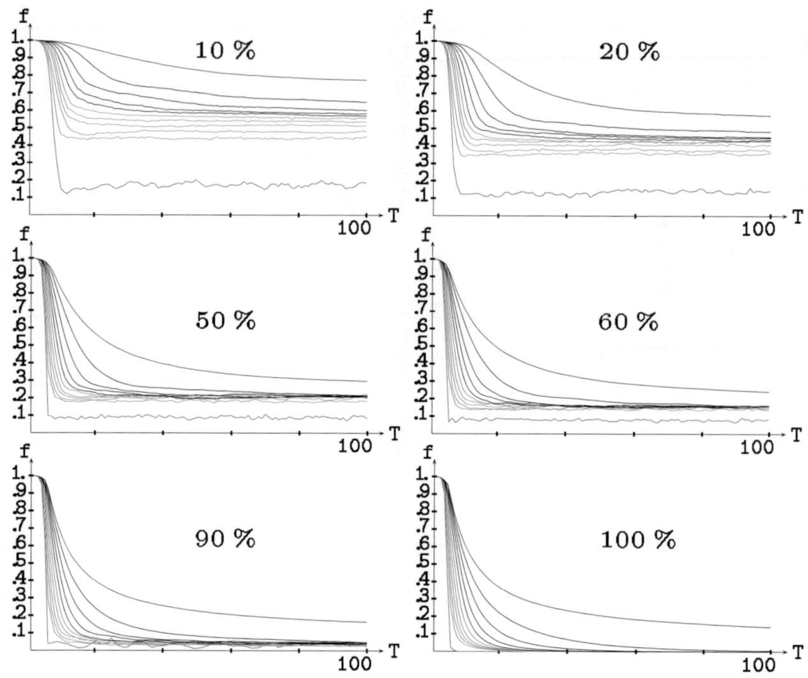

Fig. 13.16   Evolution of cooperation in the CVPD from a single defector with the percentage of rewiring indicated in each frame. Colors code as in Fig. 13.13.

as *fuzzified.*

The inhibition of the advance of cooperation in a defector context, as shown in Figs. 13.5 to 13.7 turns out fuzzier in the continuous valued PD context. Thus, for example, the *grower* in Fig. 13.18 does not extinguish at $T=6$ as it does in Fig. 13.5 .

Strategies

When the *PD* is played iteratively, the players may elaborate strategies. Reactive or one-dimensional strategies depend only on the coplayer's play in the previous round. They can be characterized by triplets $\Delta =< i, p, q >$, where i indicates the starting play, p and q are both probabilities of cooperation : after cooperation and defection on the other side respectively. The eight simple one-dimensional strategies (*strategies* for short) can thus be coded as in Table 13.5, with 1 standing for cooperation and 0 for defection :

The strategy TFT $< 1, 1, 0 >$, which cooperates on the first round

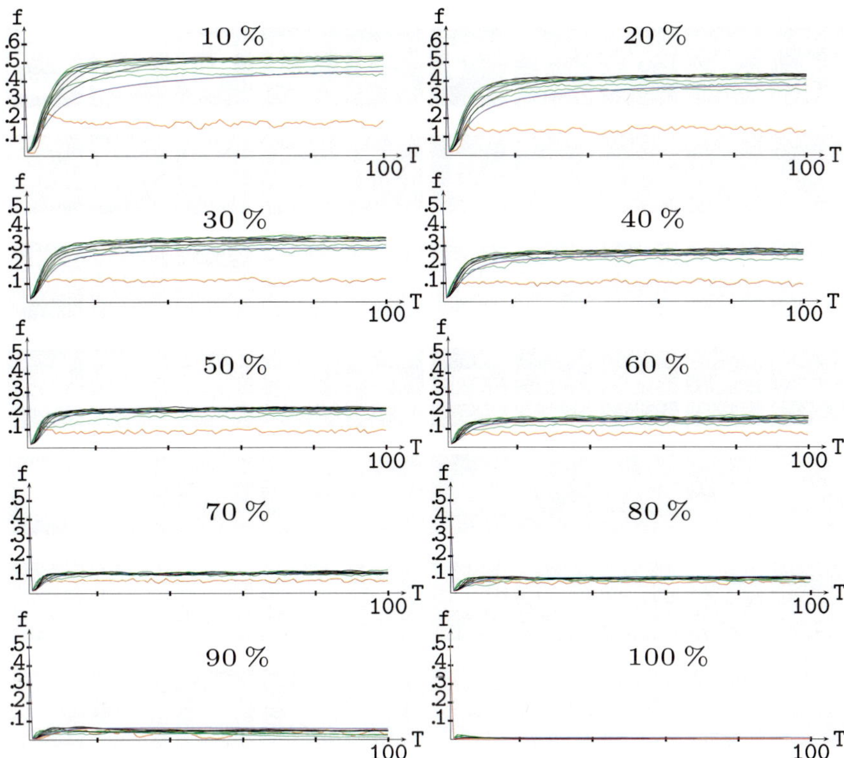

Fig. 13.17   Evolution of cooperation in the CVPD starting at random with the percentage of rewiring indicated in each frame. Colors code as in Fig. 13.13.

Table 13.5   Deterministic one-dimensional strategies.

	Moves		
Initial	After 1	After 0	
1	1	1	Always Cooperate (AllC)
1	1	0	Tit For Tat (TFT)
1	0	1	Gullibe Doormat
1	0	0	Deceptive defector
0	1	1	Suspicious Quaker
0	1	0	Suspicious TFT (STFT)
0	0	1	Suspicious Doormat
0	0	0	Always Defect (AllD)

and thereafter repeats its coplayer's play from the previous round, was the winner in the computer tournaments conducted in a world with perfect communication and errorless execution by Axelrod [70].

No memory

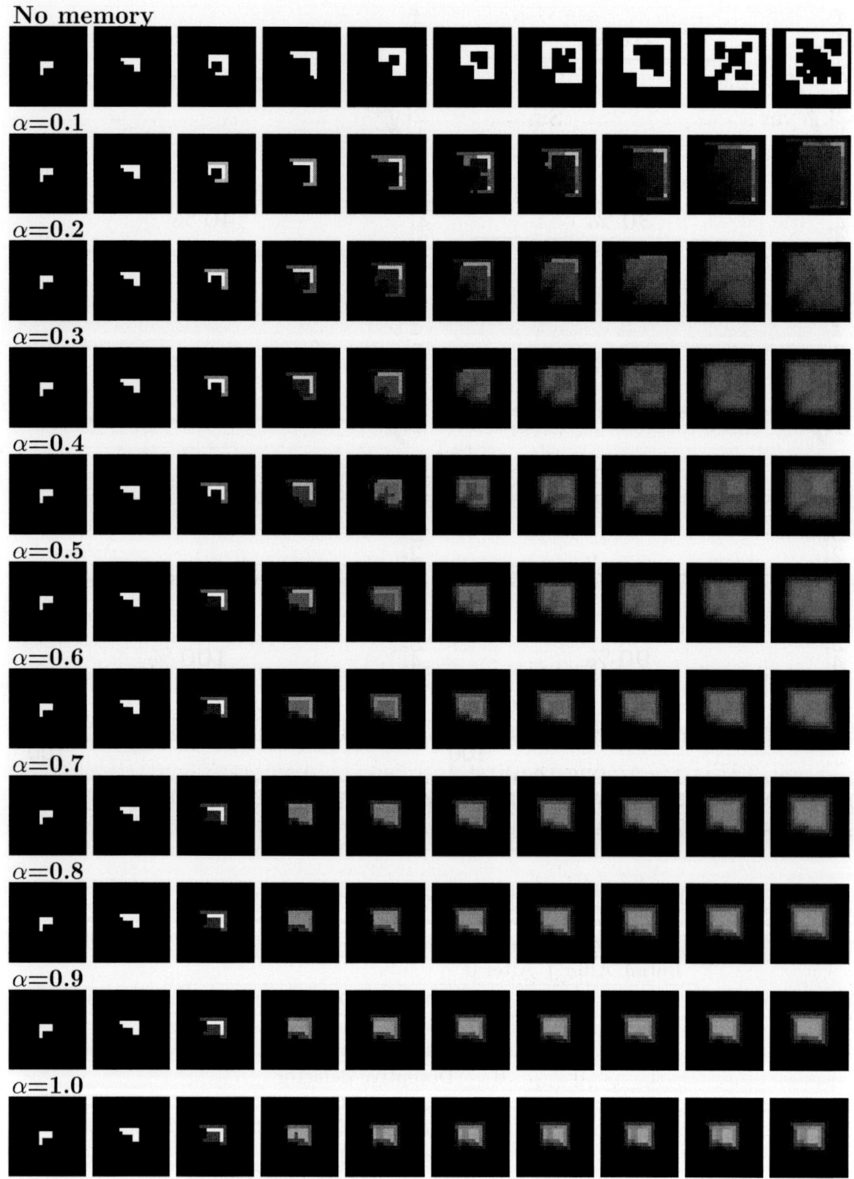

Fig. 13.18   The CVPD *grower* with memory.

Stochastic reactive strategies respond to a single previous move by a coplayer assigning probabilities for cooperation and defection. Here again strategies can be envisaged in terms of ordered triplets $\Delta =< i, p, q >$, though the components of Δ are now probabilities of cooperation: initial (i), following a cooperative move by a coplayer (p), and following a coplayer's defection (q). Simple one-dimensional strategies would remain coded as before, since the extreme values of probabilities (1 or 0) dictate the determinism. Classical TFT for example would remain $< 1, 1, 0 >$. But given a continuum of probabilities would permit the introduction of other strategies, such as the *generous* ones, which forgive defection against them with non-zero probability, i.e., $0 < q < 1$. *Generous* TFT (GTFT) $< 1, 1, 1/3 >$ proved ultimately triumphant (after an early and almost total victory of TFT) in the world of imperfect information and/or execution devised by Nowak and Sigmund in [314].

The effect of unlimited trailing memory in the SPD when players follow strategies has been studied in [58, 59]. There it is concluded that starting with an initial configuration with strategies randomly assigned historic memory: *i)* helps to preserve the initial diversity of suspicious strategies, *ii)* implies a smaller divergence of the average value of the probabilities informing the selected strategies, and *iii)* aids to the stabilization around the initial value of the overall frequency of cooperation choices.

References [58, 59] include the study of asynchronous (*continuous* time) simulations: in each elementary *step* a cell is chosen at random and immediately updated, so the model turns out to be asynchronous. Huberman and Glance [197] were among the first to raise the question about asynchrony updating in the spatial prisoners dilemma. They support the asynchronous updating in order "to mimic a real world" and argue that -when asynchrony is introduced in the model of Nowak and May [312], a fixed state is arrived at in which all the players are defecting-, so that: -results ... differ greatly when time is discrete as opposed to continuous-. Nowak *et al.* [310] reject such an idea, featuring it as an "extraordinary misapprehension" and argue that Huberman and Glance [197] only considered a specific set of parameters, and that in fact asynchronous updating does in general induce a fixed state, so that: "most of our basic conclusions are unaffected by whether we use discrete or continuous time anyway".

13.3 The structurally dynamic PD (SDPD)

Up to now, the game is based on static (crystallized) networks, i.e., the social network on which the evolution of cooperation is studied is fixed from the outset and not affected by evolutionary dynamics on top of them. However, iteration networks in the real-world are continuously evolving ones, rather than static graphs, and the preferential choice and refusal of partners plays an important role in the emergence of cooperation [65]. Instead of investigating the evolutionary games on static networks which constitute just one snapshot of the real evolving ones, recently some researchers proposed that the network structure may co-evolve with the evolutionary game dynamics [132, 133, 143, 353, 448, 447, 320–322]. A review of the work done regarding coevolutionary networks is given in [177], whereas the report by Boccaletti [86] has a broader scope, dealing with complex networks in general. A deep review of the work done on games on graphs is that of Szabo [385], whereas the compact review in [330] pays particular attention to *aging*.

In most of the models proposed, the players improve their topological position by cutting links to defectors and generating new links coupling cooperators. Both coupling and decoupling rules have a probabilistic component, so that strategies and network do not evolve at the same rate to avoid a fast freeze in the dynamics as soon as the defectors become isolated. Thus for example in [133, 448, 447] D-D links are broken with probability p. Here we renounce to any stochastic component in the coupling/decoupling transition rules and opt for full-determinism, as a primary study for a further extension with probabilistic terms embedded in the transition rules.

Thus, cooperators turning defectors (*unsatisfied* because their payoff is not the highest among their neighbors [447, 448]):

- eliminate their links with their linked defectors, and
- connect with their nearest-neighbor cooperators

The idea of decoupling defectors is fairly general, but new coupling is frequently made at random, albeit in some articles the rewiring is directed to *neighbor's neighbor*, as is done in [143] and [353], in the latter case with asynchronous updating.

Let us consider for the sake of illustration the simple case of Table 13.6, in which the initial Euclidean lattice is seeded with a single defector. After the first iteration, every player connected with the initial defector becomes unsatisfied and breaks his link to him, so that he becomes isolated at time-step T=3. Lone players will be marked in patterns of this section with empty squares. In the ahistoric model, the new defectors at T=3 in turn, break their connections with every defector and connect with every next-

Table 13.6   Co-evolving dynamics in the ahistoric model from a single defector.

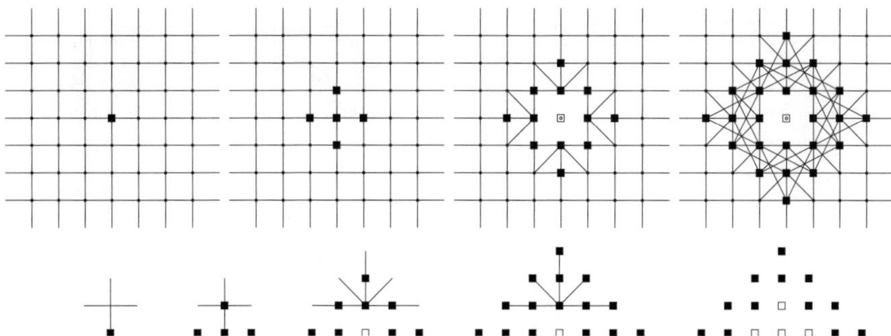

neighbour cooperator, starting at T=4 a wave of nested new connections enclosing a kernel of isolated defectors, as those patterns in Fig. 13.19. The planar representation of the web of connections used here, may lead to ambiguities, hiding those links that appear superimposed and some non-existent links may appear to be present. That is case for example of the initial neighbours of the initial defector. They would seem to continue connected at T=4 with the new defectors at T=3. But they are not, the explanation of the misleading lines lies in their new connections to next-nearest neighbours generated at T=3, as may be seen in the particular case of the northern neighbour of the initial defector, enhanced in the lower part of Table 13.6. Thus the initial neighbours of the defector at T=1 become isolated at T=5, as it is shown in Fig. 13.21 in the Appendix. The first row of patterns of this figure and that of Fig. 13.20 show the evolving patterns of choices up to T=13 starting with connectivities nine and five from a single defector.

The effect of memory in dynamic networks

Table 13.7 departs from the same scenario as Table 13.6, but deals with the co-evolving dynamics with full memory. The lower row of patterns in Table 13.7 shows the underlying values of π that drive, together with δ, the process. Thus, the initial defector gets 1.48 (=4×1.85/5) units from his encounters with his four C-neighbours at T=1, but gets no more incomes in future rounds, either because his neighbors become defectors at $T = 2$ or he turns out isolated from T=3. At variance with what happens in the ahistoric model, no player becomes a new defector at T=3 with full memory, because those that become defectors in the ahistoric context remain cooperators

Fig. 13.19   Patterns from a single defector at T=9 without memory (left) and with α=0.5 memory.

due to the fact that they have the reference of those twice cooperators, still unaffected by the initial defection, with an accumulated payoff of 2 units, higher than the maximum of 1.91 achieved by the new defectors at T=2. As a result of not being new defectors at T=3, the wiring network remains unaltered at T=4, albeit new defectors appear at this time-step as a consequence of the 3.55 payoff accumulated by the defectors up to T=3.

Let us point out here that the model proposed by [143] takes memory of past choices into account in their (asynchronous) updating of neighbours, as - if someone is picked for updating its neighbors only the most disadvantageous edge is rewired: it dismissed the link to the one who defects most times - . In the discussion of the article by Eguiluz et al.[133], when consid-

Spatial games

ering lines of research that deserve further investigations, it is stated that one possible direction is to contemplate a more complex strategy space, involving strategies dependent on past encounters, thus, ignoring payoff memory, much in the manner of the cited papers by Lindgren and Nordahl [256, 257].

Table 13.7 Co-evolving dynamics with full memory from a single defector.

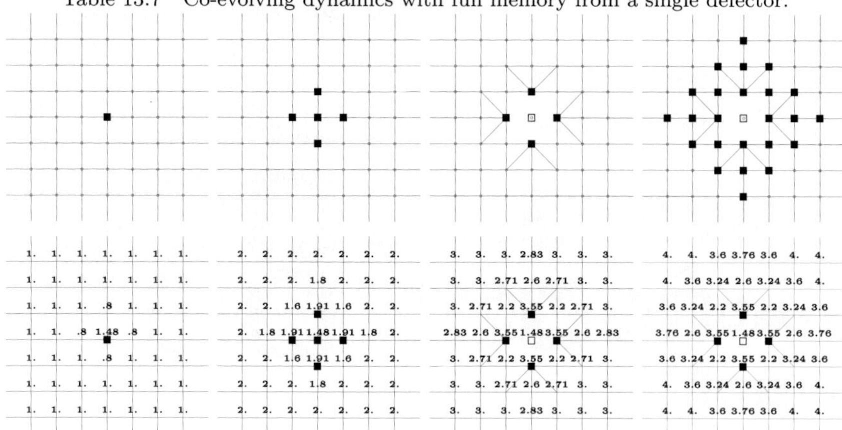

Figures 13.20 and 13.21 show the effect of memory on the evolving patterns of choices from a single defector, up to T=13. The high degree of the advance of defection with isolation becomes again apparent on these figures if $\alpha \leq 0.5$, whereas, $\alpha > 0.5$, proves a strong preventive effect on such behaviour. This is so even at low levels of α over the turning point 0.5. Thus, for example, defection is soon rejected if $\alpha=0.51$ in Fig. 13.20 and if $\alpha=0.55$ in Fig. 13.21.

For the sake of retaining the reference of the effect of memory in the structurally static scenario, the pay-off of every player at every round is divided by the number of its co-players. The model differs at this point from that in [133, 353, 448, 447], which considers absolute payoffs per player in every round, so that players may have different payoffs due solely to having a different number of neighbors. The authors argue that this approach incorporates the idea of the importance of being highly connected. This variant will be implemented in a further study of the effect of memory in the stochastic context, but for the sake of comparison with the relative-payoff model here adopted, the Fig. E.14 in the Appendix shows the effect of memory with this variant implemented starting from a single defector, i.e., the

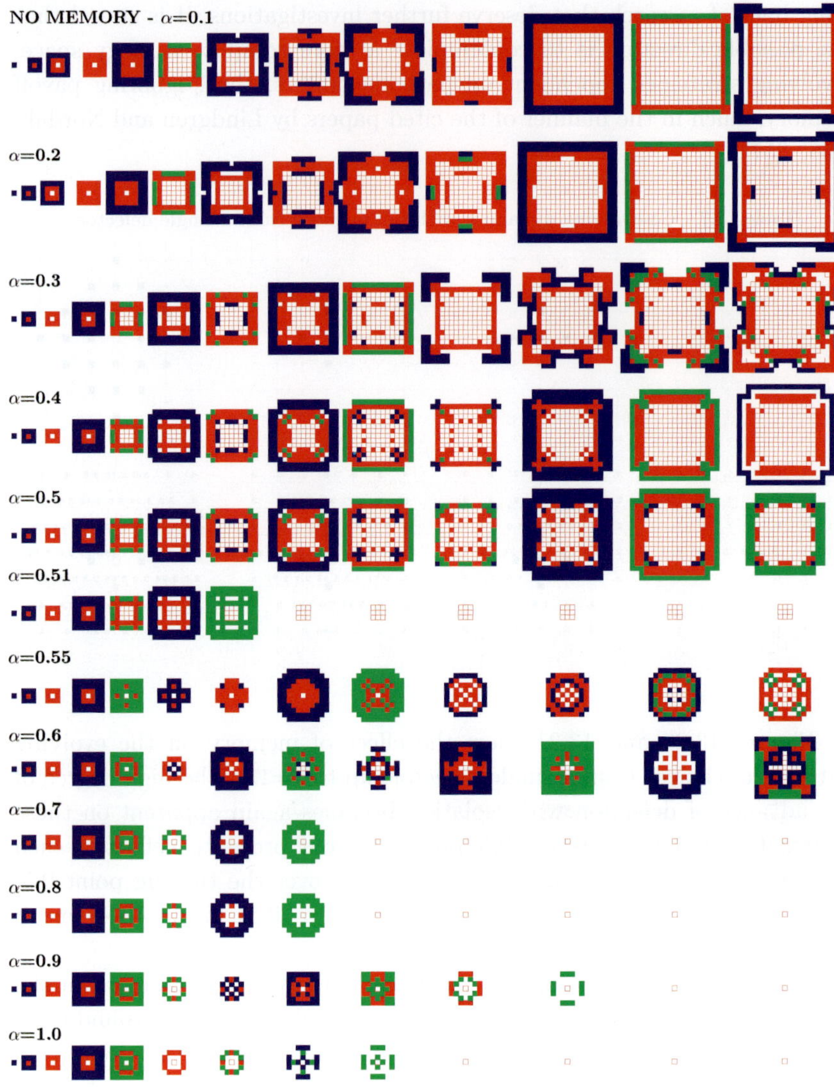

Fig. 13.20   Evolving patterns starting from a single defector in a regular lattice with
Moore neighborhood. Empty squares denote isolated cells. Evolution up to $T = 13$ in
the SDPD.

variants of Fig. 13.20 and Fig. 13.21 . Figure E.14 shows how the degradation of the initial structure via defection plus isolation advances at the
maximum speed in the ahistoric model when considering absolute payoffs

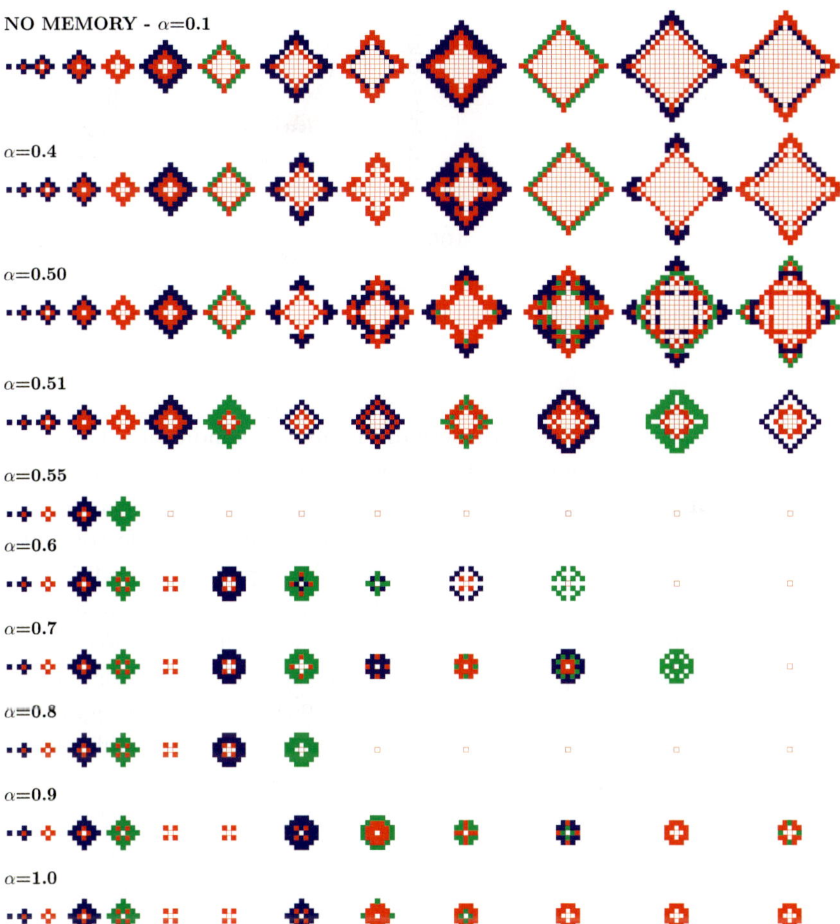

Fig. 13.21 Evolving patterns starting from a single defector in a regular lattice with von Neumann neighborhood. Evolution up to $T = 13$ in the SDPD.

in the imitation rule. Thus, no player *remorses* and returns to cooperation after defection as happens as soon as at time-step six in Fig. 13.20 and
seven in Fig. 13.21. Consequently, memory is rather ineffective in avoiding
the advance of defection/isolation up to the α=0.6 level, i.e., when it is
also operative on choices, not only in payoffs. It is at α=0.6, and not at
α=0.5, when green cells indicating D turning C appear for the first time in
Fig. E.14 .

Figure 13.22 shows the frequency of cooperators starting from a single

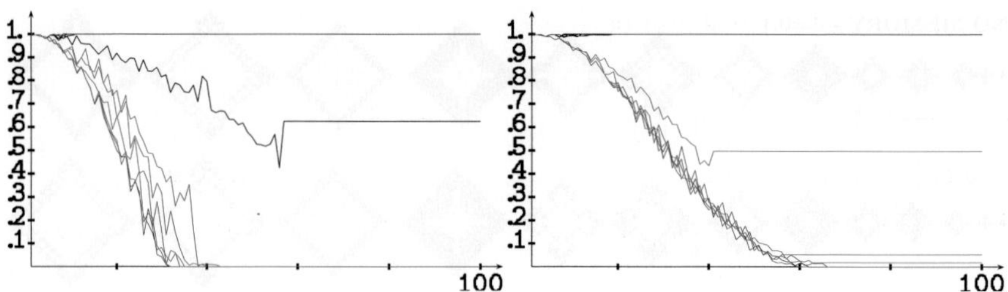

Fig. 13.22   Evolution of the frequency of cooperators from a single defector. Color code as in Fig. 13.3, with black hidden here by blue in most cases. Initial $K=9$ in the left frame and $K=5$ in the right one.

defector in a 51×51 lattice with periodic boundary conditions. Only the payoff memory, i.e., α not over 0.5, shown as green curves in figures concerning the frequency of cooperators, has the effect of only delaying the devastating effect of converting every player into an isolated defector (with *no rehab*, some singer may say), with the only exception of $\alpha=0.5$ with initial K=5 connectivity in Fig. 13.22, in which case the pattern remains unaltered soon after T=40. In contrast, incorporating memory of choices, i.e., $\alpha > 0.5$, has a dramatic preventive effect on the spread of defection, in this case with the exception of $\alpha=0.6$ with initial K=9 connectivity in Fig. 13.22, evolving by close to T=60 again towards a quiescent pattern, different to that of full cooperation.

Figure 13.23 shows the evolution of the frequency of cooperators in the scenario of Fig. 13.22 but using absolute payoffs in the imitation rule. As may be expected from Fig. E.14 in the Appendix, defection advances faster when considering absolute payoffs instead of the relative payoffs as in Fig. 13.22. Therefore, memory has more difficulties in avoiding such a *wild* tendency. In particular, memory charges under the turning point $\alpha=0.5$, thus not being effective in implementing any memory of choices, basically follow the ahistoric evolution, i.e., the green curves are almost coincident with the red-ahistoric one. Even $\alpha=0.6$ seems not be high enough: starting from K=9 cooperation finally also collapses, and starting from K=5 cooperation is led to a steady state of around 30 % of cooperation, very much under full cooperation. In any case, higher levels of memory, $\alpha \geq 0.7$, reveal full effectivity in the scenario of Fig. 13.23. No further considerations will be given in this section on the effect of memory when using absolute payoffs in the imitation rule. Suffice to say here that memory is likely to be less effective in avoiding the degradation of an initially structured cooperative lattice when using absolute payoffs than when using relative payoffs.

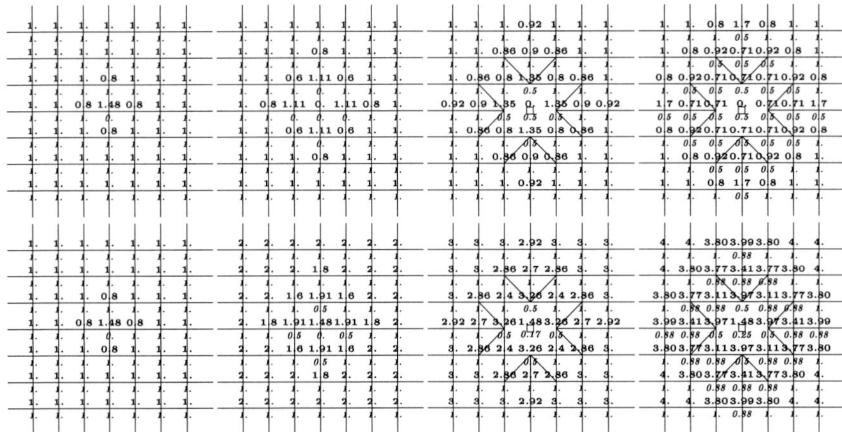

Fig. 13.23   Evolution of the frequency of cooperators in the scenario of Fig. 13.22 but using absolute payoffs in the imitation rule.

A last, but not least, remark should be made regarding the basic assumptions in this section. Moore and von Neumann neighborghoods, often referred to simply as K=9 and K=5, used here as initial ordered local wiring, carry self-interaction. But self-interaction is avoided in other works, as, to cite an example, in the group of papers [133, 448, 447]. Self-interaction may be regarded as a way to support cooperators, by giving them the extra, and maybe spurious, *reward R* income from their C-C interactions. Figure E.15 in the Appendix shows the effect of memory starting with Moore and von Neumann neighbourhoods lacking self-interaction. Contrary to expectation, starting from K=8 memory reveals to be very much effective starting from a single defector, as defection is fully rejected as soon as K=5 if $\alpha \geq 0.7$. Starting from K=4 this is achieved only with full memory. By the way, in the group of articles [133, 448, 448], self-interaction is avoided as already stated, but somehow to *compensate* the weaker background for cooperation that this implies, in the simulations reported in these articles, *i)* the values of the temptation considered are lower than the 1.85 assumed here, thus b=1.70 and 1.75 with K=8 and b=1.45 with K=4, and *ii)* the initial fraction of defectors is 40 %, instead of the 50 % used here as a rule with the only exception of the simulations in Fig. 13.25. In any case, it seems that avoiding self-interaction is not so decisive in weakening the status of cooperators as it is taking into account absolute payoffs in the imitation rule. Thus at variance with the *speed of light* advance of defection in the ahistoric scenario of Fig. E.14, in Fig. E.15 the advance of defection is constrained at some time-steps by the *repentance* of some D turning C (green cells), more abundant if α=0.5.

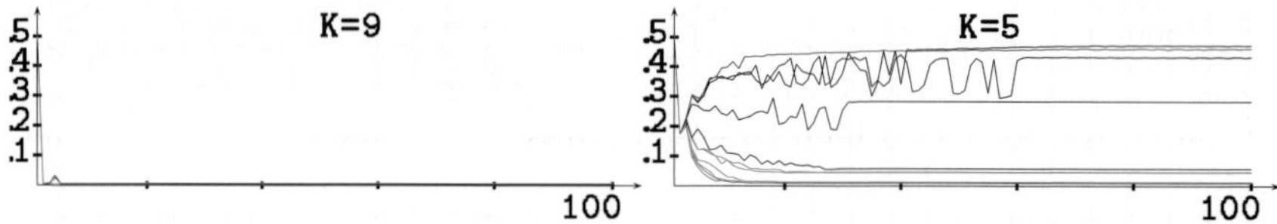

Fig. 13.24   Evolution of the frequency cooperators starting with 50% of randomly assigned defectors in 51×51 networks.

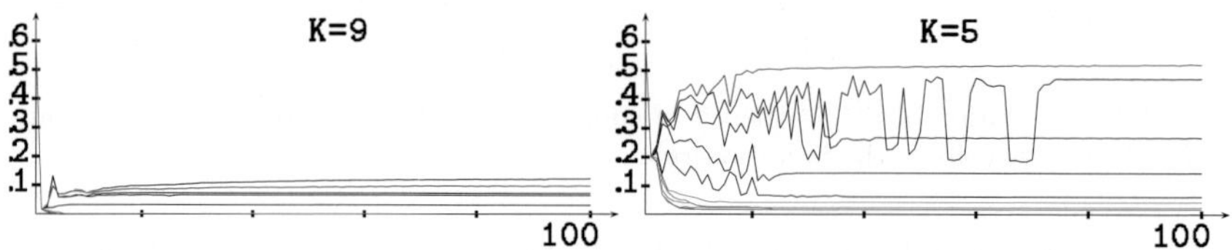

Fig. 13.25   Evolution of the frequency cooperators starting with 40% of randomly assigned defectors in 51×51 networks.

Figures 13.24 and 13.25 show the frequency of cooperators starting from 50% and 40% of randomly assigned defectors in a 51×51 lattice with periodic boundary conditions. Cooperation initially plummets in every scenario, particularly when the starting connectivity is as high as nine, in which case extinction is soon reached in the ahistoric model. Starting with K=5, cooperation decreases to near extinction in the ahistoric model, nevertheless some clusters of cooperators remain. With initial K=9, starting with 50% of defectors, the extinction of cooperation is unavoidable, but starting with a 40% of defectors, hopefully some clusters of cooperator *resist* and become the seed for a likely limited recovering of cooperation. With initial K=5, cooperation does not decrease initially so sharply, and memory turns out much more effective in recovering cooperation, particularly full memory which, even starting with a 50% of defectors, promotes a fast recovering of cooperation leading to levels over 40%. The case of memories with α in the $[0.6, 0.9]$ interval, thus black curves, appears very erratic up to their stabilization. The effect of full memory (blue curve) is here that foreseen, as it is that of maximum support of cooperation, reached after a fairly short transition period.

Figure 13.26 shows the patterns at T=100 with α=0.7 and full memories starting at random in the von Neumann's neighborhood. The wiring of the 9×9 central cells is shown in the figure. In both cases, there is a notable variability in the number and localization of the links per player. Apart from loner defectors, there are cooperators with a low number of links, cells that seem to maintain the initial NSEW neighbourhood, and *leaders*, satisfied C-agents with a high number of links, and likely high accumulated payoff. The existence of *exploiters*, i.e., defectors connected to cooperators, is rather problematic because defectors tend to be rejected and to become isolated. In the 51×51 networks starting from the Moore neighborhood, only 17 and 16 exploiters (the same except one of them) survive in the α=0.9 and full memory simulations with 50 % of initial defectors. Starting with the von Neumann neighbourhood, exploiters are more abundant: 18, 17, 17, 24, 29, 21, 118, 116, 87, 98, and 149 as α grows from 0.0 to 1.0 by 0.1 intervals. A common characteristic is that exploiters exploit a very low number of players, in most cases only one.

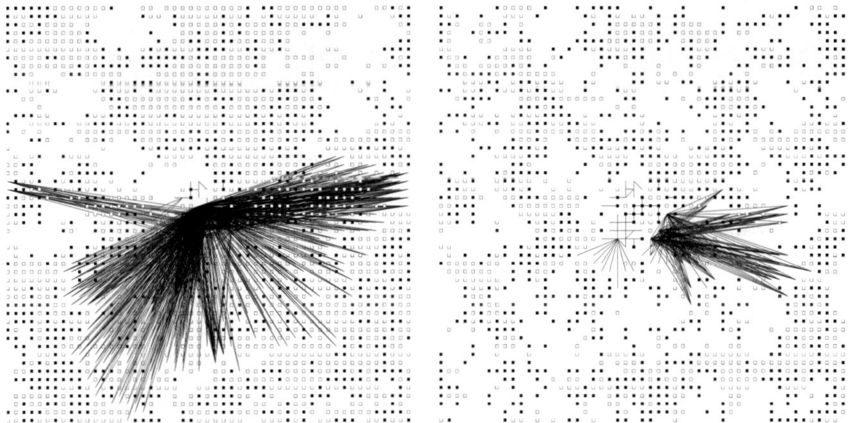

Fig. 13.26   Patterns at T=100 with α=0.7 (left) and full memories in the K=5 neighborhood of Fig. 13.24 . The wiring of the 9×9 central cells is shown in the figure.

Figure 13.27 shows a representation of the map of connectivities at T=100 in the scenario of Fig. 13.26 . It seems that the populated web of connections in the α=0.7 case in Fig. 13.26 captures the wiring network of the highly connected central cells made apparent in Fig. 13.27 . The absolute *leader* in the left snapshot of Fig. 13.27 connects to 245 players, nearly

10 % of the total number of players, whereas with full memory absolute the
leader has 157 links. The unbalanced number of links per node, together
with the presence of true hub cells detected in Fig. 13.27, would indicate
some kind of power-law like distribution in the frequency of connectivities.
The spatial location of the clusters of highly connected cells in Fig. 13.27
(and in Fig. 13.31) does not necessarily imply that they are inter-connected
after the re-definition of the network of interactions, although it is fairly
probable that they are related: most of them are cooperators not turning
defectors, thus not actively changing their neighborhood (although pas-
sively evolving it by receiving new links from *searching* D-agents), thus it
is likely that the initially connected cells remain connected in those clusters
of *leaders*.

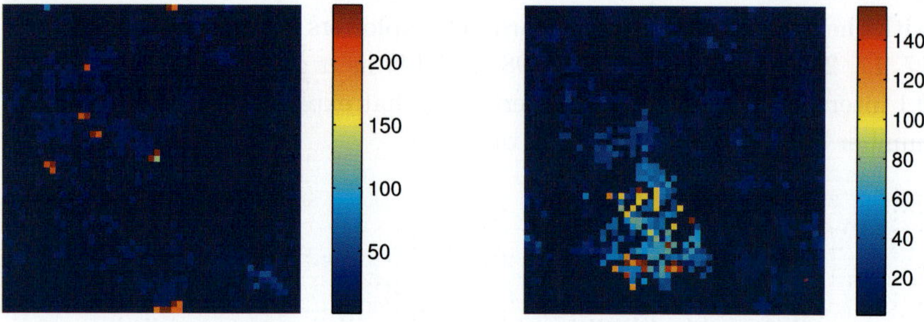

Fig. 13.27   Connectivity at T=100 in the scenarios of Fig. 13.26.

A study of the evolution of the connectivity is planned as a forthcoming
work, but their primary parameter, its mean value in the initial K=5 scenario
is shown in Fig. 13.28. In some cases, when $\alpha \in [0.7, 0.9]$ in particular, this
mean value shows an astonishingly erratic behaviour before its stabilization.
After Fig. 13.27, we envisage that the standard deviation of connectivities
will reach high values and evolve in a non-monotonic way in some α-memory
cases.

The evolution of the average payoff per node and link, and average
payoff per node in the K=5 neighborhood scenario of Fig. 13.24 is given in
Fig. 13.29. The evolution of the average payoff per node and link, top left
in Fig. 13.29, resembles that of the frequency of cooperators, i.e., the right
frame in Fig. 13.24. With the payoff matrix used in this article, to a frequency of cooperators f corresponds a expected payoff per node and link:

$$p = (f, 1-f) \begin{pmatrix} 1 & 0 \\ b & 0 \end{pmatrix} \begin{pmatrix} f \\ 1-f \end{pmatrix} = f^2 + bf(1-f).$$ Roughly, this transforma-

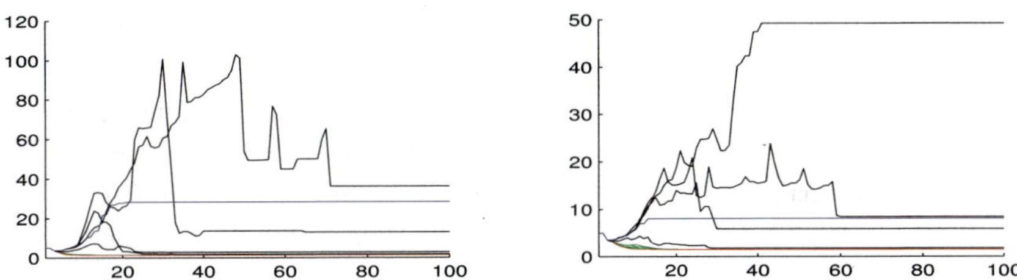

Fig. 13.28　Mean connectivity in the initial K=5 scenarios of Figs. 13.24 and 13.25. Initial 40 and 50 % of defectors correspond to left and right frames respectively.

tion from f to p operates from Fig. 13.24 to Fig. 13.29. Thus for example, initially with $f \simeq 1/2$, it is $p \simeq (1+b)/4$, thus $p \simeq 0.71$ as $b{=}1.85$.

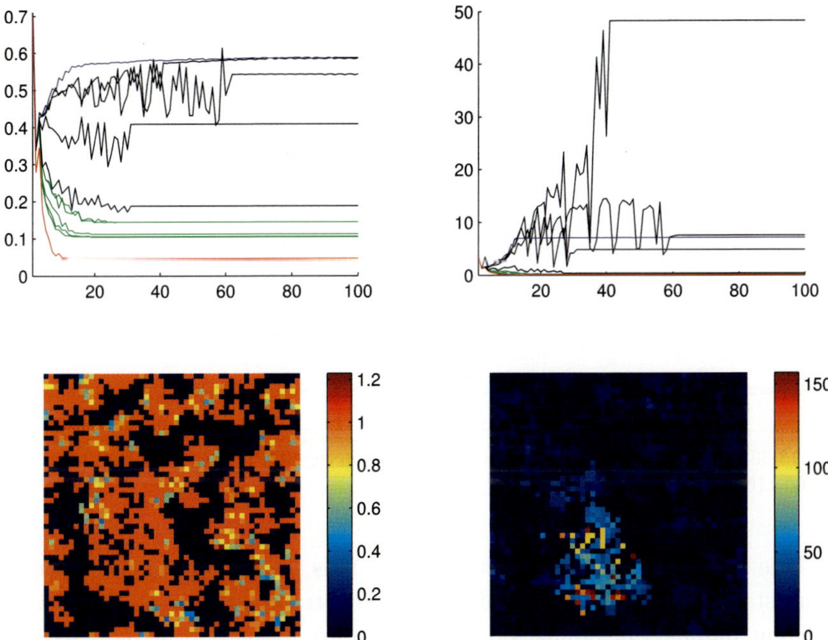

Fig. 13.29　Evolution of the average payoff per node and link (top left), of the average payoff per node (top right) in the K=5 neighborhood of Fig. 13.24, and maps of both payoff types (lower snapshots) at T=100 with full memory in this context.

The average payoff per node, thus not rating the absolute payoff per node to the number of its links, has the expected value pK, where K denotes the average connectivity. Thus the top right frame in Fig. 13.29, appears

a *composition* of the right frames of Figs. 13.24 and 13.28, departing from
0.71×5=3.51 . There is a notable correspondence in the high values of the
connectivity and total payoff in the α=0.9 simulation. Figure 13.29 shows
also the maps of both unitary and total average payoffs at T=100 with
full memory in this context. Again, there is a notable similarity between
the map of payoffs per node here, bottom right, and its corresponding
map of connectivities, the right snapshot in Fig. 13.27. The same similarity
between the maps of payoffs per node and connectivity has been found
to happen in the α=0.7 (left) scenario of Fig. 13.27. It is likely that in
the clusters of highly connected cells, some, let us say *seeds*, never turned
out to be defectors, particularly in the first iteration. Therefore they were
wired by close (in the original lattice) unsatisfied defectors that later on
became cooperators (attracted by the high payoff of the seeds), so that
many of those highly connected cells frequently hold C-C interactions. This
conjecture has been checked with the leader nodes (with 245 and 157 links)
in the snapshots of Fig. 13.27: both connect cooperators in the steady state.
In the case of the left snapshot of Fig. 13.27, the absolute leader is an
initial cooperator, whereas the leader in the simulation with full memory is
an initial defector that, rather unlikely, becomes a cooperator at the first
iteration.

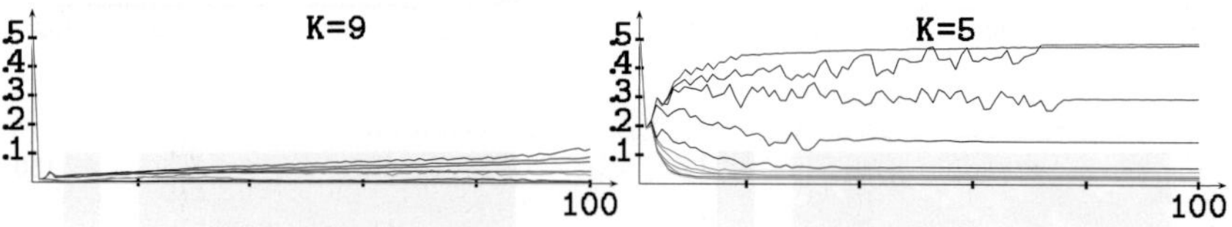

Fig. 13.30   Evolution of the frequency cooperators in the SDPD when starting at random
in 100×100 networks.

Due to the topological effects present, it is feasible that other simulations, starting with different configurations of strategies, or in networks with
different numbers of points, do not lead to full extinction in the ahistoric
context starting with K=9, or lead to extinction starting with K=5. Thus
for example, in the simulation in 100×100 lattices starting from a 50 % of
randomly assigned defectors of Fig. 13.30, when starting with the Moore
neighbourhood, cooperation recover with memory, though slightly. The
evolution starting from K=5 resembles that of its homologous in a 51×51
network in Fig. 13.24.

The evolution of the connectivity in the simulations of size 100×100 of

Fig. 13.30 is given in Fig. 13.31. As in the case of Fig. 13.28, the simulations starting with K=5 show the firing of the connectivity for α=0.9.

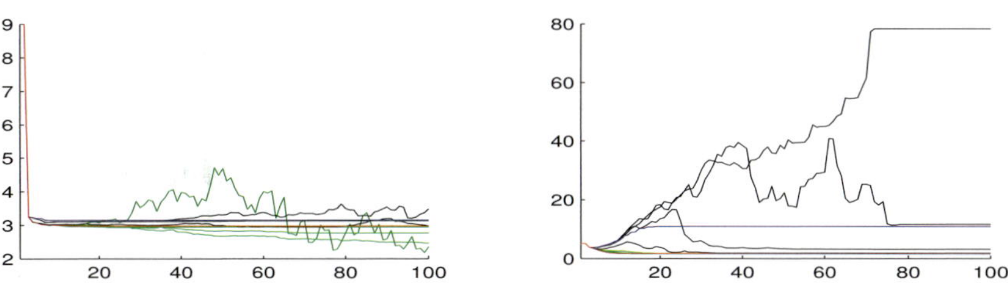

Fig. 13.31   Mean connectivity in the K=5 scenarios of Fig. 13.30 with α=0.9 (left) and full memory (right).

The map of connectivities at T=100 from K=5 with α=0.9 and full memory is shown in Fig. 13.32. The most connected hubs in Fig. 13.32 link to 806 cells in the left snapshot and to 178 in the right one. These high values expand the range in the bars for coding color in their maps of connectivities in Fig. 13.32, as happened in Fig. 13.27, masking the fact that in the snapshots many cells are isolated or lowly connected ones by amalgamating them in the darker blue color. Thus, in the snapshots of Fig. 13.32, near a 25 % of the nodes are isolated and a approximately a 20 % have only two links.

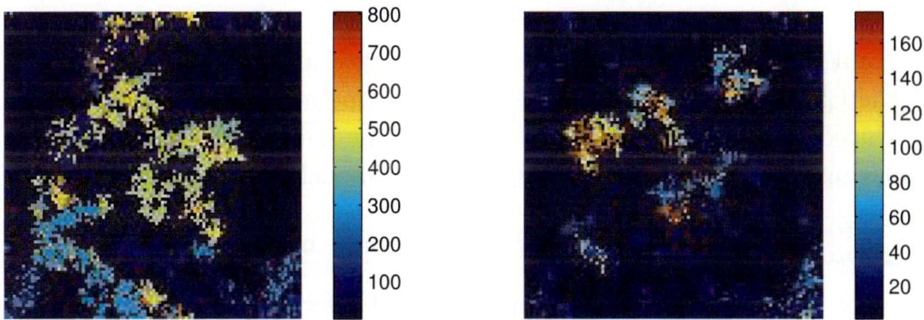

Fig. 13.32   Map of connectivity at T=100 with α=0.9 (left) and full memory (right) in the K=5 scenario of Fig. 13.30.

Degrees of cooperation in the SDPD

Generalizing the coupling/decoupling rules from the binary case, unsatisfied defectors, those with degree of cooperation decreasing at least 0.5 :

- eliminate their links with players with degree of cooperation not over 0.5,
- connect with their nearest-neighbors with degree of cooperation over 0.5 .

The SDPD admitting degrees of cooperation will be referred as CV-SDPD.

Table 13.8   Co-evolving dynamics in the CV-SDPD from a single defector. Payoffs are located over the *italiziced* degrees of cooperation in every node.

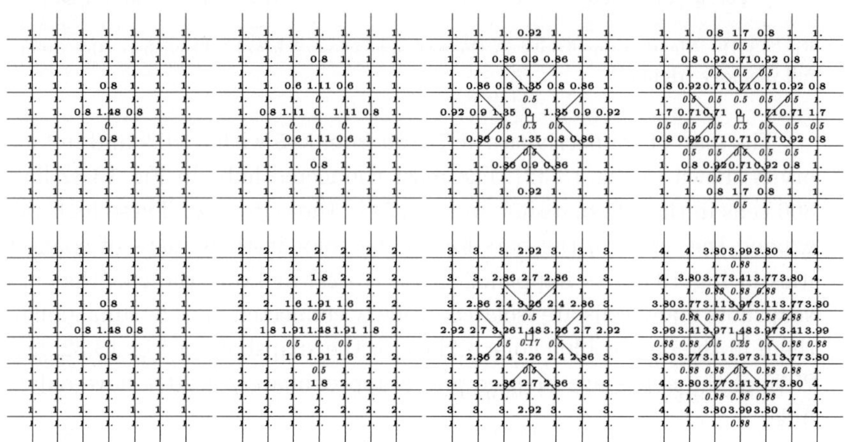

Table 13.8 starts in the same initial scenario of Table 13.7, but refers to the CV-SDPD. It shows the degrees of cooperation (x) and payoffs (p) produced at the first four stages, as well as the underlying values of π and δ driving the dynamics in the full memory model, in which case, $\delta_{i,j}^{(T)} = \left(\sum_{t=1}^{T} x_{i,j}^{(t)}\right)/(T(T+1)/2)$. Unlike in the binary model, where the initial defector never becomes cooperator, the initial defector cooperates at degree $\alpha/(1+\alpha)$ at $T=3$ (=0.5 in Table 13.8), the mean degree of cooperation up to T=2 of his (initial) neighbours which have a higher accumulated payoff $\pi^{(2)} = (4\alpha + 3b)/5$ (=1.91 in Table 13.8) than his unaltered $4b\alpha/5$ (=1.91 in Table 13.8). But this fact has no consequences in the further dynamics, as the initial defector turns out to be isolated precisely from T=3.

Figure 13.33 shows the wiring at T=9 with α=0.2 memory in the CV-SDPD starting from a single defector and initial regular connectivities nine and five. As expected, the disruption in the wiring network induced by the initial defector is smaller when starting from the lower K=5 initial connectivity.

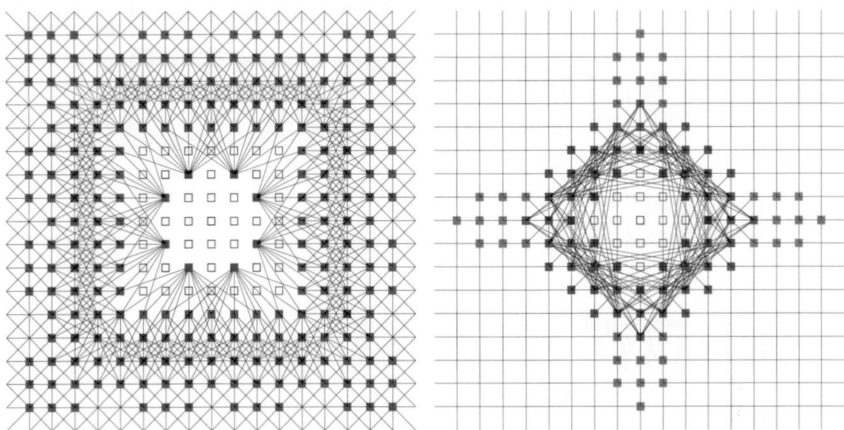

Fig. 13.33   Wiring at T=9 with α=0.2 memory in the CV-SDPD starting from a single defector with initial Moore (left) and von Neumann neighborhoods.

Figures 13.34 and 13.35 show the effect of memory on the evolving patterns of the degree of cooperation from a single full defector in the CV-SDPD up to T=13. The patterns at high levels of memory might appear as reminiscent of their equivalent in the binary case, i.e., Figs. 13.20 and 13.21, albeit there is a notable difference when turning to CV-SDPD: the advance of isolation is dramatically restrained with $\alpha \geq 0.2$. With a high memory charge, memory factor over 0.6, memory fully rejects defection starting from the Moore neighborhood in the binary context, whereas in the CV-SDPD, memory appears not so effective, and the rest of the initial defection persists over time. Figure E.16 in the Appendix shows the effect of memory with Moore neighbourhood from a single defector in the CV-SDPD, i.e., the variant of Fig. 13.34. Figure E.16 shows again how the degradation of the initial structure via defection plus isolation advancing at the *speed of light* in the ahistoric model when considering absolute payoffs in the imitation rule, is not avoided unless α is high, at least 0.3. Also in this context, high levels of memory, e.g. full memory, do not lead to full extinction of the initial defection when in the CV-SDPD context of Fig. E.16, as there is always a remaining defection around the initial defector site.

Figure 13.36 shows the evolution of the mean degree of cooperation in the CV-SDPD starting from a single defector in 51×51 lattices with initial

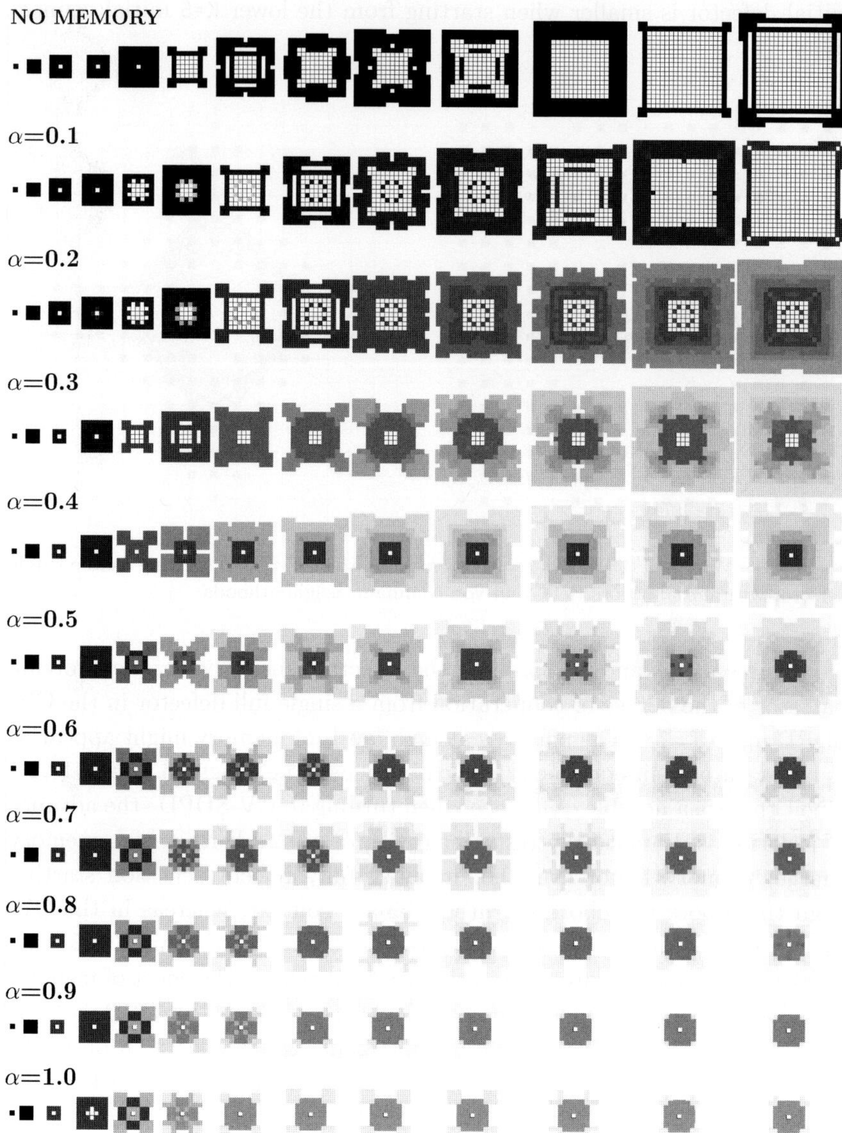

Fig. 13.34 Evolving patterns starting from a single defector up to $T = 13$ in the CV-SDPD. Initial Moore neighborhood.

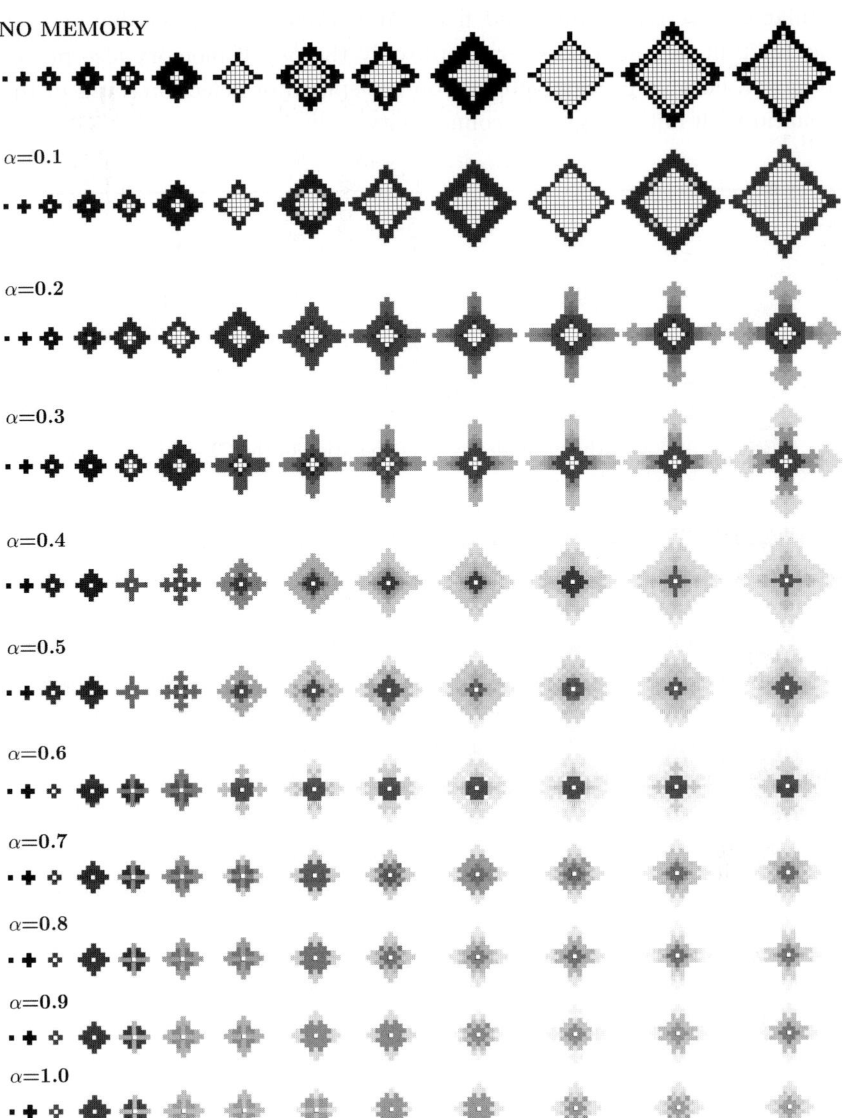

Fig. 13.35 Evolving patterns starting from a single defector up to $T = 13$ in the CV-SDPD. Initial von Neumann neighborhood.

regular connectivities nine and five. At variance with what happens in
Fig. 13.22 in the binary scenario, now only the payoff memory also proves
to have a notable preventive effect on the spread of defection, even in the
scenario with initial high K=9 connectivity.

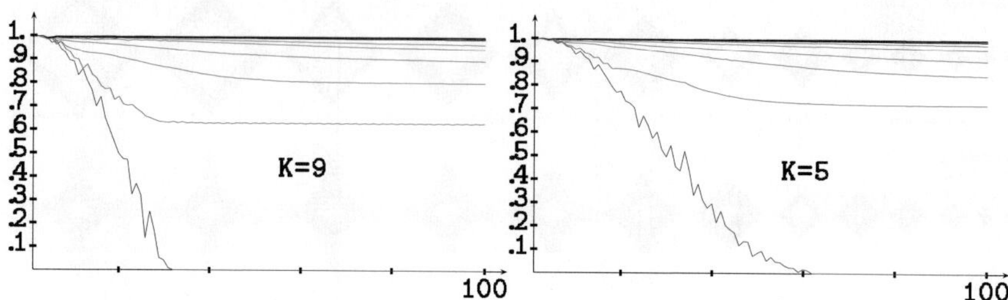

Fig. 13.36   Evolution of the mean degree of cooperation in the CV-SDPD from a single
defector in 51×51 lattices.

Figure 13.37 shows the patterns at T=100 in the scenario of Fig. 13.36
with α=0.1 memory. In Fig. 13.37 it becomes apparent how the consideration of a mere α=0.1 memory prevents the progress of the defection
and isolation effect, as the isolated cells do not invade the relatively small
51×51 network after one hundred iterations. The upper bigger snapshots
of Fig. 13.37 show the isolated cells as empty squares, making them apparent but losing the information about their degree of cooperation. The
smaller snapshots in the figure do not mark the isolated cells, so that their
degree of cooperation is made apparent giving to the patterns a higher aspect of continuity. In the patterns regarding the CV-SDPD, e.g., already
in Fig. 13.33 or in the figures in the Appendix, we opted to mark the loner
cells in the snapshots because this information is more relevant than the
degree of cooperation of the isolated cells.

Figure 13.38 shows the evolution of the mean degree of cooperation
in the CV-SDPD starting at random in 51×51 lattices and in 100×100
lattices, with initial connectivities nine and five. The case of full memory,
i.e., blue curve, starting from K=9, appears as somewhat *pathological*, as the
recovering of cooperation becomes unexpectedly slow, reaching levels below
that of the remaining models with memory. It seems that some degree of
forgetfulness helps in supporting cooperation in this scenario. This odd
phenomenon has already been traced when dealing with disordered, albeit
static, networks [21]. Figure E.17 in the Appendix, showing patterns at
T=100, makes apparent that the percolation of cooperation present with
limited memory is lost with full memory starting from K=9. The percolation

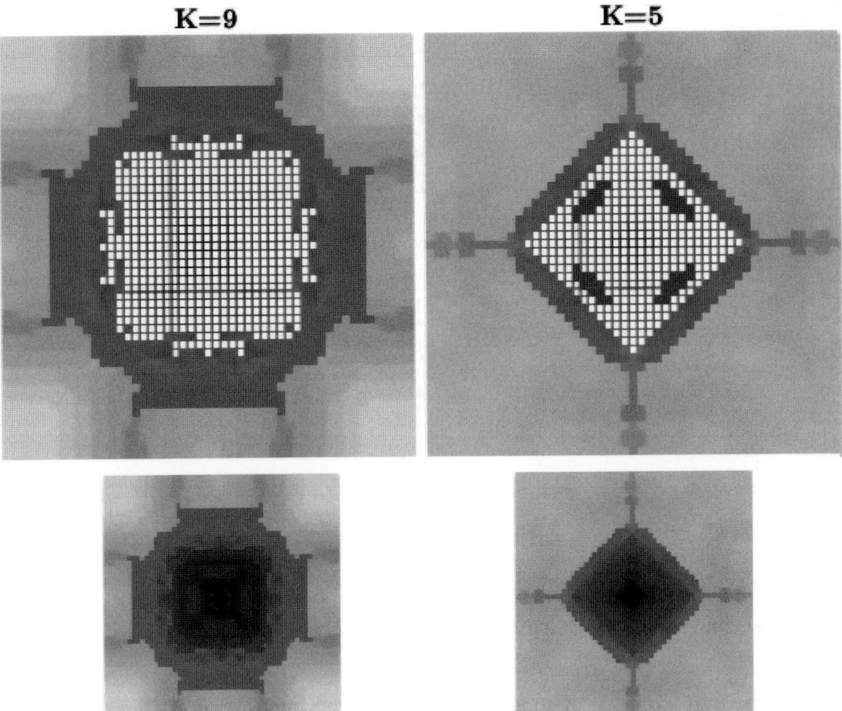

Fig. 13.37   Patterns at T=100 with α=0.1 memory in the CV-SDPD starting from a single defector in 51×51 regular lattices.

of cooperation remains with full memory starting from K=5, as shown in Fig. E.18 in the Appendix. A few small clusters of cooperators remain active with no memory in Fig. E.18, but as full cooperators are coded as blank cells, they are not easily identifiable because they are masked in the sea of isolated cells, coded as empty squares.

Figures 13.39 and 13.40 show the wiring at T=100 with α=0.5 and full memories in the CV-SDPD starting at random in the Moore and von Neumann neighborhoods. The figures exhibit a tendency to wiring remaining *local* and fairly balanced very different from the unstructured networks with true hub nodes generated in the binary model as envisaged in Fig. 13.26 when starting at random. Now, in the CV-SDPD model, the highest connected nodes link only 18 nodes in Fig. 13.39, and 11 and 13 nodes in Fig. 13.40.

Figure 13.41 shows how the wiring networks are stabilized fairly soon in the CV-SDPD. Nevertheless, the mean degree of cooperation grows, albeit

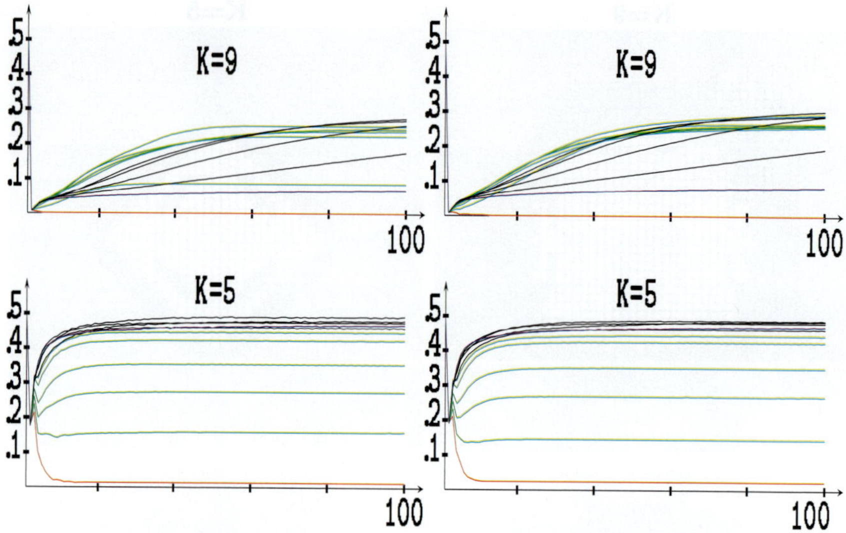

Fig. 13.38 Evolution of the mean degree of cooperation in the CV-SDPD starting at random in 51×51 lattices in the left frames, and in 100×100 lattices in the right frames. The initial connectivity (K) is indicated in each frame.

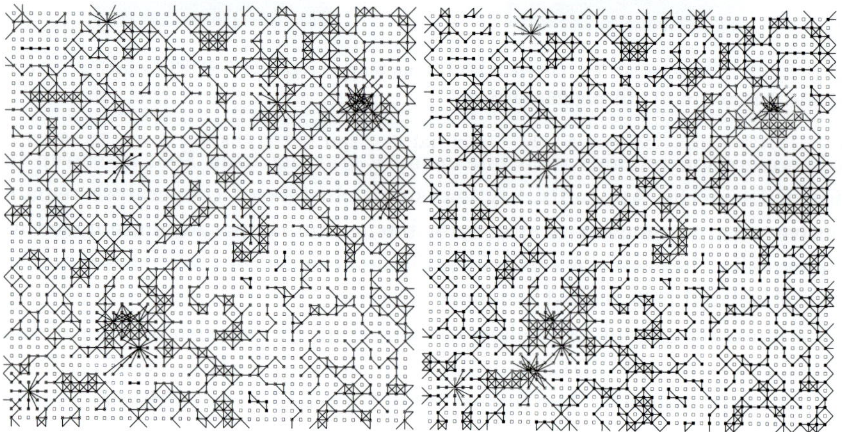

Fig. 13.39 Wiring at T=100 with α=0.5 (left) and full memories in the CV-SDPD starting at random in the Moore lattice.

slowly as shown in Fig. 13.38, after the stabilization of the networks. It seems that cooperation *percolates* across the stabilized networks of cooperators.

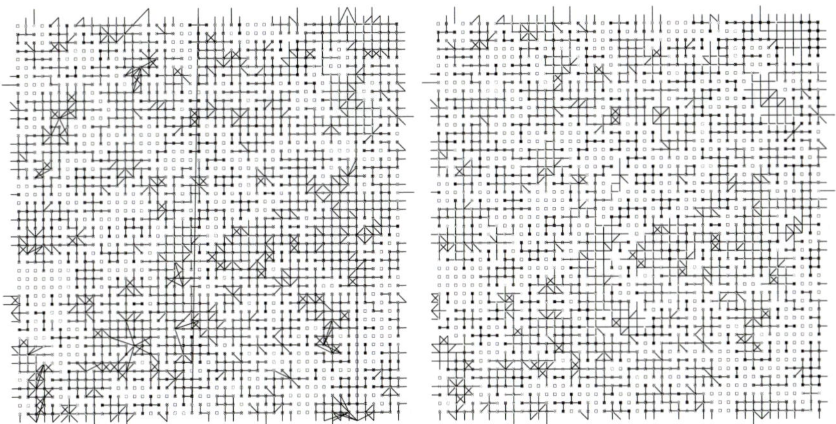

Fig. 13.40   Wiring at T=100 with α=0.5 (left) and full memories in the CV-SDPD starting at random in the von Meumann's neighborhood.

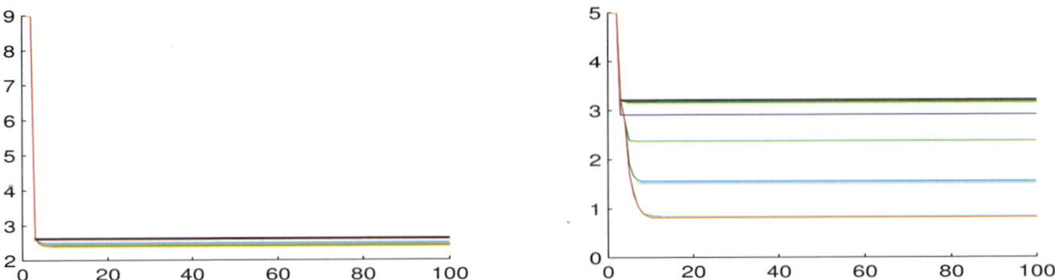

Fig. 13.41   Mean connectivity in the initial K=9 and K=5 scenarios of Fig. 13.38 with initial 51×51 lattices.

In the CV-SDPD, *exploiters* may be defined as players with degree of cooperation x under 0.5, connected to players with $x > 0.5$. Exploiters of this kind are more abundant in the CV-SDPD than in the binary model. In the 51×51 network starting from the Moore neighborhood, the number of exploiters found in the simulations here reported were: 0, 0, 144, 0, 0, 5, 175, 143, 151, 110, and 53 as α grows from 0.0 to 1.0 by 0.1 intervals. Starting with the Moore neighbourhood, exploiters are fairly numerous: 18, 60, 128, 295, 464, 484, 438, 400, 405, 404, and 402. Again, as in the binary case, the number of cooperators exploited per player is very low, in the most frequent case only one. Roughly the same proportion of defectors has been found in the 100×100 size simulations.

We consider the results reported here as a departure-base for a forth-coming study of models implementing the coevolutionary coupling and decoupling rules in a probabilistic way.

13.4 Pavlov versus anti-Pavlov (PAP) in the PD

This section considers the so called Pavlov strategy in the iterated PD game: a Pavlov player cooperates if and only if both players opted for the same alternative in the previous move. The name stems from the fact that this strategy embodies an almost reflex-like response to the payoff: it repeats its former move if it was rewarded by T or R, but switches behavior if it was punished by receiving only P or S.

By coding cooperation as 1 and defection as 0, this strategy can be formulated in terms of the choices x of Player A (Pavlov) and y of Player B as: $x^{(T+1)} = 1- \mid x^{(T)} - y^{(T)} \mid$. The Pavlov strategy has proved to be very powerful, in the sense of being successful in its contests with other strategies [314]. Let us give a simple example of this: suppose that Player B adopts an anti-Pavlov strategy (which cooperates to the extent Pavlov defects) with $y^{(T+1)} = 1- \mid 1 - x^{(T)} - y^{(T)} \mid$. Thus, in an iterated Pavlov-anti-Pavlov (PAP) contest, with $T(x,y) = \left(1 - |x - y|, 1 - |1 - x - y|\right)$, it is $T(0,0) = T(1,1) = (1,0)$, $T(1,0) = (0,1)$, and $T(0,1) = (0,1)$, so that $(0,1)$ turns out to be immutable. Therefore, in an iterated PAP contest, Pavlov will always defect, and anti-Pavlov will always cooperate. Pavlov will get the temptation and anti-Pavlov the *sucker* payoff. In other words, Pavlov has no qualms about exploiting a permanent cooperator.

Continuous-valued PAP contest

By relaxing the 0-1 assumption in the standard formulation of the PAP contest, *degrees* of cooperation can be considered in a continuous-valued scenario. Now x and y will denote the degrees of cooperation of players A and B respectively. Thus, the formulation of the PAP contest remains in the continuous-valued scenario as in the binary case, but with both x and y lying in [0,1]. The PAP contest is termed Chaotic-Dualist by Grim *et al.* [173].,

In the model allowing degrees of cooperation, not only $(0,1)$ is a fixed point, but also $T(0.8, 0.6) = (0.8, 0.6)$. An important dynamic is that of the central point, which leads to $(0,1)$ after: $(0.5, 0.5) \to (1,1) \to (1,0) \to (0,1)$. But the general tendency is not to evolve to $(0,1)$ but to (the proximity of) $(0.8,0.6)$. This is so, even starting far away from this point, as for example: $(0.1, 0.1) \to (1.0, 0.2) \to (0.2, 0.8) \to (0.4, 1.0) \to (0.4, 0.6) \to$

$(0.8, 1.0) \rightarrow (0.8, 0.2) \rightarrow (0.4, 1.0)$. Oscillators such as those *italicized,* are often generated early on.

Computer implementation of the iterated PAP tournament turns out to be fully disruptive of the theoretical dynamics (infinite precision). The errors caused by the finite precision of the computer floating point arithmetics make the final fate of every point to be $(0,1)$. This is a common problem in dynamical systems working *modulo 1,* such as the Arnold or Baker maps [128]. Without exception, not only points such as $(0.1,0.1)$ end as $(0,1)^2$, even the *theoretically* fixed point $(0.8,0.6)$ ends as $(0,1)$ in the computerized implementation[3]. Numerical results here come from a FORTRAN 90 program with *double precision* for real variables.

The effect of minimal memory on the PAP contest

Both Pavlov and anti-Pavlov strategies can be featured as ahistoric as only the last choice of both players matters in deciding the next choice. A natural way to incorporate older choices in the strategies of decision is: $m_x^{(T)} = m_x(x^{(T)}, x^{(T-1)}, ..., x^{(1)})$, $m_y^{(T)} = m_y(y^{(T)}, y^{(T-1)}, ..., y^{(1)})$. Both players are supposed to adopt the same memory mechanism: $m_x = m_y$. The PAP contest becomes in this way: $x^{(T+1)} = 1 - \mid m_x^{(T)} - m_y^{(T)} \mid$, $y^{(T+1)} = 1 - \mid 1 - m_x^{(T)} - m_y^{(T)} \mid$. The simplest historic extension results in considering only the two last choices: $m_x^{(T)} = m(x^{(T-1)}, x^{(T)}), m_y^{(T)} = m(y^{(T-1)}, y^{(T)})$. In this scenario we will consider the weighting mechanism (z stands for both x and y) : $m(z^{(T-1)}, z^{(T)}) = \frac{1}{\alpha+1}(\alpha z^{(T-1)} + z^{(T)})$, with $0 \leq \alpha \leq 1$ acting as a *memory factor,* so that $\alpha = 0$ corresponds to the ahistoric (standard) formulation: $m_x^{(T)} = x^{(T)}$, $m_y^{(T)} = y^{(T)}$, and $\alpha = 1$ to the mean memory (arithmetic mean) formulation: $m_x^{(T)} = \frac{1}{2}(x^{(T-1)} + x^{(T)})$, $m_y^{(T)} = \frac{1}{2}(y^{(T-1)} + y^{(T)})$. Initially, $m(z^{(1)}) = z^{(1)}$, so that $z^{(2)}$ is,

---

[2] Displaying numbers with three decimal digits, the period-four oscillator seems unaltered up to T=90. From this time step up to T=97 the dynamic is: $(0.801, 1.000) \rightarrow (0.801, 0.199) \rightarrow (0.398, 1.000) \rightarrow (0.398, 0.602) \rightarrow (0.797, 1.000) \rightarrow (0.797, 0.203) \rightarrow (0.406, 1.000) \rightarrow (0.406, 0.594)$.

[3] Again, displaying numbers with three decimal digits, the problem reveals at $T = 88$, and at $T = 109$ the $(0,1)$ point is reached following the series: $(0.801, 0.600) \rightarrow (0.799, 0.600) \rightarrow (0.801, 0.602) \rightarrow (0.801, 0.598) \rightarrow (0.797, 0.602) \rightarrow (0.805, 0.602) \rightarrow (0.797, 0.594) \rightarrow (0.797, 0.609) \rightarrow (0.812, 0.594) \rightarrow (0.781, 0.594) \rightarrow (0.812, 0.625) \rightarrow (0.812, 0.562) \rightarrow (0.750, 0.625) \rightarrow (0.875, 0.625) \rightarrow (0.750, 0.500) \rightarrow (0.750, 0.750) \rightarrow (1.000, 0.500) \rightarrow (0.500, 0.500) \rightarrow (1.000, 1.000) \rightarrow (1.000, 0.000) \rightarrow (0.000, 1.000)$. Note that the final steps are compulsory: $(0,1)$ is reached from $(0.5, 0.5)$.

naturally, unaffected by memory.

Table 13.9   The initial trajectories of the central point in the PAP map. **Bold** face indicate the points $(x^{(T)}, y^{(T)})$; *italicized*, the $(m_x^{(T)}, m_y^{(T)})$ values.

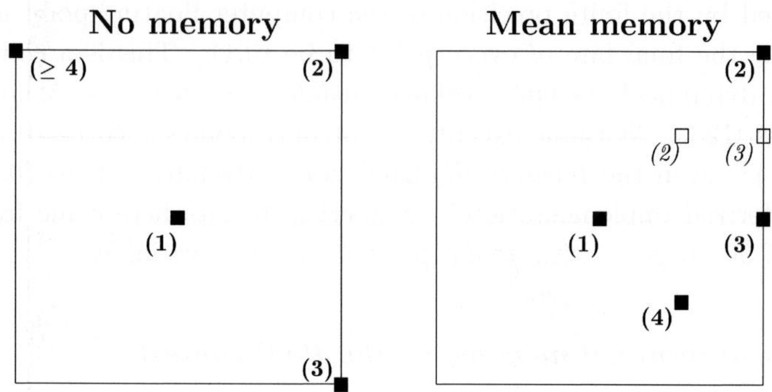

Table 13.9 shows the trajectories of the initial point (0.5,0.5) in the ahistoric, and mean memory scenarios up to $T = 4$. In the ahistoric context, the dynamics stop as early as at $T = 4$, as the fixed point (0,1) is reached at this time-step. The dynamics in the mean memory scenario is clearly different, and is featured by points closer to the initial point than in the ahistoric model.

Figures 13.42 and 13.43 show the effect of memory on the Pavlov - anti-Pavlov computerized contest up to $T = 200$, when starting from the central point $(x^{(1)} = y^{(1)} = 0.5)$ and from $x^{(1)} = y^{(1)} = 0.1$ respectively. In both cases, the ahistoric evolution ends in the fixed point (0,1): as *natural* at $T = 4$ starting from the central point, and at $T = 108$ starting from (0.1,0.1) as a result of finite precision arithmetics. There is a fairly neat parallelism in the plots obtained for every degree of memory in Figs. 13.42 and 13.43 . Thus, the *erratic* aspect of the $\alpha = 0.1$ subplots [4], the *triangular* one for $\alpha = 0.2$, the spirals for $\alpha = 0.3$, the fairly fast convergence to a point when $0.4 \le \alpha \le 0.8$, and the period-four oscillators generated in the mean memory model. In the latter case, the *square* generated starting from the central point appears as early as at T=18 .

Figure 13.44 shows the effect of $\tau{=}2$ ϵ-memory in the contexts of Figures 13.42 and 13.43, i.e. starting from (0.5,0.5) (up) and from (0.1,0.1) down.

---

[4]Which are much more *helter-skelter* that the fairly *lineal* ahistoric ones, particularly when considering the simple ahistoric trajectory from the central point.

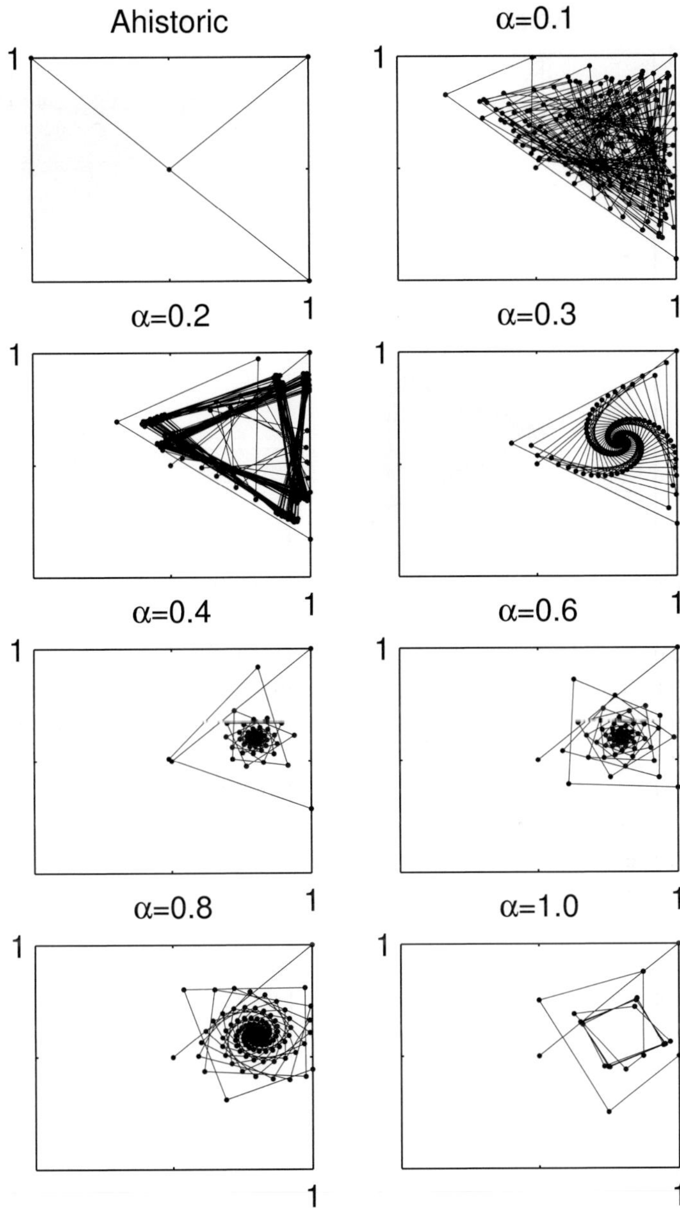

Fig. 13.42   PAP dynamics with memory starting at the central point.

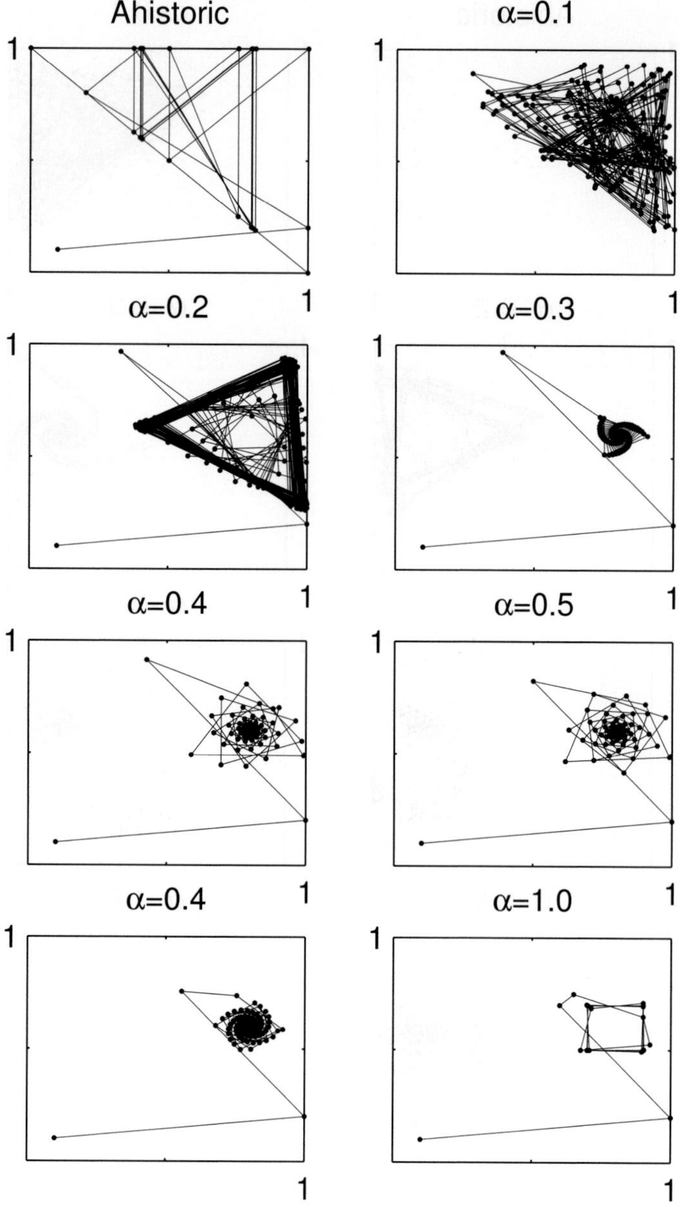

Fig. 13.43 PAP dynamics with memory starting in (0.1,0.1).

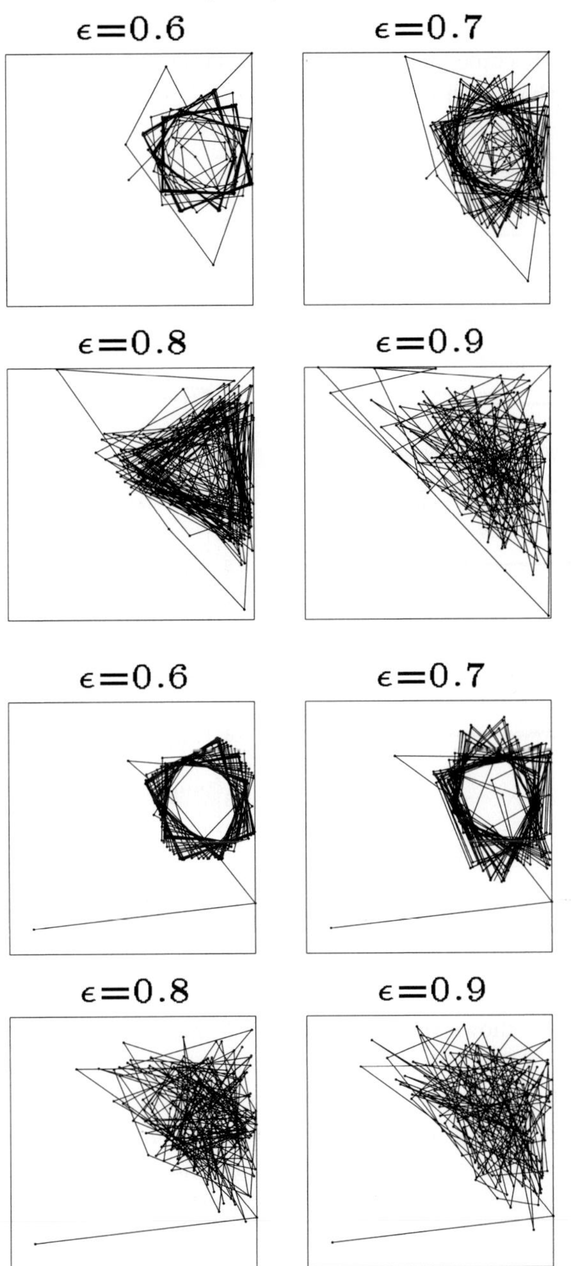

Fig. 13.44   PAP dynamics with ϵ-memory.

No structures become apparent in Fig. 13.44 as the dynamics appear much more erratic than with α-memory.

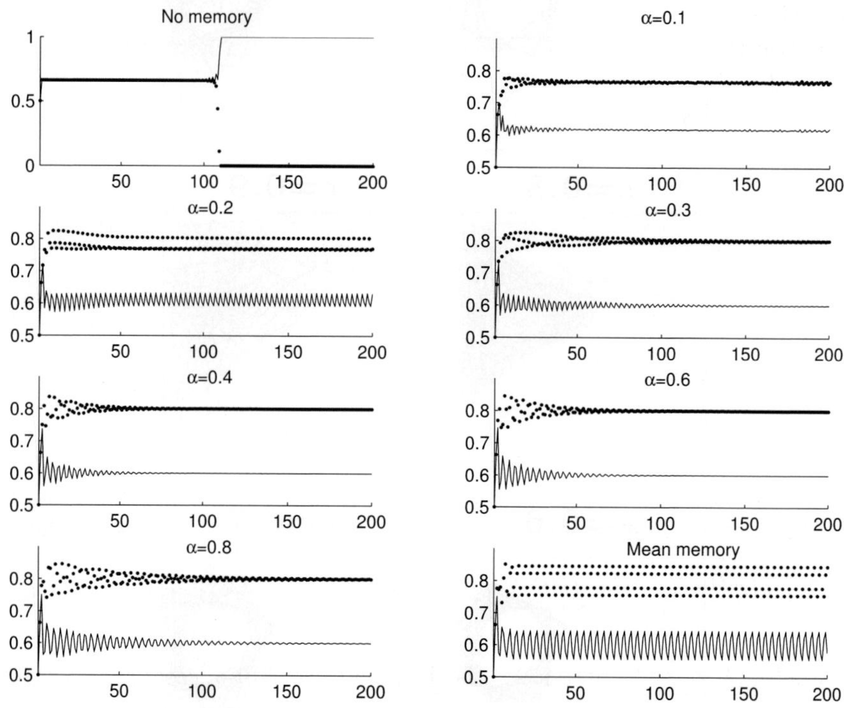

Fig. 13.45 Dynamics of the mean values of x and y starting from any of the points of the 1×1 square with different degrees of memory up to T=200. The values of $\overline{x}$ are unjoined and dot-marked; those of $\overline{y}$ are unmarked and joined.

Figure 13.45 shows the dynamics of the mean values of x and y starting from any of the 101×101 lattice points of the 1×1 square with sides divided by 0.01 intervals. The dynamics in the ahistoric context is rather striking: immediately, at $T = 2$, both $\overline{x}$ and $\overline{y}$ increase from 0.5 up to app. $0.66 \simeq 2/3$, a *generous* level of cooperation numerically coincident with the forgiveness probability (of C in response to D) found by Grim [174]. This degree of cooperation remains stable up to close to $T = 100$, when a light trembling announces an imminent symmetry-breaking with a sudden collapse of Pavlov cooperation ($\overline{x}$=0) and the corresponding firing of full cooperation of anti-Pavlov. As was illustrated starting from (0.1,0.1) and (0.8,0.6) in the previous section, finite precision arithmetics

leads every point to (0,1). The overall effect of memory is remarkable: Pavlov not only keeps a permanent mean degree cooperation but is higher than the mean degree of cooperation of anti-Pavlov, which in turn cooperates closely up to 0.6. Memory seems to lead the overall dynamics to the ahistoric (theoretically) fixed point (0.8,0.6). The $\bar{x}$ and $\bar{y}$ averages are:
$\alpha = 0.1 : (0.765, 0.617);\quad \alpha = 0.2 : (0.78, 0.61);\quad \alpha \geq 0.3 : (0.799, 0.600)$,
values not too far from the initial ahistoric, $2/3$.

Figures E.19 and E.20 in the Appendix E show the effect of different degrees of memory on the mean value of the one hundred first values of x and y starting from any of the 101×101 points of the 1×1 square mentioned above (upper set of four images) and the values of x and y at $T = 100$ (lower set of four images), just before the bifurcation in the ahistoric model. To enhance resolution, the particular point (0,1) is avoided in the presentation of the historic images of Figs.E.19 and E.20 by omitting their sides $x = 0$ and $y = 1$. The fairly structured ahistoric patterns (which resemble Vasarely's pictures) turn out to be very much altered with memory.

Figure E.21 shows the effect of memory on the mean distance to the initial point during the first one hundred time-steps. Thus,
$$\frac{1}{T} \sum_{t=1}^{T} \sqrt{\left(x^{(t)} - x^{(1)}\right)^2 - \left(y^{(t)} - y^{(1)}\right)^2} \text{ with } T = 100.$$
The (*theoretically*) fixed point (0.8,0.6) is revealed as an *attractor* in the snapshots with memory of Fig. E.21. The patterns of the distance to the initial point at $T = 100$, i.e., $\sqrt{\left(x^{(100)} - x^{(1)}\right)^2 - \left(y^{(100)} - y^{(1)}\right)^2}$, are shown in [30].

Figure E.22 shows the *escape time*: number of transitions to *escape* (for the first time) from the circle of radius one, so: $d = \sqrt{x(t)^2 + y(t)^2} > 1$. Starting from T=1, as done across this book, the number of transitions assigned to the escape time is the time step at which $d(x,y) > 1$ for the first time minus one. The ahistoric escape-time pattern [173] is not significantly altered by memory.

13.5   Other spatial games

Any two-person game can be regarded in a spatial formulation and endowed with memory as done here with the PD. So the Hawk (H)-Dove (D) [229, 286] game, a biologically interesting alternative to the PD, in which P and S have a reverse order: $S > P$. In the $H - D$ game, if both players play *Dove*, they share the resource value contested V, so both receive $R = V/2$. A *Hawk* gets all the value $V = \mathfrak{T}$ against a *Dove* who scores nothing in such an encounter ($S=0$). Mutual *Hawks* each scores $P = (V-C)/2$, with the cost of confronting C being greater than V. Thus the payoff matrices of both players are given in Table 13.10.

Table 13.10   The payoff matrices and reaction functions in the Hawk-Dove game.

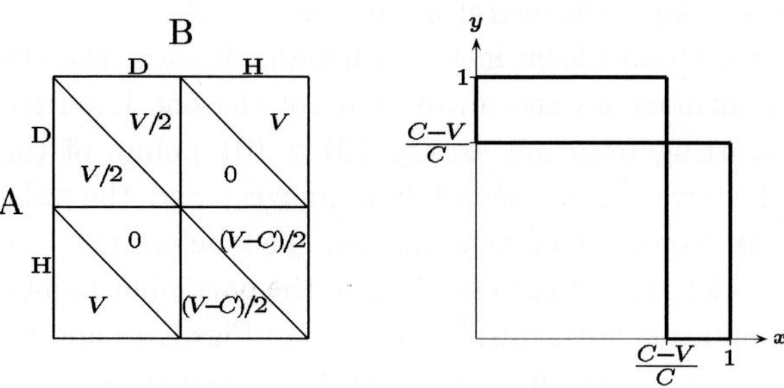

As a result of reversing the order of P and S, in the Hawk-Dove game the (H,H) pair is not in equilibrium, because D is the best reply to the co-player's H. This leads to stable coexistence of D and H in well-mixed, unstructured populations.

In the (symmetric) hawk-dove game, the expected gain when the players follow the mixed probabilistic strategies (x,1-x), and (y,1-y) are :

$$p_A(x;y) = (x, 1{-}x) \begin{pmatrix} V/2 & 0 \\ V & (V-C)/2 \end{pmatrix} \begin{pmatrix} y \\ 1{-}y \end{pmatrix} = -\frac{1}{2}(Cy+V-C)x + \left(Vy+\frac{1}{2}(V-C)(1-y)\right)$$

$$p_B(y;x) = (x, 1{-}x) \begin{pmatrix} V/2 & V \\ 0 & (V-C)/2 \end{pmatrix} \begin{pmatrix} y \\ 1{-}y \end{pmatrix} = -\frac{1}{2}(Cx+V-C)y + \left(Vx+\frac{1}{2}(V-C)(1-x)\right)$$

The pair of strategies $\big((x,1\text{-}x),(y, 1 - y)\big)$ are in Nash equilibrium if x is a best response to y and y is a best response to x. Thus, no player can benefit from changing his strategy while the other keeps his unchanged. Formally, $p_A(x,y) \geq p_A(z,y)$, and $p_B(x,y) \geq p_B(x,z)$, $\forall z$.

In game theory, the best response is the strategy (or strategies) which produces the most favorable outcome for a player, taking other players' strategies as given. Best response correspondences, also known as reaction correspondences, are not proper -functions- since functions must only have one value per argument, and some reaction correspondences may undefined. Response correspondences can be drawn for all 2×2 normal form games with a line for each player in a unit square strategy space as shown in the central panel of Table 13.10, in which vertical/horizontal lines indicate no definition. In this scenario, Nash equilibria are shown with points where the two player's correspondences agree, i.e. cross.

Thus, according to the reaction functions given in Table 13.10, the pairs of pure strategies (D,D) and (H,H) as well as the pair of mixed strategies $(x^* = 1 - V/C, y^* = 1 - V/C)$ are in Nash equilibrium, i.e., $p_A(x^*, x^*) = p_B(x^*, x^*) = V(C - V)/2C = (V/2)(1 - V/C) < V/2$. Moreover, the strategy $(1 - V/C, V/C)$ is *evolutionarily stable*, i.e., $p_A(x, x) \le p_A(y, x) \forall y$, and $p_A(x, x) = p_A(y, x) \Rightarrow p_A(x, y) > p_A(y, y), \forall y \ne x$.

A $(V=10, C=12)$ spatial simulation of the evolving patterns from a single *Hawk* irrupting in a world of *Doves* is shown in Fig. 13.46. As in the PD, historic memory leads the non-cooperative behaviour, i.e., *Hawk*, to extinction or severe restriction, unlike in the ahistoric model.

The continuous-valued version of Fig. 13.46 is given in Fig. 13.47 with the gray level corresponding to the degree of Hawk behaviour. Thus $1 - x$ and $1 - y$ in the payoff formulas:

$$p_A(x, y) = (x, 1-x) \begin{pmatrix} 5 & 0 \\ 10 & -1 \end{pmatrix} \begin{pmatrix} y \\ 1 - y \end{pmatrix} , \quad p_B(x, y) = (x, 1-x) \begin{pmatrix} 5 & 10 \\ 0 & -1 \end{pmatrix} \begin{pmatrix} y \\ 1 - y \end{pmatrix}.$$

Table 13.11 The payoff matrices and reaction functions in the Snowdrift game.

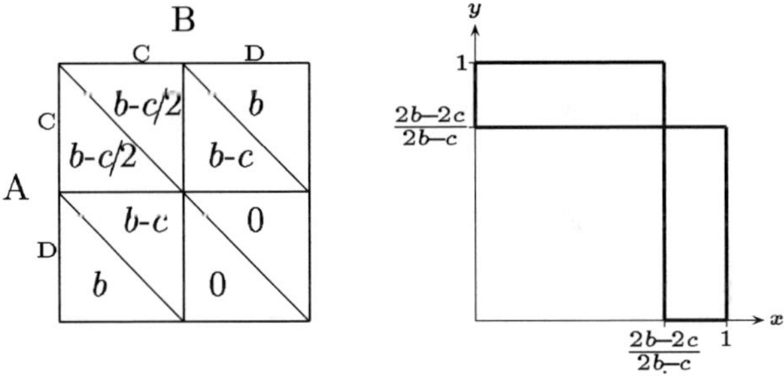

In a variant of the Hawk and Dove game, in the so called Snowdrift game (also know as the Chicken game [125]), cooperation results in a benefit b for the recipient but incurs a cost c to the donor. Mutual cooperation (sharing the cost c) pays a net benefit $R=b\text{-}c/2$, whereas mutual defection results in payoff $P=0$. With unilateral cooperation, defection yields the highest payoff $T=b$, whereas the cooperator gets $S=b\text{-}c$. Thus, the payoff matrices in the snowdrift game are given in Table 13.11.

If cost is higher than benefit (and $2b > c$), these payoffs recover the Prisoner's Dilemma, by contrast if $b > c > 0$, these payoffs generate the snowdrift game, in which the best action depends on the co-player choice: to defect if the other cooperates, but to cooperate if the co-player defects.

No memory

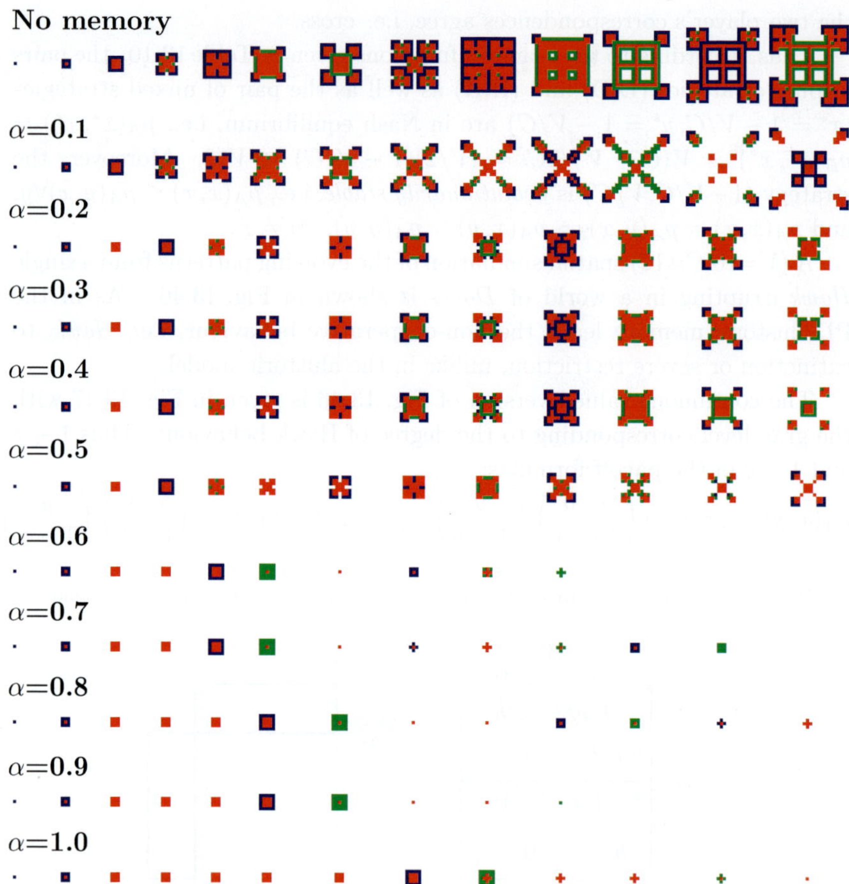

Fig. 13.46   Evolving patterns starting from a single hawk up to $T = 13$ in the Hawk and Dove spatial game. Memory factor α. Code color: Blue: Hawk (H) after Dove (D), Red: H after H, Green: D after H, blank: D after D.

The expected payoffs in the snowdrift game are:
$$p_A(x;y) = (x, 1-x)\begin{pmatrix} b - c/2 & b - c \\ b & 0 \end{pmatrix}\begin{pmatrix} y \\ 1-y \end{pmatrix} = (-(b - c/2)y + b - c)x + by,$$
and $p_B(y;x) = (x, 1-x) = (-(b - c/2)x + b - c)y + bx = p_A(y;x)$. Thus,
$x^* = y^* = \dfrac{2b - 2c}{2b - c}$ define the pair of mixed equilibrium strategies, with
$p_A(x^*, x^*) = p_B(x^*, x^*) = b\dfrac{2b - 2c}{2b - c}$. According to the replicator dynamics
[201], these probabilities coincide with the equilibrium frequency of cooperators in the snowdrift game, $f{=}1{-}r$, where $r{=}c/(2b{-}c)$ is the cost benefit ratio of mutual cooperation.

No memory

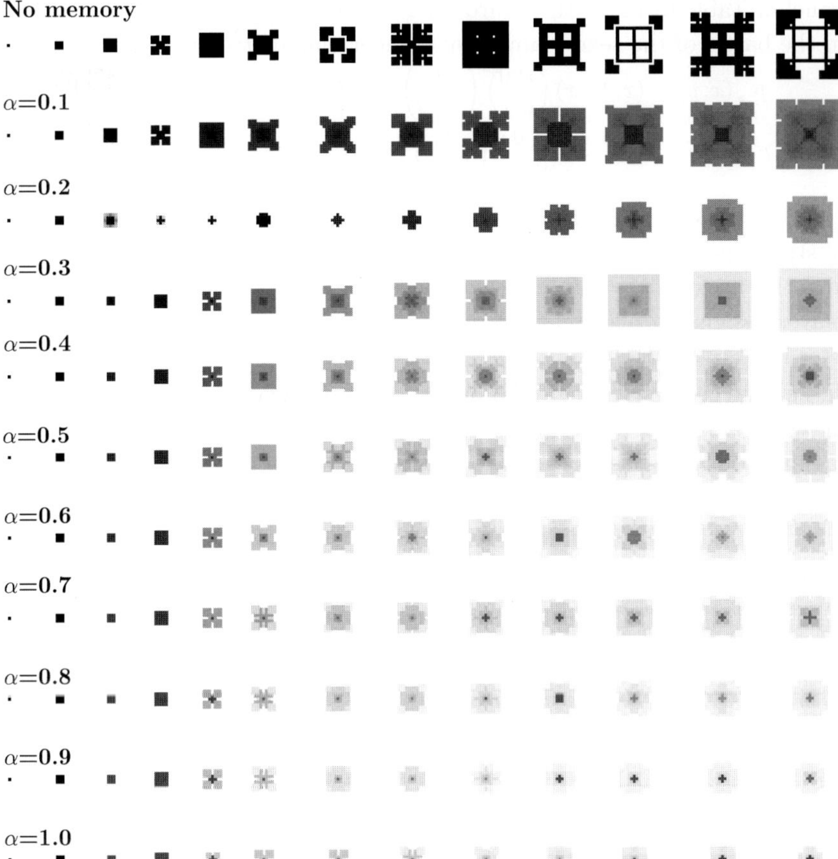

Fig. 13.47 The advance of defection up to $T=13$ in the continuous-valued Hawk and Dove spatial game starting from a single full hawk in a 13×13 lattice.

Unexpectedly, spatial structure reduces the proportion of D for a wide range of parameters in the snowdrift game as shown in [195].

Asymmetric games

The so called *battle of the sexes* is a simple example of a two-person asymmetric game [286]. In this game, the preferences of a conventional couple are assumed to fit the traditional stereotypes: the male prefers to attend a *F*ootball match, whereas the female prefers to attend a *B*allet performance. Both players (which are treated symmetrically), decide in the hope of getting together, so that their payoff matrices are given in the far left panel of Table 13.12, with rewards $R > r > 0$. Thus, the expected payoffs in the battle of the sexes game, using uncorrelated strategies are:

$$p_{\male}(x; y) = (x, 1-x) \begin{pmatrix} R & 0 \\ 0 & r \end{pmatrix} \begin{pmatrix} y \\ 1-y \end{pmatrix} = ((R+r)y - r)x + r(1-y)$$

$$p_{\female}(y; x) = (x, 1-x) \begin{pmatrix} r & 0 \\ 0 & R \end{pmatrix} \begin{pmatrix} y \\ 1-y \end{pmatrix} = ((R+r)x - R)y + R(1-x)$$

Thus, according to the reaction functions given in Table 13.12, the pairs of strategies in Nash equilibrium in the BOS are, the pure (0,0) and (1,1), and the mixed $(x^* = R/(R+r), y^* = r/(R+r))$. In the latter, every player assigns to his preferred option (the same) higher probability, i.e., $x^* = 1-y^* = R/(R+r)$. It is, $p_{\male}(x^*, y^*) = p_{\female}(x^*, y^*) = rR/(R+r) < r < R$, i.e., the geometric mean of R and r. Pareto-efficient situations are those in which it is impossible to make one player better off without necessarily making someone else worse off. Thus, only the two pure Nash equilibrium strategies (also termed *coordinated*) are Pareto efficient.

Table 13.12   The payoff matrices, reaction functions and payoff region in the *battle of the sexes* game.

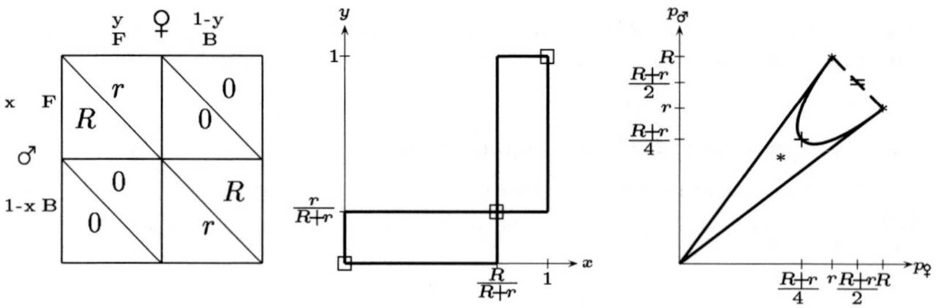

Both players get the same payoff, if $y = 1-x$, in which case, $p = (R+r)(1-x)x$. This egalitarian payoff is maximum when $x = y = 1/2$, with $p^+ = (R+r)/4$, the point marked + in the far right panel of Table 13.12. Thus, the set of payoffs which can be obtained by both players (or payoff region) is closed by the parabola passing by (R, r), (r, R), and (p^+, p^+), as shown in the payoff region panel of Table 13.12.

In a broader game scenario, a probabililty distribution $A = (a_{ij})$ assigns probability to every combination of player choices, so $A = \begin{pmatrix} a_{11} & a_{12} \\ a_{21} & a_{22} \end{pmatrix}$ in 2×2 games [318]. Thus, the expected payoffs in the BOS are:

$$p_{\sigma} = a_{11}R + a_{22}r$$
$$p_{\varphi} = a_{11}r + a_{22}R$$
. The probability distribution (strategy) A=(a_{ij}) is in *correlated equilibrium* [68, 69] if the players cannot gain by disobeying the signals given by the randomization device A. Thus, for example, in the BOS if the male player is given the F signal, by obeying it the σ-player gets: $p_{\sigma}/F = a_{11}R/(a_{11} + a_{12})$, whereas disobeying his mean payoff is: $a_{12}r/(a_{11} + a_{12})$, and correlated equilibrium demands $a_{11}R \geq a_{12}r$. Analogously, $p_{\sigma}/B = a_{22}r/(a_{21} + a_{22}) \geq a_{21}R/(a_{21} + a_{22}) \rightarrow a_{22}r \geq a_{21}R$; $p_{\varphi}/F = a_{11}r/(a_{11} + a_{21}) \geq a_{21}R/(a_{11} + a_{21}) \rightarrow a_{11}r \geq a_{21}R$; $p_{\sigma}/B = a_{22}R/(a_{12} + a_{22}) \geq a_{12}r/(a_{12} + a_{22}) \rightarrow a_{22}R \geq a_{12}r$. If $a_{12} = a_{21} = 0$ every former inequality holds, i.e., $a_{11}R \geq 0$, $a_{22}r \geq 0$, $a_{11}r \geq 0$, and $a_{22}R \geq 0$, so that $A = \begin{pmatrix} a & 0 \\ 0 & 1-a \end{pmatrix}$ is in correlated equilibrium, giving:

$$p_{\sigma} = aR + (1-a)r$$
$$p_{\varphi} = ar + (1-a)R$$
, so a convex $(0 \leq a \leq 1)$ combination of the (R,r) and (r,R) points, i.e., the segment that joints these points in the payoff region. Now it becomes accessible the payoff region limited by the parabola and the segment that joins the Pareto optimal pairs of payoffs. In this scenario both players reach a maximum egalitarian payoff $p^{=} = (R + r)/2$ (the point marked $=$ in the payoff region of Table 13.12), with $a = 1/2$, i.e., fully discarding the mutually inconvenient FB and BF combinations and adopting FF and BB with equal probability.

Correlation, namely *entanglement*, is in the core of quantum theory, thus in quantum games [135], and consequently in the quantum approach to the battle of the sexes [142, 307]. Let us point here that the study performed here differs in its the spatial component from that n-person BOS study made in [444, 445].

In the chessboard shown in the far left panel of Table 13.13, every player is surrounded by four partners (φ-σ, σ-φ), and four mates (φ-φ, σ-σ). In a CA-like implementation, every player plays with his four adjacent partners and adopts the choice of his nearest-neighbor mate (including himself) with the highest payoff. In the initial scenario of Table 13.13, every player chooses his preferred choice, except a male in the central part of the lattice that chooses *ballet*. As a result, the general income is nil with the only exception arising from the σ-ballet choice. This reports four units (assuming r=1) to the initial *deviated* male, and fires the change to *ballet* of the four males connected with the initial σ-ballet as indicated under T=2 in Ta-

  Discrete Systems with Memory

ble 13.13 . The change ♂-football to ♂-ballet advances in this way at every time-step, so that in this simple example every player will choose ballet in the long term.

Table 13.13   The battle of the sexes cellular automaton. $R = 5$, $r = 1$.

$T = 1$

♂	♀	♂	♀	♂	♀	♂	♀
♀	♂	♀	♂	♀	♂	♀	♂
♂	♀	♂	♀	♂	♀	♂	♀
♀	♂	♀	♂	♀	♂	♀	♂
♂	♀	♂	♀	♂	♀	♂	♀
♀	♂	♀	♂	♀	♂	♀	♂
♂	♀	♂	♀	♂	♀	♂	♀
♀	♂	♀	♂	♀	♂	♀	♂

F	B	F	B	F	B	F	B
B	F	B	F	B	F	B	F
F	B	F	B	F	B	F	B
B	F	B	B	B	F	B	F
F	B	F	B	F	B	F	B
B	F	B	F	B	F	B	F
F	B	F	B	F	B	F	B
B	F	B	F	B	F	B	F

0	0	0	0	0	0	0	0
0	0	0	0	0	0	0	0
0	0	0	5	0	0	0	0
0	0	5	4	5	0	0	0
0	0	0	5	0	0	0	0
0	0	0	0	0	0	0	0
0	0	0	0	0	0	0	0
0	0	0	0	0	0	0	0

$T = 2$

F	B	F	B	F	B	F	B
B	F	B	F	B	F	B	F
F	B	B	B	B	B	F	B
B	F	B	B	B	F	B	F
F	B	B	B	B	B	F	B
B	F	B	F	B	F	B	F
F	B	F	B	F	B	F	B
B	F	B	F	B	F	B	F

0	0	0	0	0	0	0	0
0	0	5	0	5	0	0	0
0	5	4	15	4	5	0	0
0	0	15	4	15	0	0	0
0	5	4	15	4	5	0	0
0	0	5	0	5	0	0	0
0	0	0	0	0	0	0	0
0	0	0	0	0	0	0	0

The ballet frequency and mean payoff per encounter (p) in the $(R = 5, r = 1)$, $(R = 2, r = 1)$, and $(R = 10, r = 1)$, battle of the sexes cellular automata starting with the same random initial configuration of choices are shown in Figs. 13.48, 13.49, and 13.50 for nine different initial configurations. All the simulations are run in a 100×100 lattice. Initially, as a result of the random assignment of choices, the frequencies of ballet choice (and that of football) are 0.5, and the mean payoffs p commence at the arithmetic mean of the payoff values, $p^+ = (R + r)/4$. Thus, 1.75, 0.75 and 2.75 respectively.

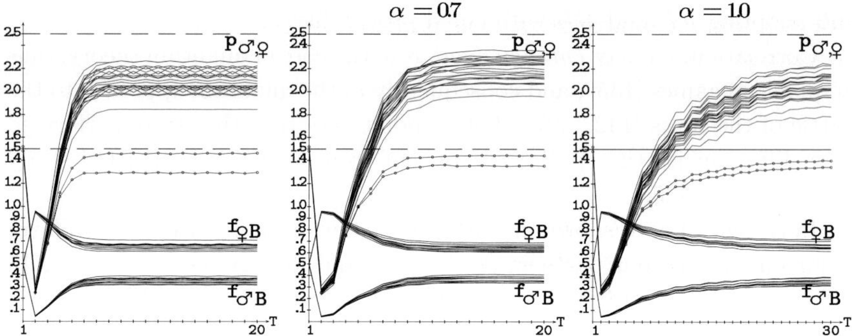

Fig. 13.48   The ballet frequency (**f**) and mean payoff per encounter (**p**) in the $(R = 5, r = 1)$ battle of the sexes cellular automaton. Ahistoric (left), $\alpha = 0.7$ (center), and full memory (right) models.

After the first round, both types of players drift to their preferred choice, and as consequence the mean payoff per encounter (p), plummets at $T = 2$. But immediately the drift to the preferred choice becomes moderated, and both payoffs recover. In the long-term the ballet frequencies stabilize at values that are not far from the B-probabilities in the mixed equilibrium strategy of a two-person game, $1 - x^* = r/(R + r), 1 - y^* = R/(R + r)$, thus, $(0.17,\ 0.83{=}5/6)$, $(0.23,\ 0.67{=}2/3)$, and $(0.09,\ 0.91{=}10/11)$ in the Figs. 13.48, 13.49, and 13.50 respectively.

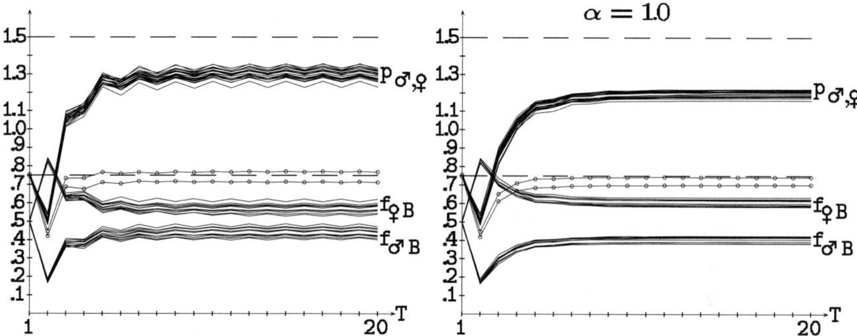

Fig. 13.49 The ballet frequency (f) and mean payoff per encounter (p) in the ($R = 2, r = 1$) battle of the sexes cellular automaton. Ahistoric (left), and full memory (right) models.

The dotted curves in Figs. 13.48, 13.49, and 13.50 show, in one of the simulations, the *theoretical* payoffs of both players in a two person game with independent strategies using as probabilities the evolving frequencies, namely:

$$p_{\male}^{(T)} = ((R + r)(1 - f_{\female B}) - r)(1 - f_{\male B}) + r f_{\female B},$$

$$p_{\female}^{(T)} = ((R + r)(1 - f_{\male B}) - R)(1 - f_{\female B}) + R f_{\male B}.$$

The actual mean payoffs of both kinds of players shown in the figures are over these *expected* values. The divergence starts modestly at $T{=}3$, and increases until the virtual stabilization of the p values. In the $R{=}2$ scenario, the divergence becomes apparent already at $T{=}3$, but in the two other higher R scenarios, the divergence progresses more slowly, although it is already appreciable in the figures at $T{=}4$.

The notable increase of the mean payoffs in the proposed cellular automaton is due to the spatial structure, which allows for the emergence of clusters of *agreement* (or cooperation), shown in Fig. E.23 as black-white (FF) and red-blue (BB) regions with interfaces of disagreement among the clusters. This notable case of self-organization of the players explains the high mean payoffs per encounter, clearly in the region accessible only by correlated strategies in the two-person game. It is remarkable that the agreement clusters appear fairly soon, so that in Fig. E.23 the pattern at $T=8$ does not qualitatively differ from that at $T=20$.

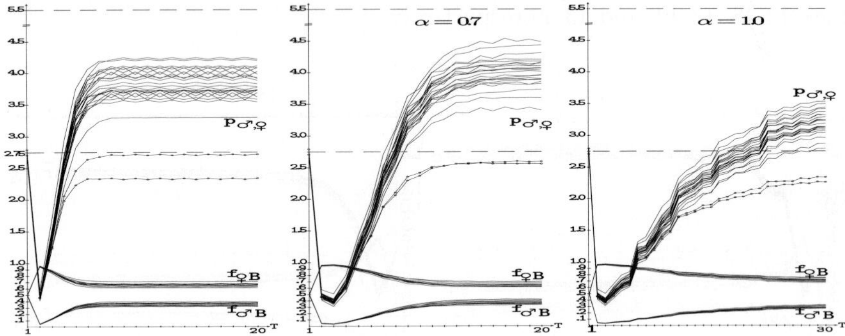

Fig. 13.50   The ballet frequency (**f**) and mean payoff per encounter (**p**) in the ($R = 10, r = 1$) battle of the sexes cellular automaton. Ahistoric (left), α=0.7 (center), and full memory (right) models.

The dynamics in higher size lattices does not qualitatively differ from that presented here in a 100×100 lattice The only difference that can be noticed is a stronger stabilization in the (almost) steady regime, i.e., with a negligible oscillation of the f and p values. Figure 13.51 shows an example in the scenario of Fig. 13.48, but in a 200×200 lattice.

Opposite to this, in small lattices the stabilization is much more problematic, and, more important, the initial configuration plays an important role in the evolutionary dynamics. Accordingly to this, the two types of curves (f and p) in the left panel of Fig. 13.52 do not amalgamate as they do in Fig. 13.48, but they cover a broad interval of values. In two of the nine simulations of Fig. 13.52 with more stabilization, there is a net drift to one of the choices, that leads in one of them to the mean payoff of the female-type to 4, and to the other one to a mean payoff of the male-type to 3.5. The first six patterns of the former, that one with drift to ballet, is shown in the last row of the right side of Fig. 13.52, where only a small cluster of FF choices (black-white colors) resists the predominant BB mutual election.

Contrary to expectations, after the previous study of the effect of memory in the spatialized prisoner's dilemma, memory does not significantly alter the dynamics of the proposed battle of the sexes cellular automaton. When dealing with the prisoner's dilemma, memory dramatically boosts

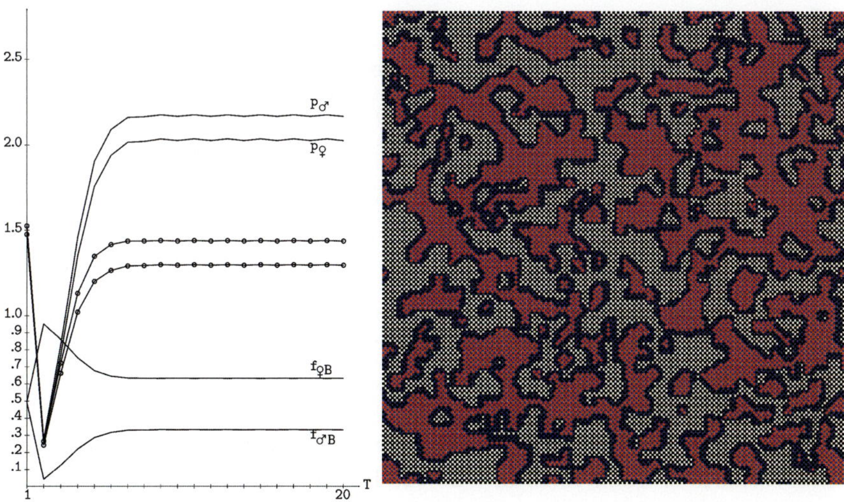

Fig. 13.51 Left: The ballet frequency (**f**) and mean payoff per encounter (**p**) in the ahistoric ($R=5, r=1$) BOS CA in a $200{\times}200$ lattice. Right: Pattern at $T=20$.

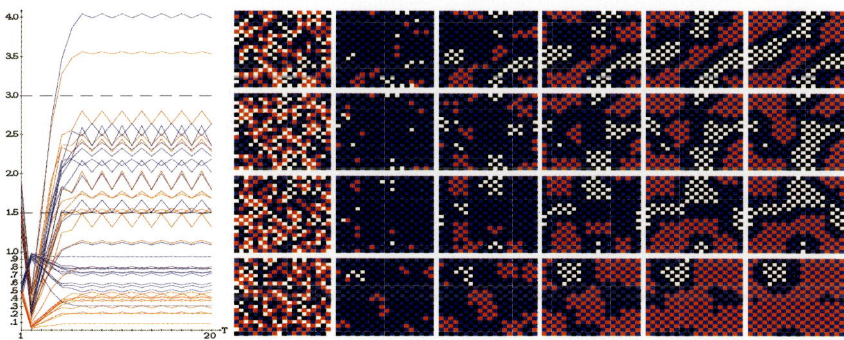

Fig. 13.52 Left: The ballet frequency (**f**) and mean payoff per encounter (**p**) in the ($R=5, r=1$) BOS ahistoric CA in a $20{\times}20$ lattice size. Right: initial patterns in four simulations.

cooperation. In the same vein, we expected here an increase in the matches with agreement (either FF or BB), as an effect of memory. But this is not so: memory tends to decrease the mean payoff, though not at a great extent, by means of a kind of inertial effect that restrains the correction of the initial drift to the preferred choices, that sooner stabilizes the agreement clusters.

As an additional example of the inertial effect of memory in the proposed

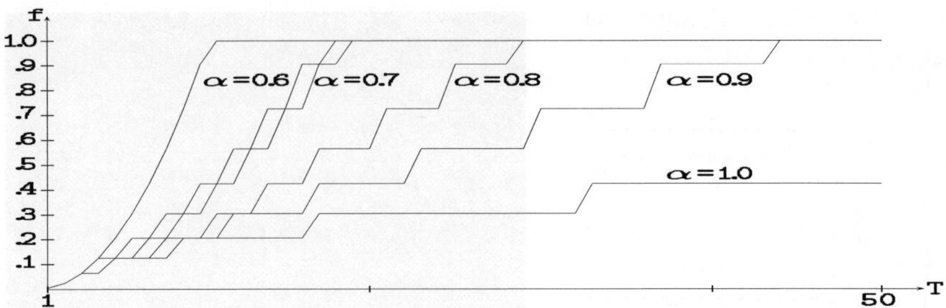

Fig. 13.53 The frequency of σB from a single σB in the inital scenario of Table 13.13 in a 20×20 lattice. Ahistoric and models with memory factor varying from 0.6 to 1.0 by 0.1 intervals.

BOS CA, Fig. 13.53 shows the frequency of σB from a single σB in the inital scenario of Table 13.13 in a 20×20 lattice. The dynamics are shown up to $T{=}50$, in the ahistoric and with the memory factor varying from 0.6 to 1.0 by 0.1 intervals. The velocity of the increase of σB frequency falls with memory, by increasing the duration of the stable periods. In the full memory context, the stable periods progressively increase their duration in such a way that the full σB-occupation is not achieved up to $T{=}513$.

Probabilistic updating

In the probabilistic updating mechanism considered from here, the individuals will play F or B with a probability proportional to the (unweighted) total payoff of F and B among their mate neighbors. Thus, denoting by $\mathcal{N}_i$ the mate neighborhood of the generic cell i, the probabilities that it is occupied by a F and B-player at time step T are:

$$P(d_i^{(T)} = F) = \frac{\displaystyle\sum_{j \in \mathcal{N}_i / d_j^{(T)} = F} p_j^{(T)}}{\displaystyle\sum_{j \in \mathcal{N}_i} p_j^{(T)}} \quad , \qquad P(d_i^{(T)} = B) = 1 - P(d_i^{(T)} = F)$$

We do not consider here asychronous updating, wich would allow to relate somehow the results presented here with other *ordering* processes in spatialized social relations. Thus, for example, the "antiferromagnetic" model proposed in [414].

Figure E.24 shows the initial patterns in the ($R{=}5, r{=}1$) probabilistic BOS cellular automaton in a simulation run in a 100×100 lattice size. The evolution of the ballet frequency and mean payoff per encounter starting at random as in the nine simulations of Fig. 13.48 is shown in Fig. 13.54

up to $T=1000$. The initial patterns in Fig. E.24 correspond to the top-left frame of Fig. 13.54.

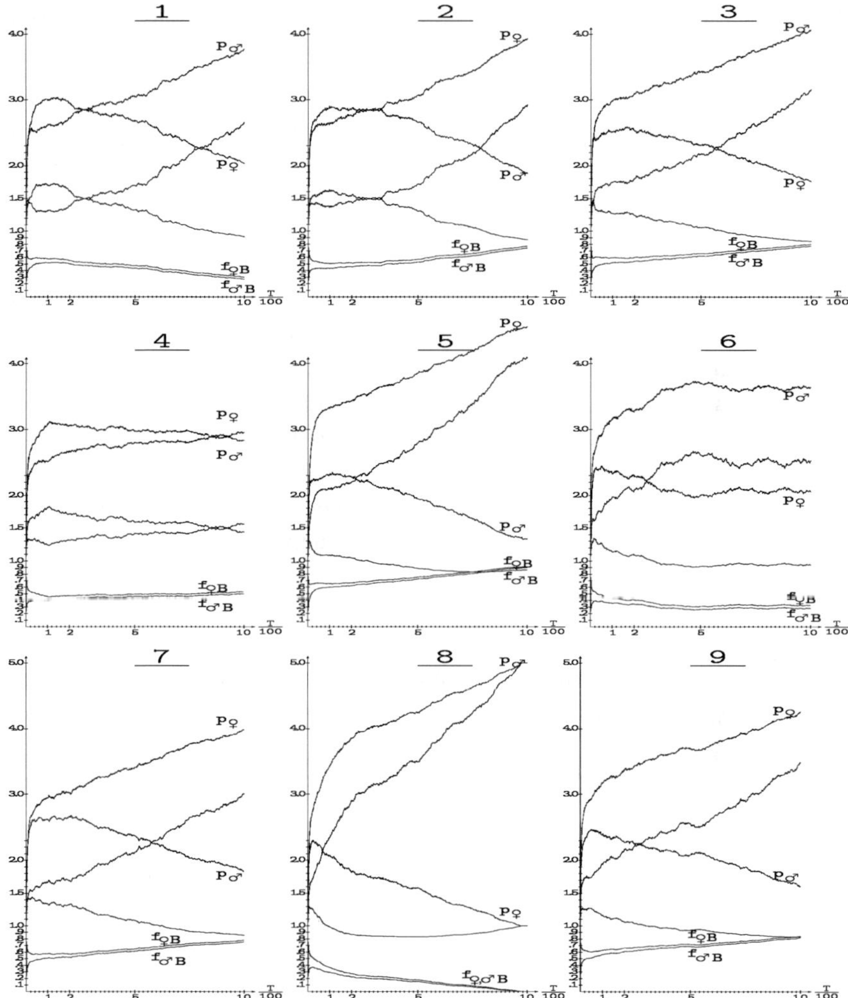

Fig. 13.54 The ballet frequency (**f**) and mean payoff per encounter (**p**) in nine simulations of the (5,1) probabilistic BOS CA Ahistoric model.

After the first round, both types of players drift to their preferred choice in the probabilistic simulations of Fig. 13.54 as happened in the deterministic simulations in Fig. 13.48, but to a lower degree. Thus, in Fig. 13.48

both $f_{♂F}$ and $f_{♀B}$ are over 0.9, whereas in Fig. 13.54 these initial frequencies reach levels lower than 0.8. After the first iteration, the frequencies of both types of choices tend to converge, as indicated in Fig. 13.54 in the *Ballet* case, where both $f_{♂B}$ and $f_{♂B}$ are almost coincident after the initial transition period, though with $f_{♀B}$ always over $f_{♂B'}$. This trend to the B and F frequencies converging in both types of players is completely absent in the deterministic model, where the frequencies are soon stabilized in rather different values as shown in Fig. 13.48 for the *Ballet* case. Although the way in which the frequencies evolve is fairly monotone, either increasing or decreasing, this is not necessarily so, as the tendency may be inverted. Thus, the simulations lebelled 1, 2 and 4 in Fig. 13.54 show fairly early swifts in the f-tendencies, that are marked by a relatively early cross in the mean payoff curves of simulations 1 and 2, whereas in the simulation lebelled 4 the p-cross is reached very late, nearly $T=1000$.

Figure E.25 shows the patterns at $T=100$, 200, 500 and 1000 in the nine probabilistic simulations of Fig. 13.54. The agreement clusters are already formed at $T=100$, and in most simulations indicate the long-term dynamics. Thus for example, the central snapshot in every frame, corresponding to the simulation lebelled 5 in Fig. 13.54, shows the expansion of the red-blue (BB) cluster, which is already predominant at $T=100$. In the same vein, the bottom-center snapshot in every frame (simulation 8 in Fig. 13.54), shows the expansion of the black-white (FF) region, already predominant at $T=100$, in this case up to the total lattice occupation at $T=1000$. As noted above, this kind of monotone dynamics is not a universal rule, and the (slightly) predominant clusters at $T=100$ in the top-left and top-central snapshots (simulations 1 and 2 in Fig. 13.54), are not predominant in the long-term evolution.

As a rule, the drift to one of the choices turns out much faster in small lattices. Thus for example, in two of the three simulations in Fig. 13.55 the lattice is soon fully occupied, either by B in the far left frame, or by F in the far right frame. But the trend to fast full occupation is not universal, as shows the simulation in the central frame, in which one the full B occupation occurs by $T=400$, after a long transient period up to $T=300$.

The probabilistic updating scheme becomes with memory:

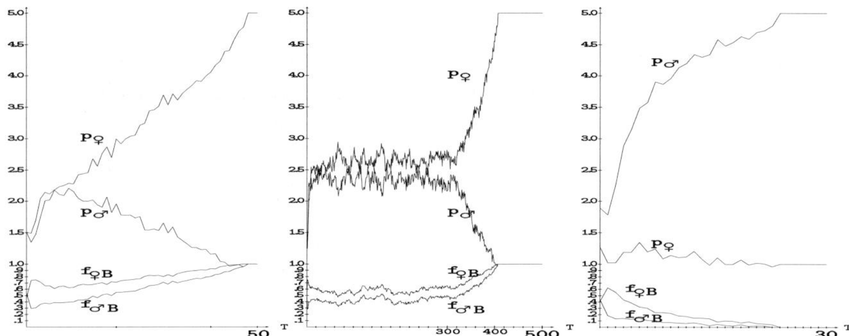

Fig. 13.55   The ballet frequency (f) and mean payoff per encounter (p) in a 20×20 lattice.

$$P(d_i^{(T)} = F) = \frac{\displaystyle\sum_{j\in\mathcal{N}_i \big/ \delta_j^{(T)}=F} \pi_j^{(T)}}{\displaystyle\sum_{j\in\mathcal{N}_i} \pi_j^{(T)}} \qquad , \qquad P(d_i^{(T)} = B) = 1-P(d_i^{(T)} = F)$$

Figure 13.56 shows the evolution of the ballet frequency and mean payoff per encounter in the nine scenarios of the probabilistic (5,1)-BOS cellular automaton of Fig. 13.54 with $\alpha=0.7$ memory. The inertial effect of memory induces a notable moderation in the trend of the $f_{\text{♀B}}$ and $f_{\text{♂B}}$ parameters either to increase or decrease, thus the evolution of the frequency curves appear much more horizontal in Fig. 13.56 compared to those in Fig. 13.54. In parallel with this kind of f-stabilization, there is a moderation in the way in which the mean payoffs per encounter diverge. Particularly appealing are the simulations lebelled 3, 7 and 8, in which ones the B-frequencies are not far off 0.5, and both mean payoffs per encounter reach fairly high values.

Figure E.26 shows the patterns at $T=1000$ in the nine scenarios of the probabilistic (5,1)-BOS CA of Fig. 13.54 with increasing values of the α memory factor. According to the $\alpha=0.7$ dynamics in Fig. 13.56, the FF and BB clusters in its corresponding panel in Fig. E.26 are not so unbalanced in size as they are in the $T=1000$ panel in Fig. E.25. In particular, not simulation leads to the full FF or BB occupation in the $\alpha=0.7$ panel of Fig. E.26.

The inertial effect of memory is more apparent as the memory charge increases. Thus for example, the dynamics of the top-left simulation with

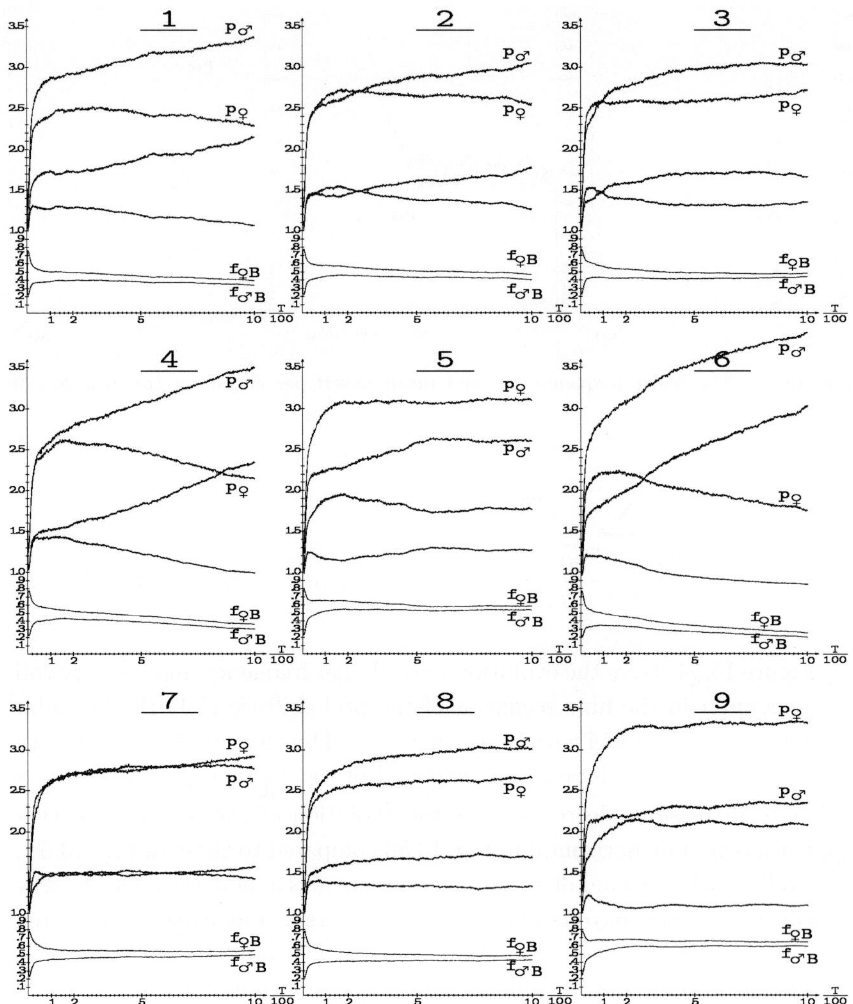

Fig. 13.56   The ballet frequency (**f**) and mean payoff per encounter (**p**) in the nine scenarios of the probabilistic (5,1)-BOS CA of Fig. 13.54 with α=0.7 memory.

α=0.7 memory in Fig. 13.56 is shown in Fig. 13.57 with higher α memory factor values. As a result of the effect of increasing the memory charge, i.e., the inertial effect, *i)* the frequencies (of Ballet) do not recover from the initial divergence at the high degree that they recover with lower memory, and *ii)* their trend to a extreme value is restrained. In parallel to that, the payoff parameters, keep lower and closer. The full memory implementation

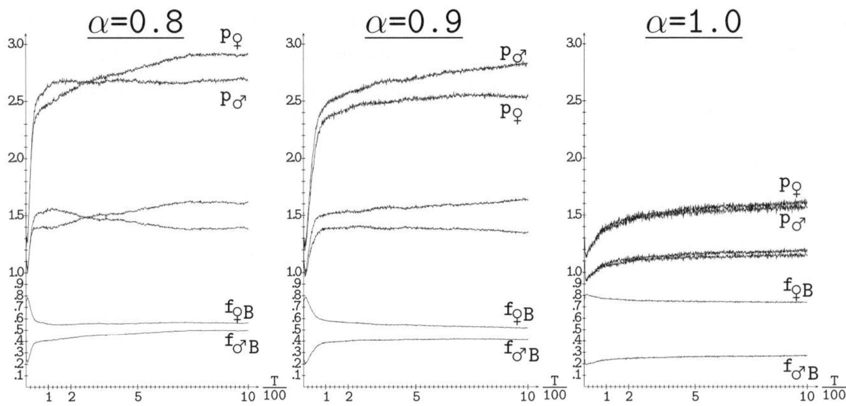

Fig. 13.57   The ballet frequency (**f**) and mean payoff per encounter (**p**) with high α memory.

turns out paradigmatic of the effect of memory, because the f-recovery is very small, and there is not any appreciable drift to extreme values. This is much in the way as in the deterministic ahistoric simulations in Fig. 13.48, albeit with lower f-recovery, and consequently lower payoffs, in the probabilistic full memory simulation compared to the deterministic ahistoric ones.

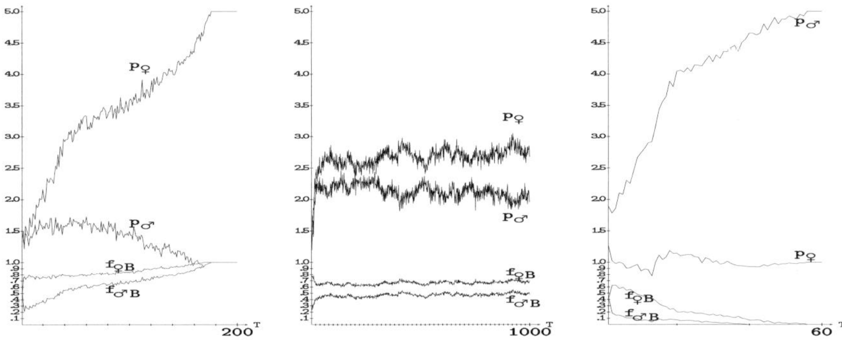

Fig. 13.58   The ballet frequency (**f**) and mean payoff per encounter (**p**) in a 20×20 lattice with $\alpha{=}0.7$ memory.

As a rule, memory notably restrains the velocity of one-choice full occupation in small lattices. This is shown in Fig. 13.58, that departs from the same scenarios of Fig. 13.55, but implementing $\alpha{=}0.7$ memory. Thus, the B-occupation in the far left frame is reached by $T{=}180$ instead of by

$T{=}50$, and that in the far right panel is delayed from $T{=}30$ up to $T{=}60$. In the simulation in the central frame, which resists the B-occupation already in the ahistoric model, there is not an apparent lattice occupation up to $T{=}1000$. On the contrary, both players get a high payoff up to this time-step, though it is likely that B will occupy the lattice in the very long term.

Appendix A

Average memory starting at random

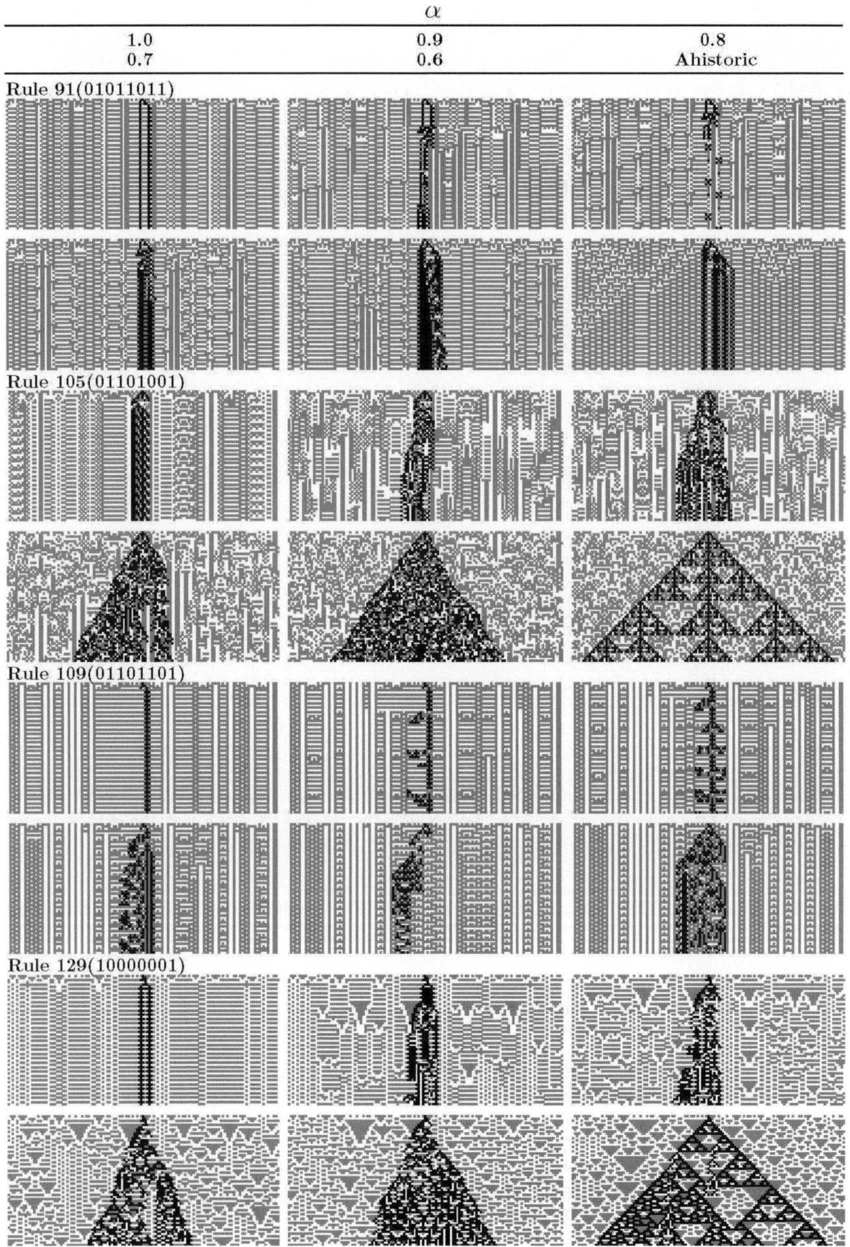

Fig. A.1   Symmetric but not quiescent elementary rules.

Rule 147(10010011)

Rule 151(10010111)

Rule 161(10100001)

Rule 165(10100101)

Rule 183(10110111)

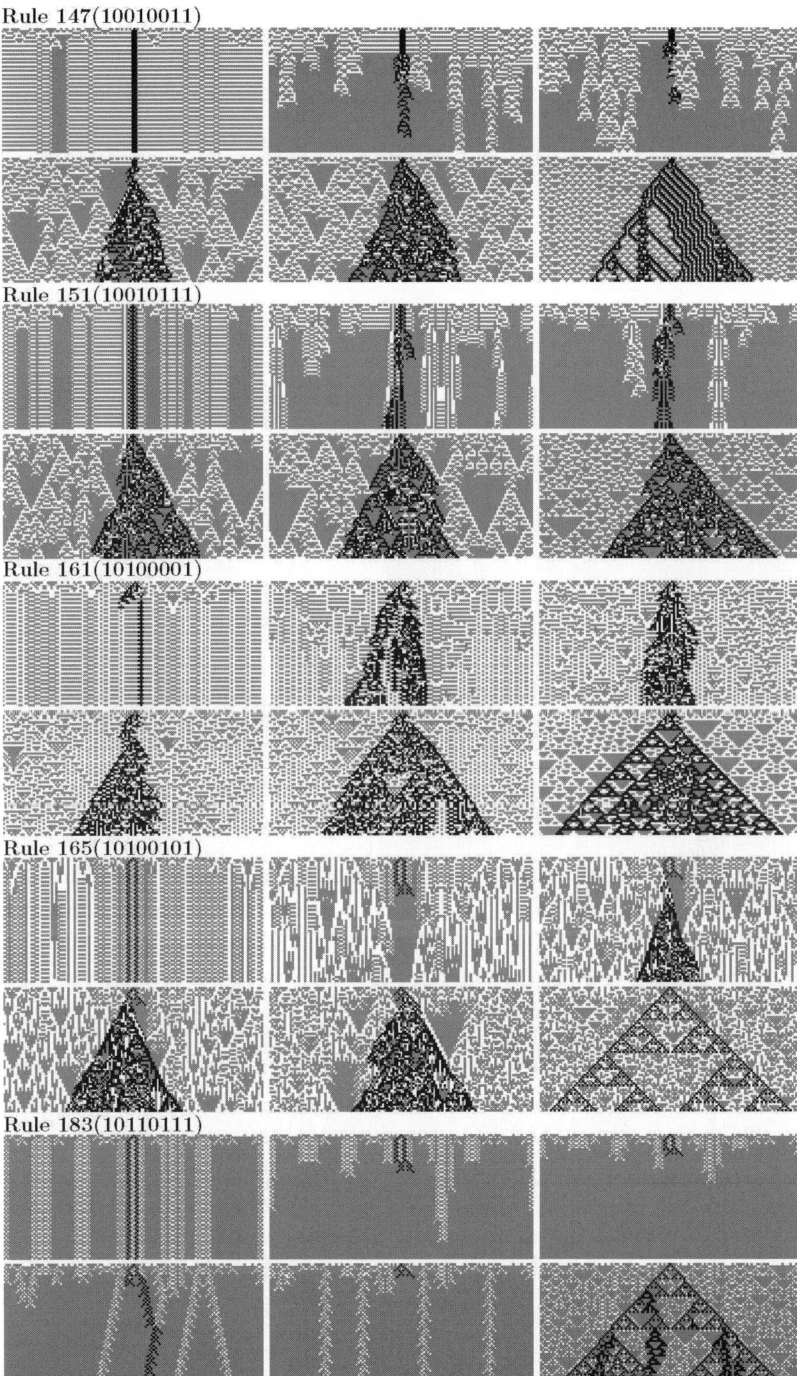

Fig. A.1 (*continued*)

Discrete Systems with Memory

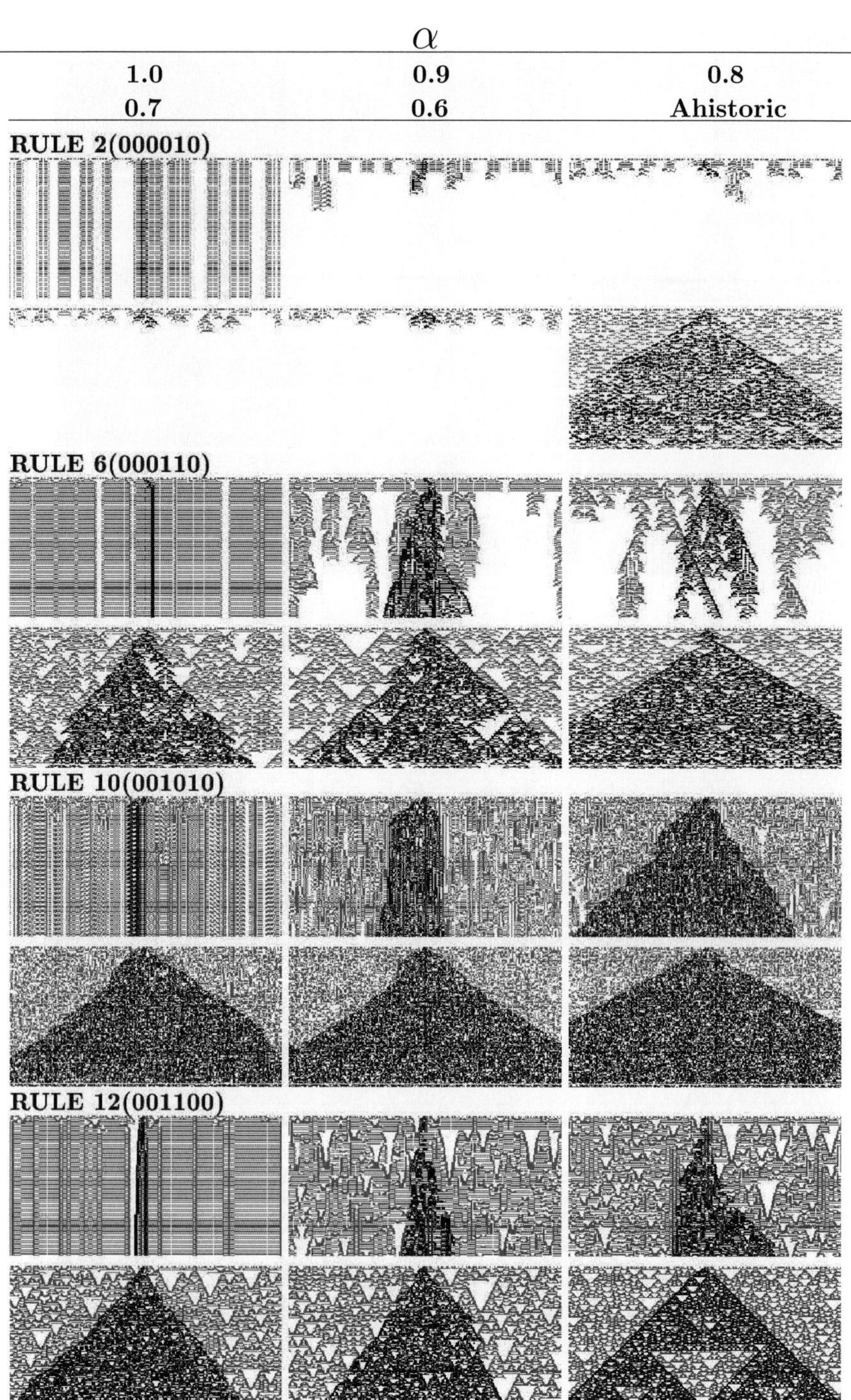

Fig. A.2   Totalistic, $k = r = 2$ CA starting at random.

RULE 14(001110)

RULE 18(010010)

RULE 20(010100)

RULE 22(010110)

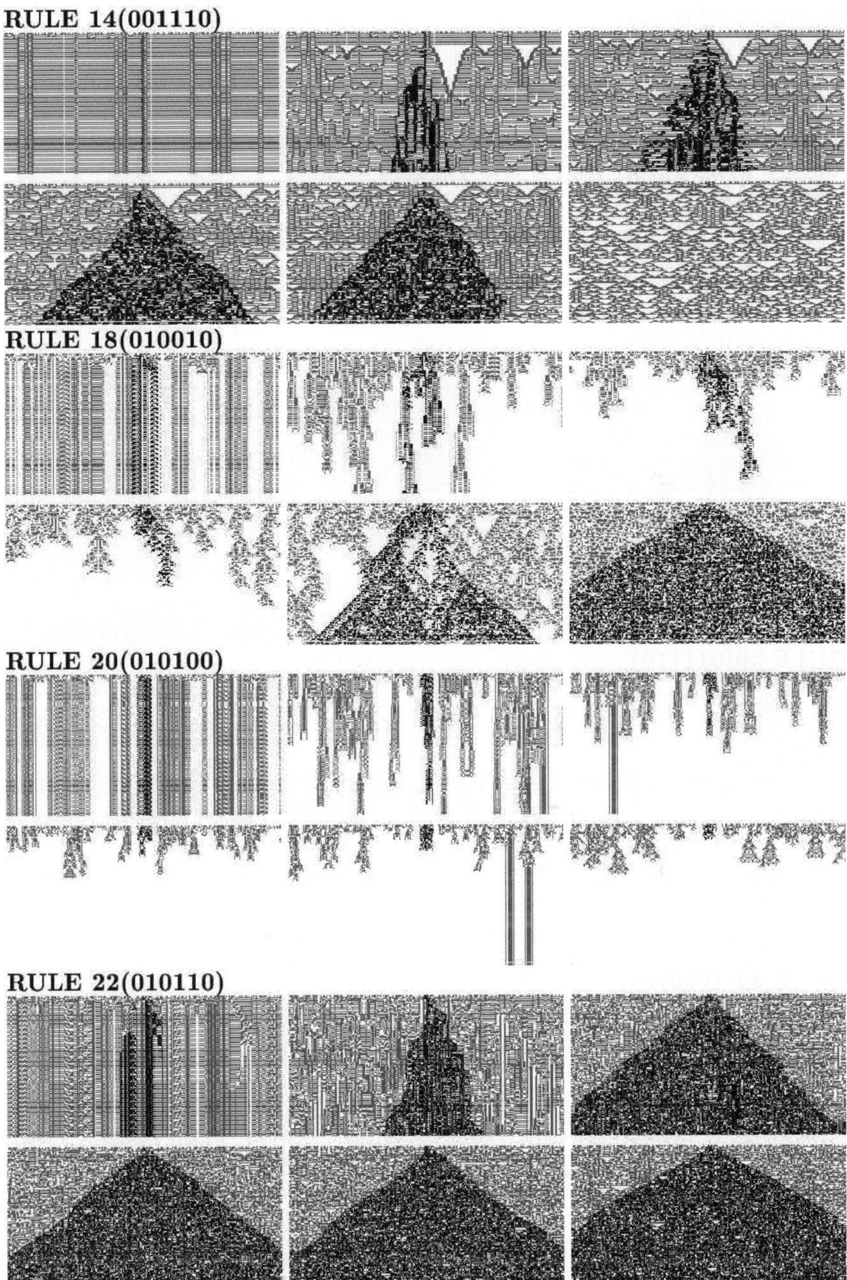

Fig. A.2   (*continued*)

RULE 26(011010)

RULE 28(011100)

RULE 30(011110)

RULE 34(100010)

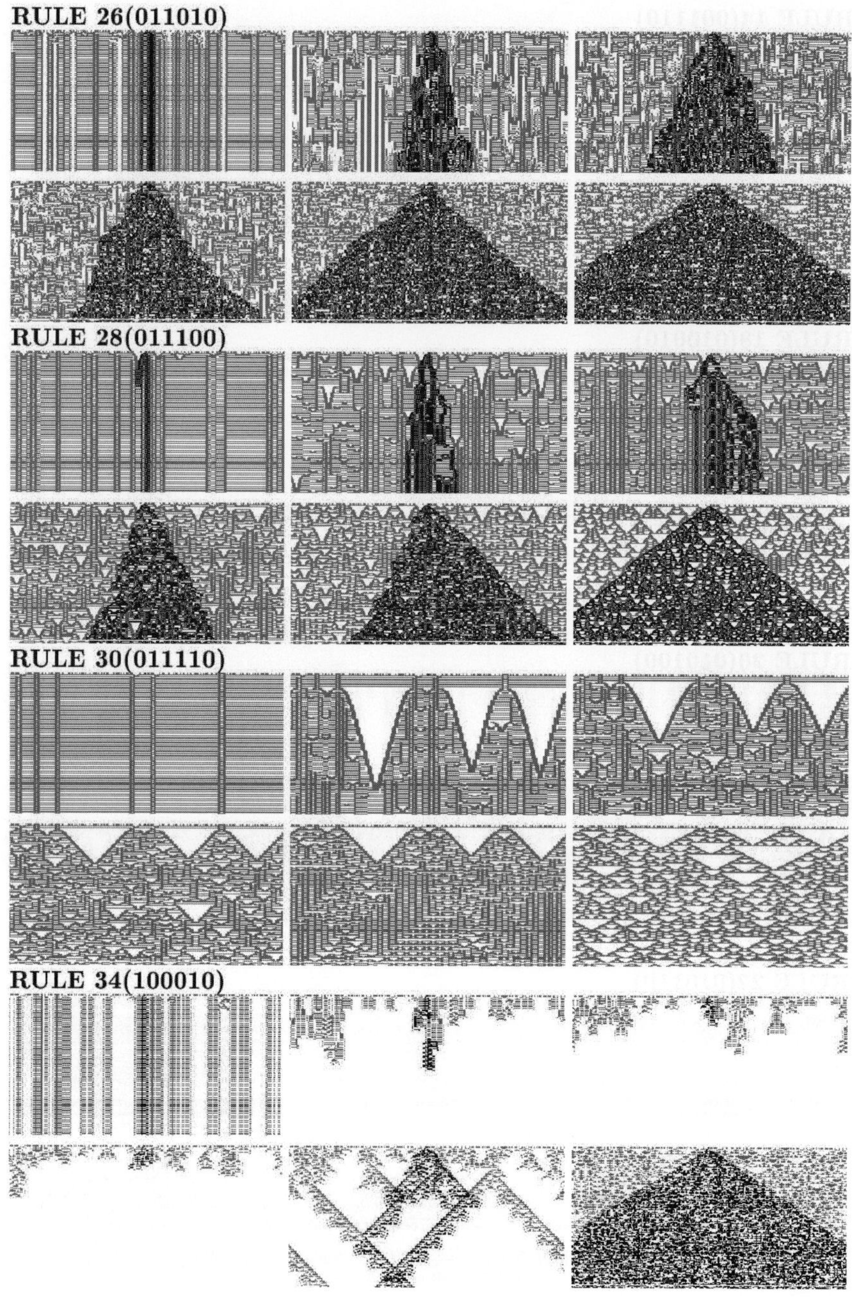

Fig. A.2   *(continued)*

RULE 38(100110)

RULE 42(101010)

RULE 44(101100)

RULE 46(101110)

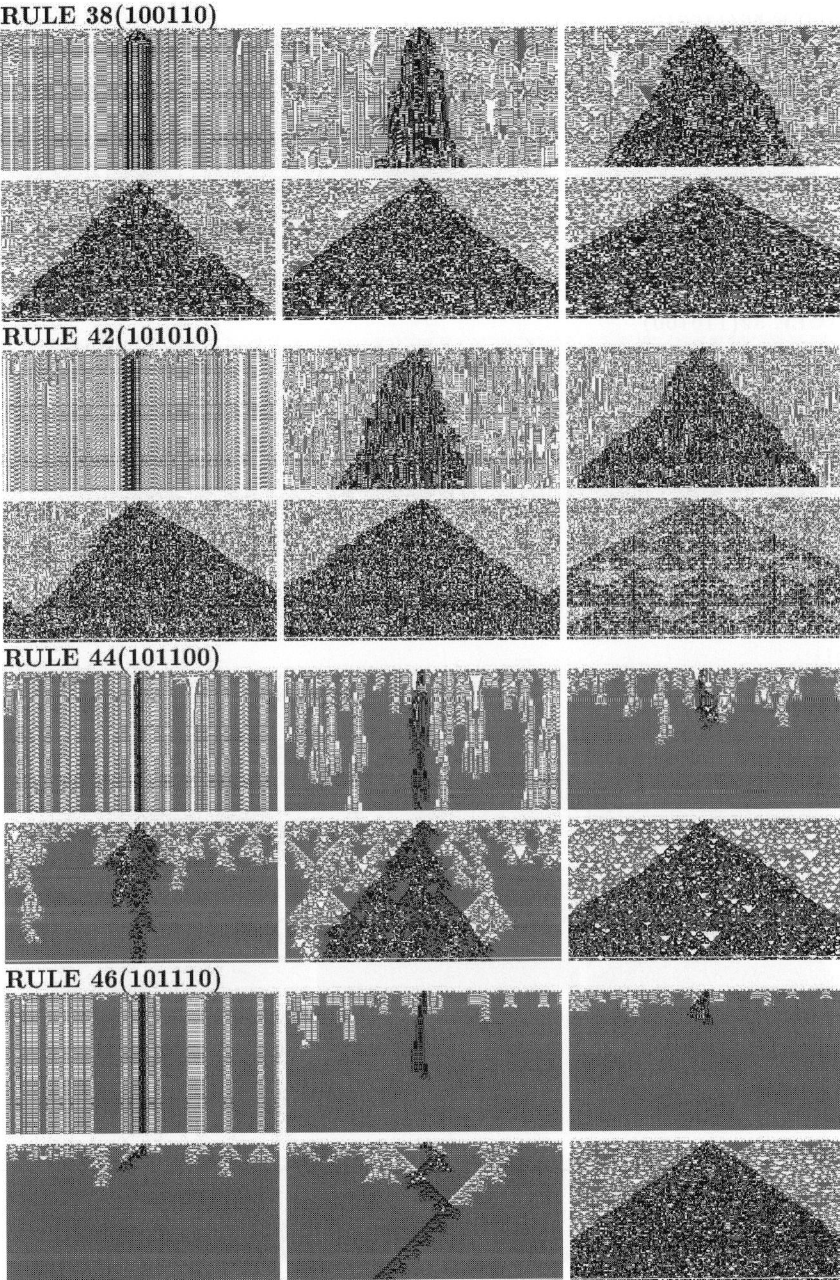

Fig. A.2 (*continued*)

RULE 50(110010)

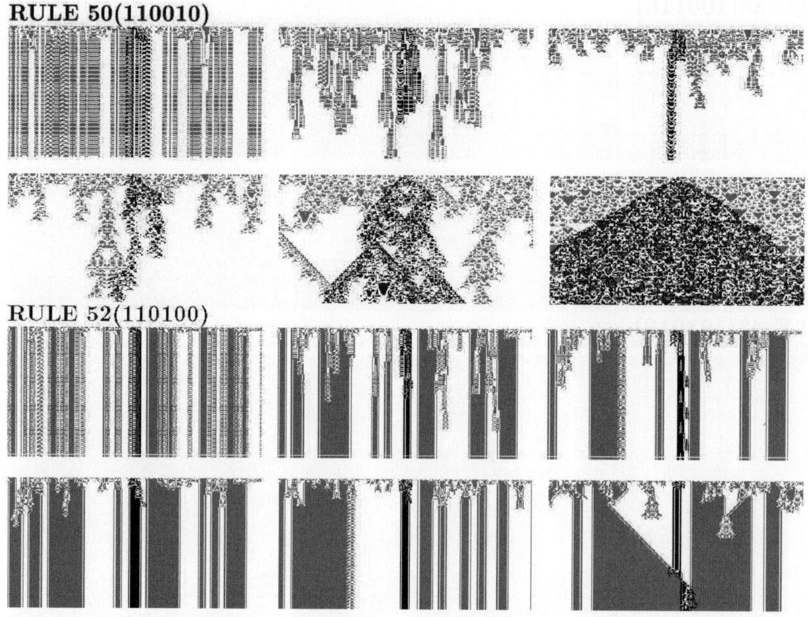

Fig. A.2   (*continued*)

RULE 42(101010)

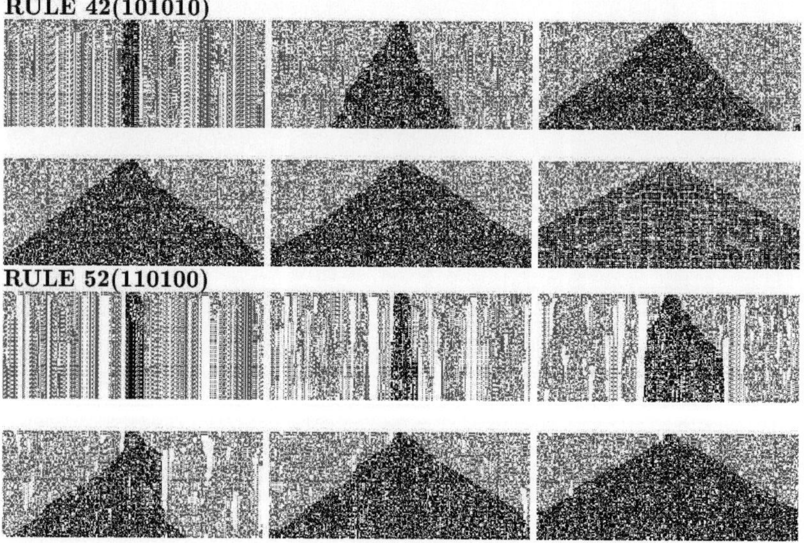

Fig. A.3   The reversible totalistic $k = r = 2$, 42 and 52 rules.

Appendix B

Dynamic with short-term memory

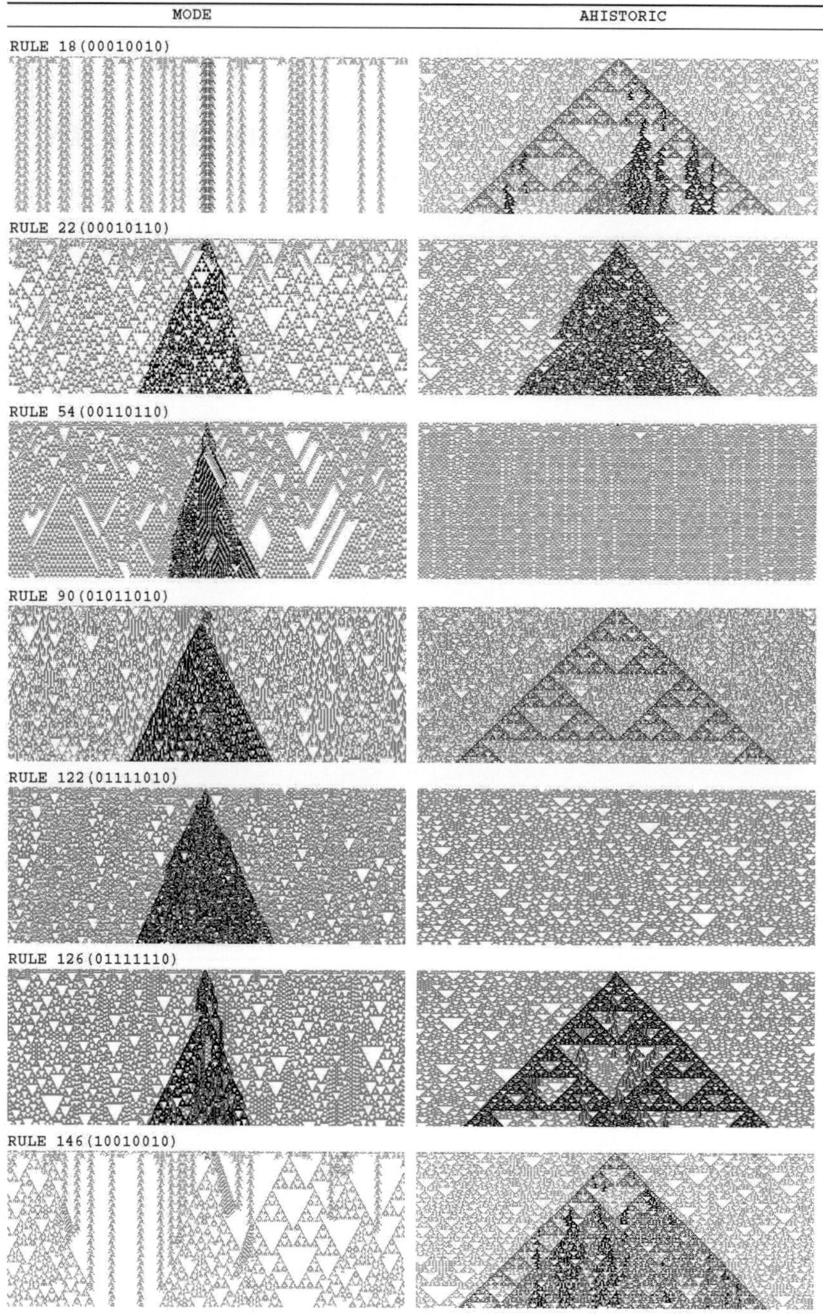

Fig. B.1   Elementary, legal rules significantly affected by the *mode* of the three-last-state memory model, starting at random.

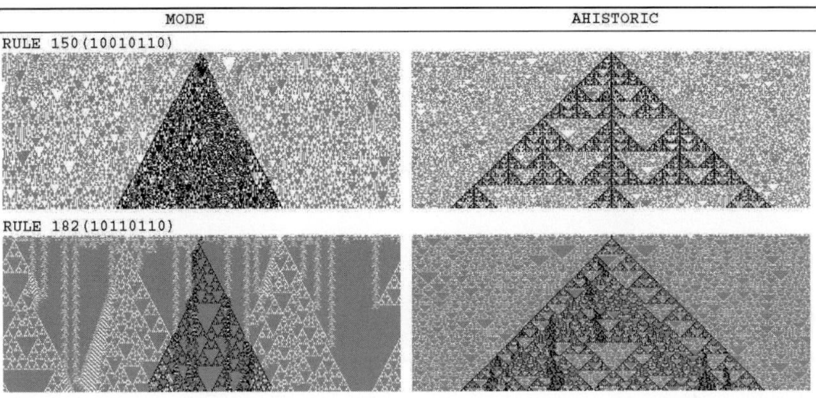

Fig. B.1 *(continued)*

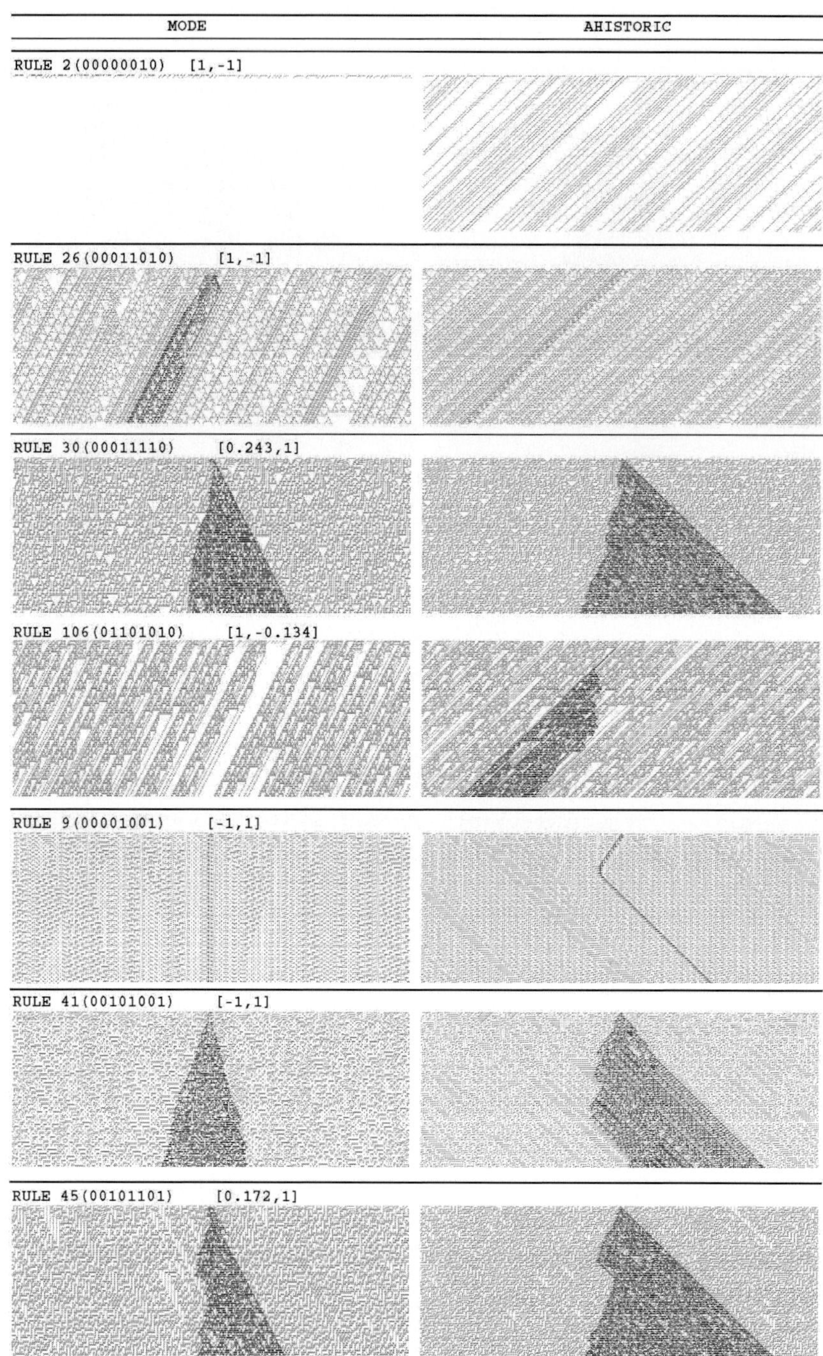

Fig. B.2 Some elementary semi-asymmetric rules significantly affected by the *mode* of the three-last-state memory in the initial scenario of Fig. B.1. The rules are grouped into equivalence classes.

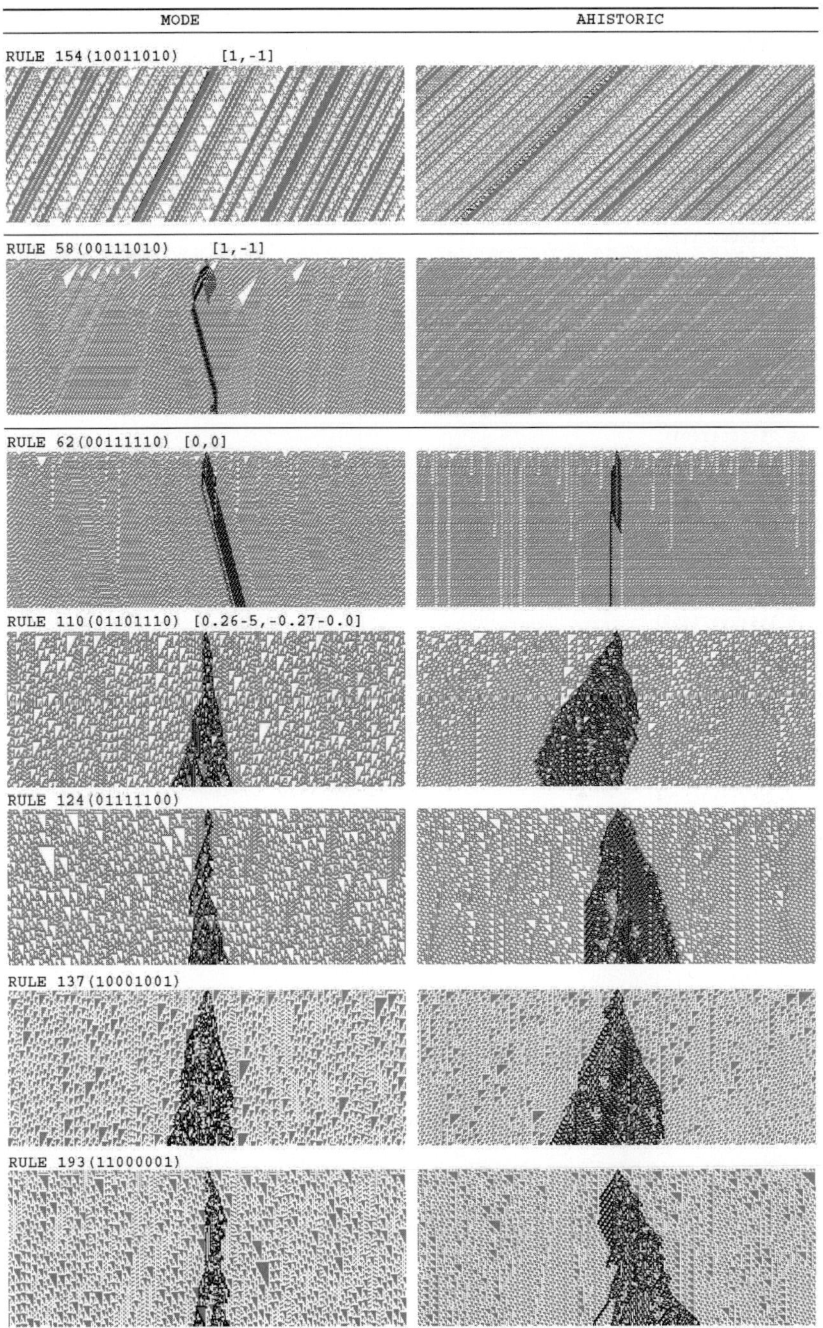

Fig. B.2   (*continued*)

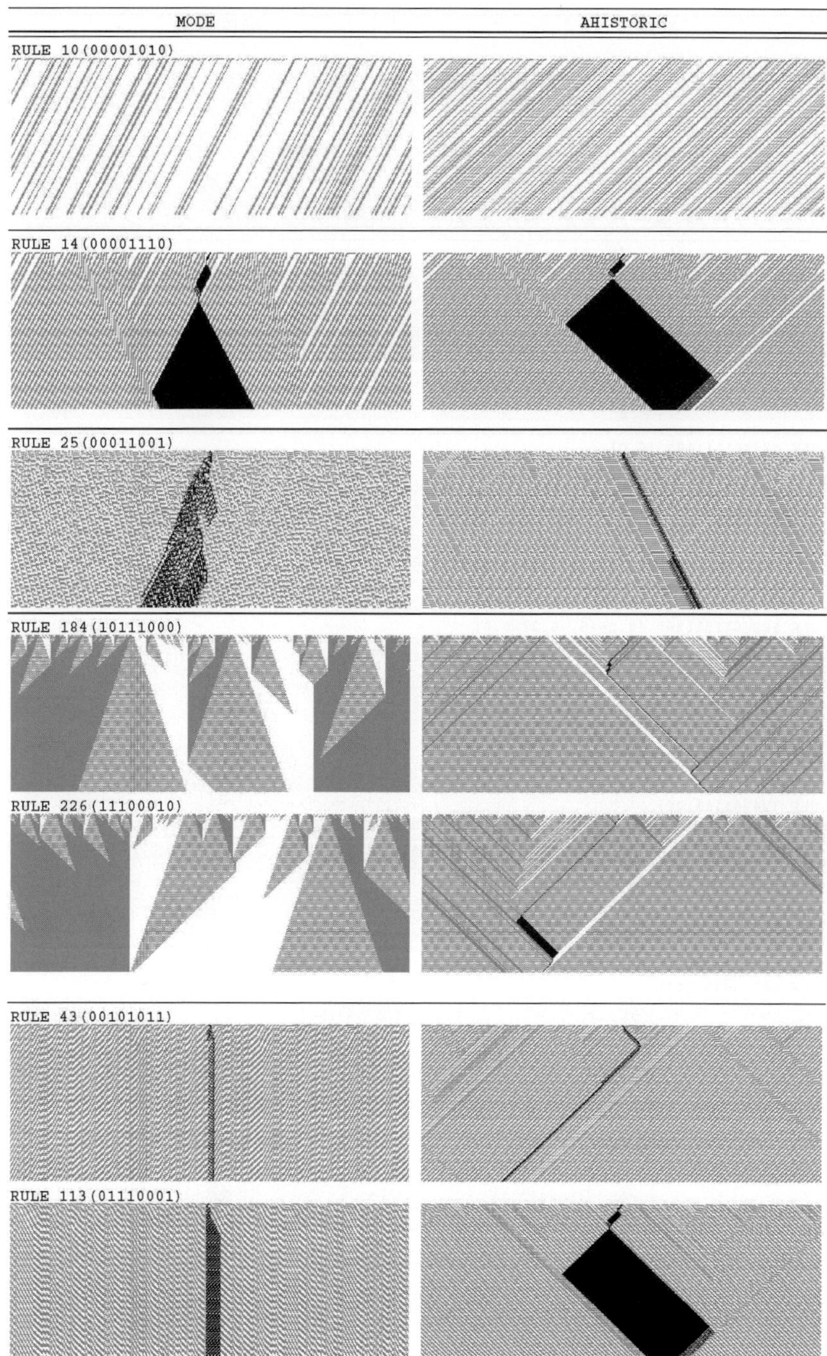

Fig. B.3   Some elementary fully asymmetric rules significantly affected by the *mode* of the three-last-state memory.

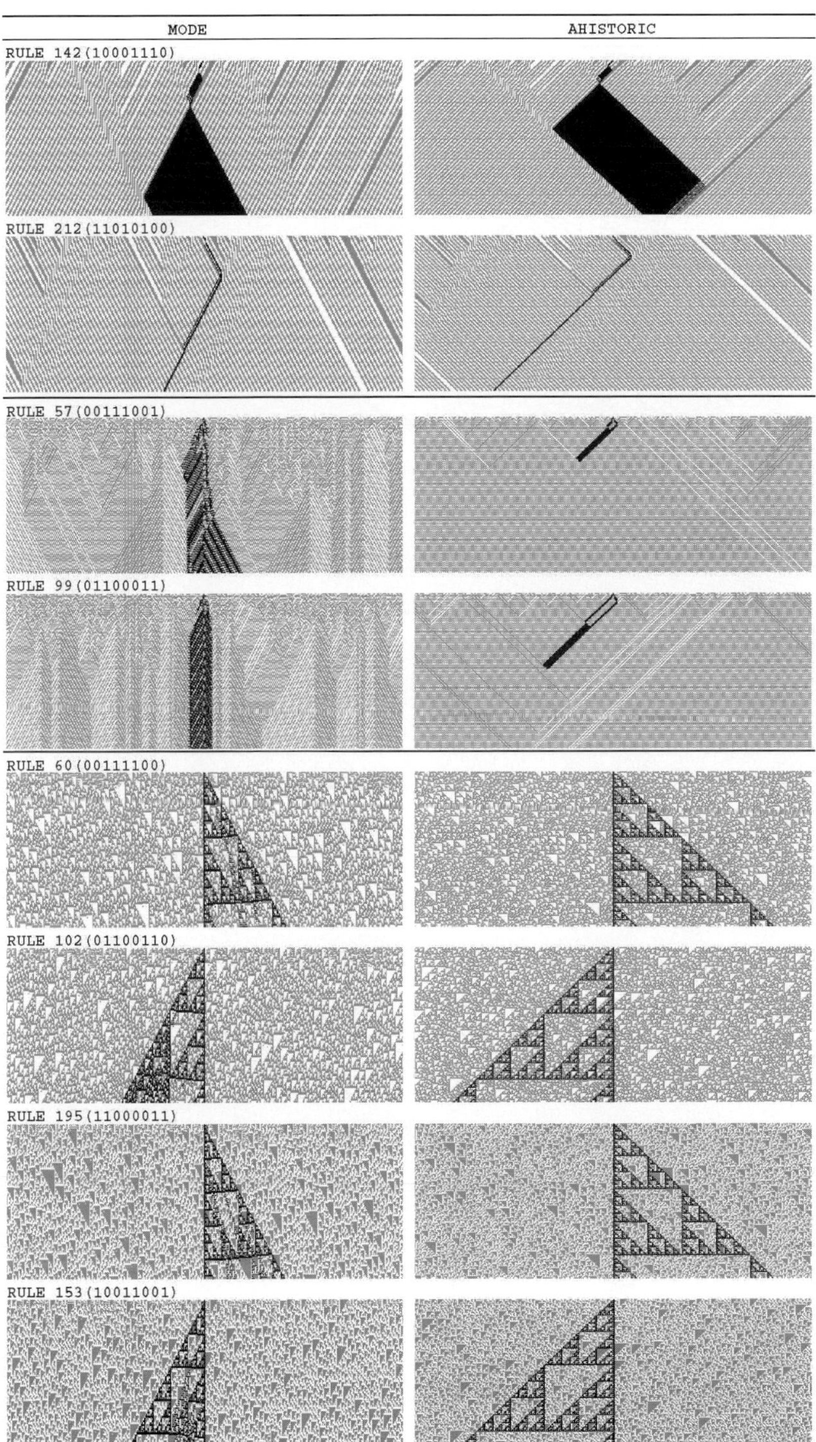

Fig. B.3 *(continued)*

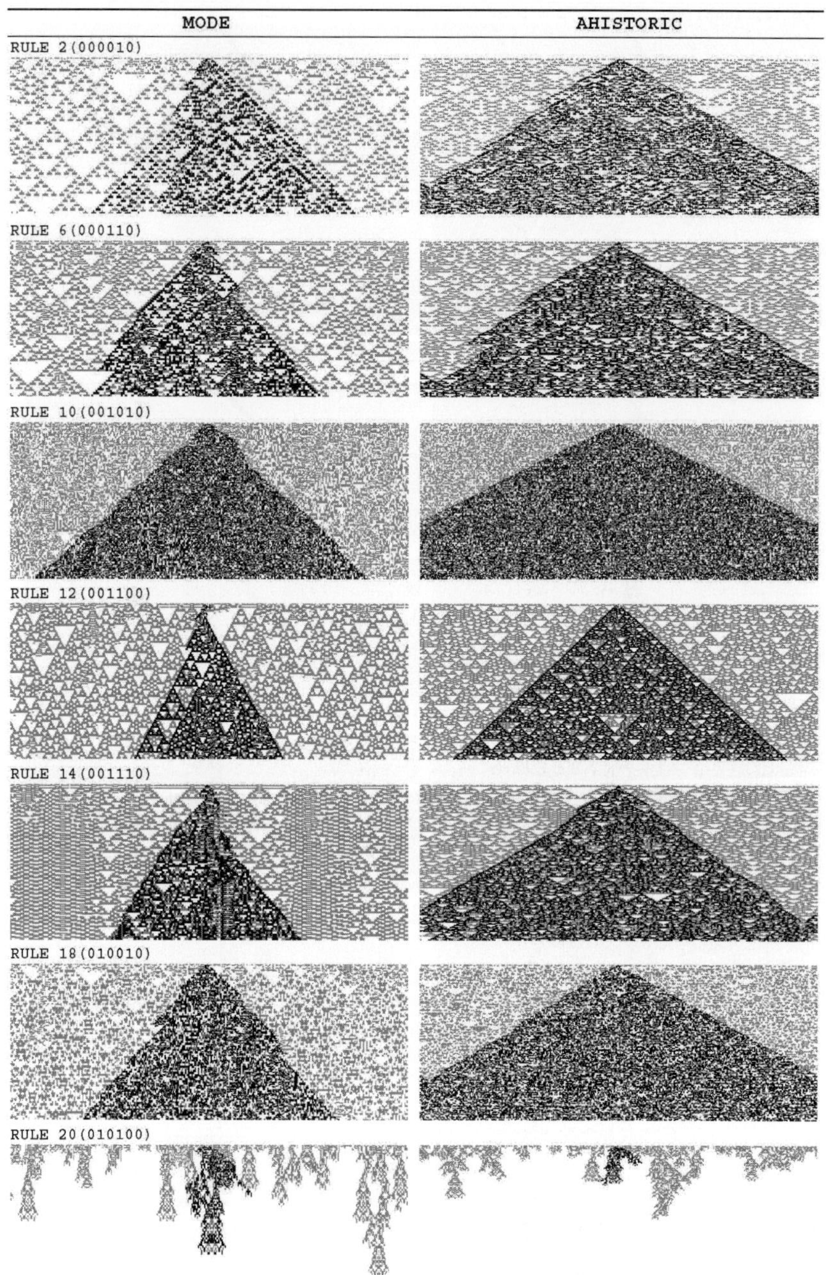

Fig. B.4   Totalistic $k = r = 2$ quiescent rules significantly affected by the *mode* of the three-last-state memory.

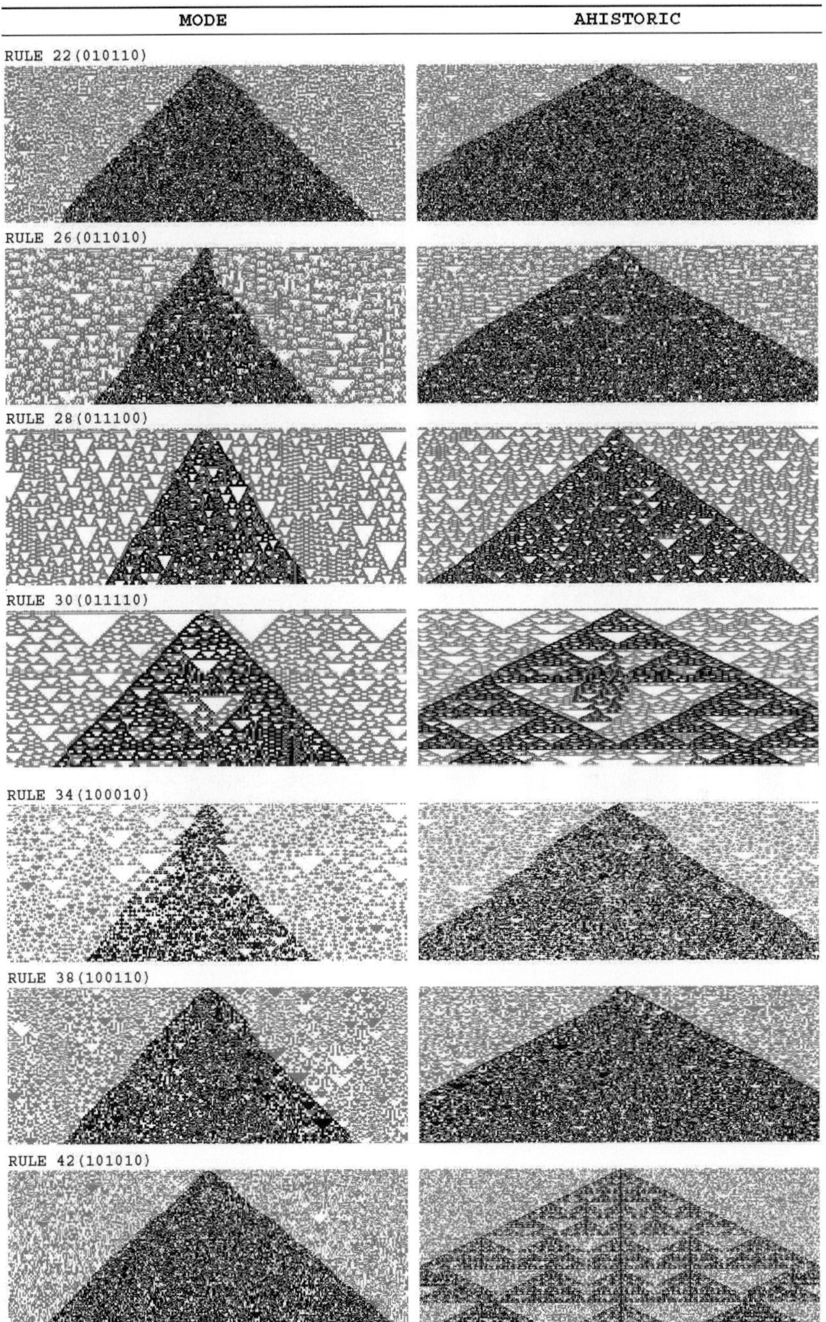

Fig. B.4 *(continued)*

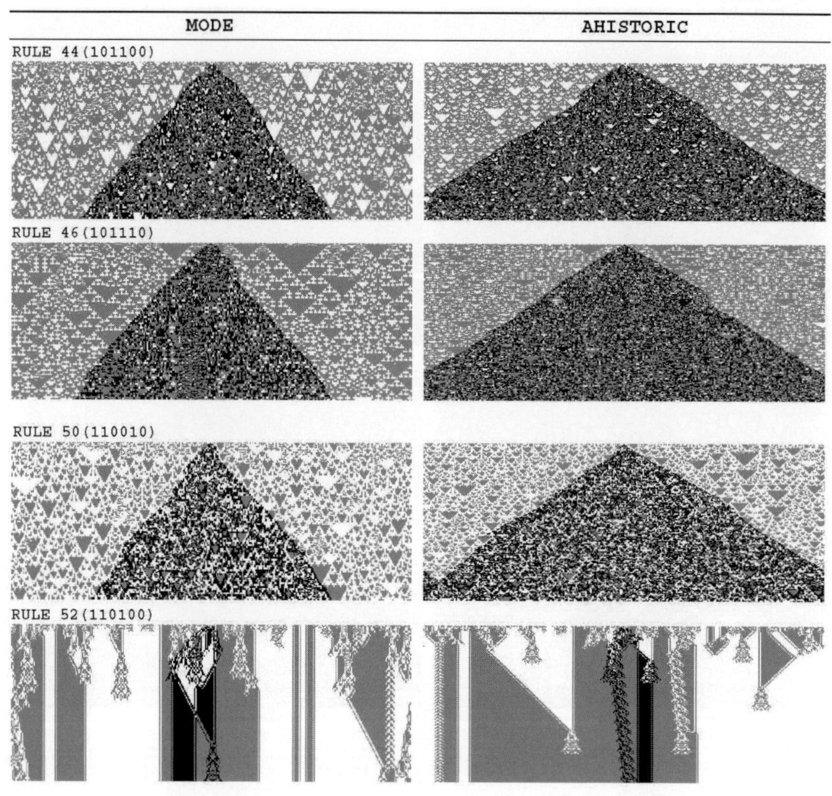

Fig. B.4   (*continued*)

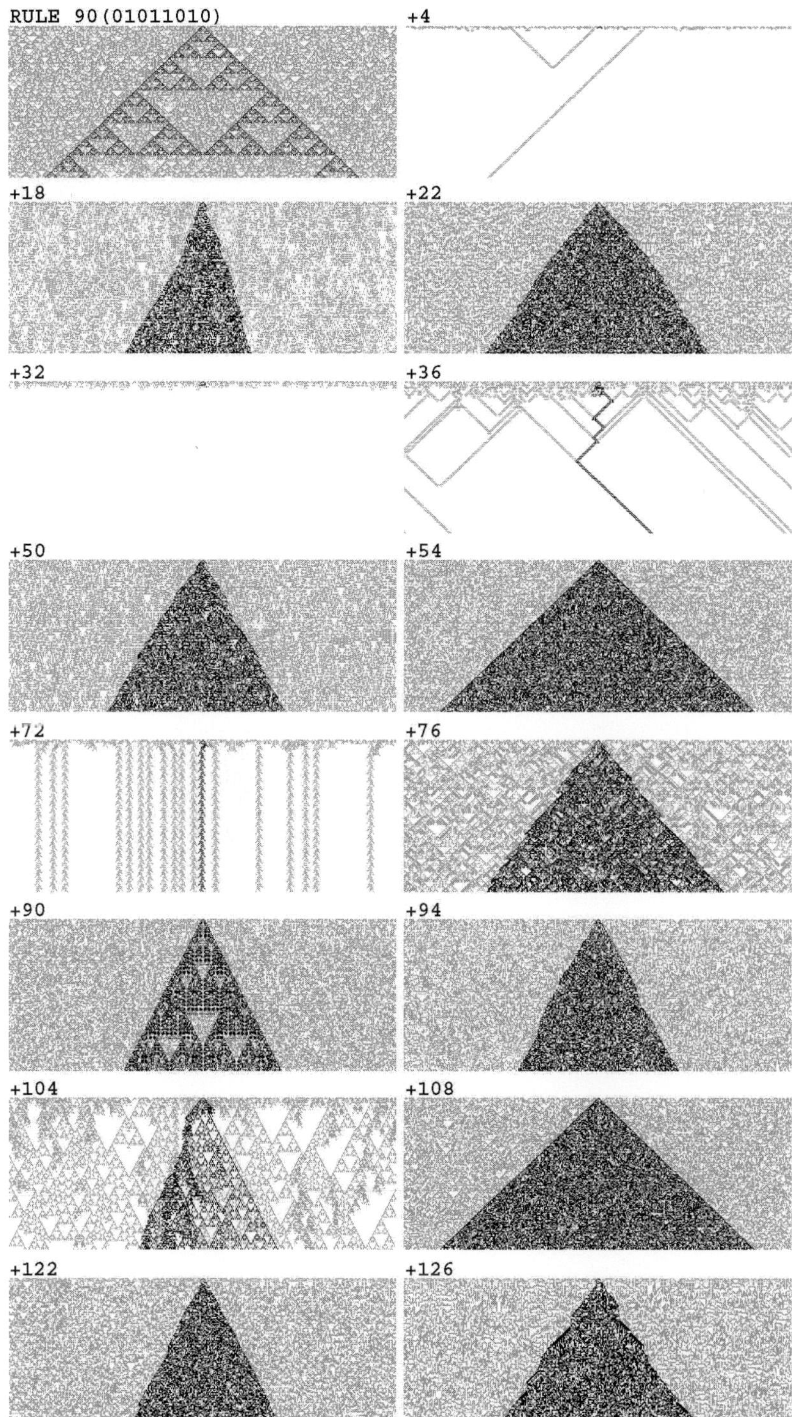

Fig. B.5   Effect of elementary legal memory rules on rule 90 starting at random.

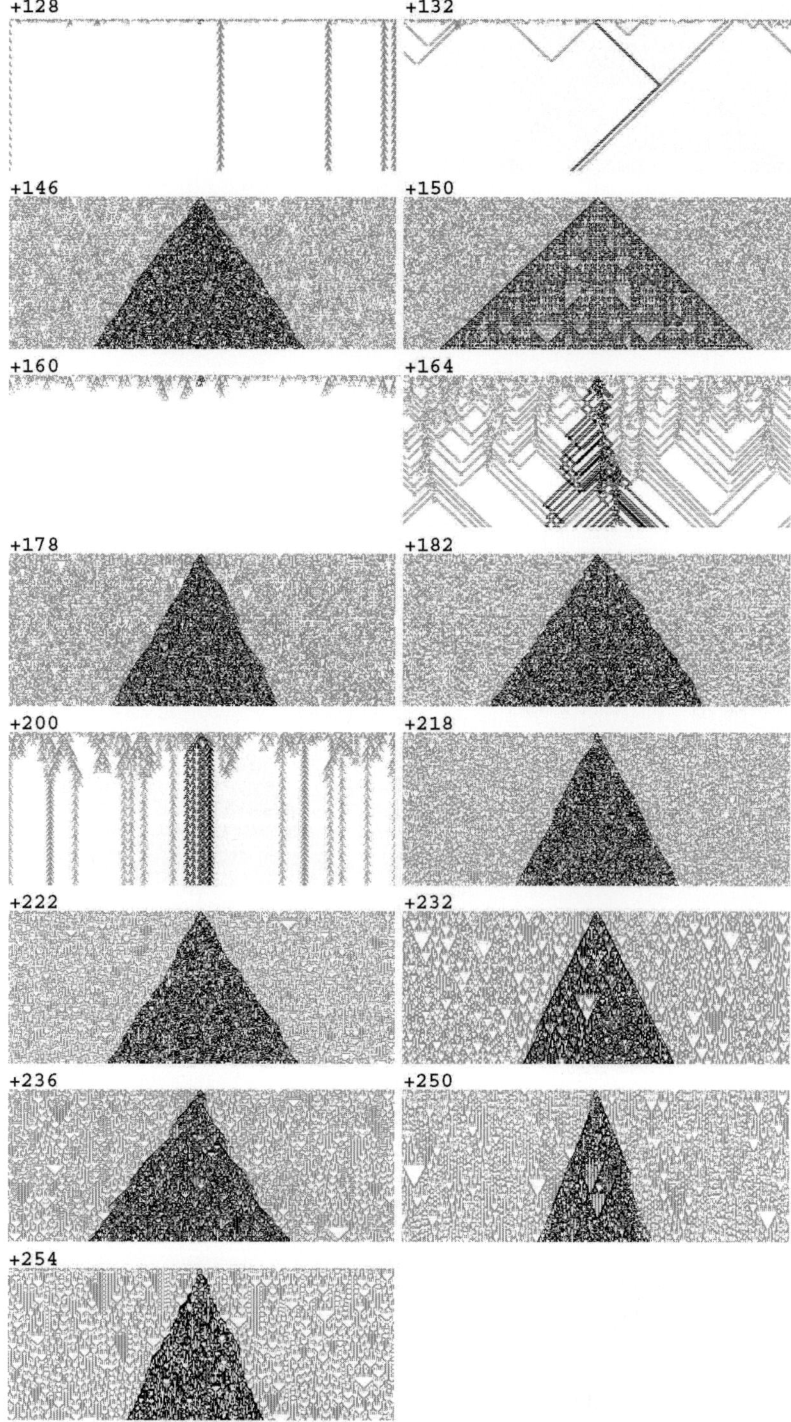

Fig. B.5   (*continued*)

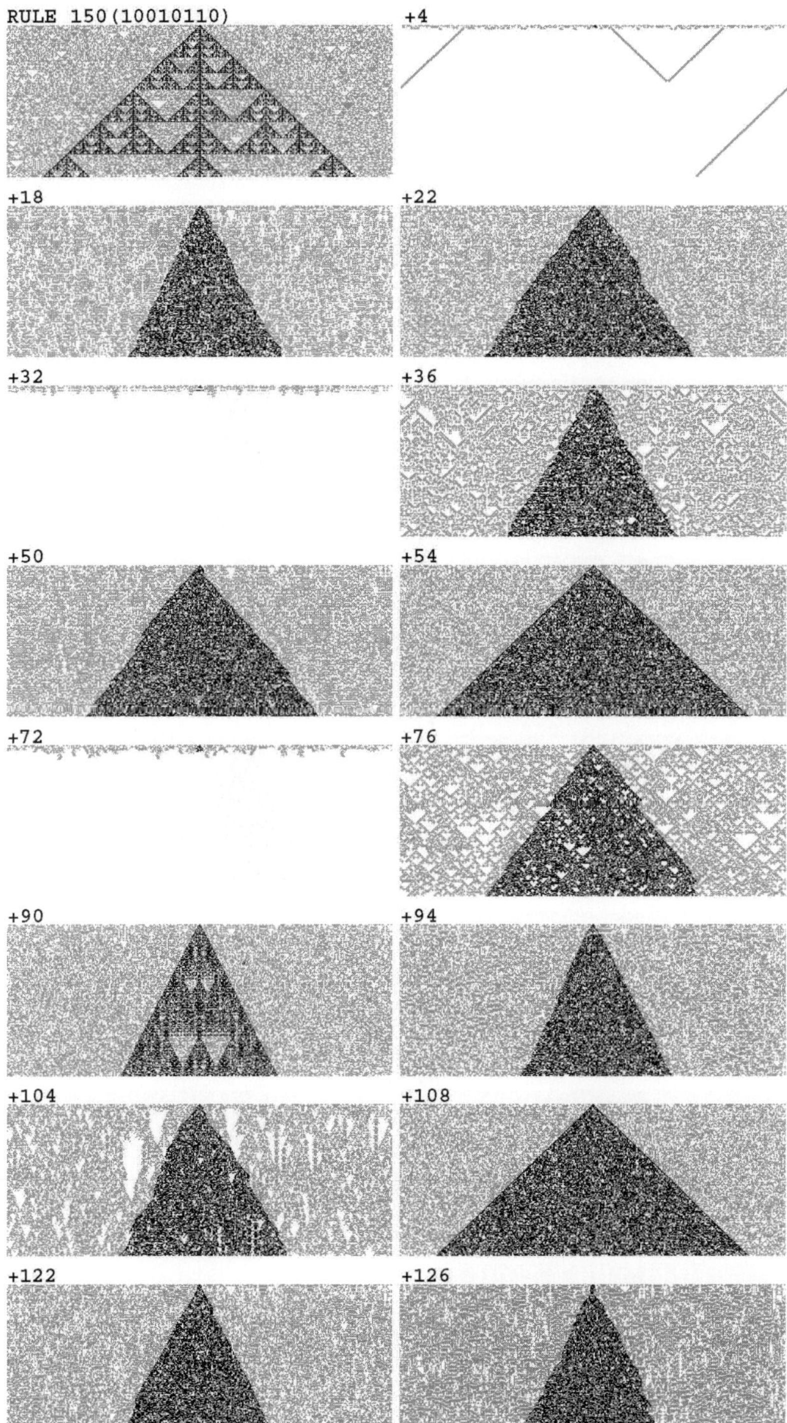

Fig. B.6   Effect of elementary legal memory rules on rule 150 starting at random.

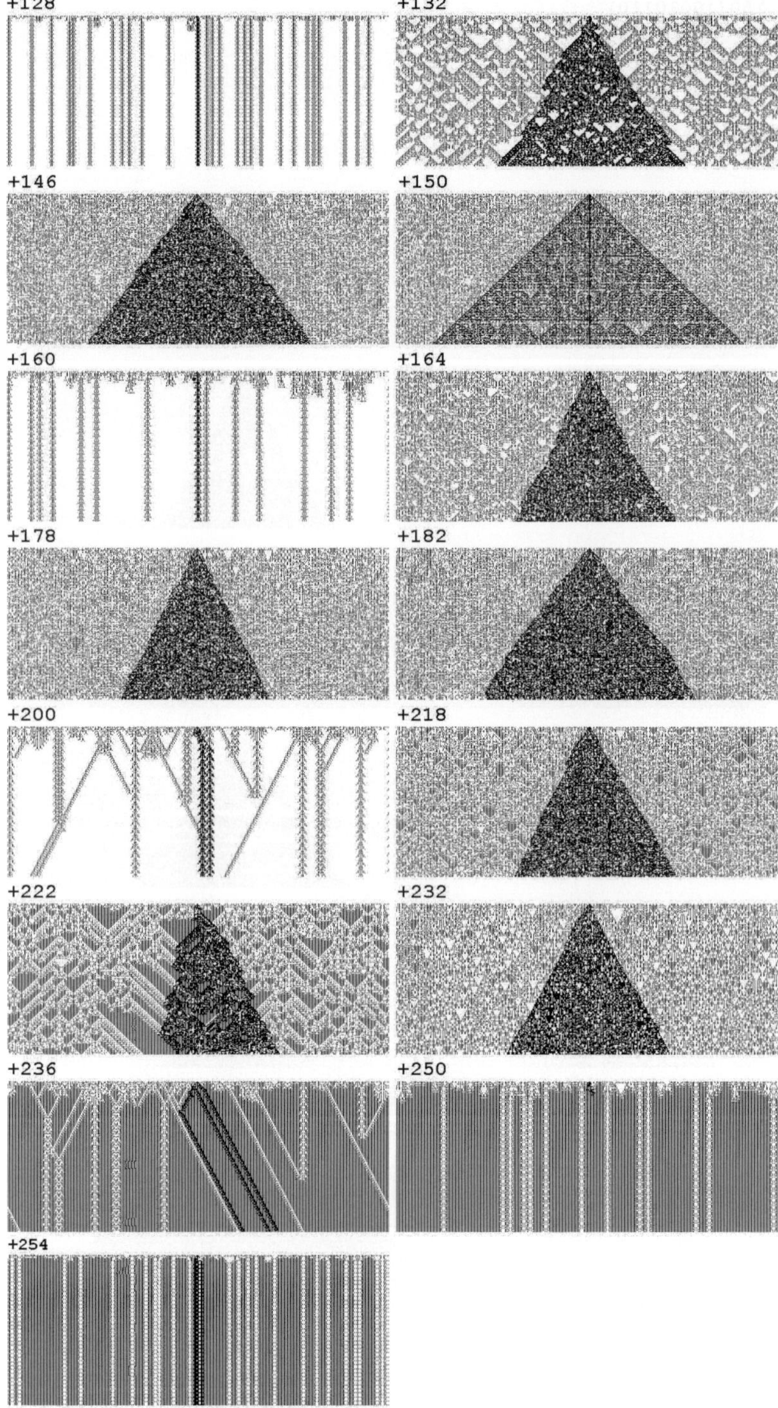

Fig. B.6   (*continued*)

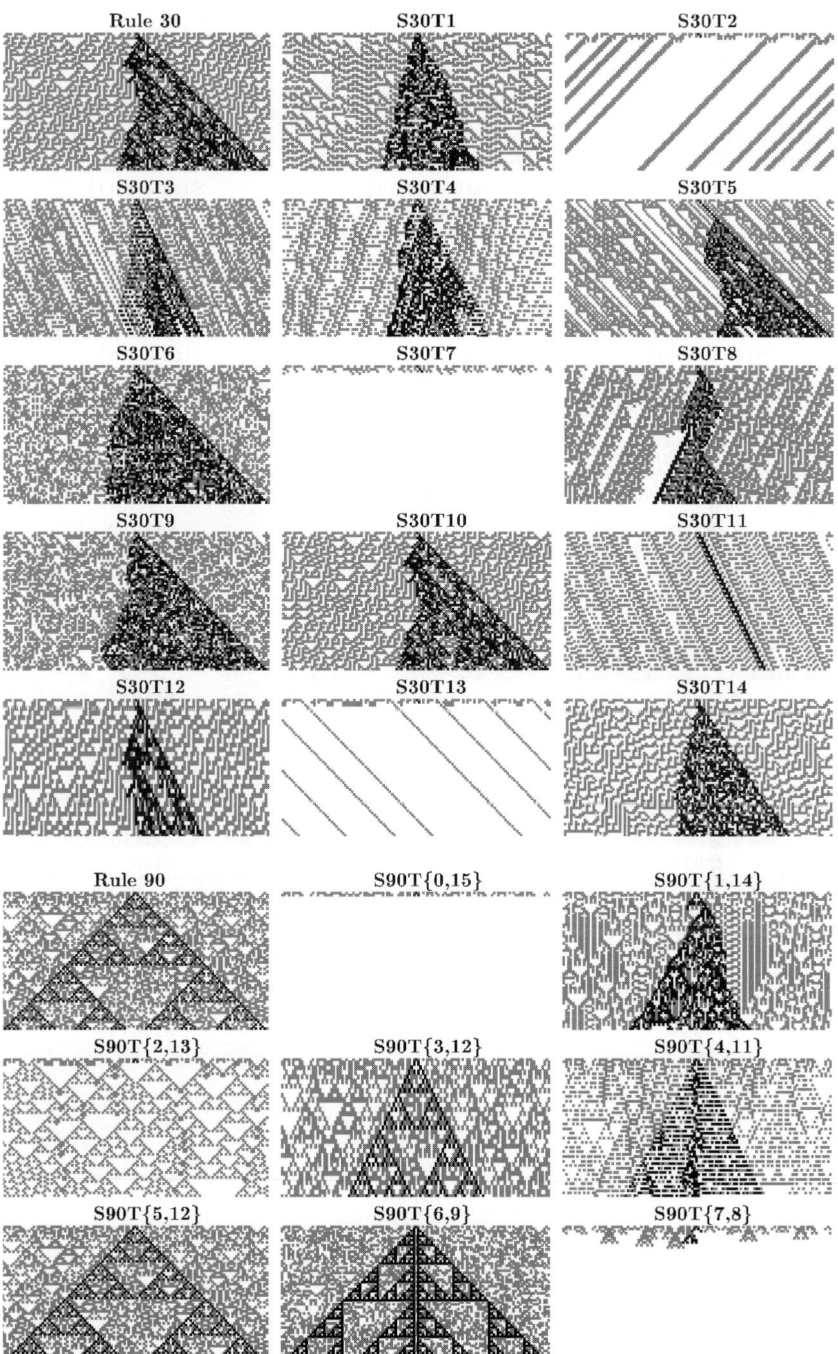

Fig. B.7   Elementary rules with two-last-state memory.

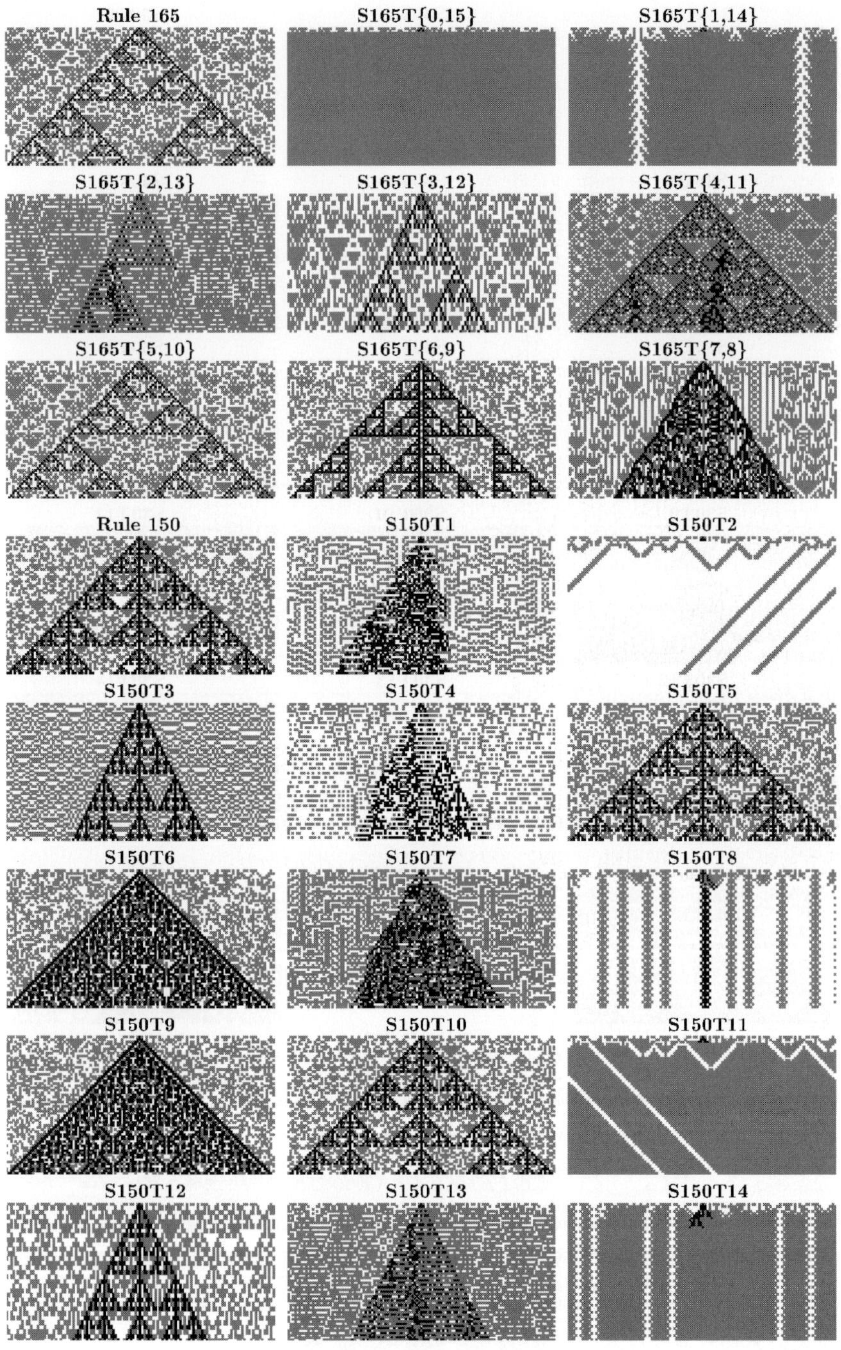

Fig. B.7   (*continued*)

Fig. B.8   Elementary rules with unlimited trailing parity memory.

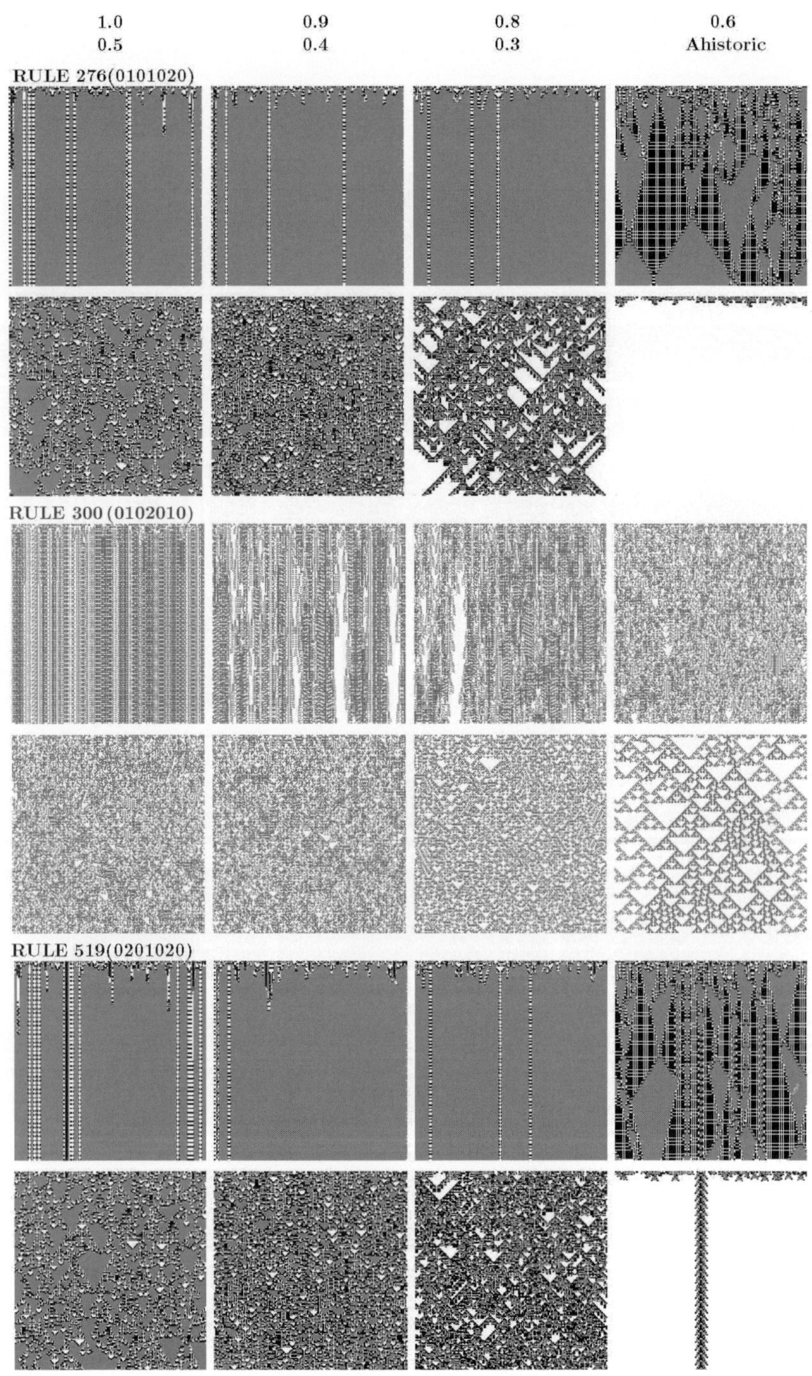

Fig. B.9　Parity $k = 3$ rules starting at random.

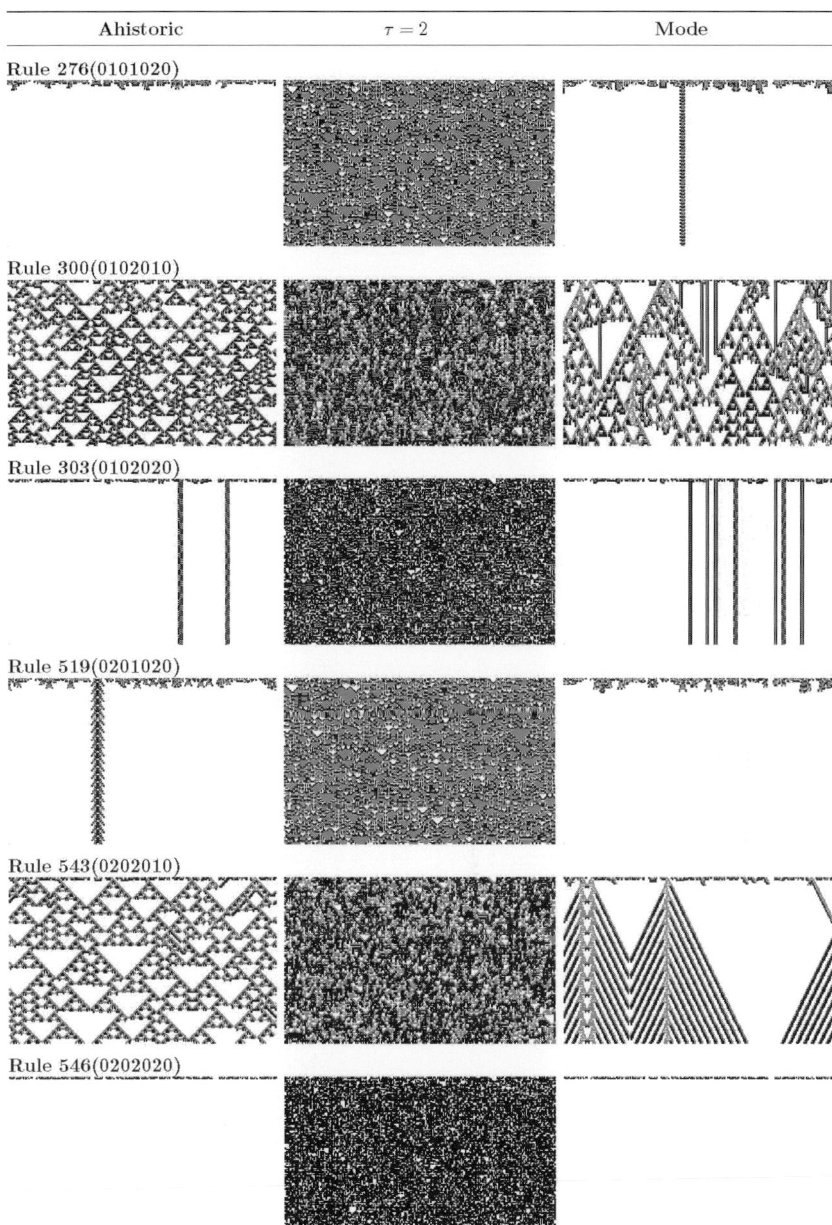

Fig. B.10 Parity $k = 3$ rules. Ahistoric model, mode of the last-three-state and average of the last-two-state memory models starting at random.

Appendix C

Heterogeneous and coupled networks

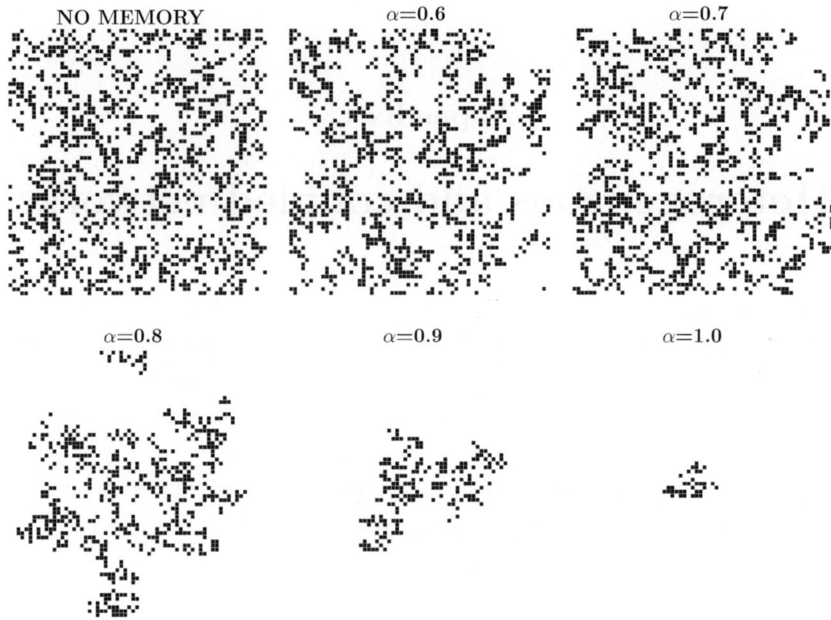

Fig. C.1   Damage at $T = 200$ in the regular graph scenario of Fig. 10.12 .

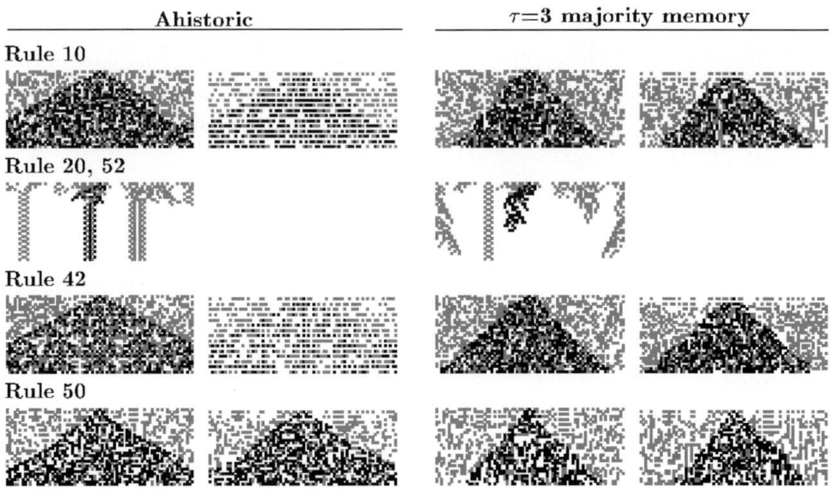

Fig. C.2   Coupled K=5 CA starting at random in the left lattice. Ahistoric and mode of the last-three-state memory models.

Ahistoric model

$\tau=3$ parity memory

Fig. C.3 Additivity of the linear rule 150 in the coupled (overlined left) and in the one-layer (right) scenarios. Ahistoric (upper group of patterns) and $\tau=3$ parity (lower group of patterns) memories.

 Discrete Systems with Memory

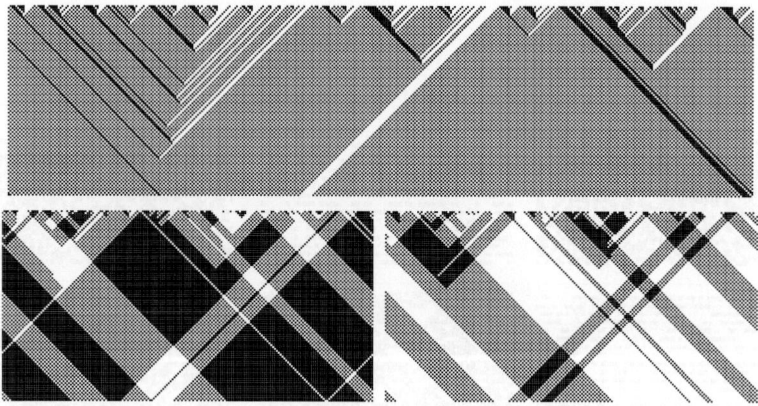

Fig. C.4   The one-layer and coupled elementary rule 226.

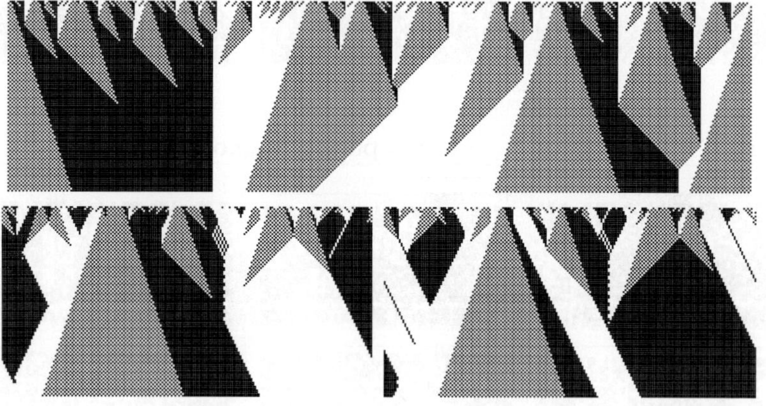

Fig. C.5   The one-layer and coupled elementary rule 226 with $\tau=3$ majority memory.

Fig. C.6   The one-layer and coupled elementary rule 226 with $\tau=3$ majority memory from $\rho_0=0.25$.

Ahistoric

$\tau=3$ majority memory

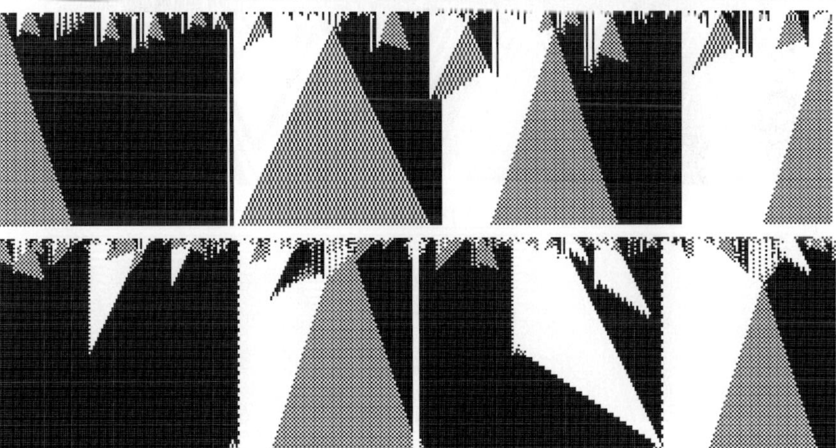

Fig. C.7 The one-layer (upper) and coupled (lower) GKL rule. Ahistoric and $\tau=3$ majority memories.

Fig. C.8   The one-layer and coupled elementary rule 99.

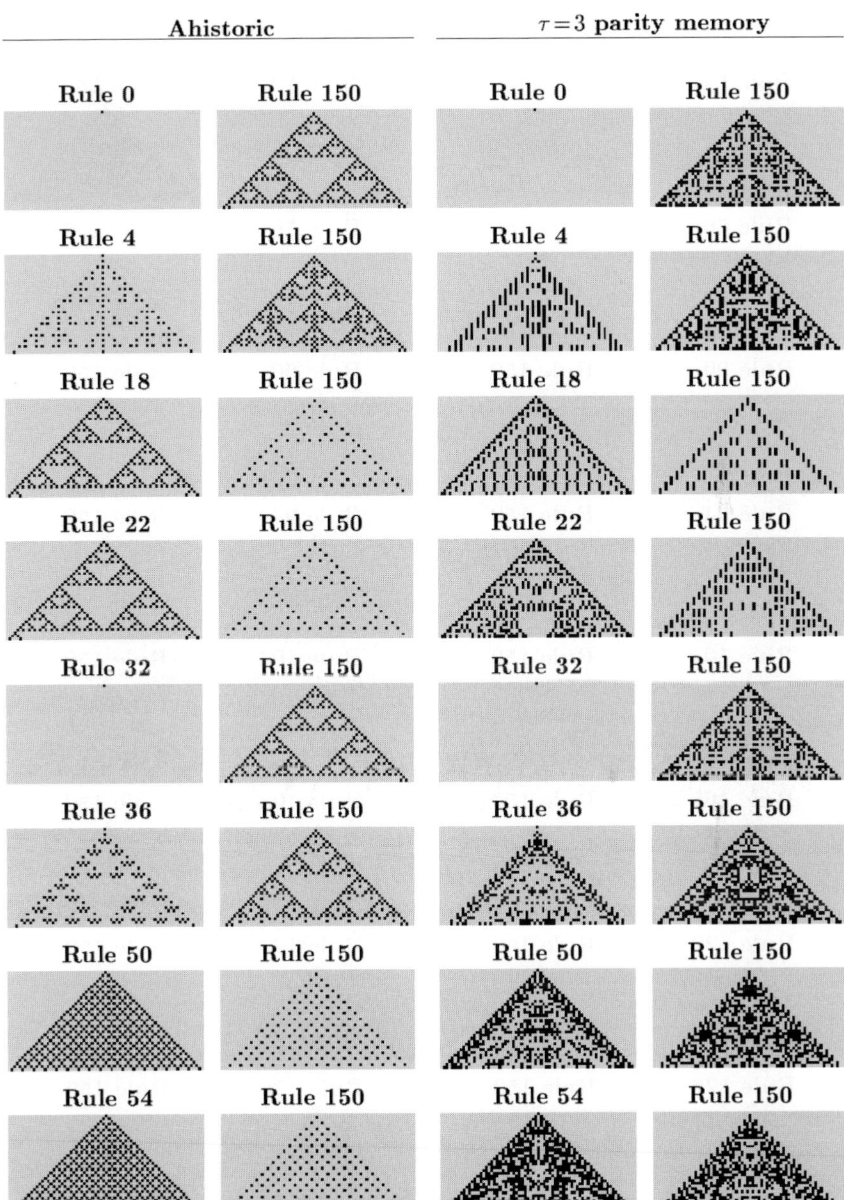

Fig. C.9 Coupled legal rules on the left layer and rule 150 on the right one. Ahistoric and parity of the last-three-state memory models.

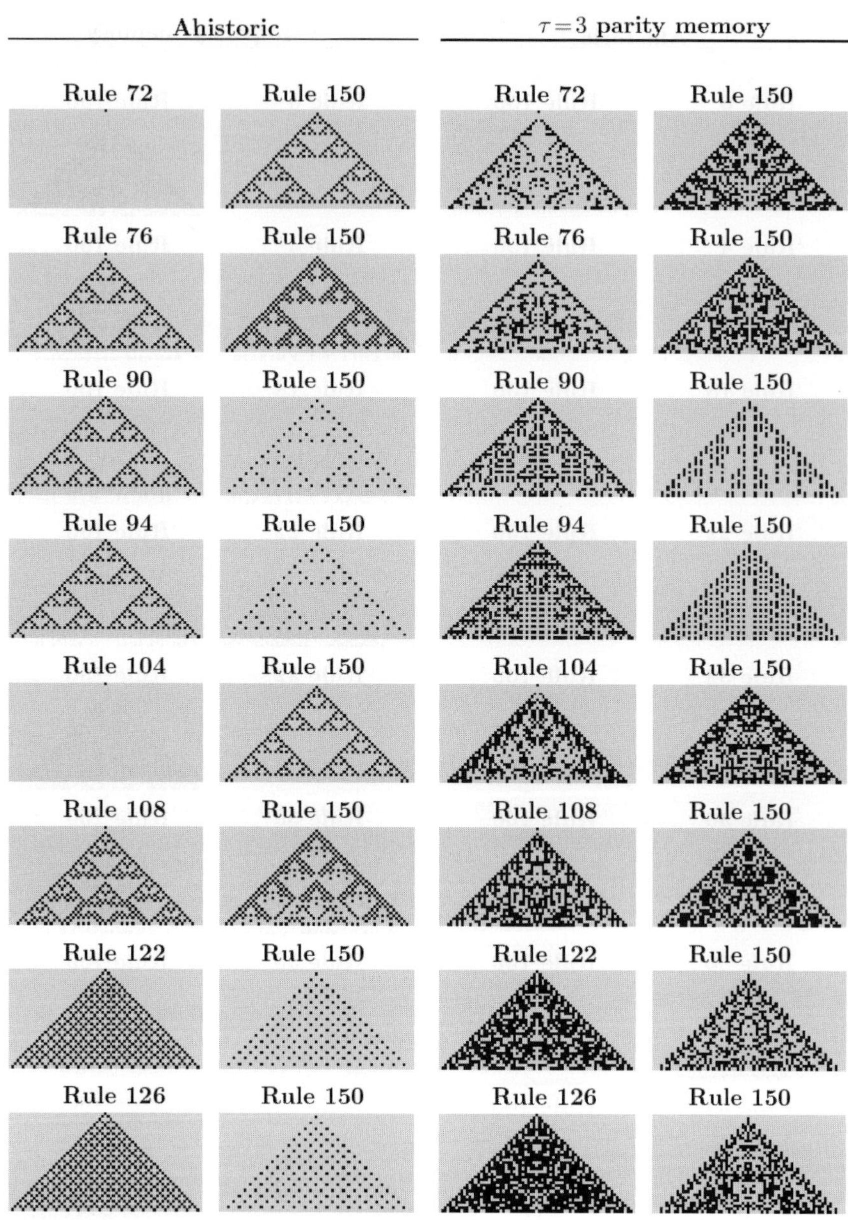

Fig. C.9   (*continued*)

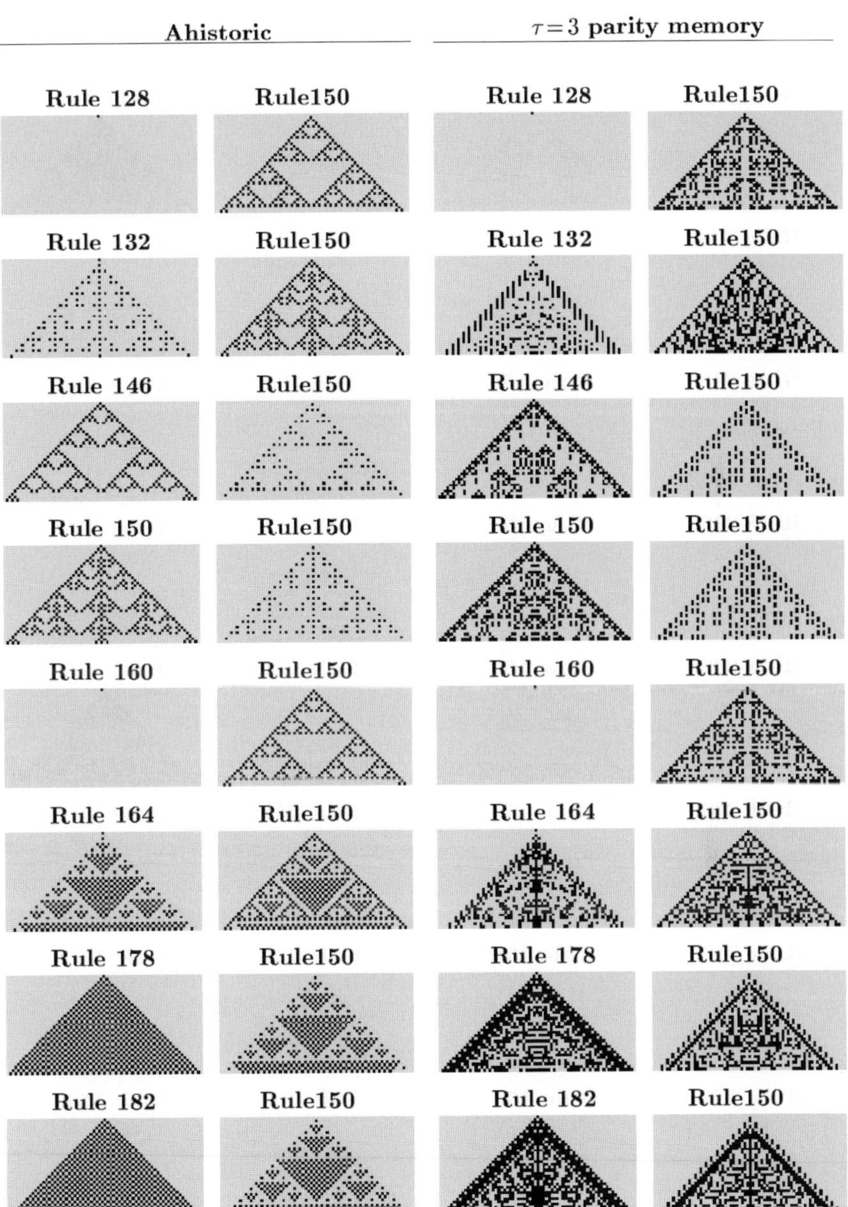

Fig. C.9   (*continued*)

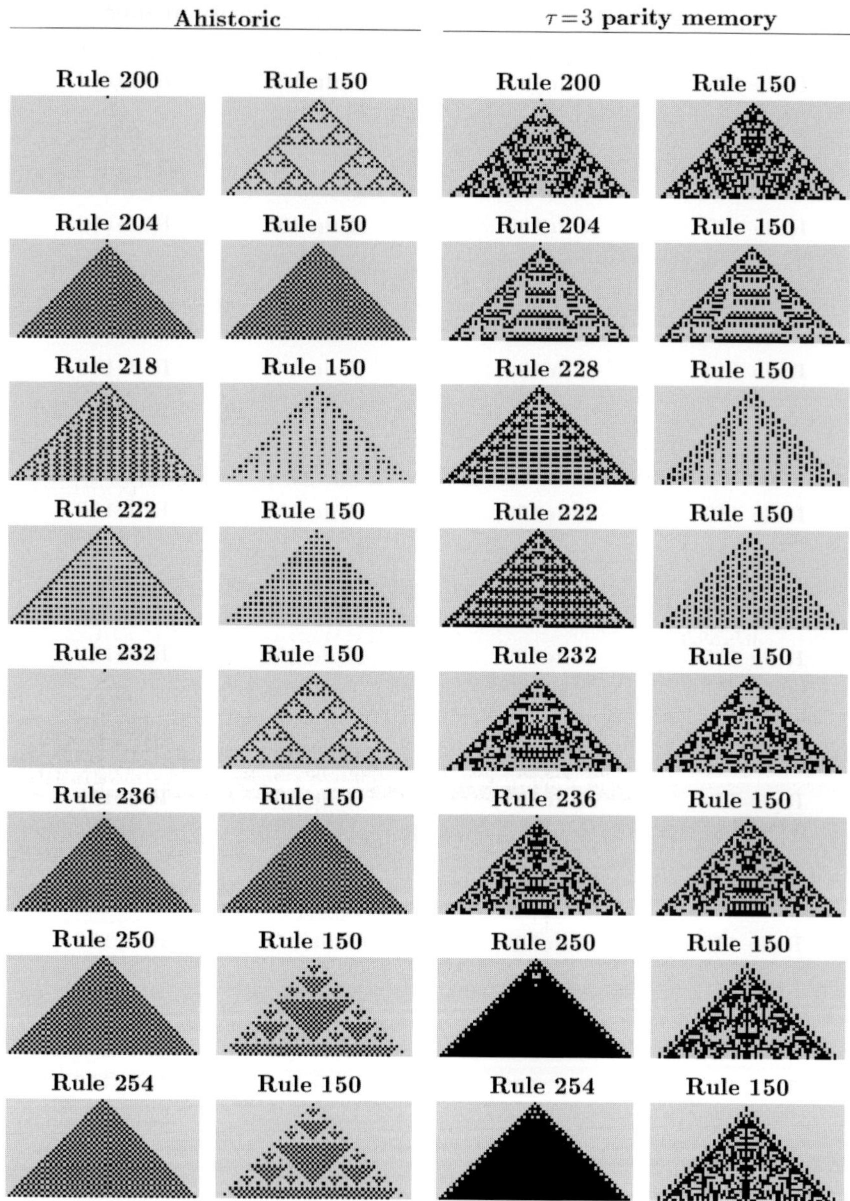

Fig. C.9   (*continued*)

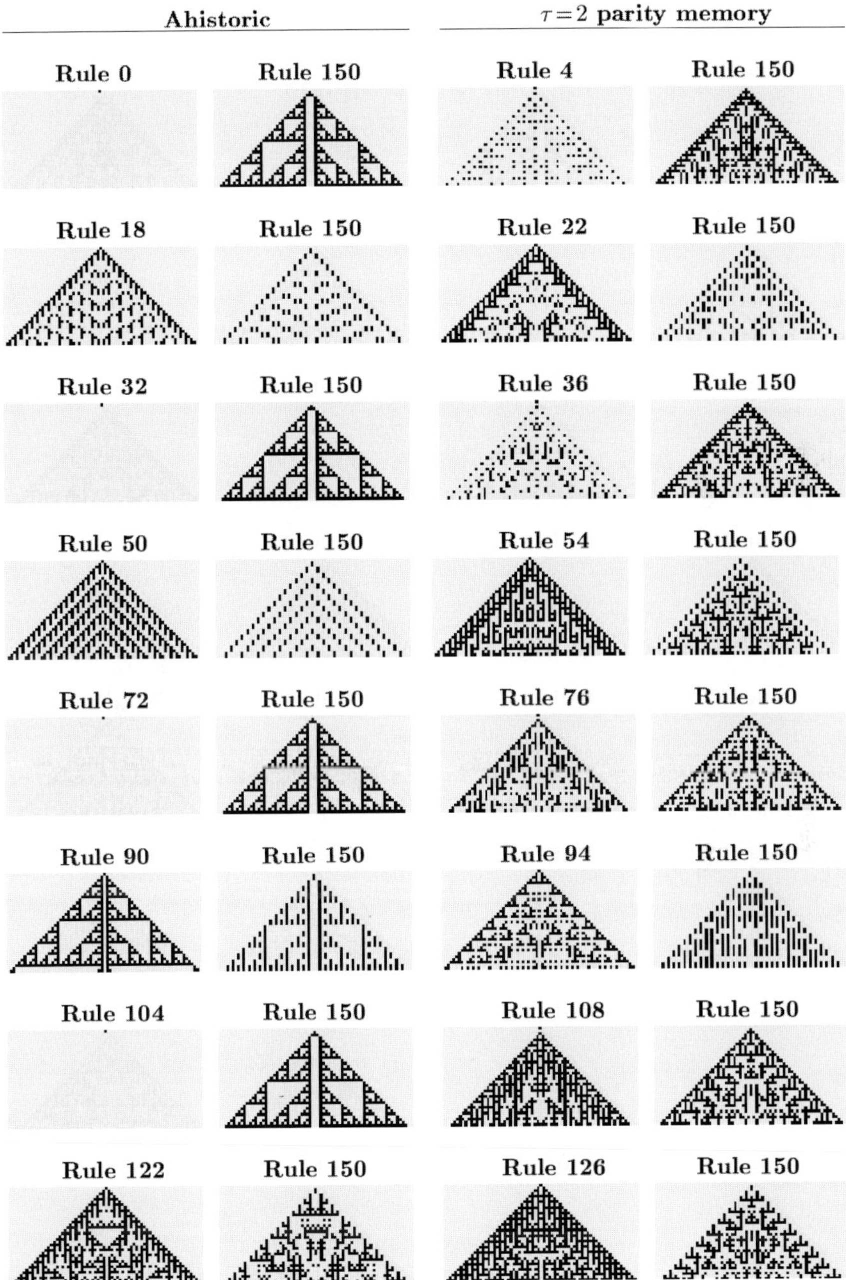

Fig. C.10   Coupled legal rules on the left layer and rule 150 on the right one. Ahistoric and parity of the last-two-state memory models.

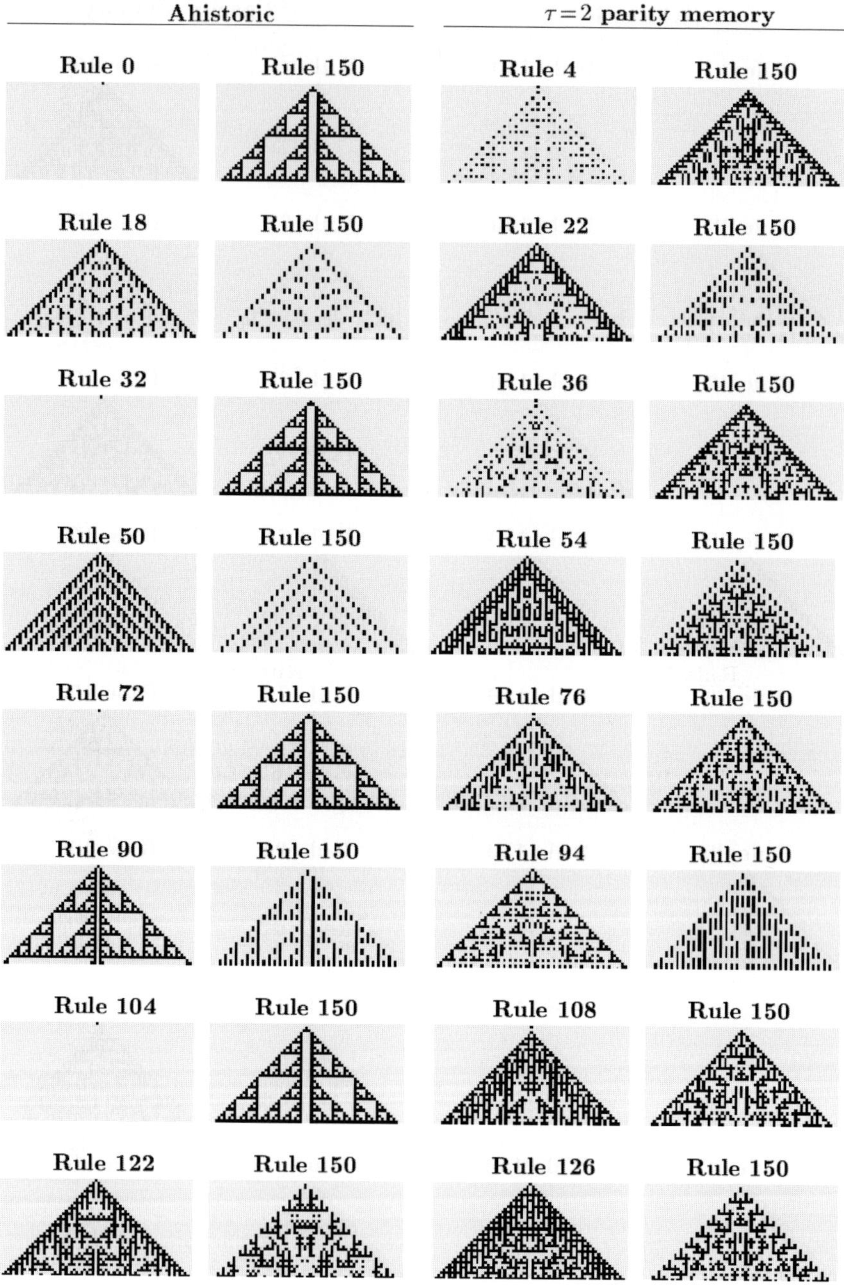

Fig. C.10   (*continued*)

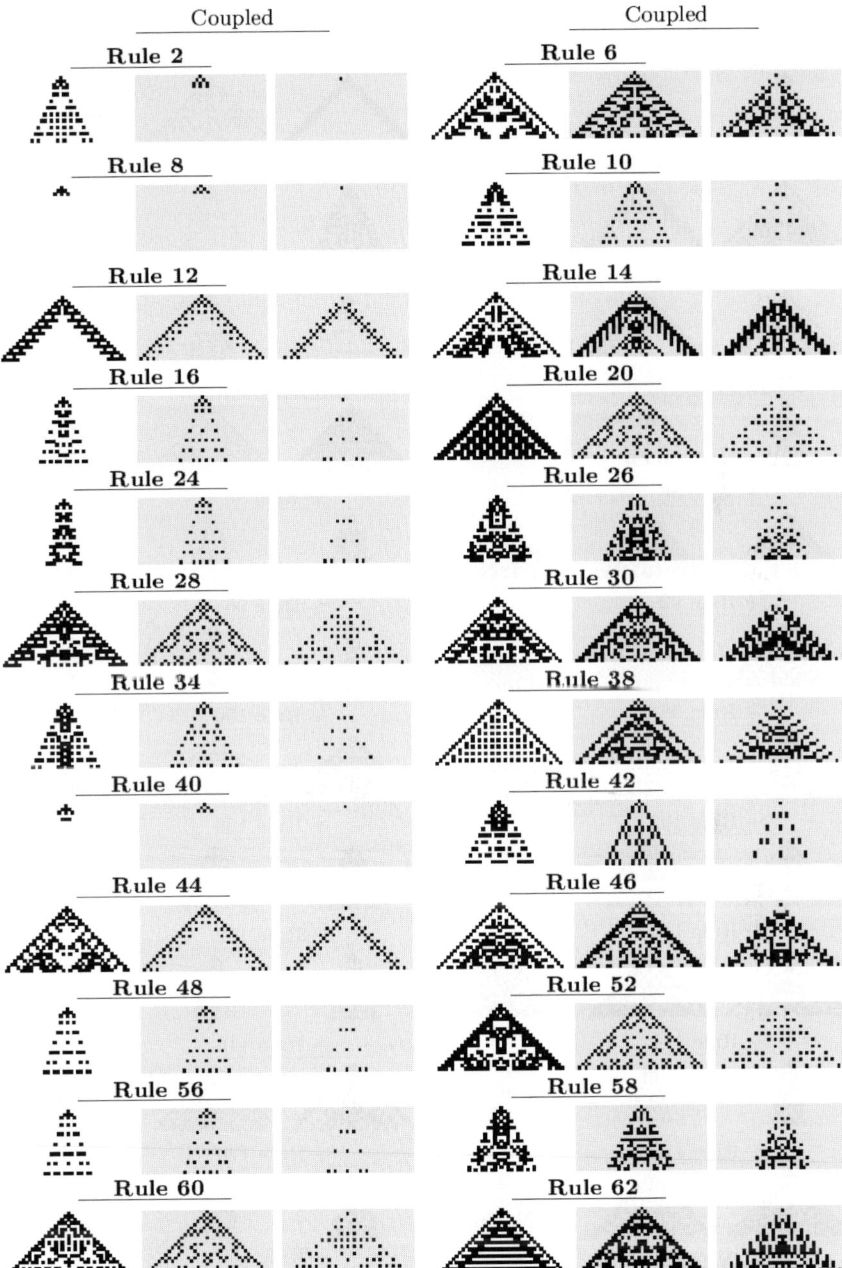

Fig. C.11   Coupled rule 150 with asymmetric quiescent rules as memory from a single seed.

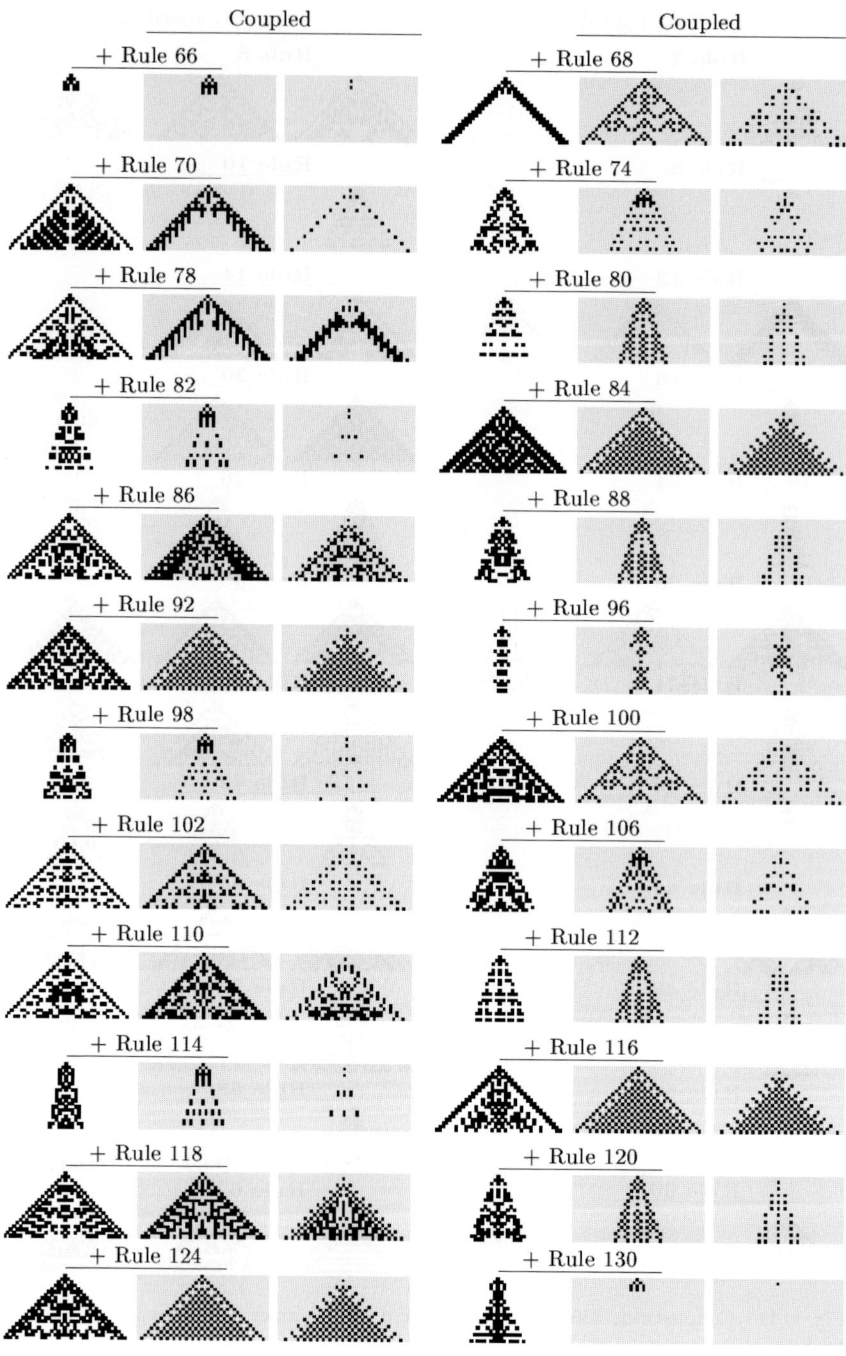

Fig. C.11   (*continued*)

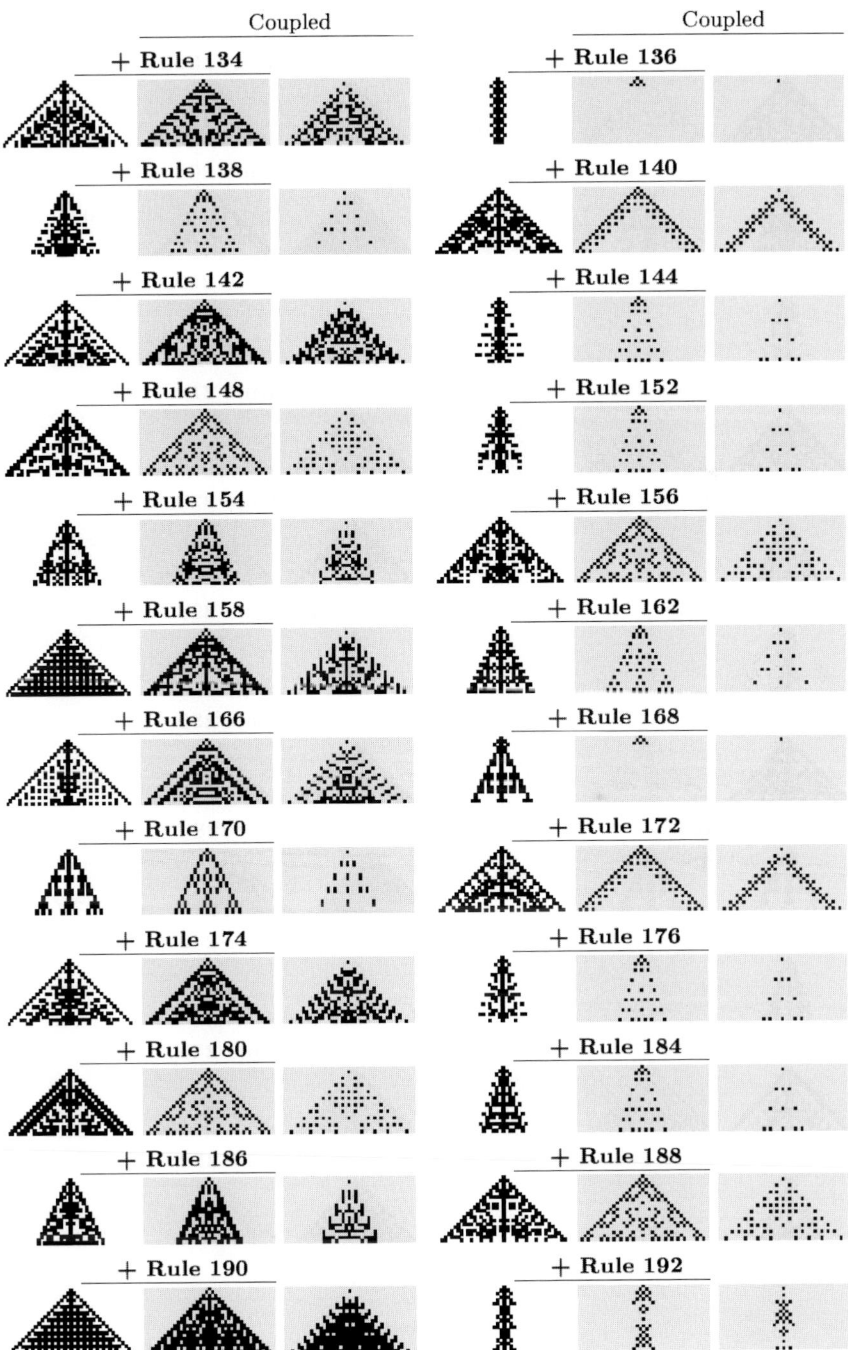

Fig. C.11 *(continued)*

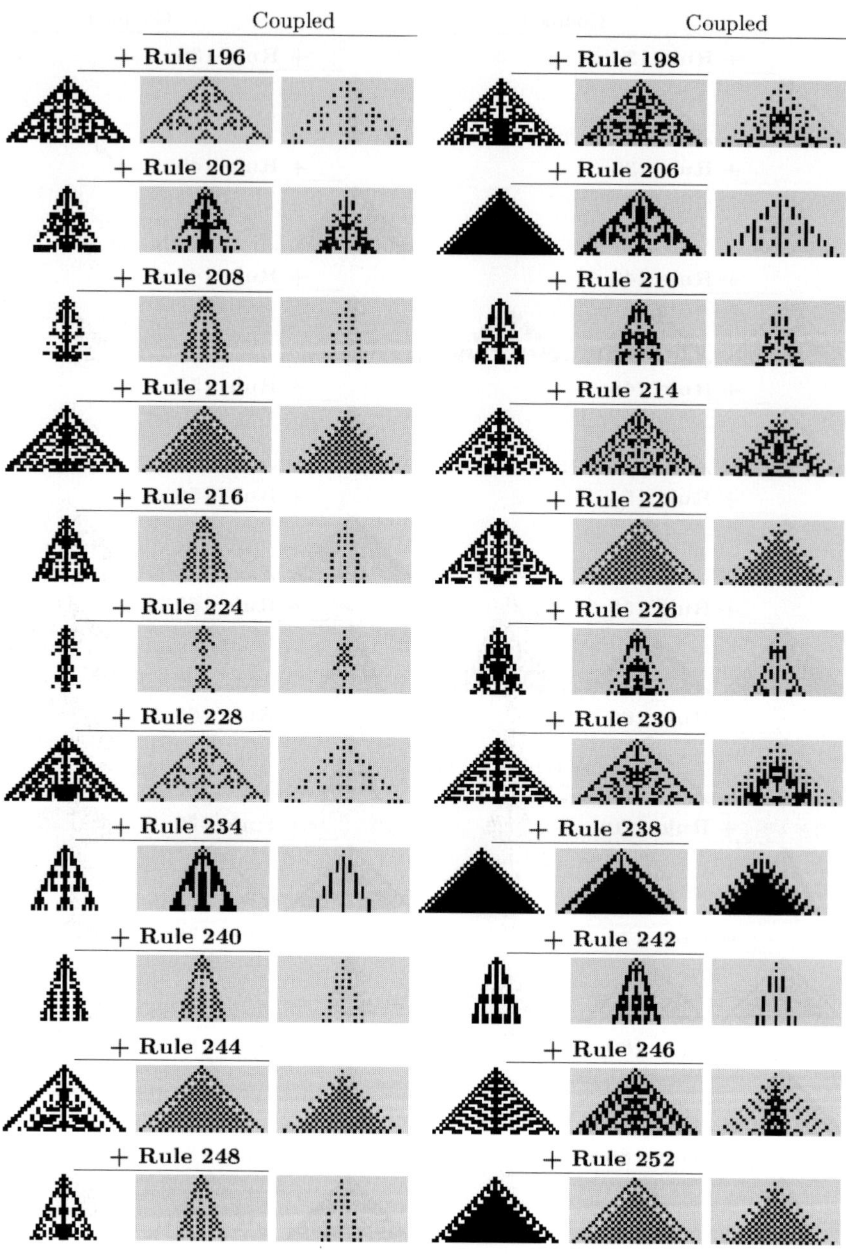

Fig. C.11   (*continued*)

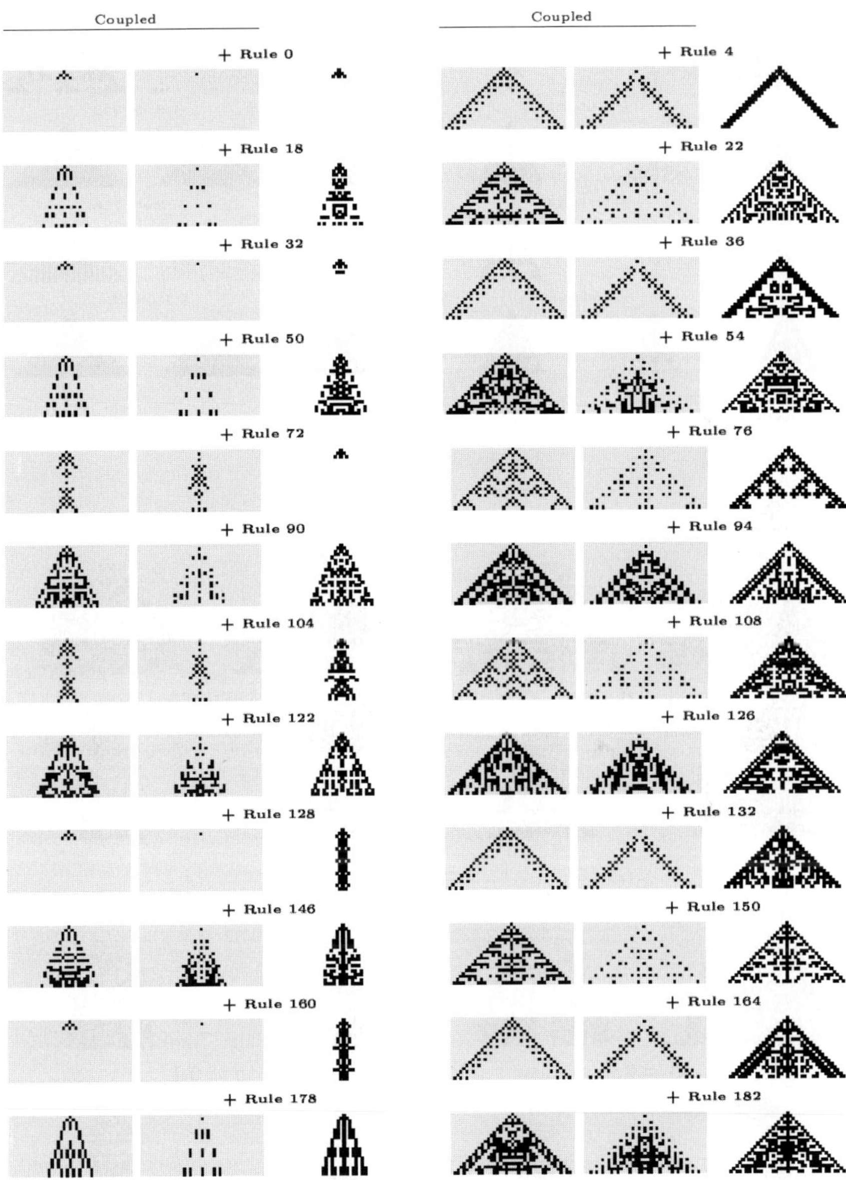

Fig. C.12   Coupled rule 150 with legal rules as memory from a single seed.

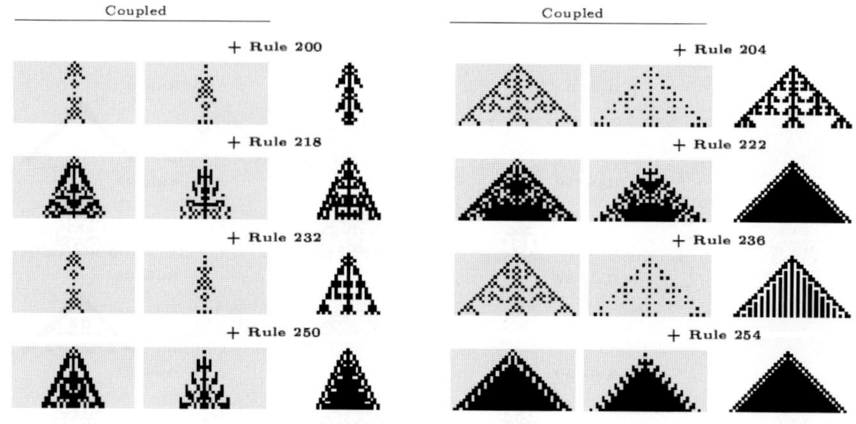

Fig. C.12   (*continued*)

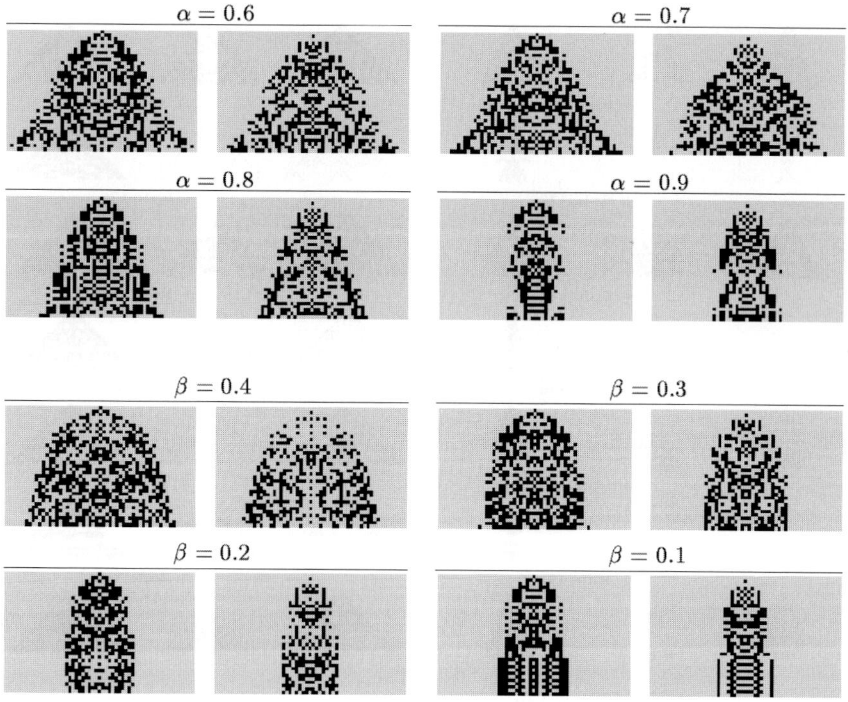

Fig. C.13   Effect of α and β memories on the coupled $K=5$ parity rule.

Appendix D

Continuous state variable

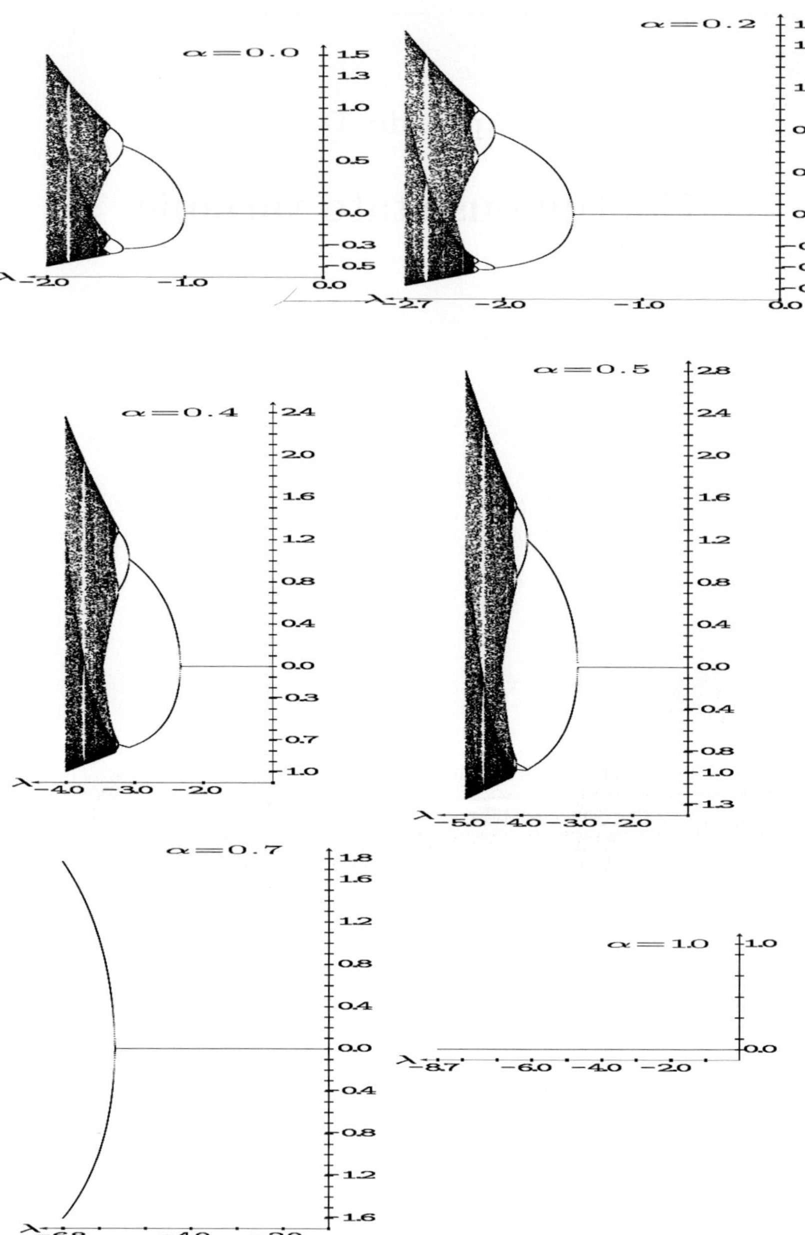

Fig. D.1   Bifurcation diagram of the logistic map with $\lambda < 0$ and α-memory.

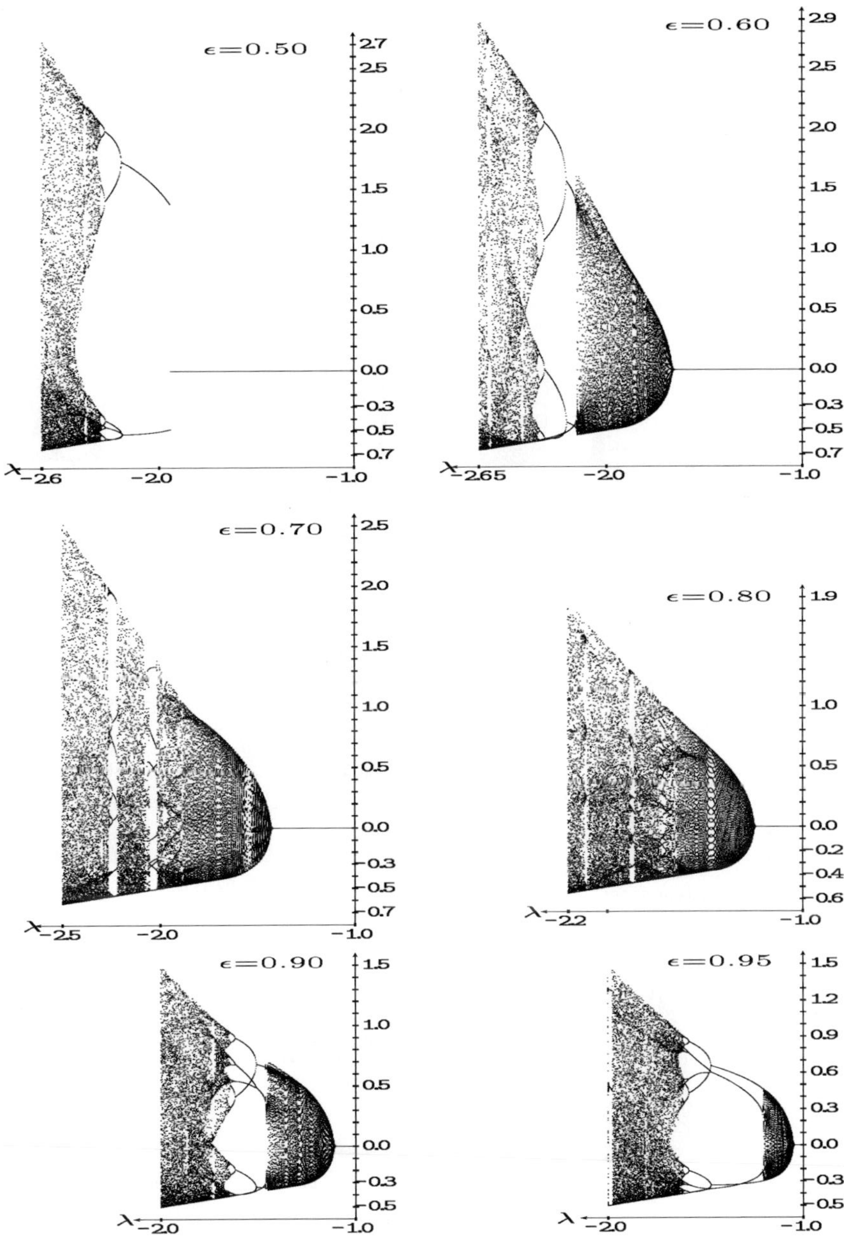

Fig. D.2 Bifurcation diagram of the logistic map with $\lambda < 0$ and ϵ-memory.

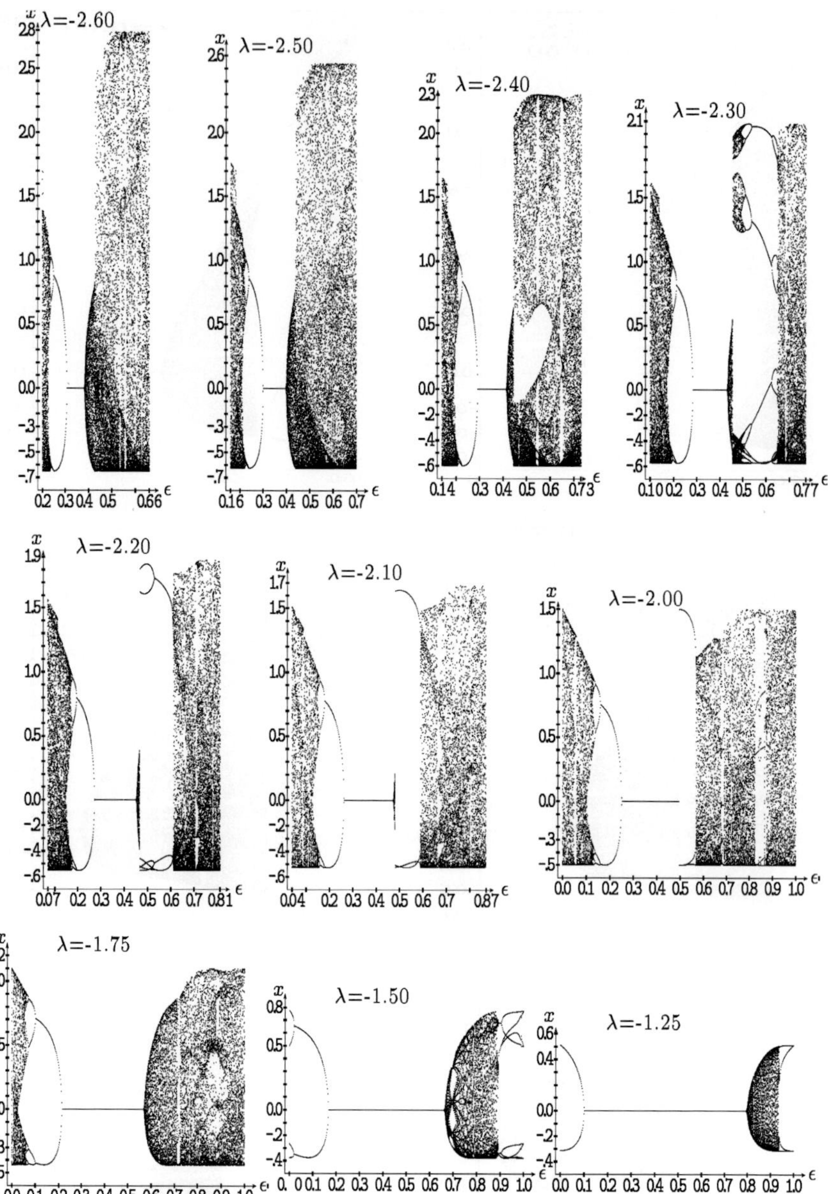

Fig. D.3   Bifurcation diagram of the logistic map with fixed $\lambda < 0$ and variable ϵ-memory.

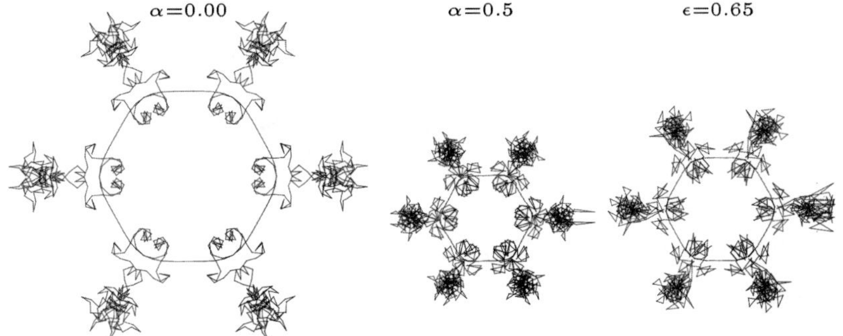

Fig. D.4 The $f(T) = T^3/1002$ curlicue with memory in z.

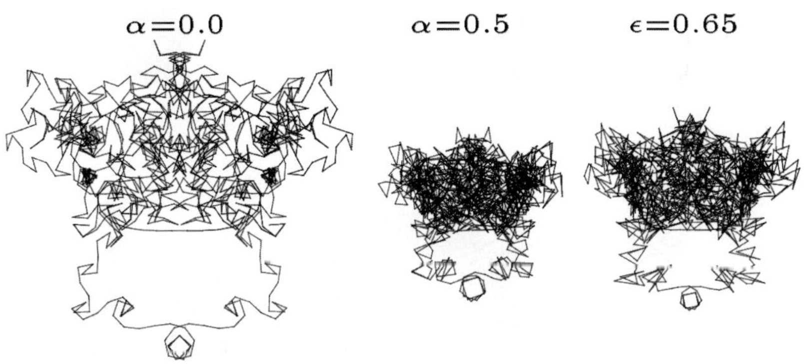

Fig. D.5 The $f(T) = T^3/1013$ curlicue with memory in z.

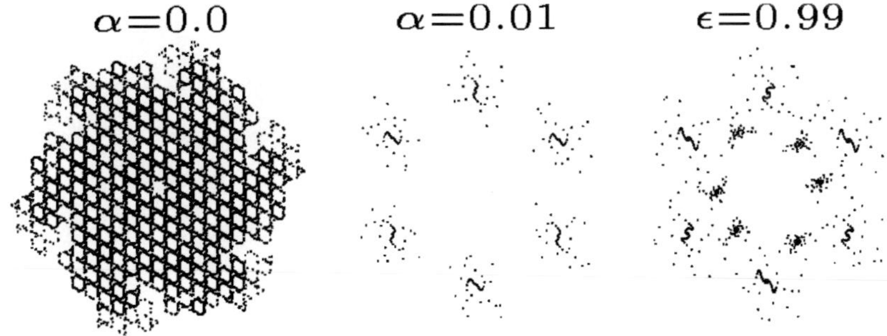

Fig. D.6 The Zaslavsky web map, with K=1.2, q=6.0. Memory in z.

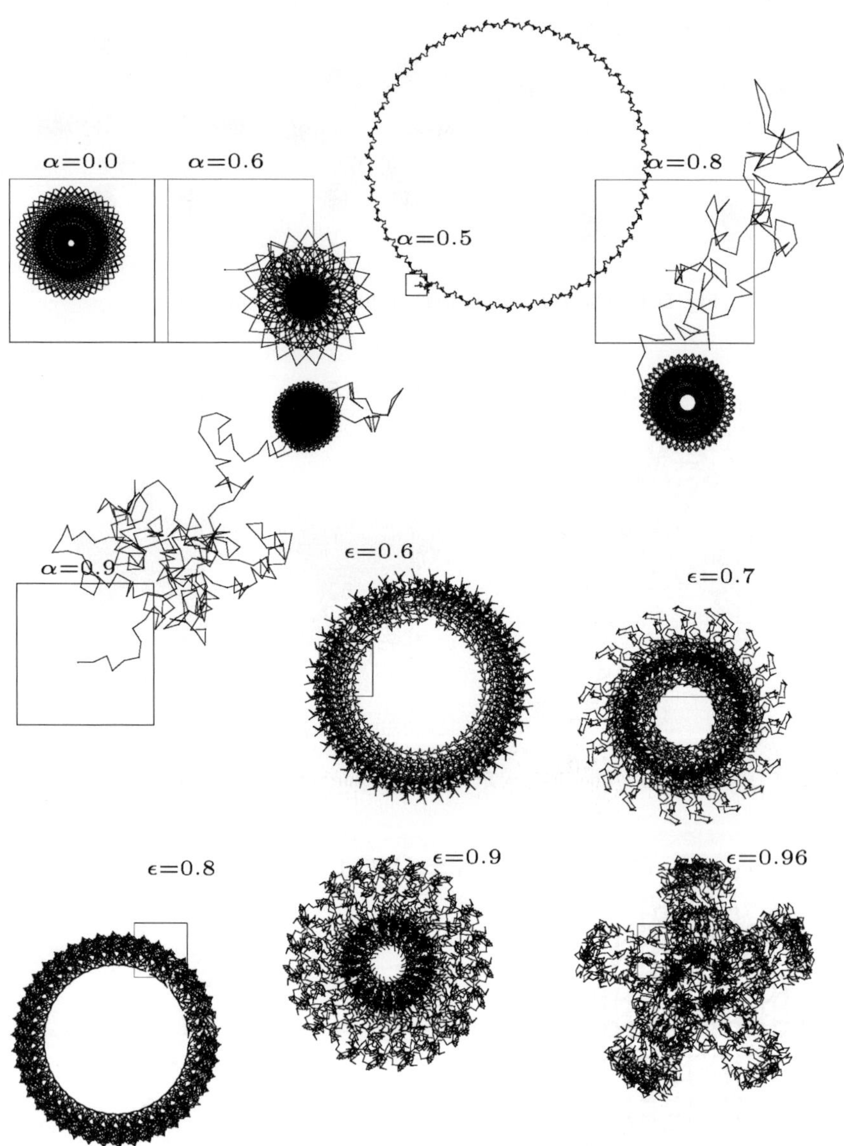

Fig. D.7   The $f(T) = T^7/1050$ curlicue with memory in f.

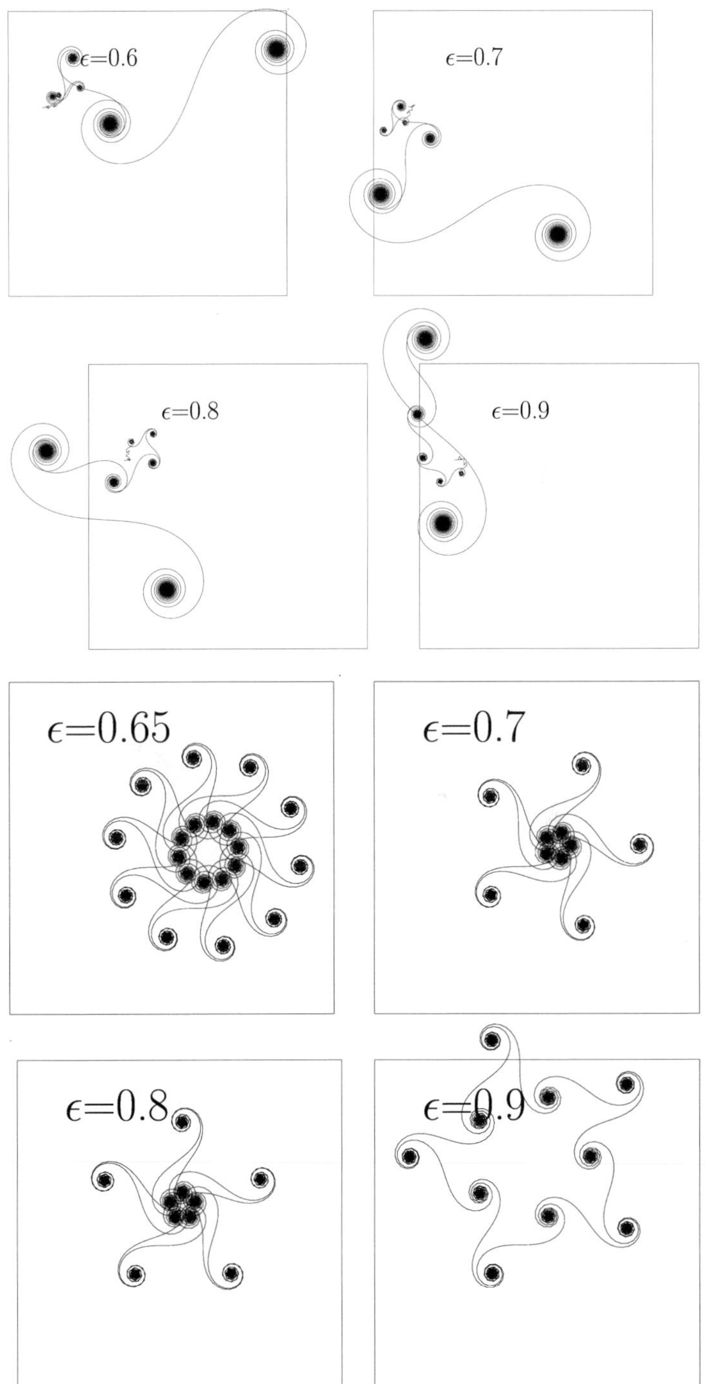

Fig. D.8   The Nessie (up) and $f(T) = T^2/321$ (down) curlicues with ϵ-memory in f.

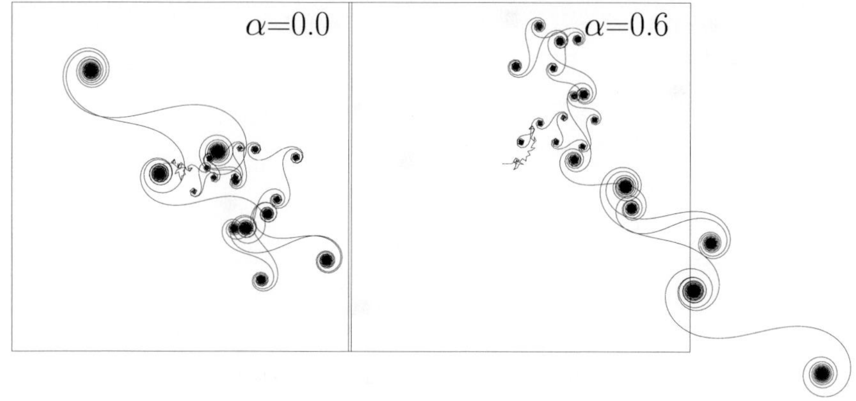

Fig. D.9   The $f(T) = (\log T)^5$ curlicue with α-memory in f.

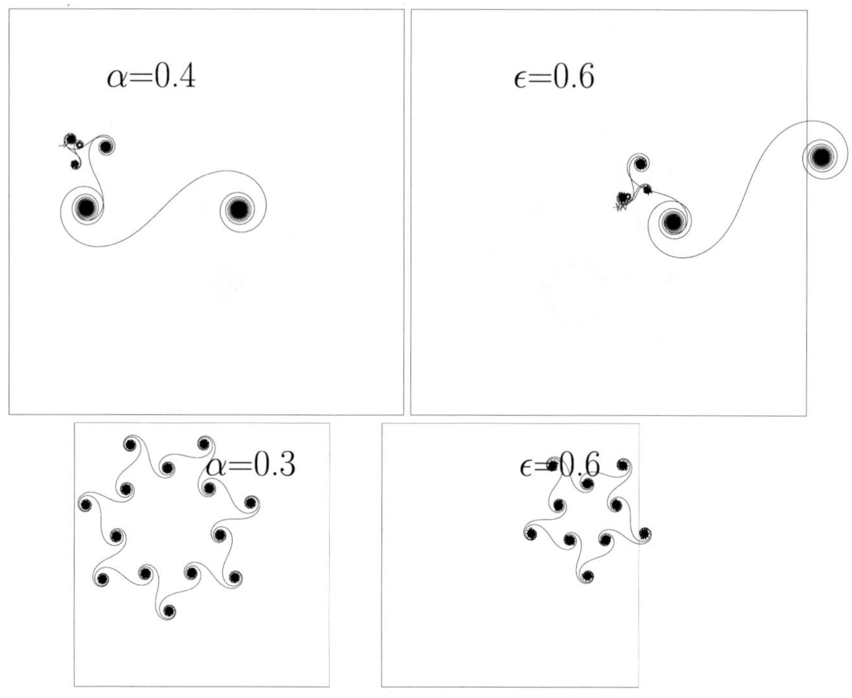

Fig. D.10   The Nessie and $f(T) = T^2/321$ curlicues with memory in z and f.

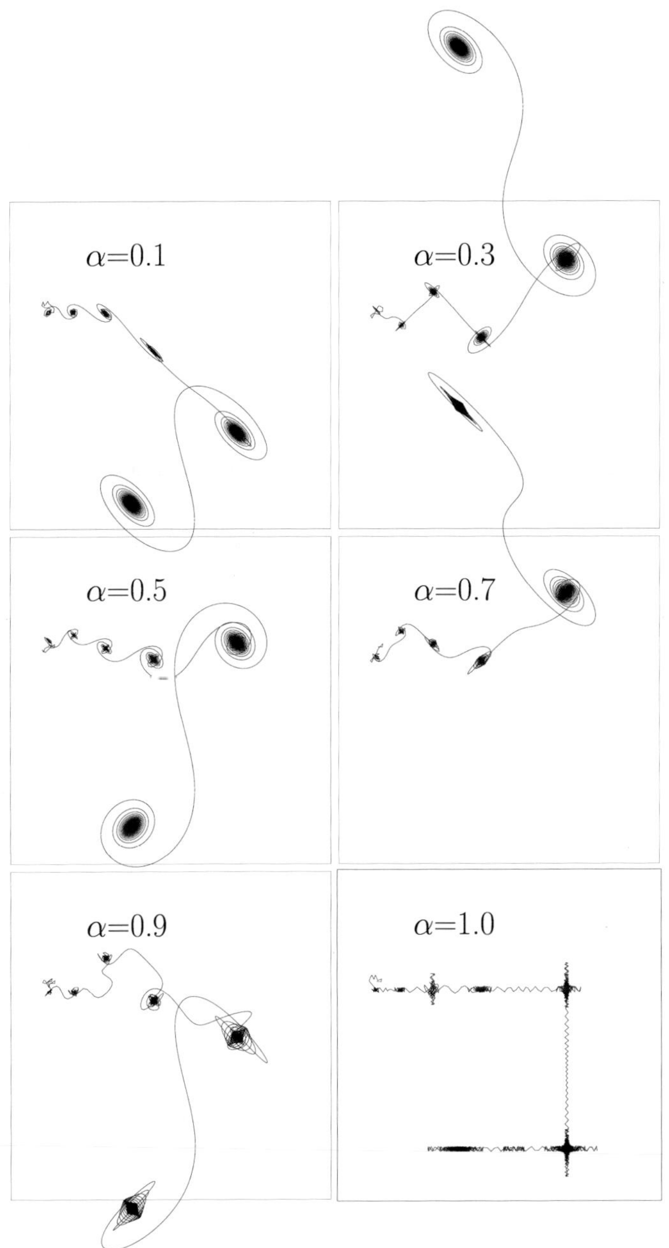

Fig. D.11   The Nessie curlicue with partial memory in f.

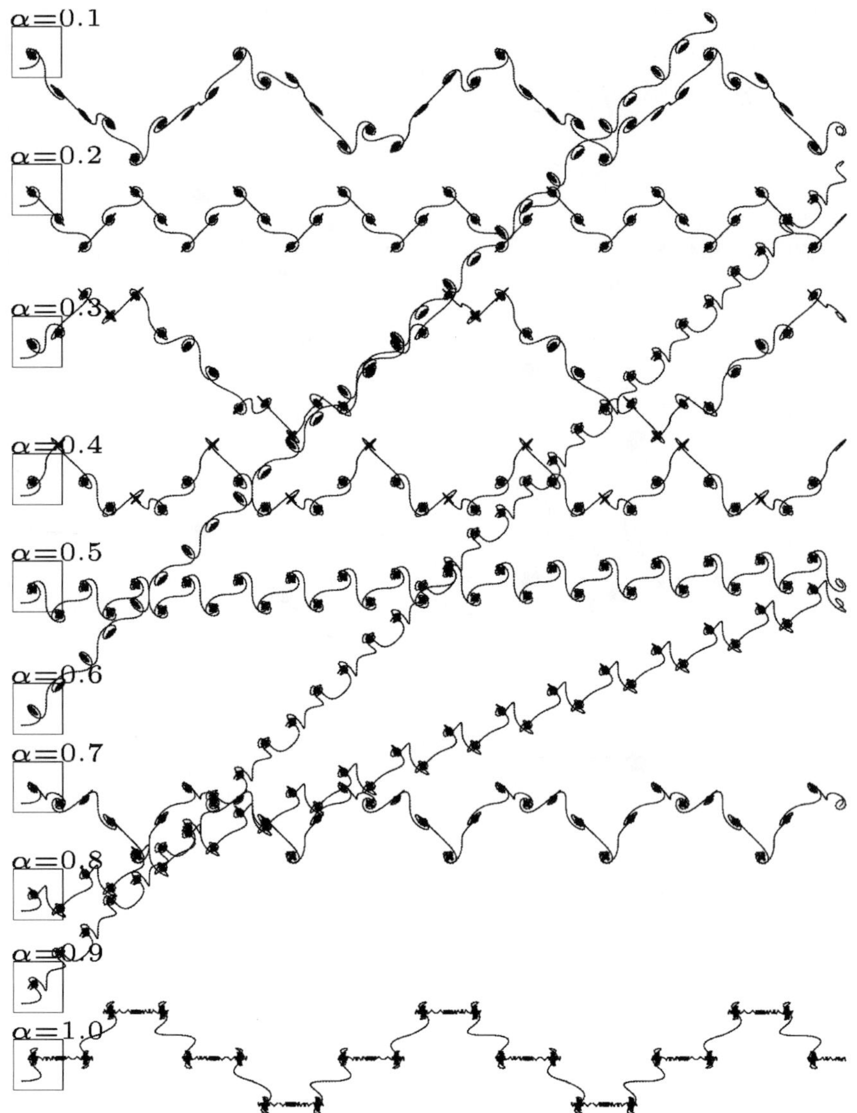

Fig. D.12   The $T^2/321$ curlicue with partial memory in f.

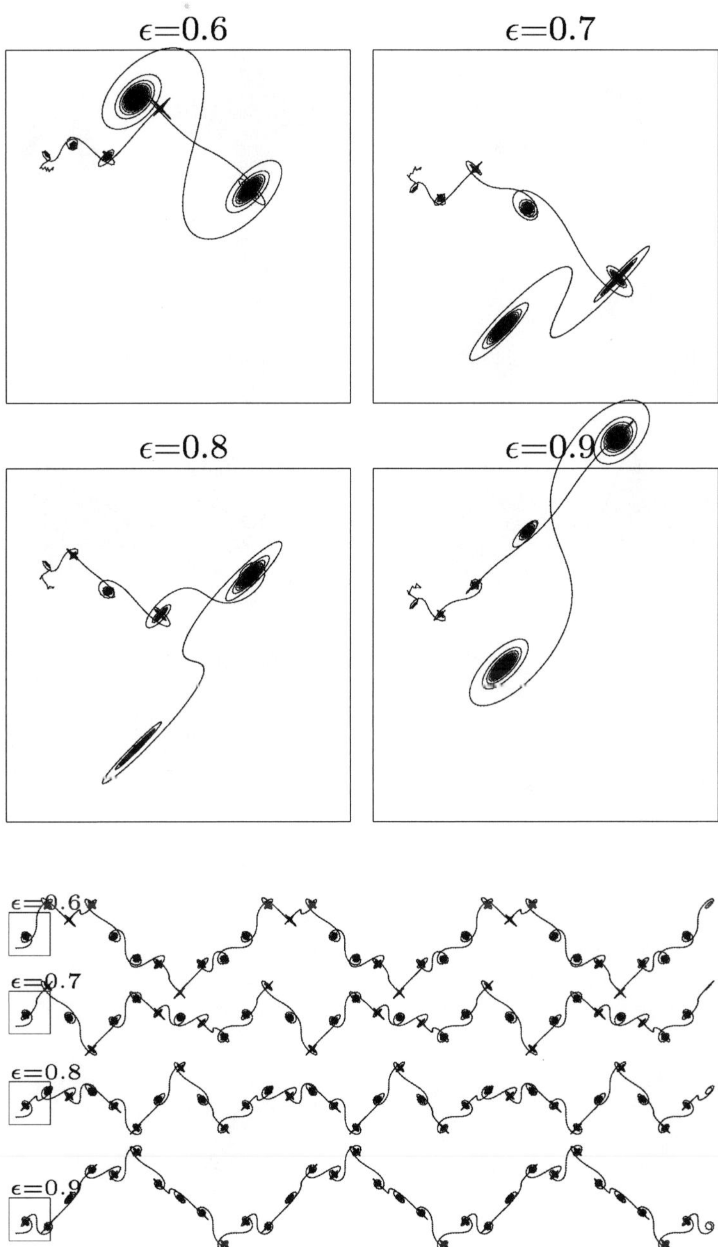

Fig. D.13   The Nessie and $T^2/321$ curlicues with partial ϵ-memory in f.

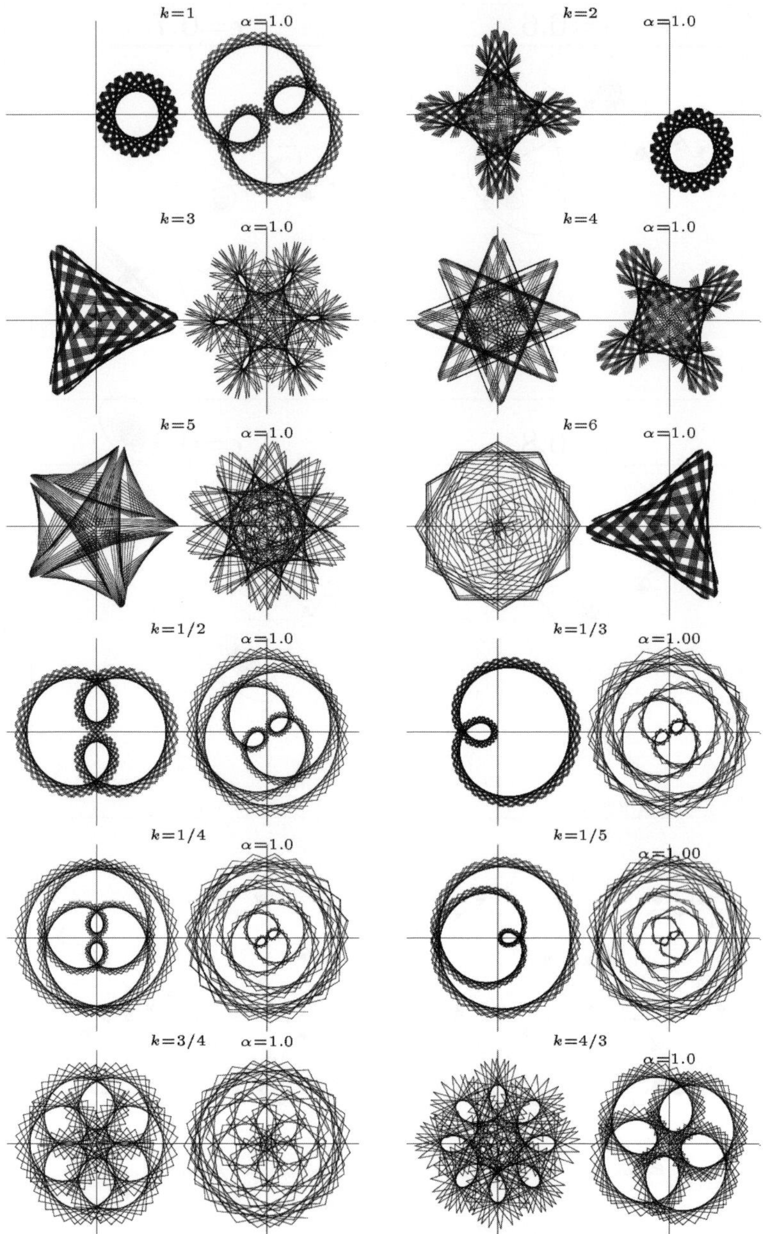

Fig. D.14   The Maurer roses from Fig. 12.30.

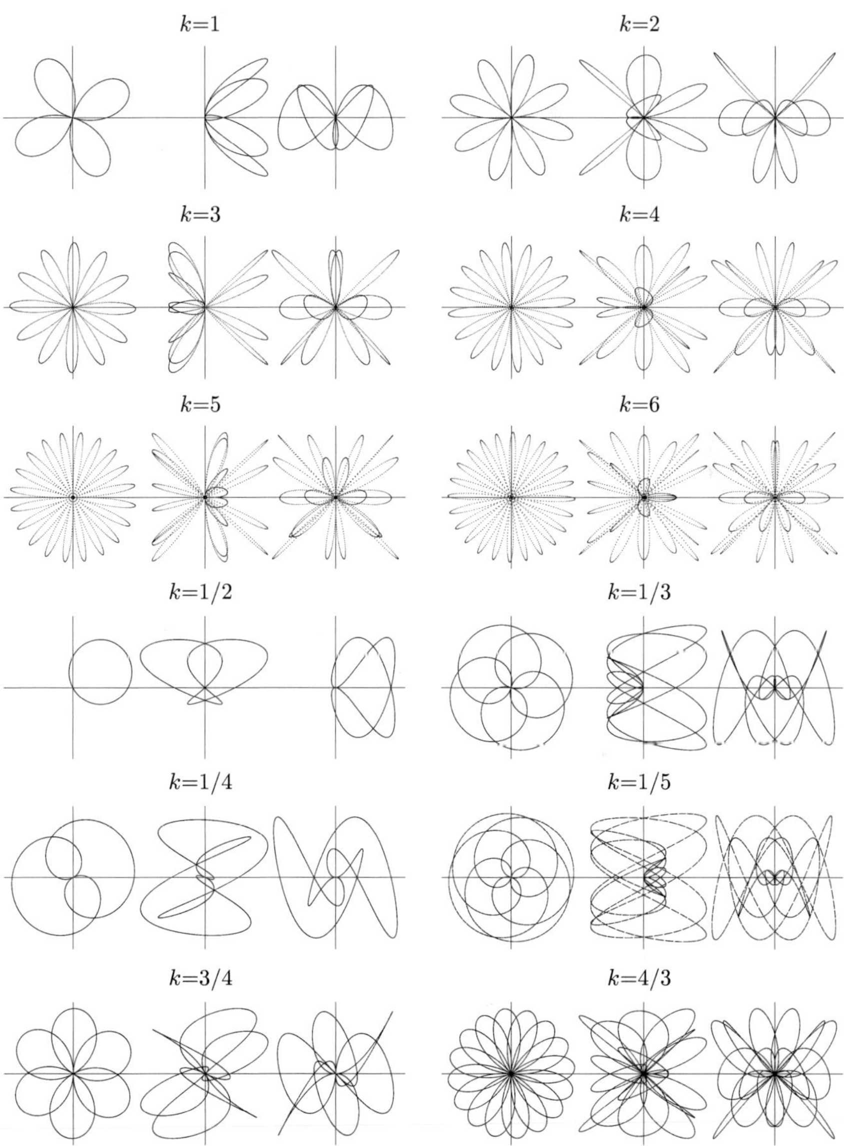

Fig. D.15   The roses from Fig. 12.30 with memory in the coordinate conversion. (Left) In both x and y, (center) only in y, (right) only in x.

Appendix E

Spatial games

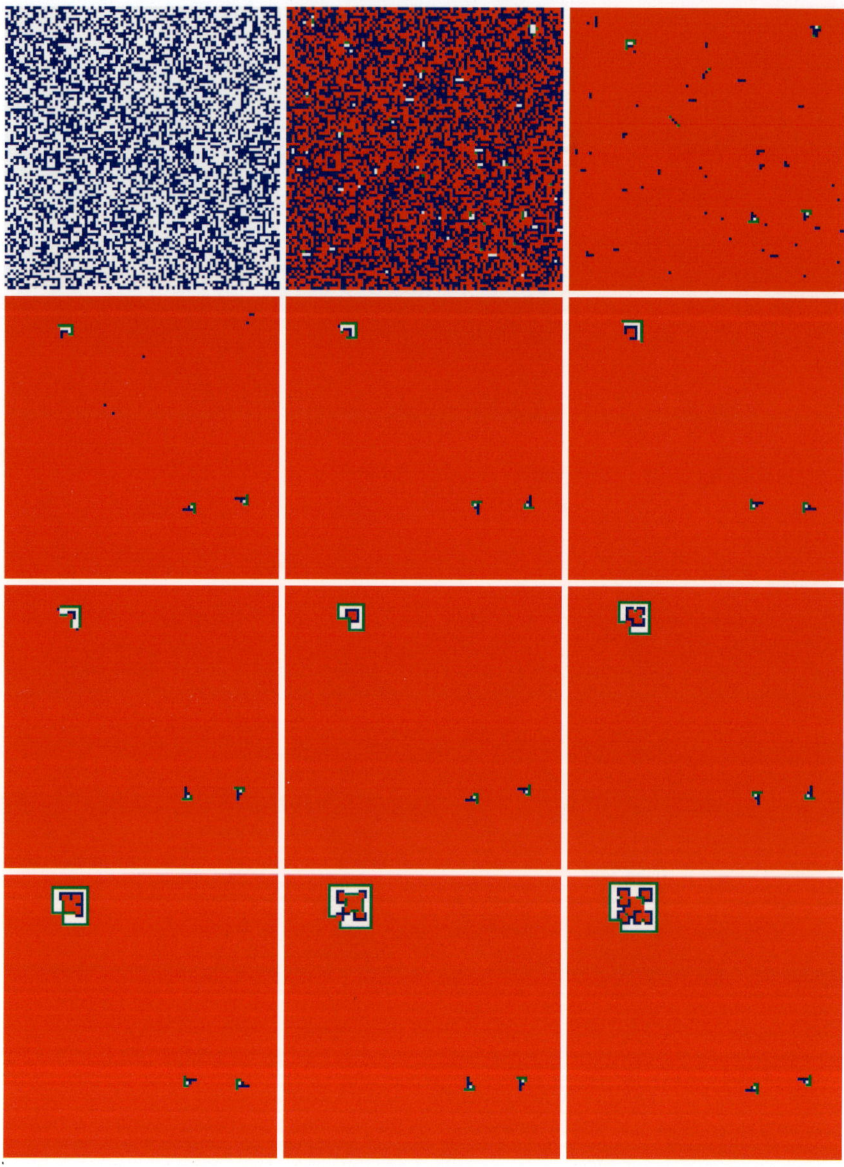

Fig. E.1   Initial patterns starting at random in the SPD. $T = 1 - 12$. No memory. $b = 1.85$.

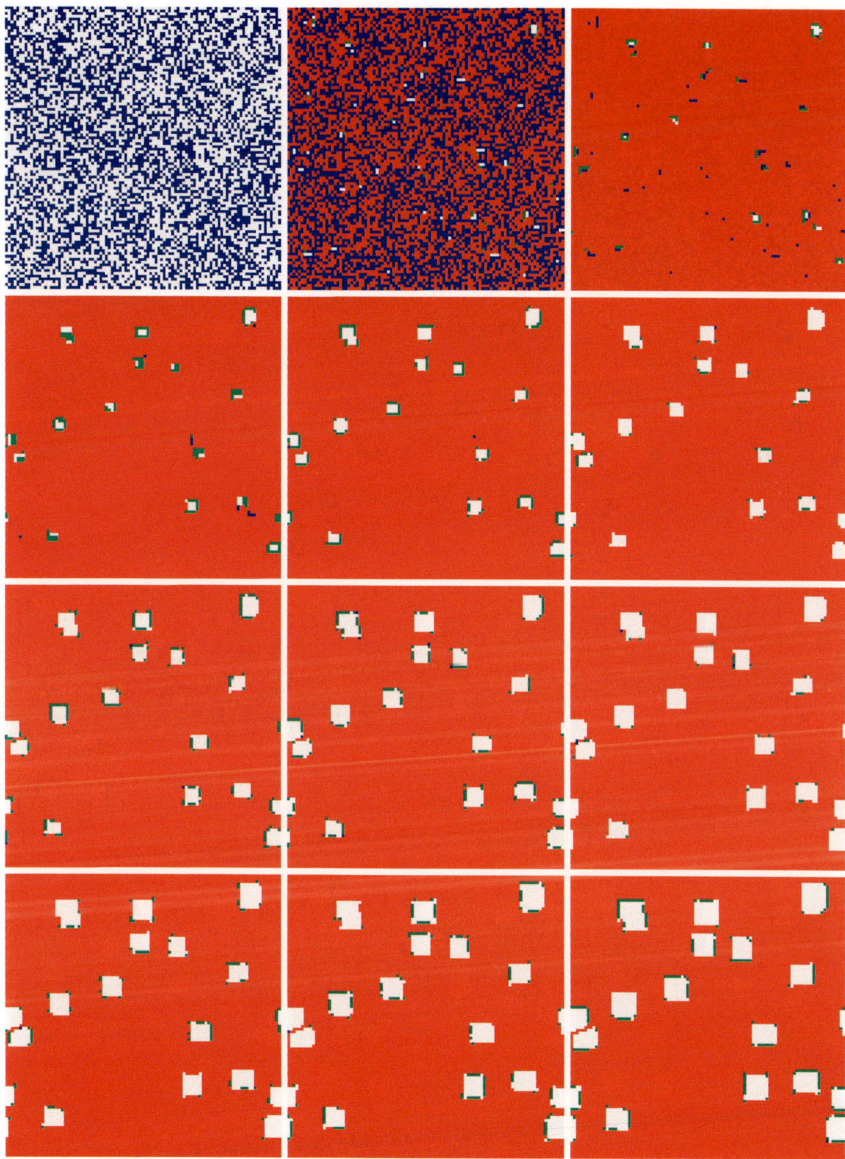

Fig. E.2   Initial patterns with α=0.9 memory starting at random in the SPD. $T = 1-12$. $b = 1.85$.

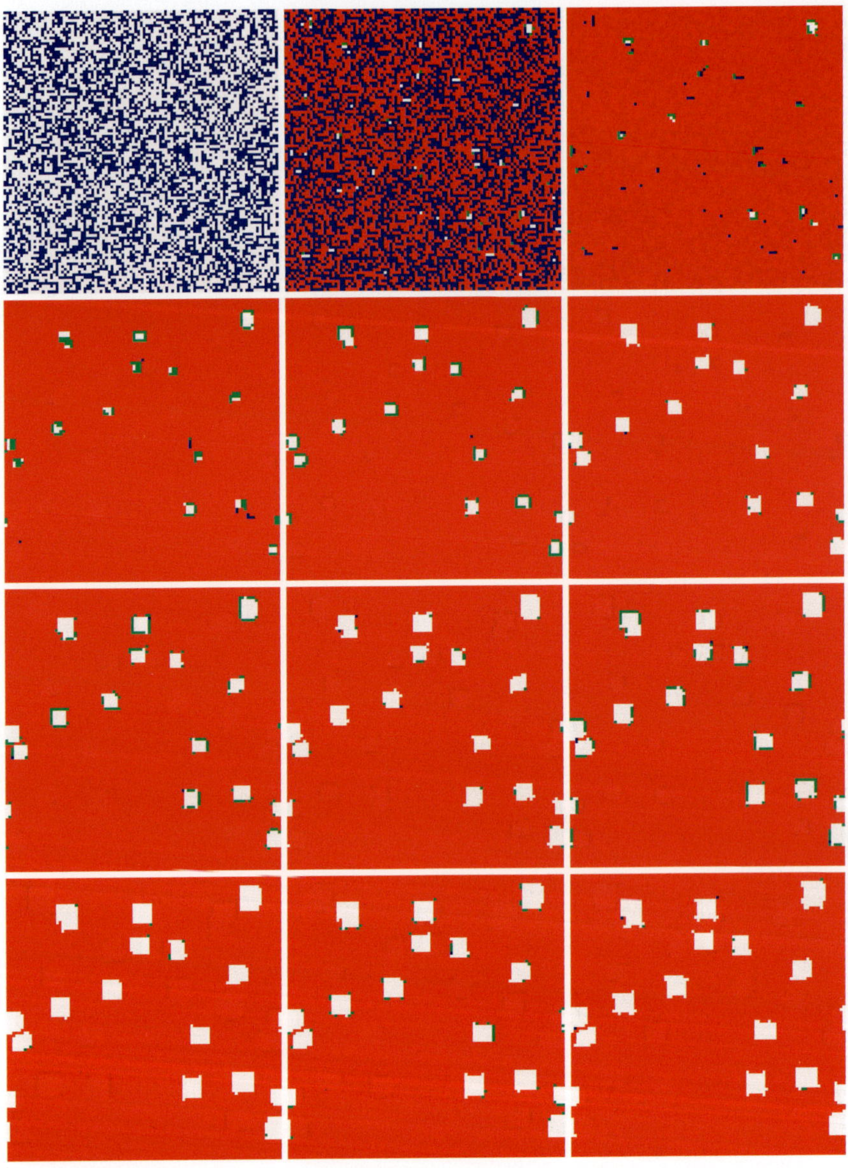

Fig. E.3   Initial patterns with full memory starting at random in the SPD . $T = 1 - 12$. $b = 1.85$.

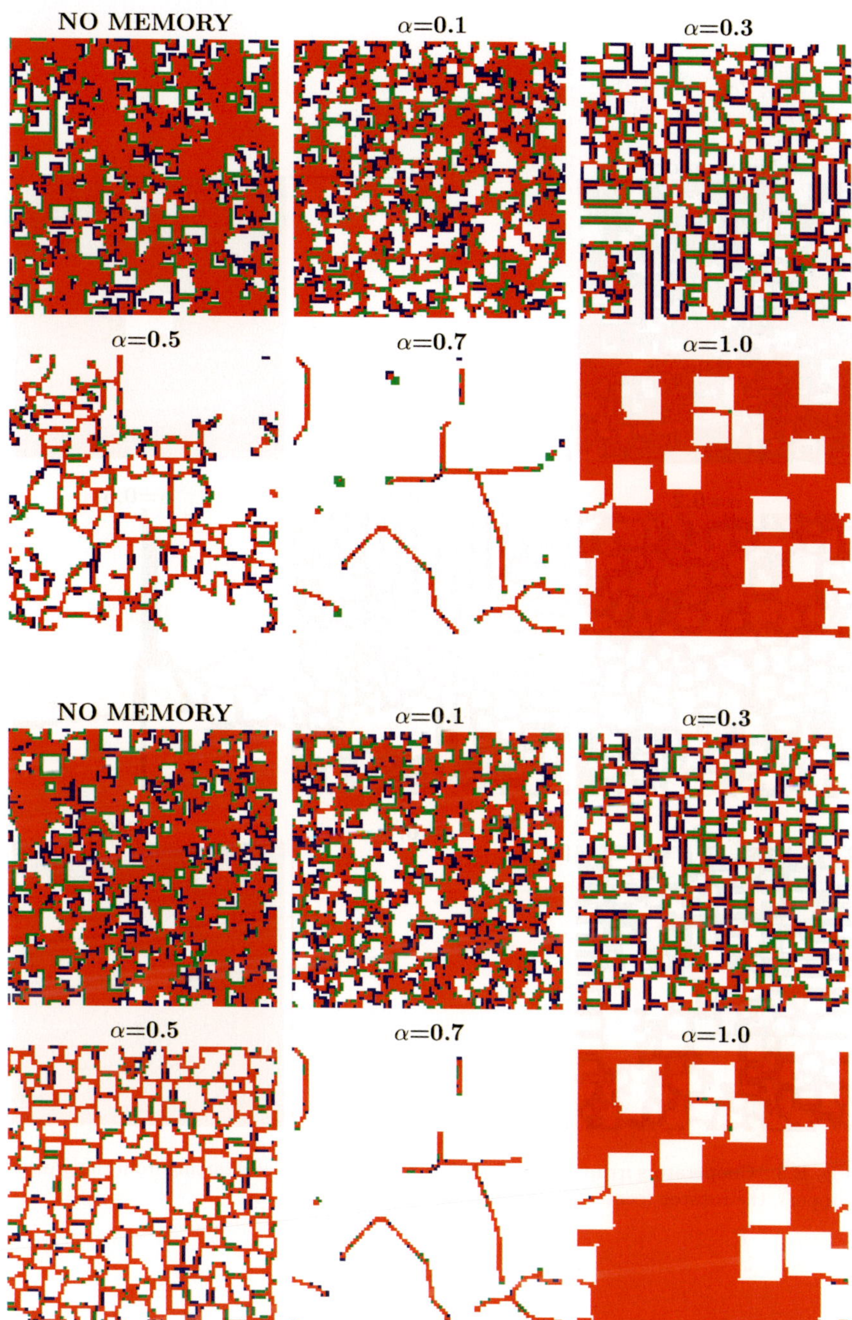

Fig. E.4   Patterns at $T=100$ (up) and at $T=200$ (down) starting at random in a 100×100 lattice.

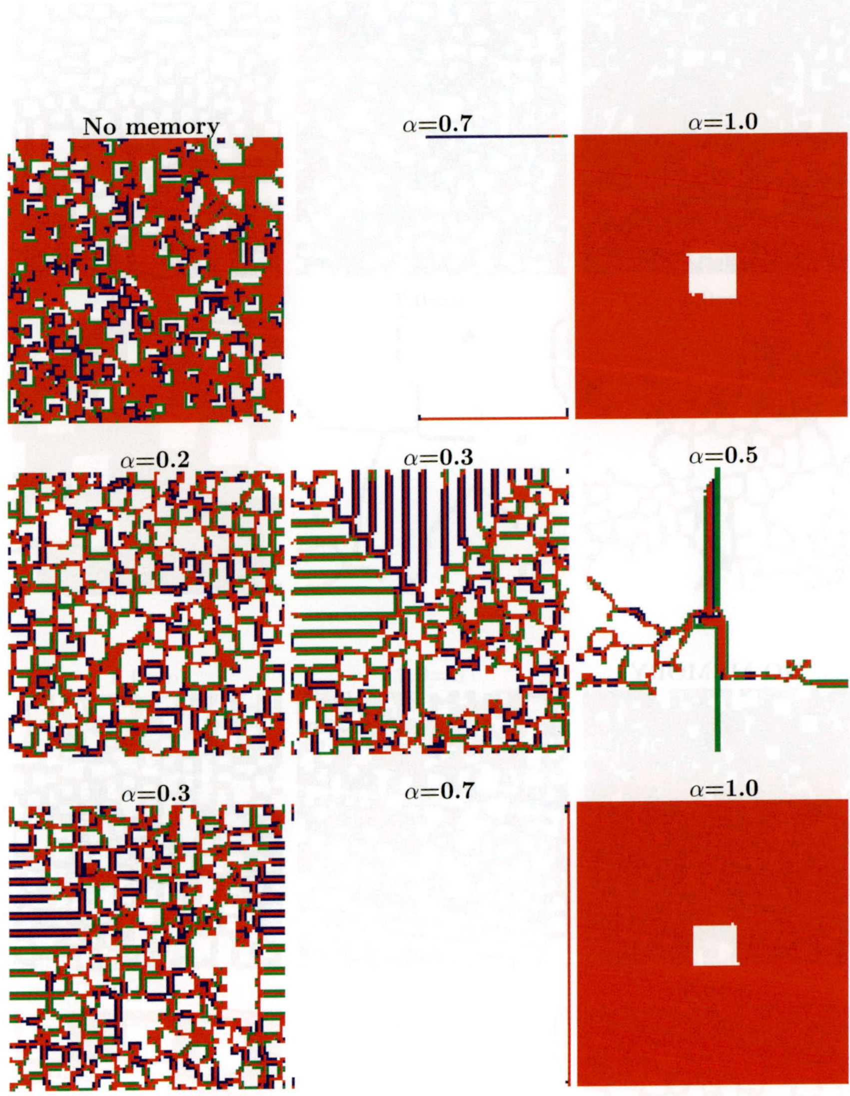

Fig. E.5 Cooperators irruption at $T=100$ in a 100×100 lattice, starting from the grower (upper), the rotator (center), and the glider (bottom).

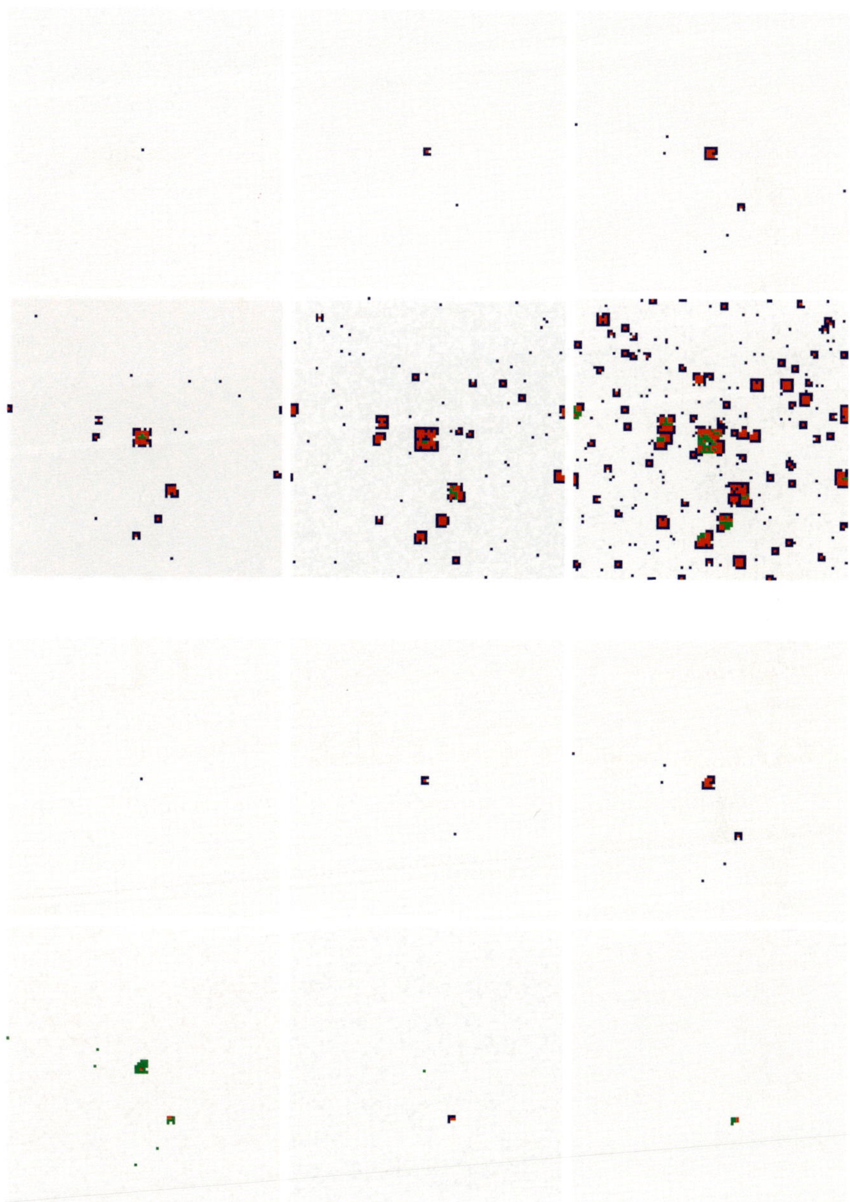

Fig. E.6 First patterns with no memory (up) and with full memory (down) from a single defector in a 100×100 network with 10 % of rewiring.

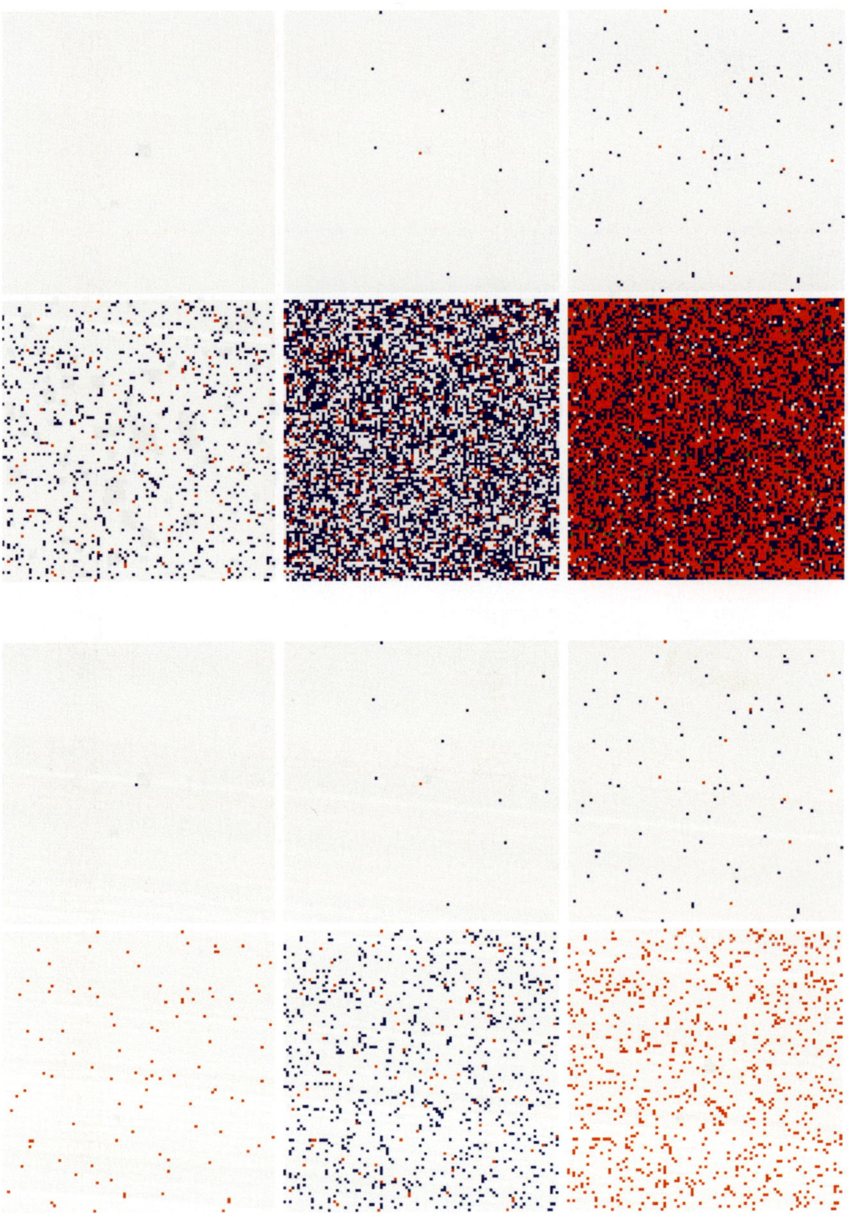

Fig. E.7 First patterns with no memory (up) and with full memory (down) from a single defector in a 100×100 network with full rewiring.

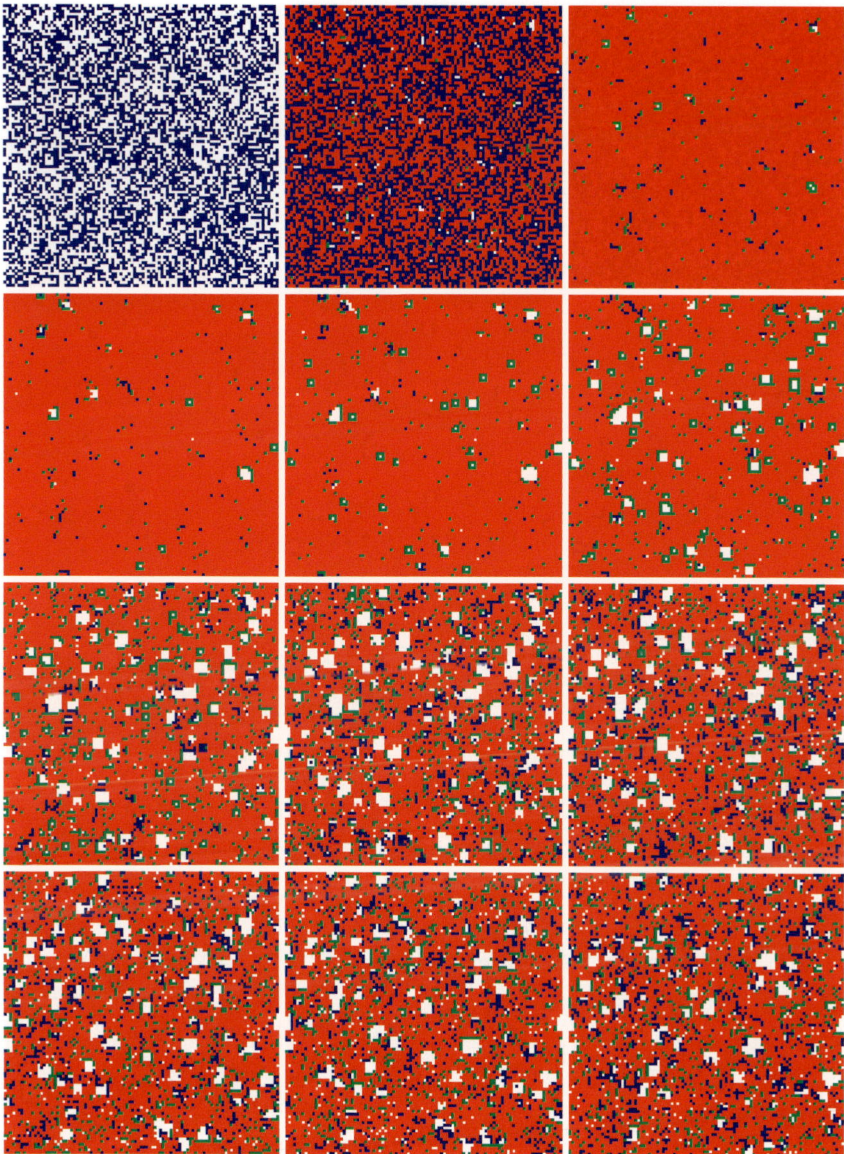

Fig. E.8 First patterns with no memory starting at random in a 100×100 network with 10 % of rewiring.

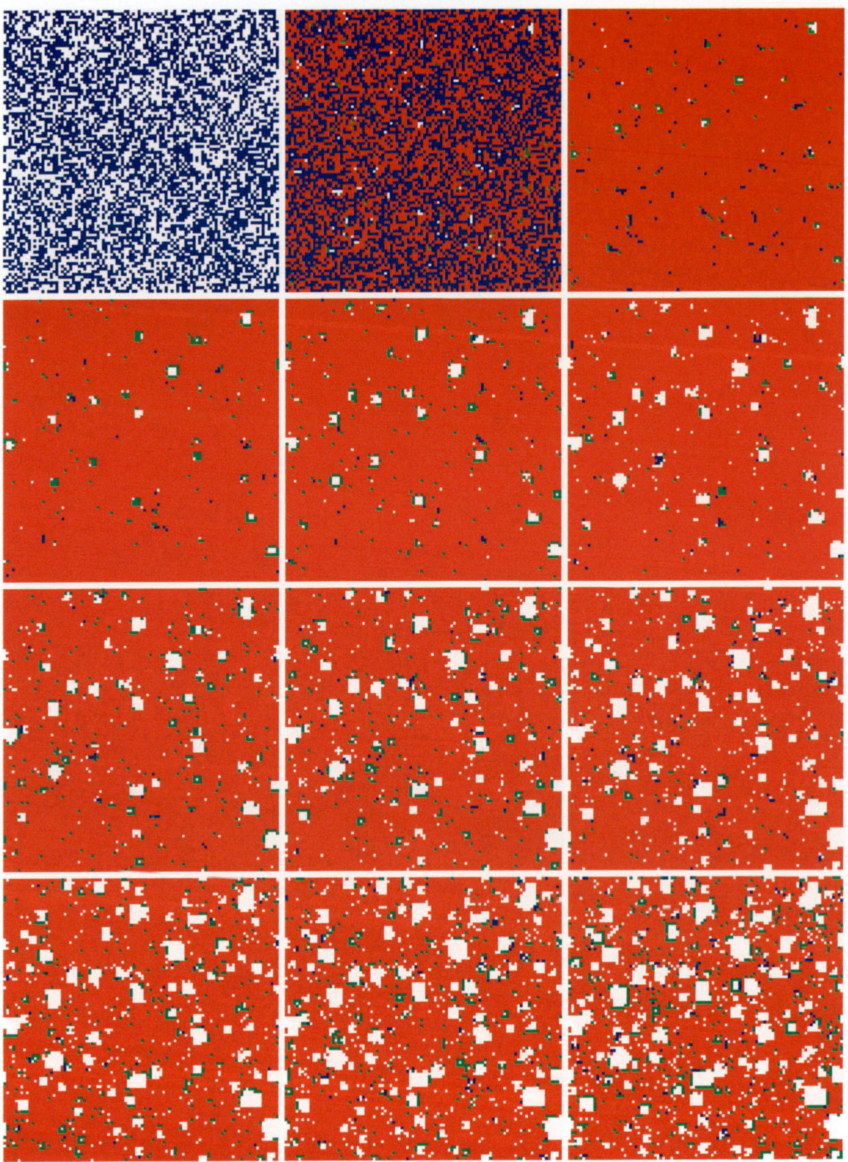

Fig. E.9   First patterns with $\alpha=0.9$ memory starting at random in a 100×100 network with 10 % of rewiring.

Unlimited trailing memory

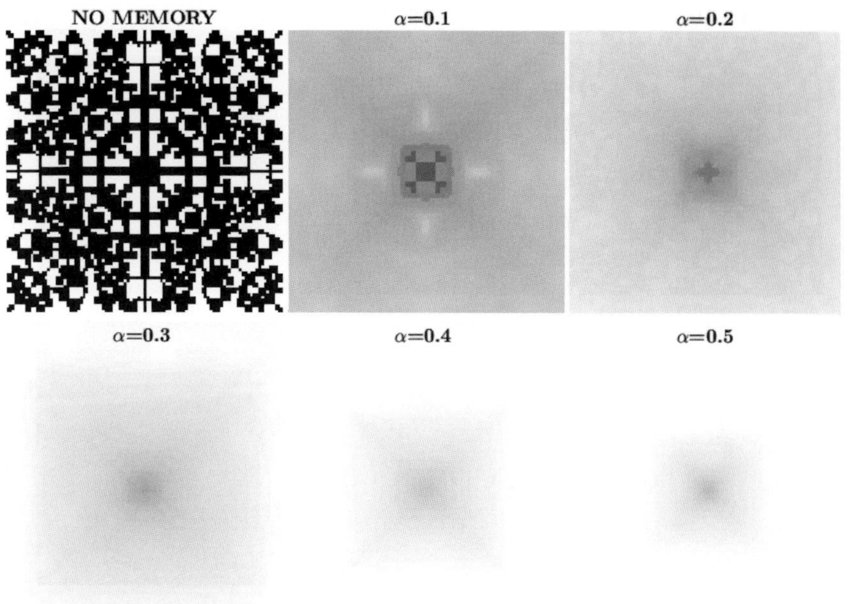

$\tau-2$ memory

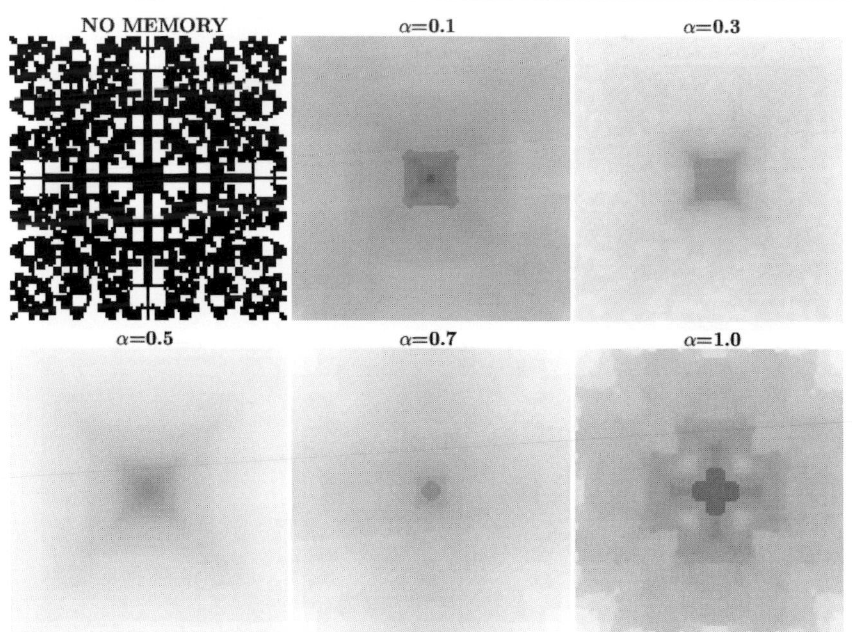

Fig. E.10 Patterns at $T=100$ starting from a single defector in a 101×101 lattice in the CVPD.

Unlimited trailing memory

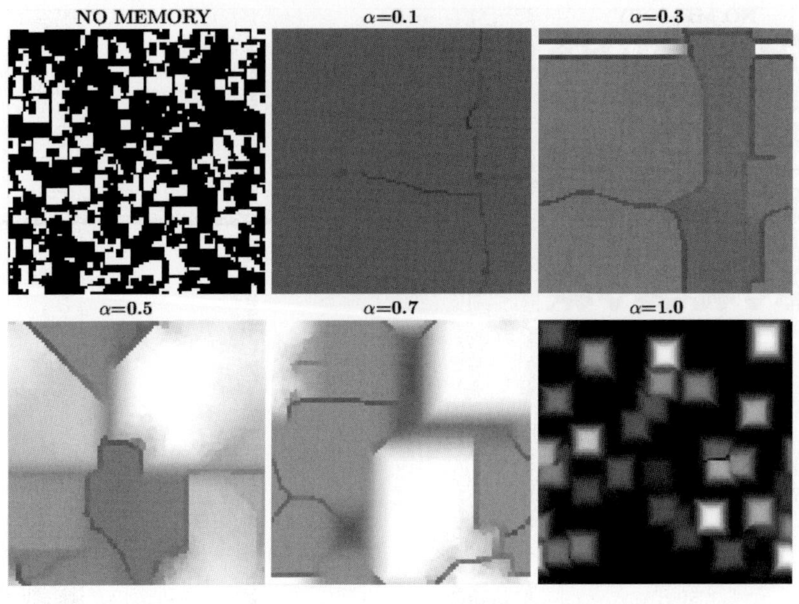

$\tau=2$ memory

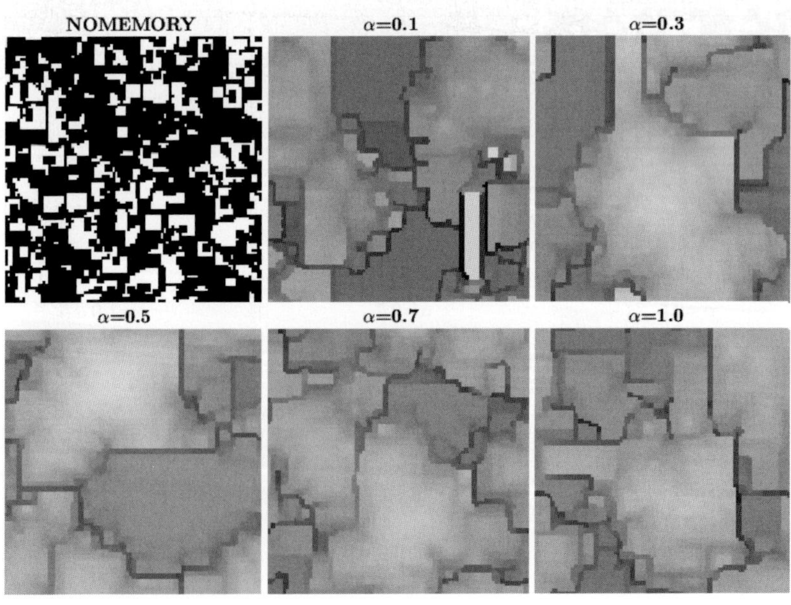

Fig. E.11   Patterns at $T=100$ starting at random in the CVPD.

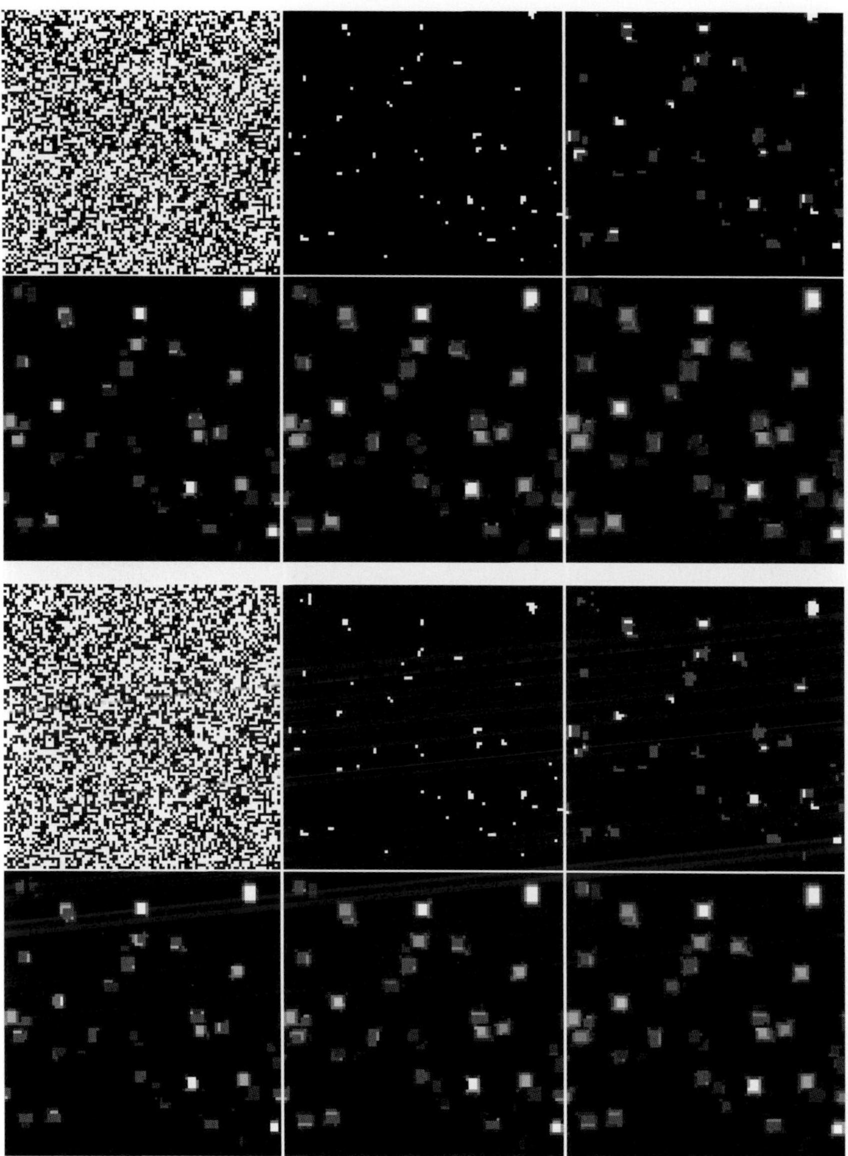

Fig. E.12   First six patterns in the CVPD with $\alpha=0.9$ (up) and full (down) memories starting at random in a 100×100 lattice.

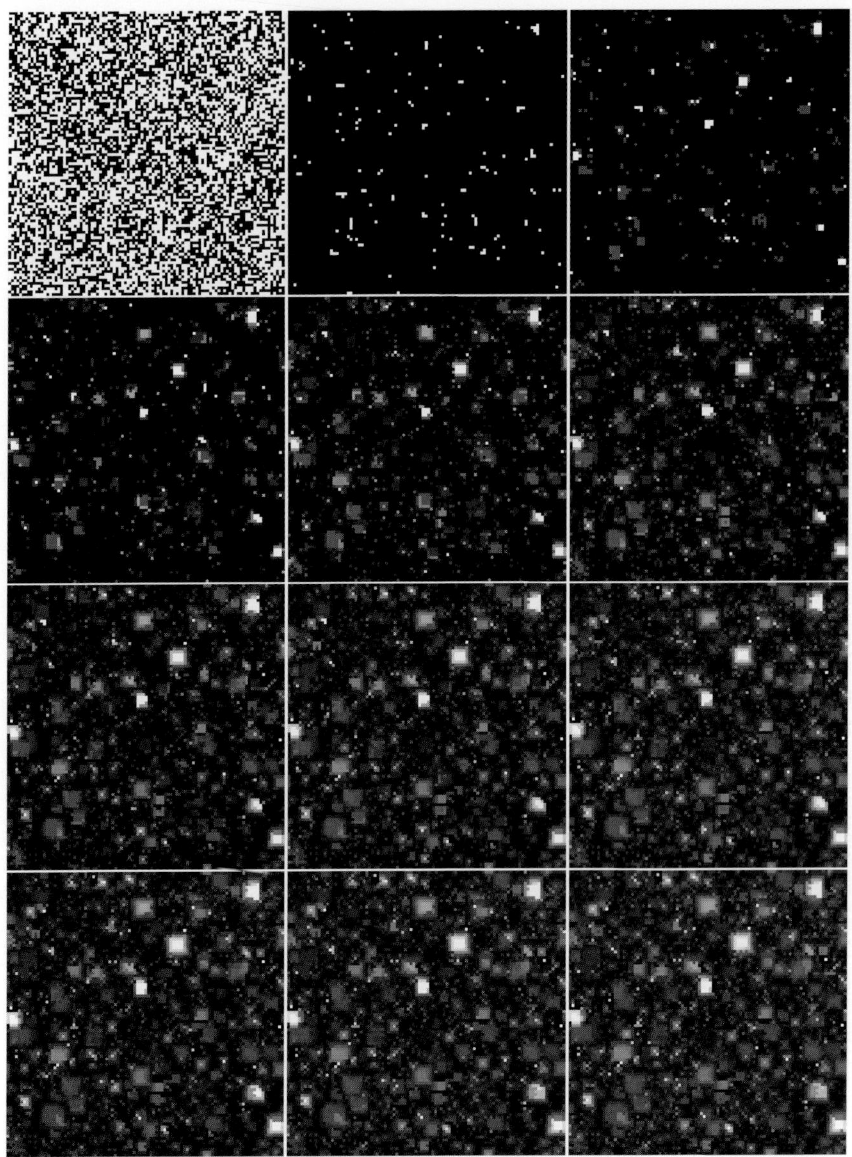

Fig. E.13   Initial patterns in the CVPD with α=0.9 memory starting at random in a 100×100 network with 10% of rewiring.

Fig. E.14   Evolving patterns in the initial scenario of Figs.13.20 and 13.21, but without rating the payoff of every player to the number of its co-players.

Fig. E.15   Evolving patterns in the initial scenario of Figs. 13.20 and 13.21 but without self-interaction.

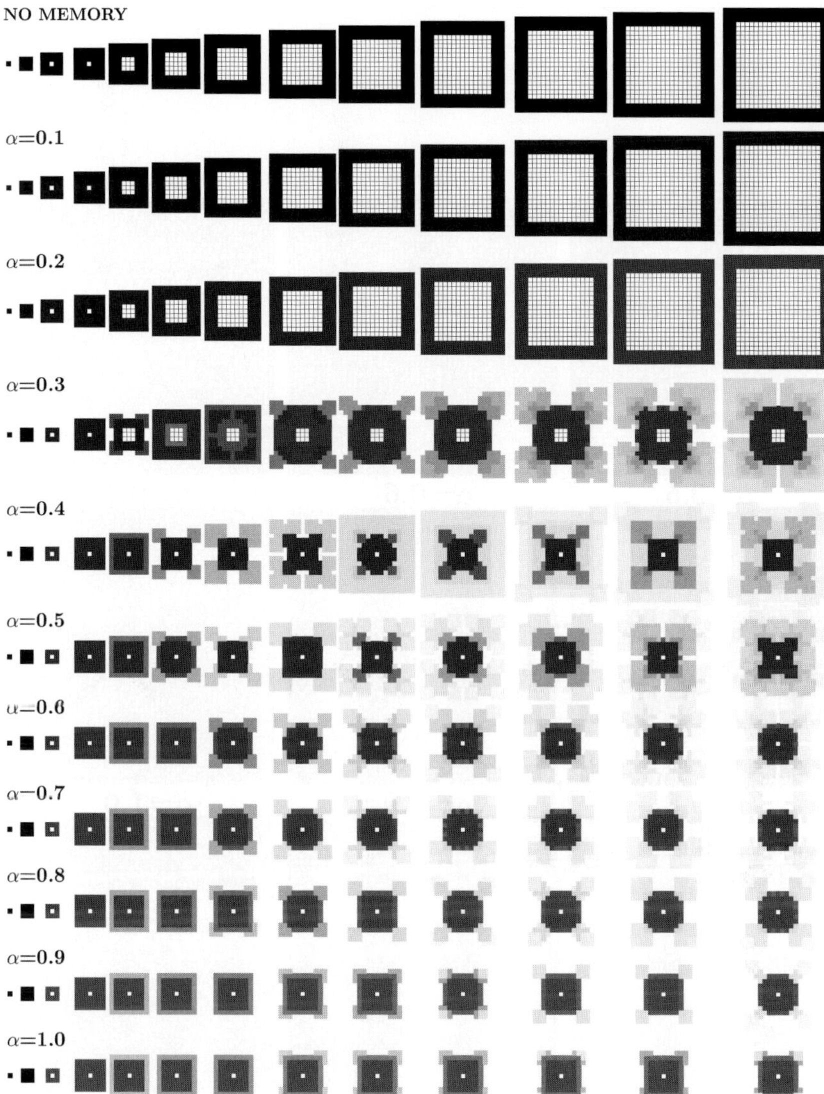

Fig. E.16 Evolving patterns in the initial scenario of Fig. 13.34 but without rating the payoff of every player to the number of its co-players.

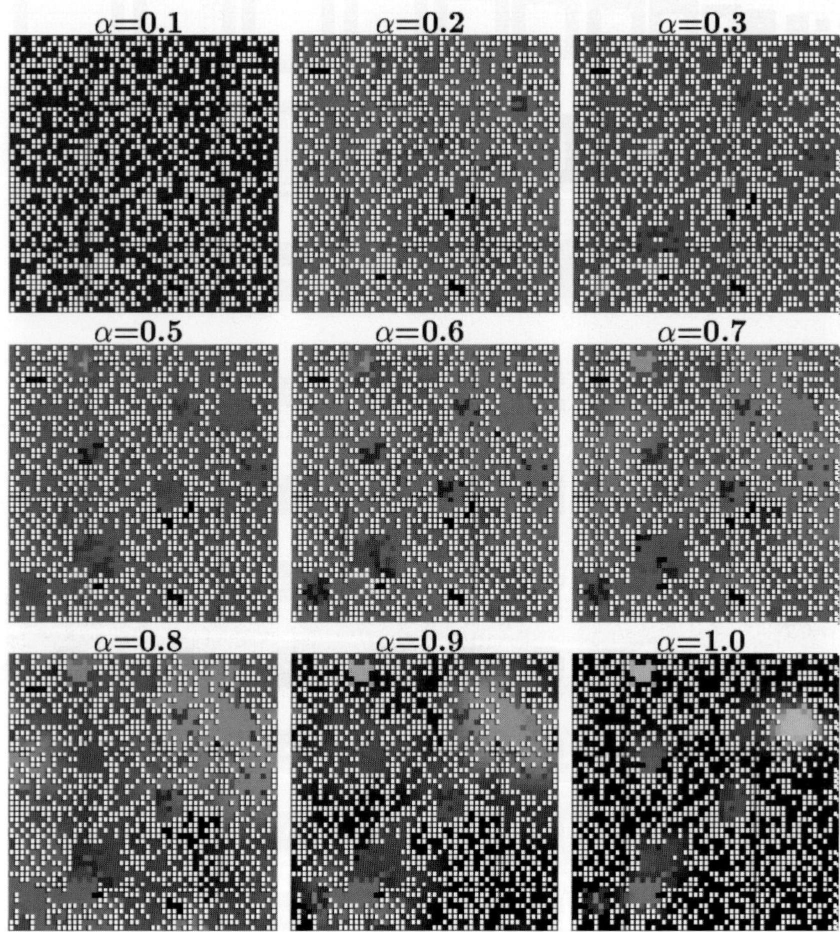

Fig. E.17   Patterns at $T=100$ in the CV-SDPD starting at random in a 51×51 lattice with Moore's neighborhood.

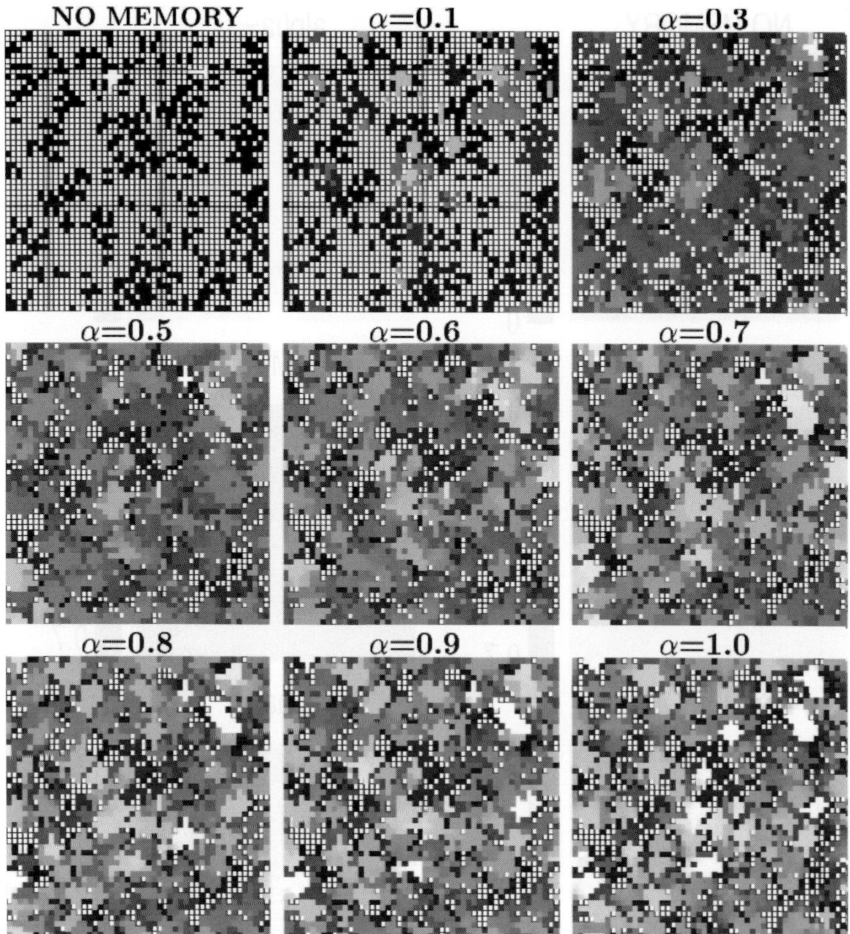

Fig. E.18 Patterns at $T=100$ in the CV-SDPD starting at random in a 51×51 lattice with von Neumann's neighborhood.

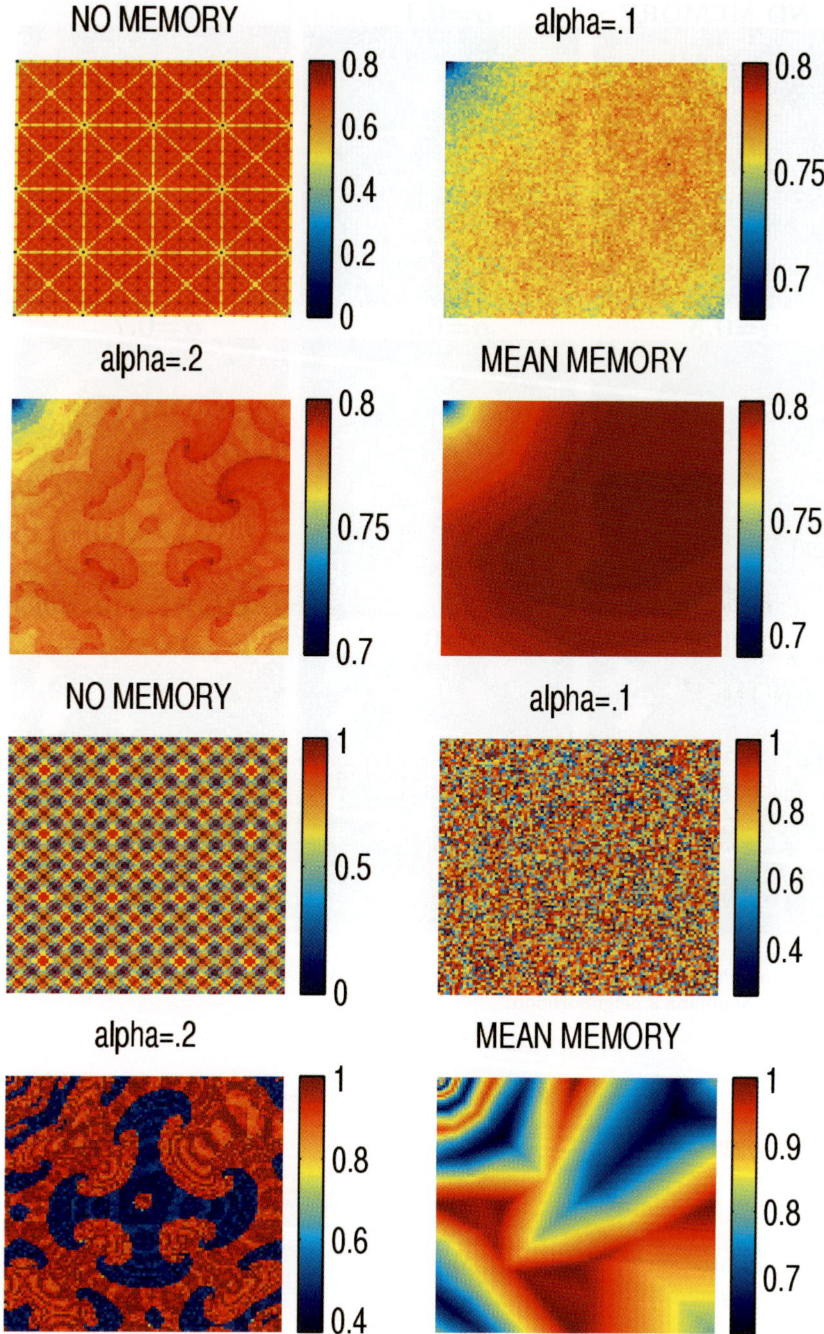

Fig. E.19   Mean value of the one hundred first values of x (upper four images), and values of x at $T = 100$ (lower four images) starting from any of the points of the 1×1 square with different degrees of memory.

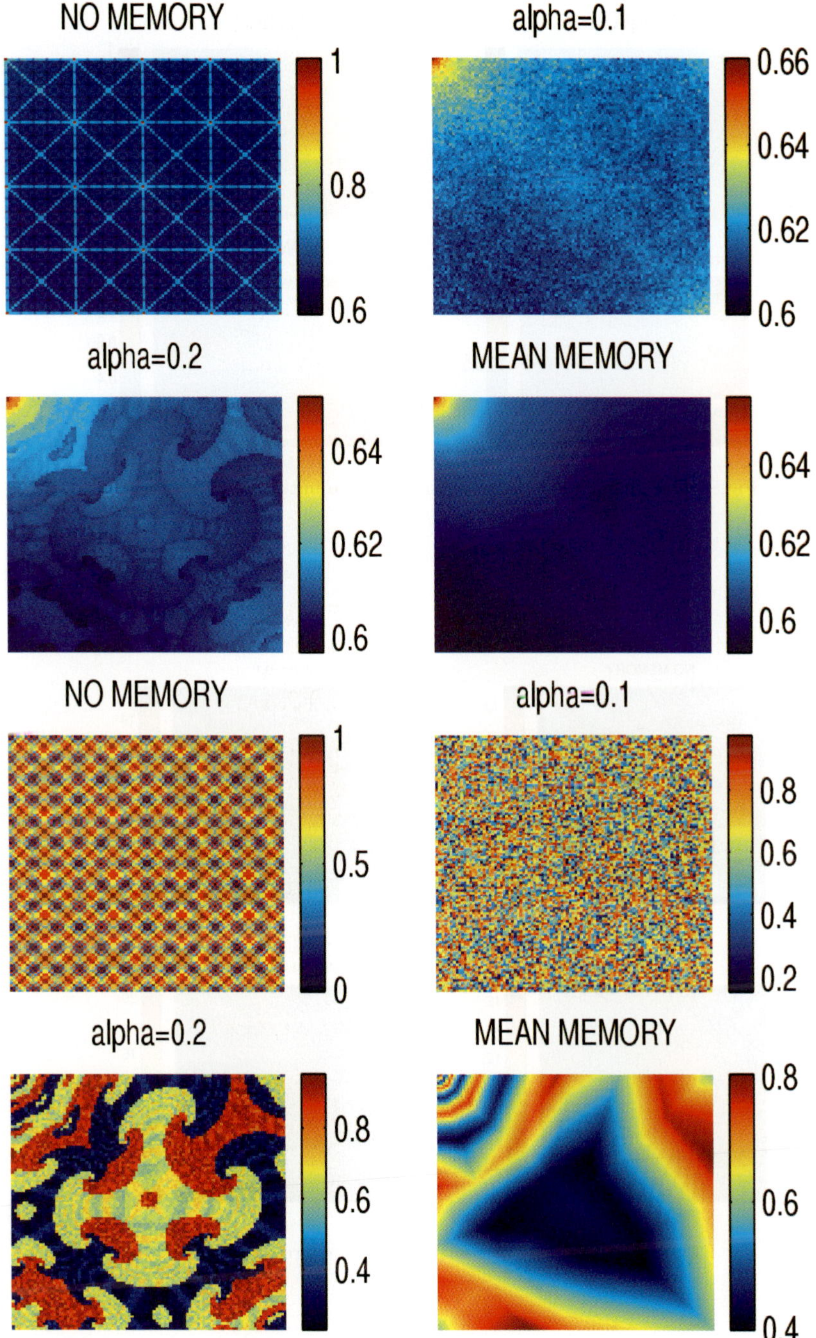

Fig. E.20   Mean value of the one hundred first values of y (upper four images), and values of y at $T = 100$ (lower four images) in the scenario of Fig. E.19.

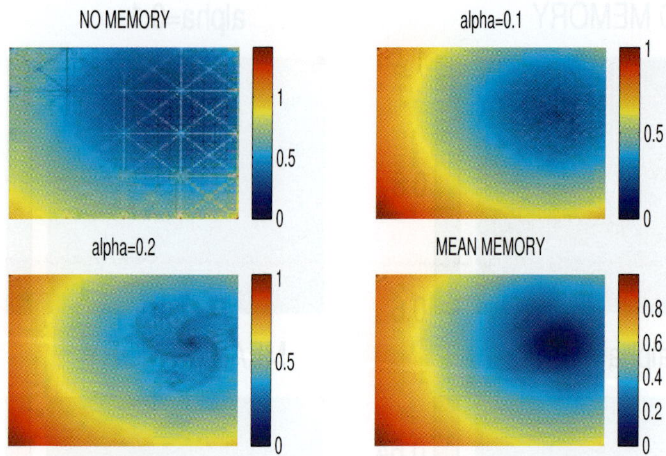

Fig. E.21　Mean value of the distance to the initial point at $T = 100$.

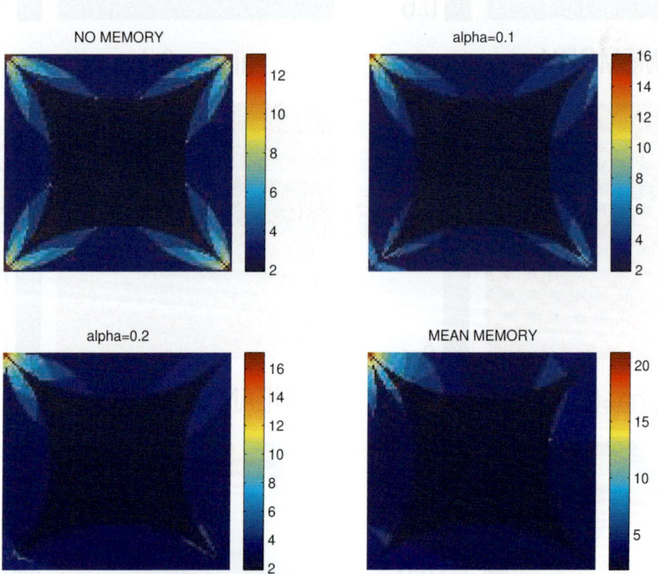

Fig. E.22　Escape time with different degrees of memory.

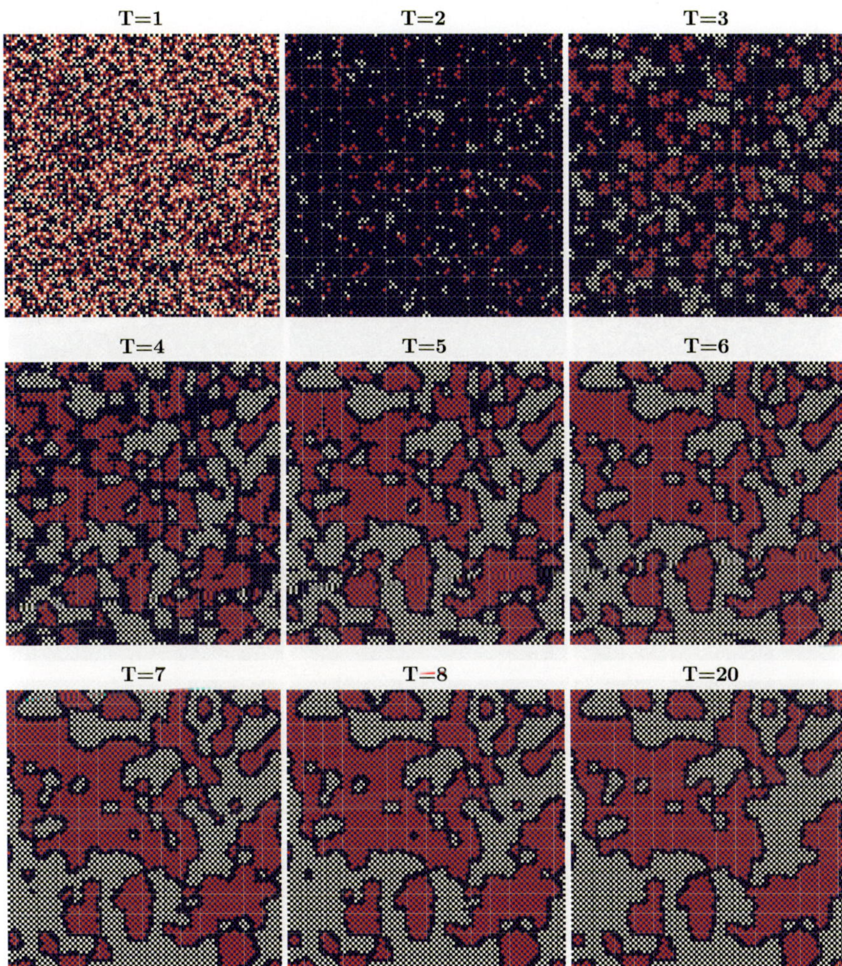

Fig. E.23   First patterns and pattern at $T=20$ in the $(R=5,r=1)$ battle of the sexes with no memory. 100×100 lattice size. Color code: $red \rightarrow \male B$, $blue \rightarrow \female B$, $black \rightarrow \male F$, $blank \rightarrow \female F$. Blue color tends to be masked by red in the agreement clusters.

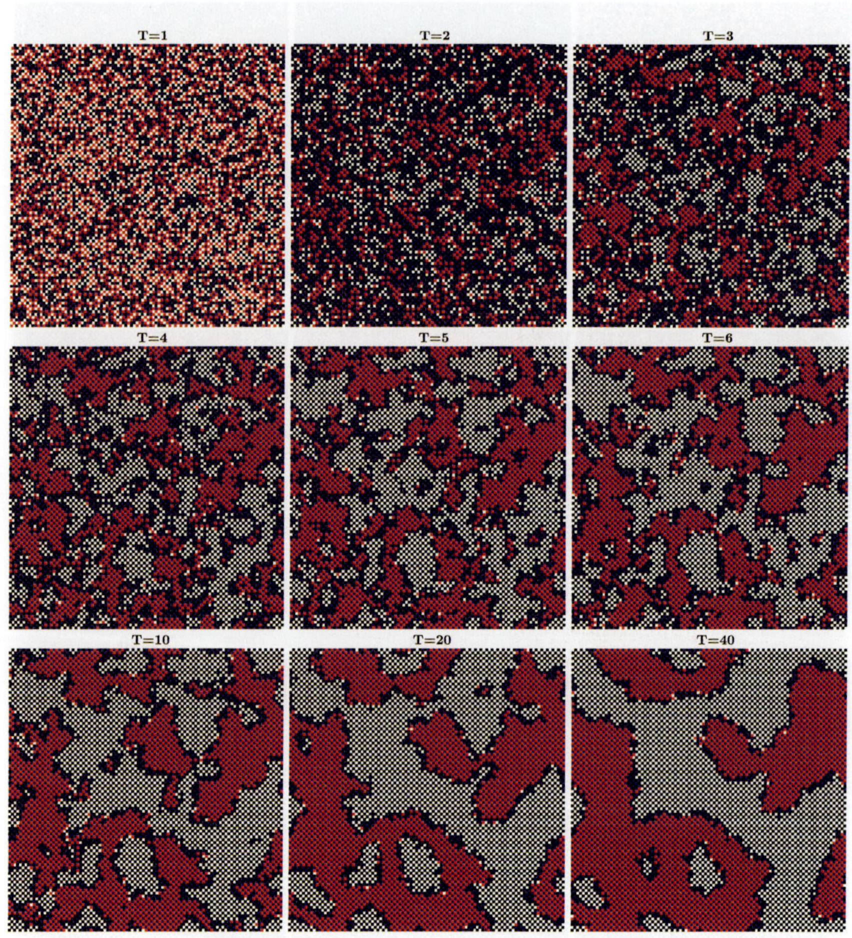

Fig. E.24   Initial patterns in a probabilistic (5,1)-BOS CA.

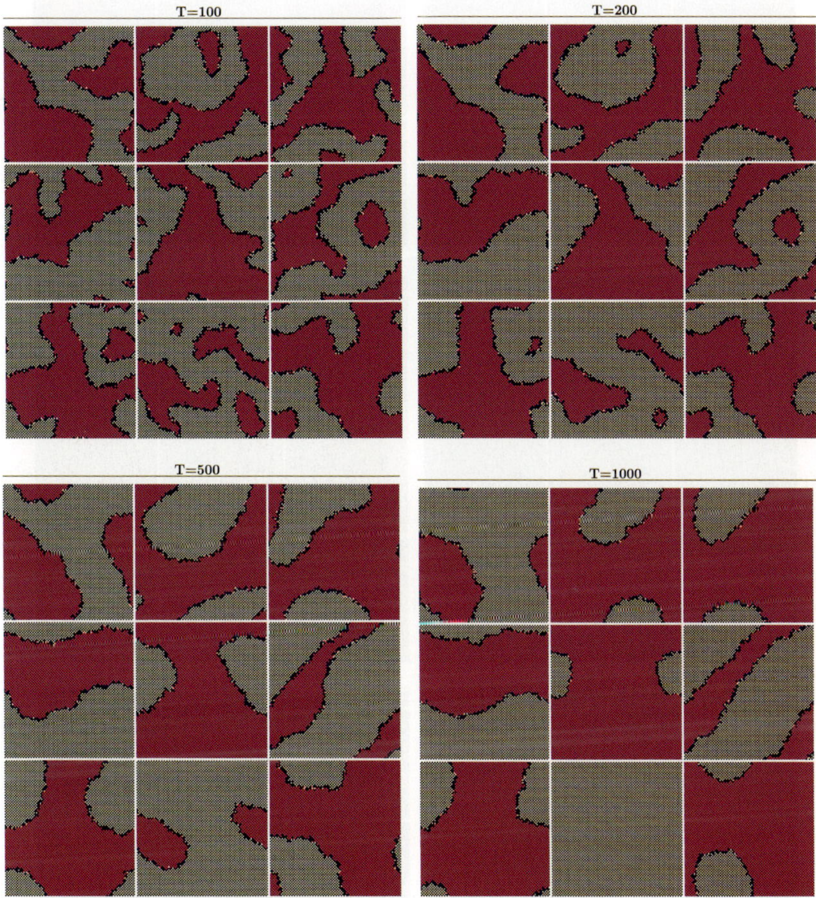

Fig. E.25   Patterns at T=100, 200, 500 and 1000 in the probabilistic (5,1)-BOS CA simulations of Fig. 13.54. Color code as in Fig. 13.48 .

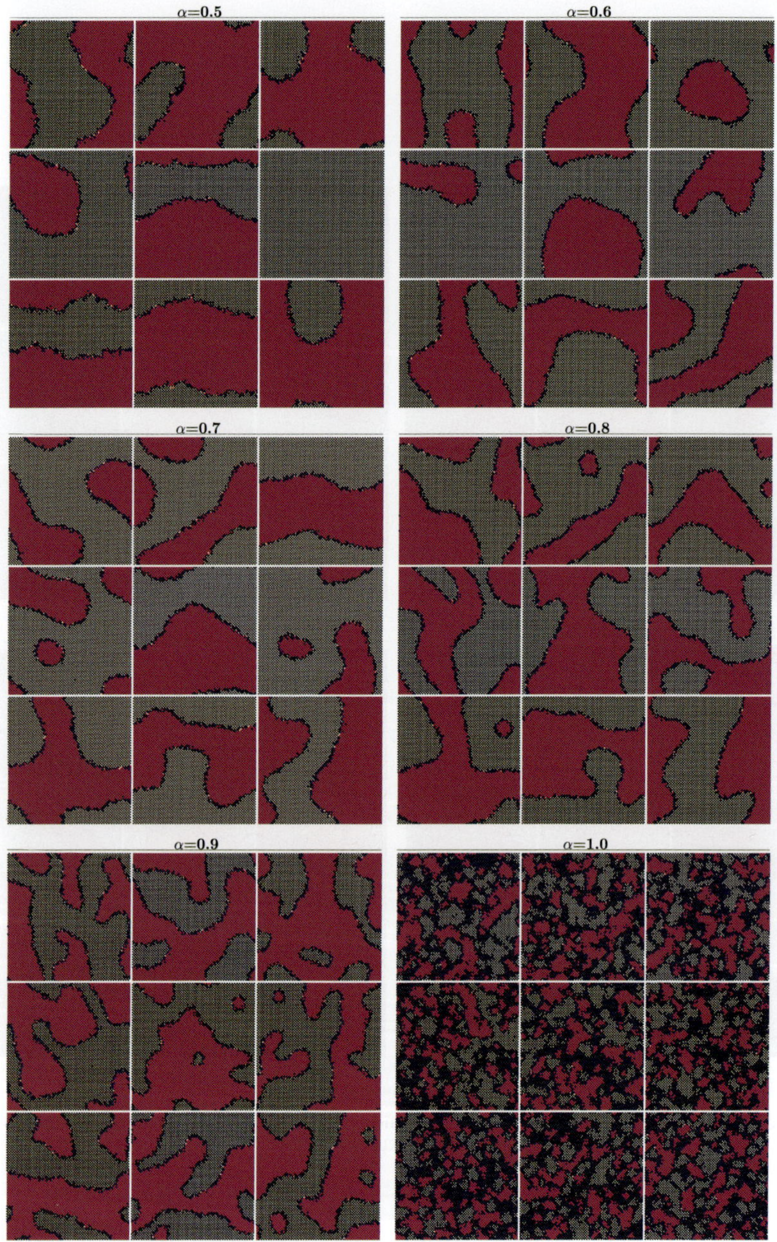

Fig. E.26   Patterns at $T{=}1000$ in the nine scenarios of the probabilistic (5,1)-BOS CA of Fig. 13.57 with increasing values of the α memory factor.

Bibliography

[1] Adamatzky,A.(2001). *Computing in Nonlinear Media and Automata Collectives.* IoP Publishing.

[2] Adamatzky A.(2009). Localizations in cellular automata with mutualistic excitation rules. *Chaos, Solitons & Fractals*, 10,981–1003.

[3] Adamatzky,A. (ed.).(2010). *Game of Life Cellular Automata.* Springer.

[4] Adamatzky,A.,Holland,O.(2002). Reaction-diffusion and ant-based load balancing of communication networks. *Kybernetes*, 31, 667–681.

[5] Adamatzky,A.,Holland,O.(1998). Phenomenology of excitation in 2-D cellular automata and swarm systems. *Chaos, Solitons & Fractals*, 9,1233–1265.

[6] Adamatzky,A.,Wuensche,A.(2007). Computing in spiral rule reaction-diffusion hexagonal cellular automaton. *Complex Systems*, 16,4.

[7] Adamatzky,A.(1994). *Identification of Cellular Automata.* Taylor and Francis

[8] Adamatzky,A.(1996). A Voronoi-like partition of lattice in cellular automata. *Math. Comput. Modeling*, 23,4,51-66.

[9] Adamatzky,A.,Wuensche,A.,De Lacy Costello,B.(2006). Glider-based computing in reaction-diffusion hexagonal cellular automata. *Chaos, Solitons and Fractals*, 27, 287–295.

[10] Adamatzky,A.(2010). On excitable β-skeletons. *J. of Computational Science*, 1,3,175-186.

[11] Ahmed,E.,Elgazzar,A.S.(2001). On some applications of cellular automata. *Physica A*, 296,529–38.

[12] Aicardi,F.,Invernizzi,S.(1992). Memory effects in discrete dynamical systems. *Int.J. Bifurcation and Chaos*, 2, 4, 815–830.

[13] Aldana,M.(2003). Boolean dynamics of networks with scale-free topology. *Physica D*, 185,1,46–66.

[14] Alexander,F.,Edrei,I.,Garrido,P.,Lebowitz,J.L.(1992). Phase transitions in a probabilistic cellular automaton : growth kinetics and critical properties. *J. Statistical Physics*, 68, 497-514.

[15] Alonso,J.,Fernandez,A.,Fort,H.(2006). Prisoner's Dilemma cellular automata revisited : evoution of cooperation under evivironmental pressure. *J. Stat, Mech.*, P06013.

[16] Alonso-Sanz,R.(2011). Extending the parameter interval in the logistic map with memory. *Int.J. Bifurcation and Chaos* (in press).

[17] Alonso-Sanz,R.(2010). The HPP rule with memory and the density classification task. *Int.J. Modern Physics C*, 21,9,1115–1118.

[18] Alonso-Sanz,R.(2010). Curlicues with memory. *Int.J. Bifurcation and Chaos*, 20,7, 2225-2240.

[19] Alonso-Sanz,R.(2010). Life with short-term memory. In Adamatzky,A.(ed.). [3] .

[20] Alonso-Sanz,R.(2009). Memory versus spatial disorder in the support of cooperation.*Biosystems*, 97,90–102.

[21] Alonso-Sanz,R.(2009). Spatial order prevails over memory in boosting cooperation in the iterated prisoner's dilemma. *Chaos*, 19,2,023102.

[22] Alonso-Sanz,R.(2009). Memory boosts cooperation in the structurally dynamic prisoner's dilemma. *Int.J. Bifurcation and Chaos*, 19,9,2899–2926.

[23] Alonso-Sanz,R.(2007). A Structurally Dynamic Cellular Automaton with Memory. *Chaos, Solitons & Fractals*, 32,1285–1295.

[24] Alonso-Sanz,R.(2007). Reversible structurally dynamic cellular automata with memory : a simple example. *J. of Cellular Automata*, 2,3,197–201.

[25] Alonso-Sanz,R.(2007). A structurally dynamic cellular automaton with memory in the triangular tessellation. *Complex Systems*, 17,1/2,1–15.

[26] Alonso-Sanz,R.(2006). The beehive cellular automaton with memory. *J. of Cellular Automata*, 1,3, 195-211.

[27] Alonso-Sanz,R.(2005). Phase transitions in an elementary probabilistic cellular automaton with memory. *Physica A*, 347,383–401.

[28] Alonso-Sanz,R.(2004). One-dimensional, $r=2$ cellular automata with memory. *Int.J. Bifurcation and Chaos*, 14,3217–3248.

[29] Alonso-Sanz,R.(2003). Reversible Cellular Automata with Memory. *Physica D*, 175,1–30.

[30] Alonso-Sanz,R.(2005). The Paulov versus anti-Paulov contest with memory. *Int.J. Bifurcation and Chaos*, 15,10,3395–3407.

[31] Alonso-Sanz,R.(1999). The historic prisoner's dilemma. *Int.J. Bifurcation and Chaos*, 9,6, 1197-1210.

[32] Alonso-Sanz,R.,Adamatzky,A.(2011). On beta-skeletons with memory. *J. Computational Sci.*, 2, 1 (in press).

[33] Alonso-Sanz,R.Adamatzky,A.(2008). On memory and structurally dynamism in excitable cellular automata with defensive inhibition. *Int.J. Bifurcation and Chaos*, 18,2,527–539.

[34] Alonso-Sanz,R.,Bull,L.(2010). One-dimensional coupled cellular automata with memory : Initial investigations. *J. of Cellular Automata*, 5,1-2, 29-49.

[35] Alonso-Sanz,R.,Bull,L.(2009). A very effective density classifier two-dimensional cellular automaton with memory. *J. of Physics A : Math. Theor.*, 42, 485101.

[36] Alonso-Sanz,R.Bull,L.(2009). On minimally coupled Boolean networks. *Int.J. Bifurcation and Chaos*, 19,4,1401–1414.

[37] Alonso-Sanz,R.,Bull,L.(2009). Elementary cellular automata with minimal memory and random number generation. *Complex Systems*, 18,2, 195-213.

[38] Alonso-Sanz,R.Bull,L.(2008). Boolean networks with memory. *Int.J. Bifurcation and Chaos*, 18,12, 3789–3799.

[39] Alonso-Sanz,R.,Bull,L.(2008). Random number generation by cellular automata with memory. *Int.J. Modern Physics C*, 19,2,351–375.

[40] Alonso-Sanz,R.,Cárdenas,J.P.(2008). On the effect of memory in one-dimensional $K=4$ automata on networks. *Physica D*, 237, 3099-3108.

[41] Alonso-Sanz,R.,Cárdenas,J.P.(2008). The effect of memory in cellular automata on *scale-free* networks: the $\overline{K} = 4$ case. *Int.J. Bifurcation and Chaos*, 18,8, 2477–2486.

[42] Alonso-Sanz,R.,Cárdenas,J.P.(2007). On the effect of memory in Boolean networks with disordered dynamics: the $K = 4$ case. *Int.J. Modern Physics C*, 18,8, 1313–1327.

[43] Alonso-Sanz,R.,Martin,M.(2006). A structurally dynamic cellular automaton with memory in the hexagonal tessellation. El Yacoubi,S., Chopard.B., Bandini,S.(eds.). *LNCS*, 4774, 30-40.

[44] Alonso-Sanz,R.,Martin,M.(2006). Elementary cellular automata with elementary memory rules in cells: the case of linear rules. *J. of Cellular Automata*, 1,1, 70-86.

[45] Alonso-Sanz,R.,Martin,M.(2004). Elementary probabilistic cellular automata with memory in cells. P.M.A. Sloot et al.(Eds.). *LNCS*, 3305, 11-20.

[46] Alonso-Sanz,R.,Martin,M.(2005). One-dimensional cellular automata with memory in cells of the most frequent recent value. *Complex Systems*, 15, 203-236.

[47] Alonso-Sanz,R.(2004). One-dimensional, $r = 2$ cellular automata with memory. *Int.J. Bifurcation and Chaos*, 14, 3217-3248.

[48] Alonso-Sanz,R.,Martin,M.(2004). Three-state one-dimensional cellular automata with memory. *Chaos, Solitons and Fractals*, 21,809-834.

[49] Alonso-Sanz,R.,Martin,M.(2003). Elementary cellular automata with memory. *Complex Systems*, 14,99–126.

[50] Alonso-Sanz,R.,Martin,M.(2003). Cellular automata with accumulative memory: legal rules starting from a single site seed. *Int.J. Modern Physics C*, 14,695-719.

[51] Alonso-Sanz,R.,Martin,M.(2003). Cellular automata with memory. In Modeling of Complex Systems. Seventh Granada Lectures P.L.Garrido,J.Marrro (eds.). *AIP Conf. Proc.*, 661,153-158.

[52] Alonso-Sanz,R.,Martin,M.(2002). One-dimensional cellular automata with memory: patterns starting with a single site seed. *Int.J. Bifurcation and Chaos*, 12, 205–226.

[53] Alonso-Sanz,R.,Martin,M.(2002). Two-dimensional cellular automata with memory: patterns starting with a single site seed. *Int.J. Modern Physics C*, 13,49–65.

[54] Alonso-Sanz,R.,Martin,M.C.,Martin,M.(2001). Historic Life. *Int. J. Bifurcation and Chaos*, 11,6,1665-1682.

[55] Alonso-Sanz,R.,Martin,M.(2006). Memory boosts cooperation. *Int.J. Modern Physics C*, 17,6,841-852.

[56] Alonso-Sanz,R.,Martin,M.(2005). The spatialized Prisoner's dilemma with

limited trailing memory. In Modeling Cooperative behavior In the Social Sciences: Eight Granada Lectures. P.L.Garrido et al.(eds.). *AIP Conf. Proc.*, 779,124–127.

[57] Alonso-Sanz,R.,Martin,M.C.,Martin,M.(2001). The effect of memory in the spatial continuous-valued prisoner's dilemma. *Int. J. Bifurcation and Chaos*, 11,8,2061-2083.

[58] Alonso-Sanz,R.,Martin,M.C.,Martin,M.(2001). The historic-stochastic strategist. *Int. J. Bifurcation and Chaos*, 11,7,2037-2050.

[59] Alonso-Sanz,R.,Martin,M.C.,Martin,M.(2001). The historic strategist. *Int. J. Bifurcation and Chaos*, 11,4,943-966.

[60] Alonso-Sanz,R.,Martin,M.C.,Martin,M.(2000). Discounting in the historic prisoner's dilemma. *Int.J. Bifurcation and Chaos*, 10,1,87–102.

[61] Alvarez,G.,Hernandez-Encinas,L.,Martin del Rey,A. (2008). A multisecret sharing scheme for color images based on cellular automata. *Information Sciences*, 178, 4382-4395.

[62] Alvarez,G.,Hernandez-Encinas,A.,Hernandez-Encinas,L.,Martin del Rey,A. (2005). A secure scheme to share secret color images. *Computer Physics Communications*, 173, 1-2,9-16.

[63] Andrade,J.S.Jr.,Herrmann,H.J.,Andrade,R.F.S.,da Silva,L.R.(2005). Apollonian Networks: Simultaneously Scale-Free, Small World, Euclidean, Space Filling, and with Matching Graphs. *Phys. Rev. Lett.*, 94, 018702.

[64] Armitage,V.,Rogers,A.(2000). Gauss sums and quantum mechanics. *J. of Physics A - Mathematical and General*, 33, 34, 5993–6002.

[65] Ashlock,D.,Smucker,M.D.,Stanley,E.A.,Tesfation,L.(1996). Preferential partner selection in an evolutionary study of Prisoner's Dilemma. *Biosystems*, 37,99–125.

[66] Astrom,J.A.,Herrmann,Timonen,J.(2000). Granular packings and fault zones. *Phys. Rev. Lett.*, 84,4, 638-641.

[67] Aubert,M.,Badoual,M.,Fereol,S.et al.(2006). A cellular automaton model for the migration of glioma cells. *Phys. Biol.*, 3, 93-100.

[68] Aumann,R.J., Subjectivity and correlation in randomized strategies, *J. of Mathematical Economics* **1**, 67-96 (1974).

[69] Aumann,R.J., Correlated equilibrium as an expression of Bayesian rationality, *Econometrica* **55**,1, 1-18 (1974).

[70] Axelrod,R.(1984). *The Evolution of Cooperation*. Basic Books, NY.

[71] Bak, P., Flyvbjerg, H,Lathrup, B.(1994). Evolution and coevolution in rugged fitness landscapes. In Langton,C. (Ed.), Artificial Life III. Addison-Wesley, 11-42.

[72] Ball,Ph.(2008). Cellular memory hints at the origin of intelligence. *Nature*, 451,7177,358.

[73] Barabási,A.-L.,Albert,R.(1999). Emergence of scaring in random networks. *Science*, 286, 5439,509-512.

[74] Bauke,H.,S.,Merten,H.S.(2004). Pseudo random coins show more heads than tails. *J. Stat. Phys.*, 114,1149-1169 (2004).

[75] Bays,C.(2009). Gliders in cellular automata. In Meyers,R.A.(ed.) [288], 4240–4249.

[76] Bidaux,R.,Boccara,N.(2000). Correlated random walks with a finite memory range. *Int.J. Modern Physics C*, 11,941-947.

[77] Bellomo,B.,Lo Franco,R.,Compagno G.(2007). Non-Markovian Effects on the Dynamics of Entanglement. *Physical Review Letters*, 99, 160502.

[78] Bennett,M.R.,Hasty,J.(2008). *E. coli* rewired. *Nature*, 452,7189,824.

[79] Berndt,B.C.,Evans,R.J.(1981). The determination of Gauss sums. *Bull. Amer. Math. Soc.*, 5,2,107-129.

[80] Berg,M. de.,Cheong.O.,Kreveld,M. van,Overmans,O.(2008). *Computational Geometry.* Springer.

[81] Berrry,M.V.,Goldberg,J.(1988). Renormalization of curlicues. *Nonlinearity*, 1,1-26.

[82] Berrry,M.V.(1999). A theta-like sum from difraction physics. *J. Phys. A : Math. and Gen.*, 32, L329-L336.

[83] Bersini,H.,Detours,V.(1994). Asynchrony induces stability in cellular automata based models. *Proc. IV Int. Conf. Artificial Life*, 382-387.

[84] Billiot J. M., Corset F., Fontenas E. (2010). Continuum percolation in the relative neighborhood graph. arXiv:1004.5292

[85] Boccara,N.(2007). Eventually number-conserving cellular automata. *Int. J. Modern Physics C* ,18,35-45.

[86] Boccaletti,S.,Latora,V.,Moreno,Y.,Chavez,M.,Huang,D.-U.(2006). Complex networks : Structure and dynamics. *Physics Reports-Rev. Sect. Phys. Lett.*, 424, 4-5,175-308.

[87] Boccara,N.,Fuks,H.(2002). Number-conserving cellular automaton rules. *Fundamenta Informaticae*, 20,1-13.

[88] Boccara,N.,Fuks,H.(1998). Cellular automaton rules conservng the number of active sites. *Journal of Physics A : Math. and Theoretical*, 31,28, 6007–6018.

[89] Blok,H.J.,Bergersen,B.(1997). Effect of boundary conditions on scaling in the "game of Life". *Physical Review E*, 55,5,6249–252.

[90] Blok,H.J.,Bergersen.B.(1999). Synchronous versus asynchronous updating in the "game of Life". *Physical Review E*, 59,4,3876-3879.

[91] Bloom,D.M.(1993). An old algoritm for the sum of integer powers. *Mathematics Magazine*, 66,5,304–305.

[92] Bonner,J.T.(2009). The social amoebae : The Biology of cellular slime models. Princeton University Press.

[93] Brauchli,K.,Lillingback,T.,Doebeli,M.(1996). Evolution of cooperation in spatially structured populations. *J. Theor. Biology*, 200, 405–417.

[94] Briesen,M.,Weimar,J.R.(2001). Distribute simulation environment for coupled cellular automata. In *PARCO 2001*. Joubert,G. *et al.* (eds). Imperial College Press.

[95] Bull, L.(1999). On the evolution of multicellularity and eusociality. *Artificial Life*, 5,1,1-15.

[96] Bull, L.,Fogarty, T.C.(1996). Artificial symbiogenesis. *Artificial Life*, 2,3,269–292.

[97] Bull, L.(1997). Evolutionary computing in multi-agent environments : partners. In *Proc. of the 7th Intern. Conf. on Genetic Algorithms.* Morgan

Kaufmann, 370-377.

[98] Calidonna,C.R.,Naddeo,A.(2006). Introducing reversibility in a high level JJL qbit model according to CAN2 paradigm. *Lect. Notes in Computer Science*, 4173, 215-221. Towards reversibility in a JJL qubit qualitative model by means of CAN2 paradigm. *Physics Letters A*, 358, 5-6, 463-469.

[99] Calidonna,C.R.,Naddeo,A.(2008). Addressing Reversibility in Quantum Devices by a Hybrid CA Approach : the JJL Case. *Int. J. of Unconventional Computing*, 4,4,315-340.

[100] Capcarrere,M.S.,Sipper,M.,Tomassini,M.(1996). Two-state, $r = 1$ cellular automata that classifies density. *Physical Reviews Letters*, 77, 4969-4971.

[101] Casti,J.L.(2002). Science is a computer program. *Nature*, 417, 381-382.

[102] Casti,J.L.,Karlquist,A.(eds.).(2003). Art and Complexity. Elsevier Science B.V..

[103] Caudle,R.M.(2006). Memory in astrocytes : a hypothesis. *Theoretical Biology and Medical Modeling*, 3,2, 1-10.

[104] Cellarosi,F.(2009). Limiting curlicue measures for theta sums. *Ann. Inst. Henri Poincare (B) Prob. Stat.*, (in press).

[105] Challet,D.,Zhang-Y.-C.(1997). Trajectory and global attractors for evolution equations with memory. *Physica A*, 246, 407–418.

[106] Challet,D.,Zhang-Y.-C.(1998). On the minority game : Analytical and numerical studies. *Physica A*, 514–532.

[107] Chepyzov,V.V.,Gatti,S.,Grasselli,M.,Miranville,A.,Pata,V.(2006). Trajectory and global attractors for evolution equations with memory. *Applied Mathematics Letters*, 19, 87-96.

[108] Chialvo,D.R.,Cecchi,A.,Magnasco,M.O.(2000). Noise-induced memory in extended excitable systems. *Physical Review E*, 61,5,5654-5657.

[109] Chong,S.W.,Yao,X.(2005). Behavioural diversity, choices, and noise in the iterated prisoner's dilemma. *IEEE Trans. Evol. Comp.*, 9, 6,540-551.

[110] Chopard,B.,Droz,M.(1998). *Cellular Automata for Modeling Physics*. Cambridge Univ. Press.

[111] Chong,S.W.,Yao,X.(2007). Multiple choices and reputation on multiagent interactions *IEEE Trans. Evol. Comp.*, 11,6 689-711.

[112] Chua,L.O.(2006). *A Nonlinear Dynamics Perspective Wolfram's New Kind Science*. World Scientific.

[113] Chua,L.O.(2008). *CNN : A Paradigm for Complexity*. World Scientific.

[114] Conway,J.H.(1982). What is Life ?. In *Winning ways for your mathematical plays*, Berkelamp,E.R.,Conway,J.H.,Guy,R.K.(eds). Academic Press, NY.

[115] Cosnard,M.,Moumida,D.,Goles,E.,St.Pierre,T.(1988). Dynamical behaviour of a neural automaton with memory. *Complex Systems*, 2,161-176.

[116] Cressoni,M.A.,da Silva,M.A.A.,Viswanathan,G.M.,(2007). Amnestically induced persistence in random walks. *Phys. Rev. Lett.*, 98,7,070603.

[117] Chruscinski,D.,Kossakowski,A.,Pascazio,S.(2009). Long-time memory in non-Markovian evolutions. arXiv:0906.5122

[118] Darwen,P.J,Yao,X.(2002). Co-evolution in iterated prisoner's dilemma with

intermediate levels of cooperation: application to missile defense. *Int. J. Computational Intelligence and Applications*, 2,1,83-107.

[119] Das,R.,Crutchfield,J.P.,Mitchell,M.,Hanson,J.E.(1995). Evolving globally synchronized cellular automata. In L.J.Eshelman (ed.). *Proc. of the 6th Int. Conf. On Genetic Algorithms*, 336-343. M.Kaufmann, San Francisco, CA.

[120] Deutsch,A.,Dormann,S.(2004). *Cellular Automaton Modelling of Biological Pattern Formation*. Birkhauser.

[121] Devaney,R.L.(1984). A piecewise linear model for the zones of instability of an area preserving map. *Physica D*, 10, 387–393 .

[122] Dekking,M.,Mendes-France,M.,Poorten,A.J.,(1982). Folds !. *Mathematical Intelligencer*, 4, 130–138, 173–181, 190–195.

[123] Devaney,R.L.(1984). The Gingerbreadmap. *Algorithm*, 3, 15–16 .

[124] Dieckman,U.,Law,R.,Metz,J.A.J.(2000). *The Geometry of Ecological Interactions. Simplifying Spatial Complexity*. Cambridge University Press. IIASA.

[125] Doebeli,M.,Hauert,C.(2005). Models of cooperation based on the Prisoner's Dilemma and the snowdrift game. *Ecology Letters*, 8,748-766.

[126] Dogaru,R.(2008). *Systematic design for emergence Cellular Neural Networks*. Springer.

[127] Domany,E.,Kinzel,W.(1984). Equivalence of cellular automata to Ising models and directed percolation. *Physical Review Letters*, 53,4, 311-314.

[128] Dorfman,J.R.(1999). *An introduction to chaos in nonequilibrium statistical mechanics*. Cambridge Lecture Notes in Physics.

[129] Dorin, A.(2000). Boolean networks for the generation of rhythmic structure. In *Proceedings of the Australian Computer Music Conference*, 38-45.

[130] Dutta, A.,Bhattacherjee,J.K. (2008). Period adding bifurcation in a logistic map with memory. *Physica D*, 237, 3153–3158.

[131] Dzwinel,W.(2010). Spatially extended populations reproducing logistic map. *Central European Journal of Physics*, 8, 1,33-41.

[132] Ebel,H.,Bornholdt,S.(2002). Coevolutionary games on nerworks. *Phys. Rev. E.*, 66, 065118.

[133] Eguiluz,V.M.,San Miguel,M.,Zimmermann,M.G,Cela-Conde,C.J. (2005). Cooperation and the emergence of role differentiation in the dynamics of social networks. *Amer. J. of Sociology*, 110, 4,977-1008.

[134] Eisele,M.(1991). Long-range correlations in chaotic cellular automata. *Physica D*, 48, 295-310.

[135] Eisert,J.,Wilkens,M.,Lewenstein,M.(1999). Quantum games and quantum strategies. *Physical Review Letters*, 83,15,3077-3080.

[136] Eslami,Z.,Ragazzi,S.H.,Ahmadabai,S,H.(2010). Secret image sharing based on cellular automata and steganography. *Pattern Recognition*, 43, 397-404.

[137] Evans,R.,Minei,M.,Yee,B.(2003). Incomplete higher order Gauss-sums. *J. Math. Anal. Appl.*, 281, 454-476.

[138] Fates,N.A.,Morvan,M.(1997). An experimental study of robustness for elementary cellular automata. *Complex Systems*, 16,1,1-19.

[139] Fay,T.H.(1989). The butterfly curve. *The American Mathematical Monthly*,

96, 5, 442–443.

[140] Fay,T.H.(1997). A study on step size. *The Mathematics Magazine*, 96, 5, 116–117.

[141] Ferrenberg,A.M.,Landau,D.P.,Wong,Y.J.(1992). Monte Carlo simulations: Hidden errors from "good" random number generators. *Physical Review Letters*, 69,23, 3382-3384.

[142] Frackiewicz,P. (2009). The ultimate solution to the quantum battle of the sexes game. *J. of Physics A -Math. and General*, 42,36,365305.

[143] Fu,F.,Chen,X.J.,Liu,L.G.,Wang,L.(2007). Promotion of cooperation induced by the interplay between structure and game dynamics. *Physica A*, 383,2,651-659.

[144] Fernández, P.,Solé, R.(2004). The role of computation in complex regulatory networks. In E. Koonin,Y. Wolf, G. Karev (Eds.) *Power Laws, Scale-free Networks and Genome Biology*. Landes.

[145] Fenton,H.F.,Evans,S.J.,Hastings,H.M.(1999). Memory in an excitable medium a mechanism for spiral wawe breakup in the low-excitability limit. *Phys. Review Letters*, 83,19,3964-3967.

[146] Feymann,R.(1996). *Feymann Lectures on Computation*. T.Hey, R.W.Allen (eds.). Perseus Pub. .

[147] Fick,E.,Fick.M.,Hausmann,G.(1991). Logistic equation with memory. *Phys. Rev. A.*, 44,4, 2469-2473.

[148] Fisk,R.,Gravner,J.,Griffeath,D.(1991). Threshold-range scaling of excitable cellular automata. *Statistics and Computing*, 1,23–29.

[149] Frean,M.(1996).The evolution of degrees of cooperation, *J.Theor. Biology*, 182, 549-559.

[150] Frean,M.R.,Abraham,E.R.(2001). A voter model of the spatial prisoner's dilemma. *IEEE Transactions on evolutionary computation*, 5,2, 117–121.

[151] Fredkin,E.(1990). Digital mechanics. An informal process based on reversible universal cellular automata. *Physica D*, 45,254-270.

[152] Freiwald,U.,Weimar,J.R.(2000). The Java based cellular automata simulation system - JCAsim. *Fut. Gen. Comp. Systems*, 18,995-1004.

[153] Fuks,H.(1997). Solution of the density classification problem with two cellular automata rules. *Phys. Rev. E*, 55,R2081-R2084.

[154] Fuks,H.,Sullivan,K.(2007). Enumeration of number-conserving cellular automata rules with two inputs. *Journal of Cellular Automata*, 2,2,141-148.

[155] Fussy,S.,Grössing,G.,Schwabl,H.(1993). Nonlocal computation in quantum cellular automata. *Physical Review A*, 48,5,3470-3479.

[156] Gabriel K. R. and Sokal R. R. (1969). A new statistical approach to geographic variation analysis. *Systematic Zoology*, 18,259,259-270.

[157] Gacs,P.,Kurdyumov,G.L.,Levin,L.A.(1978). One-dimensional uniform arrays that wash out finite islands. *Problemy Pederachi Infomatsii*, 14,92-98.

[158] Gallas,J.A.C.(1993). Simulating memory effects with discrete dynamical systems. *Physica A*, 195,417 198,339.

[159] Gallas,J.A.C.(1993). Degenerate routes to chaos. *Physical Review E*, 48,6, R4156-4159

[160] Ganguly,N.,Maji,P.,Das,A.Sikdar,B.P.,Chaudhury,P.(2002).   Generalized

multiple attractor cellular automata (GMACA) model for associative memory. *Int. J. of Bifurcation and Chaos*, 16,7,781-795.

[161] Gardner,M.(1970). The fantastic combination of John Conway's new solitaire game "life". *Scientific American*, 223,4, 120-123.

[162] Gardner,M.(1971). On cellular automat, self-reproduction, the Garden of Eden and the game "life". *Scientific American*, 224,2, 112-117.

[163] Gershenson, C.(2002). Classification of random Boolean networks. In R.K. Standish, M. Bedau, H. Abbass (Eds.) Artificial Life VIII. MIT Press, 1-8.

[164] Gershenson,C,Broekaert,J.,Aerts,D.(2003). Contextual random Boolean networks. In *Proc. of the 7th European Artificial Life Conf.*. Springer, 615-624.

[165] Giles,J.(2002). What kind of science is this ?. *Nature*, 417, 216-218.

[166] Giona,M.(1991). Dynamics and relaxation properties of complex systems with memory. *Nonlinearity*, 4, 911-925.

[167] Gonzaga de Sá,P.,Maes,C.(1992). The Gacs-Kurdyumov-Levin automaton revisited. *Journal of Statistical Physics*, 67,3/4,507-522.

[168] Grassberger,P.(1984). Chaos and diffusion in deterministic cellular automata. *Physica D*, 10,52-58.

[169] Grassberger,P.(1986). Long-range effects in an elementary cellular automaton. *J. Stat. Physics*, 45,5-6,27-40.

[170] Grassberger,P.(1999). Synchronization of complex systems with spatio-temporal chaos. *Phy. Rev. E*, 59,R2520.

[171] Gravner,J.(2009). Growth phenomena in cellular automata. In Meyers,R.A.(ed.) [288], 4497–4511.

[172] Greenberg,J.M.,Hastings,S.P.(1978). Spatial patterns for discrete models of diffusion in exciatble media. *SIAM J. Appli. Math.* ,34,515-522.

[173] Grim,P.,Mar,G.,St.Denis,P.(1998). *The Philosophical Computer*. MIT Press /Bradford Books.

[174] Grim,P.(1995). The greater generosity of the spatialized prisoner's dilemma. *Journal of Theoretical Biology*, 173, 353-359.

[175] Grinfeld,M.,Knight,P.A.,Lamba,H.(1996). On the periodically perturbed logistic equation. *J.Phys. A : Math. Gen.*, 29, 8035-8040.

[176] Grinstein,G.,Jayaprakash,C.,Yu,H.(1985). Statistical mechanics of probabilistic cellular automata. *Physical Review Letters*, 55,23, 2527-2530.

[177] Gross,T.,Blasius,B.(2008). Adaptive coevolutionary networks: a review. *J. of the Royal Soc. Interface*, 5,20,259-271.

[178] Grössing,G.,Zeilinger,A.(1988). Structures in quantum cellular automata. *Physica B*, 15,366.

[179] Grunbaum,B.,Shephard,G.C.(2009). Tilings and Patterns. *Dover Pub.*.

[180] Guan,S.U.,Zhang,S.(2003).Incremental evolution of cellular automata for random number generation. *Int. J. Modern Physics C*, 14,7, 881-896.

[181] Guan,S.U.,Zhang,S.(2003). An evolutionary approach to the design of controllable cellular automata structure for random number generation. *IEEE Trans. on Evolutionary Computation*, 7,1,23,36. A family of controllable cellular automata for pseudo-random number generation. *Int. J. Modern Physics C*, 13,8, 1047-1073.

[182] Guan,S.U.,Zhang,S.(2003). Pseudo-random number generation with self-programmable cellular automata. *IEEE Trans. on Computer-Aided Design of Integrated circuits and systems*, 23,7,1095-1101.

[183] Hardy,J.,Pazzis,O.,Pomeau,Y.(1976). Molecular dynamics of a classical lattice gas : Transport properites and time correlation functions. *Phys. Review A*, 13,5, 1949–1961.

[184] Hardy,J.,Pomeau,Y.,Pazzis,O. (1973). Time evolution of a two-dimensional model system. I. Invariant states and time correlation functions. *Journal of Mathematical Physics*, 14,12, 1746–1759.

[185] Hastings,A.(1993). Complex interactions between dispersal and dynamics - lessons from coupled logistic equations. *Ecology*, 74,5,362-1372.

[186] Hemmingsson,J.(1995). Current results on "Life". *Physica D*, 80, 151-153.

[187] Hendry,R.J.,McGlade,J.M.(1995). The role of memory in ecological systems. *Proc. R.Soc.Lond. B*, 259, 153-159.

[188] Henon,M.(1976). A two-dimensional mappping with a strange attactor. *Communications in Mathematical Physics*, 50,69-77.

[189] Hall,L.M.(1992). Trochoids, Roses and Thorns-Beyond the Spirograph. *The college of mathematics Journal*, 23,1, 20-35.

[190] Halpern,P.,Caltagirone,G.(1990). Behavior of topological cellular automata. *Complex Systems*, 4, 623-651.

[191] Halpern,P.(1989). Sticks and stones : a guide to structurally dynamic cellular automata. *Am. J. Physics*, 57,405.

[192] Harley, I.(1981). Learning the evolutionary stable strategy. *J. Theor. Biol.*, 89,611-633.

[193] Harvey, I.,Bossomaier, T.(1997). Time out of joint : attractors in asynchronous random Boolean networks. In *Proceedings of the Fourth European Artificial Life Conference*, MIT Press, 67-75.

[194] Hauert,C.,Schuster,H.G.(1997). Effects of increasing the number of players and memory steps in the iterated prisoner's dilemma, a numerical approach. *Proc. R.Soc.Lond. B*, 264, 513-519.

[195] Hauert,C.,Doebeli,M.(2001). Spatial structure often inhibits the evolution of cooperation in the snowdrift game. *Nature*, 428, 643-646.

[196] Herrmann,H.J.,Mantica,G.,Bessis,D.(1990). Space filling bearings. *Phys. Rev. Lett.*, 65,26, 3223-3226.

[197] Huberman,B.A.,Glance,N.S.(1993). Evolutionary games and computer simulations. *Proc. Natl. Acad. Sci. USA*, 90, 7716.

[198] Hillman,D.(1995). *Combinatorial Spacetimes*. Ph.D. Thesis (mathematics) Univ. of Pittsburg. hep-th /9805066.

[199] Hod,S.,Keshet,U.(2004). Phase transitions in random walk with long-range correlations. *Phys. Rev. E*, 70,1, 015104.

[200] Hoekstra,A.G.,Kroc,J,Sloot,P.M.A.(eds.).(2010). *Simulating Complex Systems by Cellular Automata*. Springer.

[201] Hofbauer,J.,Sigmund,K.(2003). *Evolutionary Games and Population Dynamics*. Cambridge University Press.

[202] Hogeweg,P.(1988). Cellular automata as a paradigm for ecological modeling. *Applied Mathematics and Computation*, 27, 81-100.

[203] Hopfield,J.J.(1982). Neural networks and physical systems with emergent collective computational abilities. *Proc. Nat. Acad. Sci. USA*, 79, 2554-2558.

[204] Hortensius,P.D.,McLeod,R.D.,Card,H.C.(1989). Parallel random number generation for VLSI systems using cellular automata. *IEEE Transactions on Computers*, 38,10, 1466-1472. Hortensius,P.D. et al.(1989). Cellular automata-based pseudo-random number generators for Built-In Self-Test. *IEEE Transactions on Computer-Aided Design*, 8,8, 842-858.

[205] Hung, Y-C., Ho, M-C., Lih, J-S.,Jiang, I-M.(2006). Chaos synchronisation in two stochastically coupled random Boolean networks. *Physics Letters A*, 356, 35-43.

[206] Ilachinsky,A.,Halpern,P.(2009). Structurally dynamic cellular automata. In Meyers,R.A.(ed.) [288], 8815–8850.

[207] Ilachinsky,A.,Halpern,P.(1987). Structurally dynamic cellular automata. *Complex Systems*, 1, 503-527.

[208] Ilachinski,A.(2001). *Cellular Automata. A Discrete Universe.* World Scientific.

[209] Ingerson,T.E.,Buvel,R.L.(1984). Structure of asynchronous cellular automata. *Physica D*, 10, 59-68.

[210] Inoue,M.,Nagadome,M.(2000). Edge of chaos in stochastically coupled cellular automata. *J. Phys. Soc. Jpn.* ,69,1-4.

[211] Isalan,M. *et al.*(2008). Evolvability and hierarchy in rewired bacterial gene networks. *Nature*, 452,7189,840-845.

[212] Ito,H.(1988). Intriguing properties of global structure in some classes of finite cellular automata. *Physica D*, 31, 318-38.

[213] Itoh,M.,Chua,L.O.(2010). Memory effects in discrete dynamical systems. *Int.J. Bifurcation and Chaos*, 19, 11, 3605-3656.

[214] Jacob,F.(1977). Evolution and tinkering. *Science*, 196,1161-1166.

[215] Jen,E.(1986). Global properties of cellular automata. *J. Stat. Physics*, 43, 1/2,219-242.

[216] Jeon,J.C.,Kim,K.W.,Yoo,K.Y.(2005). Low-complexity authentication scheme based on cellular automata in wireless network. *Lect. Notes in Computer Science*, 3794,413-421.

[217] Jimenez-Morales,F.,Crutchfield,J.P.,Mitchell,M.(2001). Evolving two-dimensional cellular automata to perform density classification: A report on work in progress. *Parallel Computing*, 27,571–585.

[218] Jin,W.,Chen,F.,Chen,G.,Chen,L.,Chen,F.(2010). Extending the Symbolic of Chua's Bernouilli-Shift rule 56. *J. of Cellular Automata*, 5,1-2,121–138.

[219] Julian,P.,Chua,L.O.(2002). Replication properties of parity cellular automata. *Int. J. Bifurcation and Chaos*, 12,3, 477-494.

[220] Kaneko,K.(1983). Transition from torus to chaos accompanied by frequency lockings with symmetry-breaking - in connection with the coupled-logistic map. *Progress of Theoretical Physics*, 69,5,1427-1442.

[221] Kaneko,K.(1986). Phenomenology and characterization of coupled map lattices. In *Dynamical Systems and Singular Phenomena*. World Scientific.

[222] Kauffman, S.A.(1969). Metabolic stability and epigenesis in randomly con-

structed genetic nets. *Journal of Theoretical Biology*, 22,437-467.

[223] Kauffman,S.A.(1984). Emergent properties in random complex automata. *Physica D*, 10, 145-156.

[224] Kauffman,S.A.(1993). *The Origins of Order: Self-Organization and Selection in Evolution*. Oxford University Press.

[225] Kauffman,S.A.(1995). *At Home in the Universe: The Search for Laws of Self-Organization and Complexity*. Oxford University Press.

[226] Kauffman,S.A.(1991). Antichaos and adaptation. *Scientific American*, August,p.64 .

[227] Kauffman S.A.,Johnsen, S.(1991). Coevolution to the edge of chaos: coupled fitness landscapes, poised states, and coevolutionary avalanches. In Langton,C. (Ed.), Artificial Life III. Addison-Wesley, 325-370.

[228] Kier.L.B.,Seybold,P.G.,Cheng,C-K.(2005). *Modeling Chemical Systems Using Cellular Automata*. Springer.

[229] Killingback,T.,Doebeli,M.(1996). Spatial evolutionary game theory: Haws and Doves revisited. *Proc. R. Soc. London B*, 263, 1135-1144.

[230] Kinzel,W.(1985). Phase Transitions of cellular automata. *Z. fur Physics B*, 58, 229-244.

[231] Kirkpatrick,D.G,Radke,J.D. (1985). A framework for computational morphology. In G. Toussaint (ed.), *Computational geometry*, 217,248. North-Holland.

[232] Knuth,D.E.(1981). *The Art of Computer Programming (Volume 2)* (2nd Edition). Addison-Wesley.

[233] Knuth,D.E.(2008). The Art of Computer Programming. 4.0 Introduction to combinatorial algorithms and Boolean functions. Addison-Wesley.

[234] Ko,J-Y.,Hung,Y-C.,Ho,M-C.,Jiang,I-M.(2008). Chaotic behaviour in the disordered cellular automata. *Chaos, Solitons and Fractals*,36,4,934-939.

[235] Kokolakis,I.,Andreadis,I.,Tsalides,Ph.(1997). Comparison between cellular automata and linear feedback shift registers based pseudo-random number generators. *Micropocessors and microsystems*, 20,643–658.

[236] Kenkre,V.M.(2007). Analytic formulation, exact solutions, and generalizations of the elephant and the Alzeimer random walks. *arXiv:0708.0034v2* .

[237] Kusch,I.,Markus,M.(1996). Mollusk shell pigmentation: cellular automaton simulations and evidence for undecidability. *J. Theoretical Biology*, 178,3, 333-40.

[238] Lakshminarayan,A.(1997). Accuracy ot trace formulas. *Pramana*, 48,2, 517-535.

[239] Land,M.,Belew,R.(1995). No perfect two-state cellular automata for density classification exists. *Physical Review Letters*, 74,25.

[240] Langer,P.,Nowak,M.A.,Hauert,C.(2008). Spatial invasion of cooperation. *J. Theor. Biology*, 250, 634–641.

[241] Larrondo,H.A,Gonzalez,C.M.,Martin,M.T.,Platino,A.,Rosso.O.A.(2005). Intensive statistical complexity of pseudorandom number generators. *Physica A*, 356, 133-138.

[242] Layman,J.W.(1992). Dynamics of multicellular automata with unbounded memory. *Complex Systems*, 6,315-331.

[243] Lawrance,J.D.(1972). *A catalog of special plane curves*. Dover.

[244] Lago-Fernandez,L.F.,Huerta,R.,Corbacho,F.(2000). Fast response and temporal coherent oscillations in small-world networks. *Phys. Rev. Lett*, 84, 2758-2761.

[245] Lebowitz,J.L.,Maes,C.,Speer,E.(1990). Statistical methods of probabilistic cellular automata. *J. Statistical Physics*, 59, 117-170.

[246] Lee,J.,Adachi,S,Peper,F.,Morita,K.(2004). Asynchronous game of life *Physica D*, 194, 369-384.

[247] Lehmer,D.H.(1976). Incomplete Gauss sums. *Mathematica*, 23, 125-135.

[248] Lentner,M.(1973). On the Sum of Powers of the First N Integers On the Sum of Powers of the First N Integers. *The American Statistician*, 27, 2, 87.

[249] Loxton,J.H.,(1983). The graphs of exponential sums. *Mathematica*, 30, 153–156.

[250] Letourneau,P-J.(2007).
http://www.wolframscience.com/conference/2006/
presentations/materials/letourneau.pdf . http://demonstrations.wolfram.
com/OuterTotalistic2DCellularAutomataWithMemory/

[251] Li,W.(1992). Phenomenology of nonlocal cellular automata. *J.Stat. Physics*, 68,3/4 ,829-882.

[252] Li X.-Y. (2004). Application of computation geometry in wireless networks. In: Cheng X., Huang X., Du D.-Z. (Eds.) Ad Hoc Wireless Networking (Kluwer Academic Publishers), 197–264.

[253] Li,W.,Nordahl,M.(1992). Transient behavior of cellular automata rule 110. *Phys. Lett. A*, 166, 5/6, 335-339.

[254] Ligget,T.M.(2004). Interacting particle systems. Springer.

[255] Lind,P.G.,Gallas,J.A.C.,Herrmann,H.J.(2004). Coherence in scale-free networks of chaotic maps. *Physical Review E*, 79,1,056207.

[256] Lindgren,K.,Nordahl,M.G.(1994). Evolutionary dynamics of spatial games. *Physica D*, 75,292-309.

[257] Lindgren,K.,Nordahl,M.G.(1995). Cooperation and community structure in artificial ecosystems. In: *Artificial Life*, 16-37. Langton,C.(ed.). The MIT Press.

[258] Liu,K-L.,H.-S.,Goan.(2007). Non-Markovian entanglement dynamics of quantum continuous variable systems in thermal environments. *Phys. Rev. A*, 76, 022312.

[259] Lloyd,A.L.(1995). The coupled logistic map: A simple model for the effects of spatial heterogeneity of populations dynamics. *J. Theoretical Bilogy*, 173, 217-230.

[260] Lockwood,E.H.(2008). *Book of curves*. Cambridge University Press.

[261] Love,P.J.,Boghosian,B.M.,Meyer,D.A.(2004). Lattice gas simulations of dynamical geometry in one dimension. *Phil. Trans. R. Soc. London A*, 362-1667.

[262] Lu,Q.,Teuscher,C.(2009). Damage spreading in spatial and small-world random Boolean networks. *Physics Review E*, (submitted).

[263] Lumer,E.D.,Nicolis,G.(1994). Synchronous versus asynchronous dynamics

in spatially extended sytems. *Physica D*, 71, 440-452.

[264] Mackiw,G.(2000). A combinatorial approach to sums of integer powers. *Mathematics Magazine*, 73,1,44–46.

[265] Magwene P. W. (2008). Using correlation proximity graphs to study phenotypic integration. *Evolutionary Biology*, 35, 191–198.

[266] Majercik,S.M.(1994). *Structurally Dynamic Cellular Automata*. Univ. of Southern Maine (MsC) .

[267] Maji,P.,Chaudhury,P.(2008). Non-uniform cellular automata based associative memory : Evolutionary design and basins of attraction. *Information Sciences*, 178,2315-2336.

[268] Mann,W.R.(1953). Mean value methods in iteration. *Proc. Amer. Math. Soc.*, 4, 506-510.

[269] Mar,G.,St.Denis,P.(1994). Chaos in cooperation : continuous-valued prisoner's dilemmas in infinite-valued logic. *Int. J. of Bifurcation and Chaos*, 4,4,943-958.

[270] Maranon,G.A.,Encinas,L.H.,del Rey,A.M.(2005). A new secret sharing scheme for images based on additive 2-dimensional cellular automata. *Lect. Notes in Computer Science*, 3522, 411-418.

[271] Marklof,J.(1999). Limit theorems for theta sums. *Duke Mathematical Journal*, 97,1, 127-153.

[272] Margolus,N.(1984). Physics-like models of computation. *Physica D*, 10, 81-95.

[273] Marr,C.,Hutt,M-T.(2008). Outer-totalistic cellular automata on graphs. *Physics Letters A*, 373,5,546-549.

[274] Marti,A.A.,Masoller,C.(2006). Chaotic maps coupled with random delays : Connectivity, topology and network propensity to synchronization. *Physica A*, 371,104–107.

[275] Martinez,G.J.,Adamatzky,A,McIntosh,H.V.(2008). On the Representation of Gliders in Rule 54 by De Bruijn and Cycle Diagrams. *LNCS*, 5191, 83-91.

[276] Martinez,G.J,Adamatzky,A.,Alonso-Sanz,R.(2010). Complex dynamics of cellular automata emerging in chaotic rules. *Int.J. Bifurcation and Chaos*, (in press)

[277] Martinez,G.J,Adamatzky,A.,Alonso-Sanz,R.,Seck-Touh-Mora,J.C. (2009). Complex dynamics emerging in Rule 30 with majority memory. *Complex Systems*, 18,3, 345-365.

[278] Martinez,G.J,Adamatzky,A.,Seck-Touh-Mora,J.C.,Alonso-Sanz,R.(2010). How to make dull cellular automata complex by adding memory : Rule 126 case study. *Complexity*, 15,6,34-39.

[279] Martin del Rey,A.,Mateus,J.P.,Rodriguez-Sanchez,G.(2005). A secret sharing scheme based on cellular automata. *Applied Mathematics and Computation*, 170,2,1356-1364.

[280] Masoller,C.,Marti,A.A.(2005). Random delays and the synchronization of chaotic maps. *Phys. Rev. Letters*, 94, 134102.

[281] Mason,D.R.,C.,Hudson,T.S.,Sutton,A.P.(2005). Fast recall of state-history in kinetic Monte Carlo simulations using the Zobrist Key. *Computer Physics Communications*, 165,1,37-48.

[282] Masuda,N.,Aihara,K.,T.(2203). Spatial prisoner's dilemma optimally played in small-world networks. *Phsysics Letters A*, 313, 55–61.

[283] Matula D. W. and Sokal R. R. (1980). Properties of Gabriel graphs relevant to geographic variation research and clustering of points in the plane. *Geogr. Anal.*, 12, 205-222.

[284] Maurer,P.(1987). A rose is a rose ... . *The Amer. Math. Monthly*, 94,7, 631–645.

[285] Maynard-Smith,J.(1968). *Mathematical ideas in Biology.* Cambridge University Press.

[286] Maynard-Smith,J.(1982). *Evolution and the theory of games.* Cambridge University Press.

[287] May,R.(1976). Simple mathematical models with very complicated dynamics. *Nature,* 261,459.

[288] Meyers,R.A.(ed.)(2009). *Encyclopedia of complexity and systems sciences.* Springer.

[289] Méndez,V.,Fedotov,S.,Hortshemke,W.G.(2007). Effect of memory kernels on the speed of reaction-diffusion fronts. *Europhysics Letters,* 77,58006-10.

[290] Metzler,W.,Beau,W.,Frees,W.,Ueberla,A.(1987). Symmetry and self-similarity with coupled logistic maps. *Zeitschrift fur Naturforschung,* 42,3, 310-318.

[291] Mitchell,M.(2009). *Complexity: A Guided Tour.* Oxford Univ. Press. USA.

[292] Moore,R.R.,Poorten,A.J.(1989). On the thermodinamics of curves and other curlicues. *Proceedings of Conf. on Geometry and Physics,* Camberra. Barber,M.N.,Murray,M.K.(eds.)

[293] Moreira,A.A.,Mathur,A.,Diermeier,D.,Amaral,L.A.N.(2004). Efficient system-wide coordination in noisy environments. *Proc. Nat. Acad. Sci. USA,* 101,33,12085–12090.

[294] Morelli,L.G.,Zanetti,D.H.(2001). Synchronization of Kaufmann networks. *Physical Review E,* 63, 036204-10.

[295] Morelli,L.G.,Zanetti,D.H.(1998). Synchronization of stochastically coupled cellular automata. *Physical Review E,* 58,1, R8-R11.

[296] Morgado,F.,Ciesla,M.,Longa,L.,Oliveira,F.A.(2007). Synchronization in the presence of memory. *Europhysics Letters,* 79,1,10002-7.

[297] Morvai,G.,Weiss,B.(2007). On estimating the memory for finitarily Markovian processes. *Ann. Inst. H.Poincaré Probab. Statist.,* 43,1, 15-30. arXiv:0712.0105v1.

[298] Motoike,I.N.,Yoshikawa,K.,Iguchi,Y.,Nataka,S.(2001). Real-time memory on and excitable field. *Phys. Review E,* 63, 036220-4.

[299] Mukherji,A.,Rajan,V.,Slagle,V.(1996). *Nature,* 379,125.

[300] Muhammad R. B.A. (2007). distributed graph algorithm for geometric routing in ad hoc wireless networks. *J. Networks,* 2,49–57.

[301] Murguia,J.S.,Perez-Terrazas,J.E.,Rosu,H.C.(2010). Improvement and anlysis of a pseudo-random bit generator by means of cellular automata. *Int. J. Modern Physics C,* 21,6, 741-756.

[302] Murguia,J.S.,Perez-Terrazas,J.E.,Rosu,H.C.(2009). Multifractal properties of elementary cellular automata in a discrete wavelet approach of MF-DFA.

Europhisics Letters, 87, 28003.

[303] McIntosh,H.V.(2009). *One Dimensional Cellular Automata*. Old City Pub. .

[304] McGettrick,M.(2009). *One dimensional quantum walks with memory.* ArXiv.

[305] Nandi,A. Dutta, A.,Bhattacherjee,J.K. (2005). The phase modulated logistic map. *Chaos*, 15, 023107.

[306] Nandi,S. Kar,B.K.,Pal Chaudhury,P.(1994). Theory and applications of cellular automata in cryptography. *IEEE Transactions on Computers*, 43-12, 1346–1357.

[307] Nawaz,A.,Toor,A.H. (2004). Dilemma and quantum battle of sexes *J. of Physics A -Math. and General*, 37,15, 4437-4443.

[308] von Neumann, J.(1966). *The Theory of Self-Reproducing Automata*. University of Illinois.

[309] Nowak,M.A.(2006). *Evolutionary dynamics : Exploring the Equations of Life*. Harvard University Press.

[310] Nowak,M.A.,Bonhoefer,S.,May,R.M.(1994). Spatial games and the maintenance of cooperation. *Proc. Nat. Acad. of Sciences*, 91,4877-81.

[311] Nowak,M.A.,Bonhoeffer,S.,May,R.M.(1994). More spatial games. *Int. J. Bifurcation and Chaos*, 4,1,33-56.

[312] Nowak,N.M.,May,R.M.(1992). Evolutionary games and spatial chaos. *Nature*, 359, 826-829.

[313] Nowak,M.A.,May.M.(1993). The spatial dilemmas of evolution,*Int. J.Bifurcation and Chaos*,3,11, 35-78.

[314] Nowak,M.A.,Sigmund,K.(1993). A strategy of win-stay, lose-shift that outperforms tit-for-tat in the prisoner's dilemma game. *Nature*, 364,56-58.

[315] Nowak,M.A.,Sigmund,K.(1988). Evolution of indirect reciprocity by image scoring. *Nature*, 393, 573–577.

[316] Oliveira,G.M.B.,Siqueira,S.R.C.(2006). Parameter characterization of two-dimensional cellular automaton rule space. *Physica D*, 217,1, 1-6.

[317] Oprisan,S.O.(2002). Optimization of the memory weighting function in stochastic functional self-organized sorting performed by a team of autonomous agents. *Complex Systems*, 13, 205-225.

[318] Owen,G.(1995). *Game Theory*. Academic Press.

[319] Owens,G.(1992). Sum of powers of integers. *Mathematics Magazine*, 65,1,38–40.

[320] Pacheco,J.M.,Traulsen,A.,Nowak,M.A.(2006). Active linking in evolutionary games. *J. of Theoretical Biology*, 243,437-443.

[321] Pacheco,J.M.,Traulsen,A.,Nowak,M.A.(2006). Coevolution of strategy and structure in complex networks with dynamical linking. *Phys. Rev. Letters A*, 97,25,258103.

[322] Pacheco,J.M.,Traulsen,A.,Ohtsuki,H.,Nowak,M.A.(2008). Repeated games and direct reciprocity under active linking. *J. of Theoretical Biology*, 250,723-731.

[323] Paraan,F.N.C.,Esguerra,J.P.(2006). Exact moments in a continuous time random walk with complete memory of its story. *Phys. Rev. E.*, 74,3,032101.

[324] Paris,R.B.(2009). The asymptotics of a new exponential sum. *J. of Com-*

putational and Applied Mathematics, 223,1,314-325

[325] Paris,R.B.(2008). An asymptotic approximation for incomplete Gauss sums. II *J. of Computational and Applied Mathematics*, 212,1, 16-30.

[326] Paris,R.B.(2005). An asymptotic approximation for incomplete Gauss sums. *J. of Computational and Applied Mathematics*, 180, 2, 461-477.

[327] Paris,R.B.(2003). On the asymptotic connection between two exponential sums. *J. of Computational and Applied Mathematics*, 157,2, 297-308.

[328] Paul,J.L.(1971). On the Sum of the kth Powers of the First n Integers. *The American Mathematical Monthly*, 78, 3, 271-272.

[329] Peak,D.,West,J.D.,Messinger,S.M.,Mott,K.A.(2004). Evidence for complex, collective dynamics and emergent, distributed computation in plants. *Proc. Nat. Acad. Sci. USA*, 101,4, 918-922.

[330] Perc,M.,Szolnoki,A.(2010). Coevolutionary games - A mini review. *Biosystems*, 99,2,109-125.

[331] Petersen,N.K.,Alastrom,P.(1997). Phase transitions in an elementary probabilistic cellular automaton. *Physica A*, 235, 473-485.

[332] Pitsianis,N.,Tsalides,Ph.,Bleris,G.L.,Thanailakis,.A.,Card,H.C.(1989). Deterministic one-dimensional cellular automata. *J. Statistical Physics*, 56,99-112.

[333] Poundstone,W.(1985). *The Recursive Universe*. William Morrow, NY.

[334] Prado,C.P.C.,Bosco,F.(2007). Dynamics of moving agents with memory. *Physica A*, 374, 134-151.

[335] Qin,S-M,Chen,Y,Zao,X-L.,Shi,J.(2008). Effect of memory on the prisoner's dilemma game. *Physical Review E*, 78,4, 041129.

[336] Quick,T.,Nehaniv,C.,Dautenhahn,K.,Roberts,G.(2003). Evolving embedded genetic regulatory Network-driven control systems. In *Proceedings of the Seventh European Artificial Life Conference*. Springer, 266-277.

[337] Quieta,M.T.,Guan,S.U.(2005). Configurable cellular automata for pseudo-random number generation. *Int. J. Modern Physics C*, 16,7, 1051-1073.

[338] Raftery,A.E.(1985). A model for high-order Markov chains. *J.R.Statist. Soc. B.*, 47,3, 528-539.

[339] Rapaport,D.C.(2004). *The Art of Molecular Dynamics*. Cambridge University Press.

[340] Reynaga,R.,Amthauer,E.(2003). Two-dimensional cellular automata of radius one for density classification task $\rho = 1/2$. *Pattern Recognition Letters*, 24,2849-2856.

[341] Requardt,M.(1998). Cellular networks as models for Plank-scale physics. *J. Phys. A*, 31,7797; (2006). The continuum limit to discrete geometries, arxiv.org/abs/math-ps/0507017.

[342] Requardt,M.(2006). Emergent properties in structurally dynamic disordered cellular networks. *J. of Cellular Automata*, 2,4,273-289.

[343] Rohlf,T.,Jost,C.(2009). A new class of cellular automata: How spatio-temporal delays affect dynamics and improve computation. In *European Conf. on Complex Systems 2009*.

[344] Rohlf,T.,Tsallis,C.(2007). Long-range memory elementary 1D cellular automata: Dynamics and nonextensivity. *Physica A*, 379, 2, 465-470.

[345] Ros,H.,Hempel,H.,Schimansky-Geier,L.(1994). Stochastic dynamics of catalytic CO oxidation on Pt(100). *Physica A*, 206, 421-440.

[346] Rucker,R.(2005). *The Lifebox, the seashell and the soul.* Basic Books.

[347] Rujan,P.(1987). Cellular automata and statistical mechanical models, *J. Statistical Physics*, 49, 1,2 139-222.

[348] Sabatelli,L.,Richmond,P.(2003). Phase transitions, memory and frustration in a Snzajd-like model with synchronous updating. *Int. J. Modern Physics C*, 14,9, 1223-1229.

[349] Saigusa,T.,L.,Tero,A.,Nakagaki,.T.,Kuramot,Y.(2008). Amoebae anticipate periodic events. *Phys. Rev. Lett.*, 100,018101.

[350] Sanchez,J.R.,Alonso-Sanz,R.(2004). Multifractal Properties of R90 Cellular Automaton with Memory. *Int.J. Modern Physics C*, 15, 1461.

[351] Sanchez,J.R.(2003). Multifractal characteristic of linear one-dimensional cellular automata. *Int.J. Modern Physics C*, 14,4, 491-499.

[352] Santi P. (2005). Topology Control in Wireless Ad Hoc and Sensor Networks. Wiley.

[353] Santos,F.C.,Pacheco,J.M.,Lenaerts,T.(2006). Cooperation prevails when individuals adjust their social ties. *PLoS Comput. Biol.*, 2, 10,e140, 1284-1291.

[354] Saidami,S.(2004). Self-Reconfigurable robots topodynamic. *Proc. 2004 IEEE Int.Conf. on Robotics & Automation*, 2883-7.

[355] Schiff,J.L.(2008). *Cellular automata : a discrete view of the world.* Wiley.

[356] Shackleford,B.,Tanaka,M.,Carter,R.J.,Snider.G.(2002). FPGA implementation of neighborhood-of-four CA random number generation. FPGA02, 106-112. Random number generators implemented with neighborhood-of-four, non-locally connected CA. *IEICE Trans. on Fundamentals of Electronics Comm. and Comp. Sci.*, 12, 2612-2623.

[357] Shapley,L.S.(1953). Stochastic games. *Proc. Nat. Acad. Sciences*, 39:1095-1100.

[358] Schnegg,M.,Stauffer,D.(2007). Dynamics of networks and opinions. *Int. J. Bifurcation and Chaos*, 17, 2399-2409.

[359] Shi,X-M.,Shi,L.,Zhang,J-F.(2010). Opinion evolution based on cellular automata rules in small world networks. *Chin. Phys. B*, 19,3, 038701-7.

[360] Schultz,H.J.(1980). The Sum of the kth Powers of the First n Integers *The American Mathematical Monthly*, 87, 6, 478-481.

[361] Schulz,M.,Trimper,S.,Zabrocki,K.(2007). Spatiotemporal memory in a diffusion-reaction system. *J. Physics A : Math. and Theor.*, 40,3369-3378.

[362] Schütz,G.M.,Trimper,S.(2004). Elephants can always remember : long-range memory effects in a non-Markovian random walk. *Phys. Rev. E*, 70, 045101.

[363] Schönfisch,B.,Roos,A.(1999). Synchronous and asynchronous updating in cellular automata. *Biosystems*, 51, 123-143.

[364] Serra, R., Villani, M.,Agostini,L.(2004). On the dynamics of random Boolean networks with scale-free outgoing connections. *Physica A*, 339, 665-673.

[365] Sheng,Z-H,Hou,Y-Z,Wang,X-L.,Liu,J-G.(2007). The evolution of coopera-

tion with memory, learning and dynamic preferential selection in spatial prisoner's dilemma game. *Int. J. of Nonlinear Sci. and Numerical Simulat.*, 96,012107.

[366] Shepard,M.(1997). A rose is a rose is a rose . . . . *The College Math. Journal*, 28,1, 55–56.

[367] da Silva,M.A.A.,Cressoni,J.C.,Viswannathan,G.M.(2006). Discrete-time non-Markovian random walks : The effect of memory limitations on scaling. *Physica A*, 364, 70-78.

[368] Sinai,Y.G.(2008). Limit theorems for trigonometric sums. Theory of curlicues. *Russ. Math. Surveys*, 63,6, 1023-1029,

[369] Sipper,M.(1997). *Evolution of Parallel Cellular Machines* Springer.

[370] Sipper,M.(1996). Co-evolving non-uniform cellular automata to perform computations. *Physica D*, 92, 193-208.

[371] Sipper,M.,Tomassini,M.(1996). Generating parallel random number generators by cellular programming. *Int. J. Modern Physics C*, 7,2, 181-190.

[372] Smale,S.(1980). The prisoner's dilemma and dynamical systems associated to non-cooperative games. *Econometrica*, 48,1617-1633.

[373] Smolin,L.(2004). Atoms of space and time. *Scientific American*, January.

[374] Smolin,L.(2008). Big bang or big bounce ?. *Scientific American*, October.

[375] Sokal R. R. and Oden N. L. (2008). Spatial autocorrelation in biology 1. Methodology. *Biological Journal of the Linnean Society*, 10,199–228.

[376] Song W.-Z., Wang Y., Li X.-Y. (2004). Localized algorithms for energy efficient topology in wireless ad hoc networks. In: *Proc. MobiHoc*, (May 24-26, 2004, Roppongi, Japan).

[377] Stanislavsky,D.(2006). Long-term memory contribution as applied to the motion of discrete dynamical systems. *Chaos*, 16, 034105.

[378] Sridharan M. and Ramasamy A. M. S. (2010). Gabriel graph of geomagnetic Sq variations. *Acta Geophysica*, `10.2478/s11600-010-0004-y`

[379] Stauffer,D.(1994). Evolution by damage spreading in Kauffman model. *J. Stat. Physics*, 74, 5/6, 1293-1299.

[380] Stauffer,D.(1994). *Introduction to percolation Theory*. Taylor & Francis.

[381] Stewart,I.(2004). *Another fine maths you've got me into*. Dover.

[382] Stone,C.,Bull,L.(2009). Solving the density classification task with cellular automaton rule 184 and memory. *Complex Systems*, 18,3.

[383] Stone,C.,Bull,L.(2009). Evolution of cellular automata with memory : The density classification task. *Biosystems*, 92, 108-116.

[384] Svozil,K.(1986). Are quantum fields cellular automata ?. *Phy. Lett. A*, 119,41, 153-156.

[385] Szabo,G.,Fáth,G.(2007). Evolutionary games on graphs. *Physics Reports*,446, 97-216.

[386] Tachikawa,M.(2009). A mathematical model for period-memorizing behaviour in *Physarium* plasmodium. *J. Theor. Biology*, 263,4,449-495.

[387] Tsalides,Ph.,York T.A.,Thanailakis,A.(1991). Pseudorandom number generators for VLSI systems based on linear cellular automata. *IEE Proceedings-E*, 138, 4, 241-249.

[388] Toffoli,T.(1977). Computation and construction universality of reversible

cellular automata. *J. Comput. System Sci.*, 15, 213-31.

[389] Toffoli,T.(1984). Cellular automata as an alternative to (rather than an approximation of) differential equations in modeling physics. *Physica D*, 10, 117-27.

[390] Toffoli,T.(1994). Occam, Turing von Neumann, Jaynes: How much can you get for how little ?. In Di Gregorio,S.,Spezzano,G.(eds.). ACRI-94. ENEA, Italy.

[391] Toffoli,T.(1990). *Encyclopedia of Physics*. R.G.Lerner and G.L.Trigg (eds). p.196. VCH Pub. NY

[392] Toffoli,T.,Margolus,N.(1987). *Cellular Automata Machines*. MIT Press.

[393] Toffoli,T.,Margolus,N.(1990). Invertible cellular automata: a review. *Physica D*, 45 229-253.

[394] Tomassini,M.,Sipper,M.,Zolla,M.,Perrenoud,M.(1999). Generating high-quality random numbers in parallel by cellular automata. *Future Generation Computer Systems*, 16, 291-305.

[395] Tomassini,M.,Sipper,M.,Perrenoud,M.(2000). On the generation of high-quality random numbers by two-dimensional cellular automata. *IEEE Transactions on Computers*, 49, 10, 1146-1151.

[396] Tomassini,M.,Sipper M.Zolla and M.Perrenoud, Generating high-quality random numbers in parallel by cellular automata. *Future Generation Computer Systems*, 16, 10, 291-305 (1999).

[397] Toroczkai Z.,Guclu H.(2007). Proximity networks and epidemics. *Physica A*, 378, 68.

[398] 't Hooft,G.(1988). Equivalence relations between deterministic and quantum mechanical systems. *J. Statistical Physics*, 53,(1/2),323-344.

[399] Ungethüm,A.,Lammering,R.(2007). Simulation of temperature and stress induced phase transformation of shape memory alloys using cellular automaton and the finite element method. In *Proc. 1st seminar on The Mechanics of Multifunctional Materials*. J.Schröeder *et al.* (eds.).

[400] Vichniac,G.(1984). Simulating physics with cellular automata. *Physica D*, 10, 96-115.

[401] Vichniac,G.Y.,Tamayo,P.,Hartmann,H.(1986). Annealed and quenched inhomogeneous cellular automata (INCA). *J. Stat. Physics*, 45,5/6 ,875-883.

[402] Villani,M.,Serra,R.,Ingrami,P.,Kauffman,S.A.(2006). Coupled random Boolean networks forming an artificial tissue. In *Proc. of the 7th Int. Conf. on Cellular Automata for Research and Industry*. Springer, 548-556.

[403] Voorhes,V.(2006). Emergence and induction of cellular automata rules via probabilistic reinforcement paradigms. *Complexity*, 11,3,44-57.

[404] Voorhes,V.(2006). Induction of cellular automata rules I: a reinforcement scheme. *Int. J. Unconventional Computing*, 2,2,91-127.

[405] Voorhes,V.,Arthur,R.,Keeker,T.(2006). Probabilistic induction of cellular automata rules II: Probing CA rule space. *Int. J. Unconventional Computing*, 2,3,195-229.

[406] Voorhes,V.(2005). Emergence of cellular automata rules through fluctuation enhancement. *Int. J. Unconventional Computing*, 1,1,69-100.

[407] Wan P.-J., Yi C.-W.(2007). On the longest edge of Gabriel Graphs in

wireless ad hoc networks. *IEEE Trans. on Parallel and Distributed Systems*, 18, 111–125.

[408] Wang,W-X.,D.J.,Ren-J.,Chen,G.,Wang,B-H.(2006). Memory-based snowdrift game on networks. *Physics Rev. E*, 74, 056113.

[409] Watanabe D. (2005). A study on analyzing the road network pattern using proximity graphs. *J. of the City Planning Institute of Japan*, 40,133–138.

[410] Watanabe D. (2008). Evaluating the configuration and the travel efficiency on proximity graphs as transportation networks. *Forma*, 23,81-87.

[411] Watts,D.J.(1999). *Small Worlds*. Princeton University Press.

[412] Watts,D.J.,Strogatz,S.H.(1998). Collective dynamics of small-world networks. *Nature*, 393, 440-442.

[413] Weimar,J.R.(2001). Coupling microscopic and macroscopic cellular automata. *Parallel Computing*, 27, 601-611.

[414] Weisbuch,G.,Stauffer,D.(2007). "Antiferromagnetism" in social relations and Bonabeau model. *Physica A* **384**, 542-548.

[415] Weisstein,E.W.(2009). *CRC Encyclopedia of Mathematics I*, p. 836. Chapman and Hall.

[416] Wilde,R.E.,Singh,S.(1998). *Statistical Mechanics*. Wiley.

[417] Wiesner,E.(2009). Quantum cellular automata. In Meyers,R.A.(ed.) [288].

[418] Willeboordse,F.H.(2003). The spatial logistic map as a simple prototype for spatiotemporal chaos. *Chaos*, 13,2, 533–540.

[419] Wissel,C.(2000). Grid-based models as tools for ecological research. In *The Geometry of Ecological Interactions*, Dieckmann *et al.*(eds).

[420] Wolfram,S.(1983). Statistical mechanics of cellular automata. *Reviews of Modern Physics*, 55,601-644.

[421] Wolfram,S.(1984). Algebraic properties of cellular automata. *Physica D*, 10,1-35.

[422] Wolfram,S.(1986). Random sequence generation by cellular automata. *Advances in Applied Mathematics*, 7, 123-169.

[423] Wolfram,S.(1984). Universality and complexity in cellular automata. *Physica D*, 10,1-35.

[424] Wolfram,S.,Martin,O.,Odlyzko,A.M.(1984). Algebraic properties of cellular automata. *Communications in Math. Physics*, 93,219–258.

[425] Wolfram,S.(1994). *Cellular Automata and Complexity*. Addison-Wesley.

[426] Wolfram,S.(2002). *A new kind of science*. Wolfram Media.

[427] Wolf-Gladrow,D.A.(2000). *Lattice-gas Cellular Automata and Lattice Boltzmann Models*. Springer.

[428] Wolz,D.,Oliveira,P.P.B.(2008). Very effective evolution techniques for searching cellular automata spaces. *J. of Cellular Automata*, 3,1, 289-312.

[429] Wuensche,A.,Lesser,M.(1992). *The Global Dynamics of Cellular Automata*. Addison-Wesley.

[430] Wuensche,A.(1994). The ghost in the machine: basins of attraction of random Boolean networks. In Artificial Life III, Langton,C.(ed.), 465-501.

[431] Wuensche,A. DDLab: Discrete Dynamics Lab. www.ddlab.com

[432] Wuensche,A.(2005). Glider dynamics in 3-value hexagonal cellular automata: the beehive rule. *Int. J. of Unconventional Computing*, 1, 375-398.

[433] Wuensche,A.,Adamatzky,A.(2006). On spiral glider-guns in hexagonal celular automata : activator-inhibitor paradigm. *Int. J. Modern Physics C*, 17,7,1009–10026.

[434] Xiao-Ming,S.,Lun,S.,Jie-Fang,Z,X.,Fengyu,L.(2010). Opinion evolution based on cellular automata rules on small networks. *Chinese Physics B*, 9,3,038701.

[435] Xuelog,Z.,Qianmu,L.,Manwu,X.,Fengyu,L.(2005). A symmetric cryptography based on extended cellular automata. *IEEE Int. Conf. on Systems, Mand and Cybernetics*, 1,10-12.

[436] Yadav,A.,Fedotov,S.,Méndez,V.,Horsthemke,W.(2007). Propagating fronts in reaction-transport systems with memory. *Physics Letters A*, 371,374-378.

[437] Yao,X.,Darwen,P.J.(1999). How important is your reputation in a multiagent environment *Proc. IEEE Conf. Systems, Man and Cybernetics*, IEEE Prees, II-575-580.

[438] Yates,R.C.(1941). The trisection problem. *National Mathematics Magazine*, 15,4, 191–202.

[439] Yonac,M.,Yu,T.,Eberly,J.H.(2006). Sudden death of entanglement of two Jaynes-Cummings atoms. *J. Phys. B : At. Mol. Opt. Phys.*, 39,15, S621-S625.

[440] Zamora,R.R.,Vergara,S.V.C.(2004). Using de Bruijn diagrams to analyze 1d cellular automata traffic models. LNCS, 3305, 306–315.

[441] Zabolyzky,J.G.(1988). Critical properties of rule 22 elementary cellular automata. *J. Stat. Physics*, 50,5-6 ,1255-1262.

[442] Zarnitsyna,V.I.,Huang,J.,Zhang,F.,Chien,Y-H,Leckband,D.,Zhu,C.(2007). Memory in receptor ligand-mediated cell adhesion. *Proc. Nat. Acad. Sci. USA*, 104,46, 19037-19042.

[443] Zaslavsky,G.M.,Sagdeev,R.Z.,Usikov,D.A.,Chernkov,A.A.(1991). Weak chaos a and quasi-regular pattens. Cambrige University Press.

[444] Zhao,J.,Szilagyi,N.,Szidarovsky,F. (2008). An n-person battle of sexes game. *Physica A*, 387,3669-3677.

[445] Zhao,J.,Szilagyi,N.,Szidarovsky,F. (2008). An n-person battle of sexes game – a simulation study. *Physica A*, 387,3668-3788.

[446] Zia,L.(1991). Using the Finite Difference Calculus to Sum Powers of Integers. *The College Mathematics Journal*, 22, 4, 294-300.

[447] Zimmermann,M.G.,Eguiluz,V.M.(2005). Cooperation, social networks, and the emergence of leadership in a prisoner's dilemma with adaptive local interactions. *Physical Review E*, 72,056118.

[448] Zimmermann,M.G.,Eguiluz,V.M.,San Miguel,M.(2004). Coevolution of dynamical states and interactions in dynamic networks. *Physical Review E*, 69,065102-6.

List of Figures

List of Tables

Index